Lecture Notes in Computer Science 13292

More information about this series at https://link.springer.com/bookseries/558

Pierre Schaus (Ed.)

Integration of Constraint Programming, Artificial Intelligence, and Operations Research

19th International Conference, CPAIOR 2022
Los Angeles, CA, USA, June 20–23, 2022
Proceedings

 Springer

Editor
Pierre Schaus 🄳
UCLouvain
Louvain-la-Neuve, Belgium

ISSN 0302-9743   ISSN 1611-3349 (electronic)
Lecture Notes in Computer Science
ISBN 978-3-031-08010-4   ISBN 978-3-031-08011-1 (eBook)
https://doi.org/10.1007/978-3-031-08011-1

This Springer imprint is published by the registered company Springer Nature Switzerland AG
The registered company address is: Gewerbestrasse 11, 6330 Cham, Switzerland

Preface

This volume contains the papers that were presented at the 19th International Conference on the Integration of Constraint Programming, Artificial Intelligence, and Operations Research (CPAIOR 2022), held in Los Angeles, USA, as a hybrid physical/virtual conference.

The conference received a total of 73 submissions, including 60 regular papers and 13 extended abstract submissions. The regular papers reflect original unpublished work, whereas the extended abstracts contain either original unpublished work or a summary of work that was published elsewhere. Each regular paper was reviewed by at least three Program Committee members. The reviewing phase was followed by an author response period and a general discussion by the Program Committee. The extended abstracts were reviewed for appropriateness for the conference. At the end of the review period, 28 regular papers were accepted for presentation during the conference and publication in this volume, and 13 abstracts were accepted for a short presentation at the conference.

In addition to the regular papers and extended abstracts, three invited talks, whose abstracts can be found in this volume, were given by André Cire (University of Toronto, Canada), Carla Gomez (Cornell University, USA), and Vinod Nair (Deepmind, UK).

The conference program included a Master Class on the topic "Bridging the Gap between Machine Learning and Optimization", organized by Adam Elmachtoub and Elias Khalil, with invited talks by Brandon Amos (Facebook AI Research, USA), Priya Donti (Carnegie Mellon University, USA), Paul Grigas (University of California, Berkeley, USA), Tias Guns (KU Leuven, Belgium), Bryan Wilder (Carnegie Mellon University and Harvard University, USA), Angela Zhou (University of California, Berkeley, and University of Southern California, USA). Of the regular papers accepted to the conference, a committee - comprising of myself and Mark Wallace (Monash University, Australia) - selected for the Best Paper Award the paper "Shattering Inequalities for Learning Optimal Decision Trees" by Justin Boutilier, Carla Michini, and Zachary Zhou and selected for the Best Student Paper Award the paper "A MinCumulative Resource Constraint" by Yanick Ouellet and Claude-Guy Quimper.

We acknowledge the main local organizers Bistra Dilkina, Sven Koening, and Pheve Vayanos (University of Southern California, USA).

We also acknowledge the generous support of our sponsors including, at the time of writing, the USC Daniel J. Epstein Department of Industrial and Systems Engineering, the USC Epstein Institute, the USC-Meta Center for Research and Education in AI and Learning, Gurobi, Google, Nextmv, Kinaxis, The Optimization Firm, Lindo, Cosling, and Springer.

April 2022

Pierre Schaus

Organization

General Chairs

Bistra Dilkina	University of Southern California, USA
Sven Koenig	University of Southern California, USA
Phebe Vayanos	University of Southern California, USA

Program Chair

Pierre Schaus	UCLouvain, Belgium

DEI Chair

Phebe Vayanos	University of Southern California, USA

Master Class Chairs

Adam Elmachtoub	Columbia University, USA
Elias Khalil	University of Toronto, Canada

Sponsorship Chair

Thiago Serra	Bucknell University, USA

Program Committee

Beste Basciftci	University of Iowa, USA
Chris Beck	University of Toronto, Canada
Nicolas Beldiceanu	LS2N, IMT Atlantique, France
David Bergman	University of Connecticut, USA
Armin Biere	Albert-Ludwigs-Universität Freiburg, Germany
Quentin Cappart	Ecole Polytechnique de Montréal
Carlos Cardonha	University of Connecticut, USA
Mats Carlsson	RISE, Sweden
Andre Augusto Cire	University of Toronto, Canada
Simon de Givry	MIAT INRAE, France
Emir Demirović	Delft University of Technology, The Netherlands
Guillaume Derval	Université de Liège, Belgium
Pierre Flener	Uppsala University, Sweden

Tias Guns	Vrije Universiteit Brussel, Belgium
Emmanuel Hebrard	LAAS-CNRS, France
John Hooker	Carnegie Mellon University, USA
Serdar Kadioglu	Brown University, USA
Roger Kameugne	University of Maroua, Cameroon
George Katsirelos	MIA Paris, INRAE, AgroParisTech, France
Joris Kinable	Amazon, USA
Zeynep Kiziltan	University of Bologna, Italy
Christophe Lecoutre	CRIL, Université d'Artois, France
Jiaoyang Li	University of Southern California, USA
Andrea Lodi	Cornell Tech, USA
Michele Lombardi	DISI, University of Bologna, Italy
Pierre Lopez	LAAS-CNRS, Université de Toulouse, France
Arnaud Malapert	Université Côte d'Azur, CNRS, I3S, France
Ciaran McCreesh	University of Glasgow, UK
Laurent Michel	University of Connecticut, USA
Nysret Musliu	TU Wien, Austria
Margaux Nattaf	G-SCOP, Grenoble INP, France
Barry O'Sullivan	University College Cork, Ireland
Marie Pelleau	Université Côte d'Azur, CNRS, I3S, France
Laurent Perron	Google, France
Gilles Pesant	Polytechnique Montréal, Canada
Claude-Guy Quimper	Laval University, Canada
Jean-Charles Regin	University Nice-Sophia Antipolis CNRS, I3S, France
Andrea Rendl	Satalia, Austria
Louis-Martin Rousseau	CIRRELT, Canada
Elina Rönnberg	Linköping University, Sweden
Domenico Salvagnin	University of Padua, Italy
Pierre Schaus	UCLouvain, Belgium
Thomas Schiex	INRAE, France
Paul Shaw	IBM, France
Mohamed Siala	INSA Toulouse and LAAS-CNRS, France
Helmut Simonis	University College Cork, Ireland
Christine Solnon	INSA Lyon, France
Willem-Jan Van Hoeve	Carnegie Mellon University, USA
Hélène Verhaeghe	Polytechnique Montréal, Canada
Petr Vilím	CoEnzyme, Czech Republic
Mark Wallace	Monash University, Australia
Roland Yap	National University of Singapore, Singapore

Additional Reviewers

Matteo Cacciola	Polytechnique Montréal, Canada
Shao-Hung Chan	University of Southern California, USA
JingkaiChen	Massachusetts Institute of Technology, USA
Timothy Curry	University of Connecticut, USA
Julien Ferry	LAAS-CNRS, France
Emilio Gamba	Vrije Universiteit Brussel, Belgium
Tobias Geibinger	TU Wien, Austria
Yong Lai	Massachusetts Institute of Technology, USA
Jayanta Mandi	Vrije Universiteit Brussel, Belgium
Florian Mischek	TU Wien, Austria
Pierre Montalbano	MIAT INRAE, France
Maxime Mulamba	Vrije Universiteit Brussel, Belgium
Magnus Rattfeldt	Tomologic, Sweden
Philippe Refalo	IBM, France

Abstract of Keynote Speakers

Decision Diagrams for Deterministic and Stochastic Optimization

Andre A. Cire

University of Toronto, Canada
andre.cire@Rotman.utoronto.ca

Abstract. In this talk we will discuss alternative solution techniques for discrete and stochastic optimization based on decision diagrams (DDs). A DD, in our context, is a graph-based extended formulation of an optimization problem that exposes network structure, leading to novel bounding and branching mechanisms that complement classical model-based approaches. We will investigate the principles of DD modeling for combinatorial problems and develop the intrinsic connections between DDs and (approximate) dynamic programming. We will then leverage links with mathematical programming and polyhedral theory to propose stronger formulations, cutting-plane methods, and new decomposition approaches for difficult combinatorial and stochastic discrete problems. The talk will highlight examples in routing, scheduling, and planning, while also emphasizing new applications and future research in the area.

Combining Reasoning and Learning for Discovery

Carla P. Gomez

Computer Science Department, Cornell, USA
gomes@cs.cornell.edu

Abstract. Artificial Intelligence (AI) is a rapidly advancing field inspired by human intelligence. AI systems are now performing at human and even superhuman levels on various tasks, such as image identification and face and speech recognition. The tremendous AI progress that we have witnessed in the last decade has been largely driven by deep learning advances and heavily hinges on the availability of large, annotated datasets to supervise model training. However, often we only have access to small datasets and incomplete data. We amplify a few data examples with human intuitions and detailed reasoning from first principles and prior knowledge for discovery. I will talk about our work on AI for accelerating the discovery for new solar fuels materials, which has been featured in Nature Machine Intelligence, in a cover article entitled, Automating crystal-structure phase mapping by combining deep learning with constraint reasoning [1]. In this work, we propose an approach called Deep Reasoning Networks (DRNets), which seamlessly integrates deep learning and reasoning via an interpretable latent space for incorporating prior knowledge. and tackling challenging problems. DRNets requires only modest amounts of (unlabeled) data, in sharp contrast to standard deep learning approaches. DRNets reach super-human performance for crystal-structure phase mapping, a core, long-standing challenge in materials science, enabling the discovery of solar-fuels materials. DRNets provide a general framework for integrating deep learning and reasoning for tackling challenging problems. For an intuitive demonstration of our approach, using a simpler domain, we also solve variants of the Sudoku problem. The article DRNets can solve Sudoku, speed scientific discovery [2], provides a perspective for a general audience about DRNets. DRNets is part of a SARA, the Scientific Reasoning Agent for materials discovery [3]. Finally, I will also talk about the effectiveness of a novel curriculum learning with restarts strategy to boost a reinforcement learning framework [4]. We show how such a strategy can outperform specialized solvers for Sokoban, a prototypical AI planning problem.

References

1. Chen, D., et al.: Automating crystal-structure phase mapping by combining deep learning with constraint reasoning. Nat. Mach. Intell. **3**(9), 812-822 (2021)
2. Fleischman, T.: DRNets can solve Sudoku, speed scientific discovery, Cornell Chronicle, 17 September 2021. https://news.cornell.edu/stories/2021/09/drnets-can-solve-sudoku-speed-scientific-discovery
3. Ament, S., et al.: Autonomous materials synthesis via hierarchical active learning of nonequilibrium phase diagrams. Sci. Adv. **7**(51), eabg4930 (2021)
4. Feng, D., Gomes, C.P., Selman, B.: A novel automated curriculum strategy to solve hard sokoban planning instances. Adv. Neural Inf. Process. Syst. **33**, 3141–3152 (2020)

Deep Learning and Neural Network Accelerators for Combinatorial Optimization

Vinod Nair

DeepMind
vinair@google.com

Abstract. Deep Learning has been used to construct heuristics for challenging combinatorial optimization problems. It has two advantages: a) expressive neural network models can learn custom heuristics from data by exploiting the structure in a given application's distribution of problem instances, and b) once learned, the models can be executed on accelerators such as GPUs and TPUs with high throughput for computations such as matrix multiplication. In this talk we'll present two works that illustrate these advantages. In the first work, we apply Deep Learning to solving Mixed Integer Programs by learning a Branch and Bound variable selection heuristic and a primal heuristic. We show results on datasets from real-world applications, including two production applications at Google. In the second work, we use a class of neural networks called Restricted Boltzmann Machines to define a stochastic search heuristic for Maximum Satisfiability that is well-suited to run on a large-scale TPU cluster. Results on a subset of problem instances from annual MaxSAT competitions for the years 2018 to 2021 show that the approach achieves better results than competition solvers with the same wall-clock budget across all four years.

Contents

A Two-Phase Hybrid Approach for the Hybrid Flexible Flowshop with Transportation Times

Eddie Armstrong[1], Michele Garraffa[2,3], Barry O'Sullivan[2,3], and Helmut Simonis[2,3(✉)]

[1] Johnson and Johnson Research Centre, Limerick, Ireland
[2] Confirm SFI Research Centre for Smart Manufacturing, Limerick, Ireland
`helmut.simonis@insight-centre.org`
[3] School of Computer Science and IT, University College Cork, Cork, Ireland

Abstract. We present a two-phase heuristic approach for the Hybrid Flexible Flowshop with Transportation Times (HFFTT) which combines a metaheuristic with constraint programming (CP). In the first phase an adapted version of a state-of-the-art metaheuristic for the Hybrid Flowshop [15] generates an initial solution. In the second phase, a CP approach reoptimizes the solution with respect to the last stages. Although this research is still in progress, the initial computational results are very promising. In fact, we show that the proposed hybrid approach outperforms both the adapted version of [15] and earlier CP approaches.

Keywords: Metaheuristics · Constraint Programming · Scheduling · Hybrid Flowshop

1 Introduction

Real world scheduling problems are generally tackled by means of two different types of approaches. On one hand, mixed integer linear programming (MILP) and constraint programming (CP) approaches put the focus on modelling the scheduling problem, instead of developing a solution algorithm from scratch. This approach shortens the development time and provides a high level of flexibility in the case where the problem formulation needs to be adjusted to take into account new characteristics. In that case, an optimization specialist just needs to adapt the MILP/CP model, then a modern solver will provide high quality solutions by exploiting many years of algorithmic improvements. On the other hand, metaheuristic algorithms have been widely studied, since they can exploit problem-specific properties to perform the search very efficiently. In many cases, this leads to achieving high quality results, at the cost of a higher development time and a reduction in the flexibility/generality of the solution approach.

The combination of both approaches, so called hybrid heuristics, has been receiving significant attention from the research community since the early 2000 s

© Springer Nature Switzerland AG 2022
P. Schaus (Ed.): CPAIOR 2022, LNCS 13292, pp. 1–13, 2022.
https://doi.org/10.1007/978-3-031-08011-1_1

(see [2] for a survey about the topic). Most hybrid heuristics rely on efficient MILP solvers, due to their use of powerful mathematical programming techniques. However, the performance of MILP solvers is typically poor in cases where the model has a weak linear relaxation, e.g. due to the use of big-M variables to represent logical constraints. MILP-based hybrid heuristics are generally known as "matheuristics" [6,9], and they have been widely used for scheduling problems [3–5,8]. On the other hand, CP solvers rely on the expressive power of global constraints, and on the effectiveness of propagation algorithms and automatic search heuristics. One of the best known commercial CP solvers, CP Optimizer by IBM [7], offers support for efficiently solving many types of scheduling problems. It provides a model-and-run paradigm, which is quite simple to master and is more generic than the one provided by MILP solvers, since non-linearities and logical constraints can be easily included.

In this paper we propose a hybrid heuristic for a real-world scheduling problem, which combines a metaheuristic with a local search procedure relying on CP Optimizer. Hybrid heuristics based on CP for scheduling problems are not very common, but they have been considered [12,14]. The motivation behind hybridizing CP with another type of solution approach is usually to exploit the complementarity of the two in order to achieve a better performance than each of the two approaches separately. The problem considered in this study is the Hybrid Flexible Flowshop with Transportation Times (HFFTT) which arises in modern production facilities and has been recently introduced [1]. The problem is an extension of the Hybrid Flowshop Problem (HFP) [13] and of the Hybrid Flexible Flowshop (HFF) [10] where transportation times between the machines for the different production steps of each job are assumed to be non-negligible.

The HFFTT is defined as follows. Let J be a set of jobs, M a set of machines and S a set of production stages. We denote as $p_{j,s}$ the processing time of a job $j \in J$ to complete a stage $s \in S$. All jobs complete the stages in the same order but some of the stages may be skipped by some jobs. We indicate with $S_j \subseteq S$ the subset of the production stages performed by job $j \in J$, while we indicate with $J_s \subseteq J$ the set of all jobs that complete a stage $s \in S$. The successor stage of a stage $s \in S$ with respect to a job $j \in J$ is denoted as $succ(s, j)$. The transportation time required to move a job $j \in J$ from a machine $m \in M_s$ to another machine $m' \in M_{s'}$, where $s, s' \in S$ and $s' = succ(s, j)$, is represented by $\delta_{m,m'}$. Finally, the problem objective is to minimize the makespan.

The paper is organized as follows. Section 2 describes the CP model of the problem using the global constraints available in CP Optimizer. Section 3 describes an iterated greedy method, IGT_NEH, which is an adaptation of a state-of-the-art metaheuristic for the HFP [15]. Section 4 presents a novel two-phase hybrid approach to solve the HFFTT. Section 5 presents a computational assessment of the different approaches, using the benchmarks defined in [1].

2   The Constraint Programming Model

This section presents a CP model of the problem, analogous to the model based on interval variables presented in [1], and was encoded by using the OPL API of CP Optimizer [7]. The model is based on the following variables:

- Optional interval variables $tm_{m,j}$ for each $j \in J$ and $m \in M$;
- Interval variables $ts_{s,j}$ for each $j \in J$ and $s \in S_j$;
- Integer variables $machine_{s,j}$ for each $j \in J$ and $s \in S_j$, with feasible values in the range $\{1, \cdots, |M|\}$.

The variables $tm_{m,j}$ are optional interval variables representing the execution of a job on a certain machine. Given a stage $s \in S$ and a job $j \in J$, only one of these variables is active, which is constrained to be equal to $ts_{s,j}$. The variables $machine_{s,j}$ are linked to the machine used to performed a job $j \in J$ at stage $s \in S$.

The CP model of the HFFTT is as follows:

$$\min \max_{s \in S, j \in J_s} endOf(ts_{s,j}) \tag{1}$$

subject to:

$$(machine_{s,j} = m) \implies presenceOf(tm_{m,j}) \quad \forall j \in J, s \in S_j, m \in M_s \tag{2}$$

$$alternative(ts_{s,j}, \{tm_{m,j} : m \in M_s\}) \quad \forall j \in J, s \in S_j \tag{3}$$

$$endBeforeStart(tm_{m,j}, tm_{m',j}, \delta_{m,m'}) \quad \forall j, s \in S_j, m \in M_s, m' \in M_{succ(s,j)} \tag{4}$$

$$endBeforeStart(ts_{s,j}, ts_{succ(s,j),j})) \quad \forall j \in J, s \in S_j : succ(s,j) \neq \emptyset \tag{5}$$

$$noOverlap(\{tm_{m,j} : j \in J\}) \quad \forall m \in M \tag{6}$$

$$cumulative(\{ts_{s,j} : j \in J_s\}, |M_s|) \quad \forall s \in S \tag{7}$$

The objective (1) indicates that we minimize the makespan. Constraints 2 deal with assigning a job $j \in J$ to one machine $m \in M_s$ at each stage $s \in S_j$, and setting the corresponding interval variable $tm_{m,j}$ to active. Constraints 3 link the interval variables $tm_{m,j}$ and the interval variables $ts_{s,j}$. Constraints 4 indicate that a job $j \in J$ can start being processed by a machine $m' \in M_{succ(s,j)}$ after being completed by the previous one $m \in M_s$ and spending $\delta_{m,m'}$ time units for the transportation. Constraints 5 are flowshop constraints, meaning that each job $j \in J$ can perform the next stage $succ(s,j) \in S_j$ after completing the previous one $s \in S_j$. Constraints 6 state that each machine can process one job at a time. Constraints 7 are redundant, requiring that the maximum number of jobs performing simultaneously at stage $s \in S$ is equal to the number of machines $|M_s|$ available at that stage.

3   Metaheuristic Approach

State-of-the-art metaheuristic approaches for the HFP, e.g. [15], can be easily adapted to solve instances of the HFFTT. These approaches are based on forward scheduling. Given a certain jobs permutation γ, they follow these two steps:

- Assign the jobs by following the order in γ, to the earliest available machine (first stage);
- Assign the earliest available job to the earliest available machine (each of the other stages).

A random choice is taken in case of ties. In such a way, a feasible solution to the HFP can be generated given a reference permutation. The adaptation needed to use forward scheduling on the HFFTT is to consider the transportation times. We do not consider the transportation time when we compute the earliest available job, since the job is available once it has been processed by a certain machine. However, we consider the transportation time when performing the machine assignment. In this case, we do not evaluate the machines according to their earliest available time because the job may not be able to reach the machine at that time. We rank the machines according to the maximum between the:

- Earliest time when the machine is available;
- Earliest time when the job can reach the machine.

This change allows us to replicate approaches based on forward scheduling on the HFFTT.

   In the following, we describe our adaptation of the approach denoted as IGT (Iterated Greedy with fixed temperature T) [15]. We denote our approach as IGT_NEH. The most important changes with respect to IGT are:

- IGT_NEH considers transportation times when applying forward scheduling;
- IGT_NEH computes the initial solution using a different procedure.

   Section 3.1, Sect. 3.2 and Sect. 3.3 describe the main components of the IGT_NEH heuristic, while its structure is discussed in Sect. 3.4.

3.1   Computation of the Initial Solution

The NEH heuristic [11] is one of the most common constructive heuristics used for scheduling problems. It takes its name from the three authors who proposed it for the first time (Nawaz, Enscore and Ham). The approach is quite generic: only the evaluation strategy and the initial sorting criteria change from one problem to another. First, the sum of the processing times on all stages, indicated as $TP_j = \sum_{s \in S} p_{j,s}$, is computed for each $j \in J$. The jobs are then sorted by decreasing order of TP_j, in order to have a good job reference permutation γ. The first job γ_1 of the permutation is selected to establish a partial solution of length one. Then the other jobs in γ are sequentially inserted into the output permutation γ^{out} one by one. At the i-th iteration, the job γ_i is chosen and tentatively inserted into all the i possible positions of γ^{out}, it is then inserted at the position resulting in the best makespan value. Each evaluation of a permutation is performed by means of forward scheduling and considering the makespan as the objective. Once the n-th iteration is reached, the solution constructed is provided as an output.

We now briefly discuss why we used NEH as a method to compute initial solutions, instead of the approach denoted as GRASP_NEH in [15]. The computational complexity of NEH and GRASP_NEH is different ($O(n^3m)$ vs $O(n^4m)$). In our preliminary computational experiments, we explored both variants in the metaheuristic approach. According to our results, spending too much time on computing the initial solution affected the results in large instances ($n \geq 200$), while the two approaches showed quite similar performance in smaller instances. For this reason, we decided to use NEH as a method to compute the initial solution.

3.2   Local Search Approaches

The local search moves used in IGT_NEH are guided by a reference jobs permutation. They are the same as used in IGT, with the only difference that forward scheduling takes into account the transportation times as explained at the beginning of Sect. 3. These procedures are denoted by RIS (Referenced Insertion Scheme) and RSS (Referenced Swap Scheme). Given a reference permutation γ and an initial permutation γ^{in} – the job permutation used to generate the initial solution – we iteratively select a job from γ. In RIS, we remove the job from γ^{in} and re-insert it in the best position. In RSS, we try to swap the selected job with the others in γ^{in} and choose the swap leading to the best objective. The authors in [15] discuss the fact that a reference permutation associated with a high-quality solution can improve the performance of the local search moves. Both local search routines stop whenever n iterations with no improvements are performed. The best solution found is then provided as an output.

3.3   Deconstruction and Reconstruction

The deconstruction-reconstruction procedure, denoted as DEC_REC, is straightforward. First, dS random jobs are removed from the initial permutation γ, then they are re-inserted one by one at the best possible position. Again, the only difference with IGT is that the transportation times are taken into account when applying forward scheduling.

3.4   General Structure of IGT_NEH

The IGT_NEH procedure is an iterated greedy procedure, which starts from an initial solution and iteratively applies some local search moves, followed by a shaking step to escape local minima. First, an initial solution is computed by using the function NEH. Then, a while loop iterates until a time limit $\frac{T_{max}}{2}$ is reached. The loop starts with the deconstruction-reconstruction procedure DEC_REC, where dS jobs are removed from the current permutation and re-inserted to optimality. Afterwards, one of the local search moves RIS and RSS is chosen, the first one with probability jP and the second one with probability $1 - jP$. If the solution computed by the local search is better than the solution considered at the beginning of the current iteration, the next iteration continues the search from that solution. Otherwise, the output of the local search is considered as a starting point of the next iteration with probability $\exp(-(f(\Pi') - f(\Pi))/g(\mathcal{I}, \tau P))$, where $g(\mathcal{I}, \tau P) = \frac{\sum_{j \in J} \sum_{s \in S} p_{j,s}}{|J||S|10} \times \tau P$. The best solution found is updated whenever is necessary and it is provided as an output when the time limit is reached. The pseudocode of the procedure is given in Algorithm 1.

Algorithm 1. The IGT_NEH approach

Function IGT_NEH **is**

    Input: Time limit T_{max}, Parameters dS, τP, jP

    Output: Solution Π_{best}

    $\Pi, \gamma \leftarrow$ NEH();

    $\Pi^{best} \leftarrow \Pi$;

    $\gamma^{best} \leftarrow \gamma$;

    while *time limit T_{max} not reached* **do**

        $\gamma'' \leftarrow$ DEC_REC(γ, dS);

        $r \leftarrow$ random value between 0 and 1;

        if $r < jP$ **then**

            $\Pi', \gamma' \leftarrow$ RIS(γ^{best}, γ'');

        else

            $\Pi', \gamma' \leftarrow$ RSS(γ^{best}, γ'');

        if $f(\Pi') < f(\Pi)$ **then**

            $\Pi \leftarrow \Pi'$;

            $\gamma \leftarrow \gamma'$;

            if $f(\Pi') < f(\Pi^{best})$ **then**

                $\Pi^{best} \leftarrow \Pi'$;

                $\gamma^{best} \leftarrow \gamma'$;

        else

            if $r < \exp -(f(\Pi') - f(\Pi))/g(\mathcal{I}, \tau P)$ **then**

                $\Pi \leftarrow \Pi'$;

                $\gamma \leftarrow \gamma'$;

    return Π^{best}, γ^{best};

---

4   Hybrid Approach

`IGT_NEH` implements a very different type of search compared to CP-based approaches. In fact, reference-based heuristics generate a new solution from scratch each time a different job permutation is considered, which occurs many times in the different local search moves. This is the reason why these local search moves are not suitable for embedding into a CP system. Moreover, reference-based heuristics do not perform any local search that is devoted to optimizing the makespan by purely modifying how jobs are scheduled at the last stages. As an alternative, CP-based approaches branch on specific decisions, taken at each stage and consist of assigning a job to a machine such that the job starts on a certain time point.

The idea of the proposed hybrid approach is to exploit `IGT_NEH` to quickly find high quality solutions, while CP is used to intensify the search over the last stages of the schedule. The approach is based on two phases. The first phase consists of running `IGT_NEH` until the time limit $\frac{T_{max}}{2}$ is reached. Hence, we consider the CP model of the problem and we impose that the solution is identical to the one computed at the first phase up to stage ρ. The second phase consists of solving the resulting CP model, which basically re-optimizes the best solution found at the first phase, with respect to the stages that are successive to stage ρ. The model is solved using CP Optimizer, and its black-box search routine. This simplifies development compared to using the best performing solver in [1], which is SICStus Prolog, for which we would have to write a new specific search routine. The time limit of the second phase is $\frac{T_{max}}{2}$, such that the overall time limit of the approach is equal to T_{max}. Figure 1 depicts the structure of the proposed approach, which we denote as `HYBRID`.

5   Computational Experiments

This section describes the experiments conducted to assess the performance of `IGT_NEH` and `HYBRID`. The instances considered are the ones in [1], which refer to a realistic lane scenario in industry with 8 stages, where Stages 4 and 8 may be skipped by some jobs.

All experiments were performed on an Intel(R) Xeon(R) E5620 processor running at 2.40 GHz with 32 GB of memory. Please note that this machine is slightly slower than the machine used for the experiments in [1]. `IGT_NEH` was run by using the same parameter configuration as the one used in [15] ($dS = 2$, $\tau P = 0.5$, $jP = 0.4$). For the sake of performing a fair comparison, `HYBRID` is run in single-thread mode. Table 1 and Table 2 show the results obtained when the overall time limit is set to $T_{max} = 300$ s both for `IGT_NEH` and `HYBRID`. The values of each row of both tables are averages over 25 instances and 5 seeds.

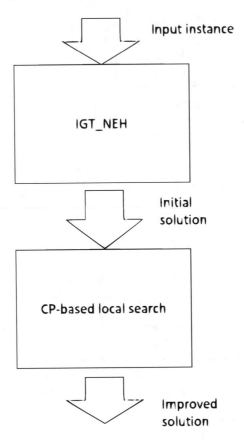

Fig. 1. Structure of the approach HYBRID.

Please note that a dash indicates that the corresponding experiments have not been conducted. Table 1 includes:

- the average objective value (obj value),
- the number of runs where an improvement was achieved with respect to IGT_NEH run for one half of the time limit (imp.),

for IGT_NEH run for 300 s, and HYBRID run for 300 s with $\rho \in \{0, 4, 6\}$. Only the average objective values are reported for NEH, and for IGT_NEH run for 150 s. Furthermore, the last two columns of both tables indicate the best results obtained with the different CP approaches published in [1], and the average of the lower bounds of the instances; see [1] for a description of the procedure used to compute these bounds.

Table 1. Average objective value and number of improved solutions per approach on different instance sizes, with the time limit $T_{max} = 300\,$s.

	NEH	IGT_NEH $\frac{T_{max}}{2}$	IGT_NEH T_{max}		HYBRID $T_{max}, \rho = 0$		HYBRID $T_{max}, \rho = 4$		HYBRID $T_{max}, \rho = 6$			
n	obj	obj	obj	imp.	obj	imp.	obj	imp.	obj	imp.	CP21	LB
20	66.72	62.78	62.76	3	62.75	4	62.77	2	62.776	1	**62.72**	61.88
25	68.84	64.32	64.26	7	64.26	8	**64.16**	6	64.30	3	**64.16**	62.84
30	72.52	66.81	66.68	16	66.74	8	**66.59**	26	66.78	4	66.68	64.12
40	78.2	72.37	72.26	14	72.18	21	**71.74**	66	72.12	40	72.56	65.32
50	85.42	78.94	78.81	21	78.31	20	**77.72**	82	78.12	43	78.4	67.24
100	115.56	109.78	109.51	29	-	-	**108.90**	74	109.26	54	113.04	94.72
200	176.24	171.69	170.08	73	-	-	**169.8**	87	170.98	60	176.72	153.08
300	239.12	238.54	**234.21**	93	-	-	235.24	113	240.96	75	240.96	214.96
400	298.82	298.82	298.54	32	-	-	**296.1**	98	298.07	61	303.16	275.36

Table 2 shows the average percentage (optimality) gap from the lower bound for each approach. It is interesting to note that, as with the previous CP results, the largest gap occurs for 50 or 100 jobs, while the optimality gap shrinks again for larger problem sizes.

The first key point that we can notice from Table 1 and Table 2 is that HYBRID is the best performing approach for most of the instance sizes. The best of its versions is the one with $\rho = 4$ stages for all the instances, but the ones that are very small ($n = 20$). Please note that in those instances the results published in [1] are slightly better, probably because of a more favorable computational setting (more powerful CPU, use of multiple threads). We can also note that IGT_NEH achieves a better result than HYBRID only for $n = 300$.

Table 3 provides an explanation for this, showing the average sum of squared deviation of the achieved objective values of runs performed with different seeds. This value is generally quite low, but very high for $n = 300$ and again very low for $n = 400$. For $n = 300$, the time limit stopped the search before the different runs converge to a common value, while for $n = 400$ there was not even enough time to find initial, but very variable, improvements in the local search routine over NEH.

Table 4 shows some experiments, performed with just one seed for an increased time limit, showing that HYBRID finds even better solutions given more time, and still outperforms IGT_NEH for the same time limit. Table 3 shows that we achieve a much lower average sum of squared deviations for $n = 300$, when the time limit is set to $T_{max} = 2400\,$s. Thus, an extension of the time limit helps improve the quality of the solutions provided by the proposed approaches and reduces the variance in the results. A closely related research question is to establish what is the best way to split the time limit in HYBRID, which we will cover in our future studies.

Figure 2 shows how the approaches perform on the first instance with $n = 100$ when different time limits are used, with the exception of NEH which is executed once, until the procedure is completed. We notice that HYBRID outperforms IGT_NEH after 200 s and maintains its lead for larger time limits. Figure 3 shows

Table 2. Average percentage gap from the lower bound per approach on different instance sizes, with the time limit $T_{max} = 300$ s.

n	NEH	IGT_NEH $\frac{T_{max}}{2}$	IGT_NEH T_{max}	HYBRID $T_{max}, \rho = 0$	HYBRID $T_{max}, \rho = 4$	HYBRID $T_{max}, \rho = 6$	CP21
20	7.25%	1.44%	1.40%	1.40%	1.41%	1.43%	**1.34%**
25	8.72%	2.30%	2.22%	2.20%	**2.06%**	2.26%	2.06%
30	11.58%	4.02%	3.84%	3.93%	**3.71%**	3.98%	3.84%
40	16.47%	9.74%	9.60%	9.50%	**8.94%**	9.43%	9.98%
50	21.28%	14.82%	14.68%	14.14%	**13.48%**	13.93%	14.23%
100	18.03%	13.72%	13.51%	-	**13.02%**	13.31%	16.21%
200	13.14%	10.84%	10.00%	-	**9.85%**	10.47%	13.38%
300	10.10%	9.89%	**8.22%**	-	8.62%	10.79%	10.79%
400	7.85%	7.85%	7.76%	-	**7.00%**	7.62%	9.17%

Table 3. Average sum of squared deviation of the objective for IGT_NEH runs with different seeds on different instance sizes, with time limits $T_{max} = 300$ s and $T_{max} = 2400$ s.

n	IGT_NEH ($T_{max} = 300$)	IGT_NEH ($T_{max} = 2400$)
20	0.064	-
25	0.272	-
30	0.336	-
40	0.464	-
50	0.544	-
100	0.88	-
200	8.624	0.624
300	38.56	3.616
400	0.688	5.588

Table 4. Average objective values and gaps obtained with IGT_NEH and HYBRID on different instance sizes, with time limit $T_{max} = 2400$ s.

	IGT_NEH (T_{max})		HYBRID($T_{max}, \rho = 4$)		
n	Obj value	Gap	Obj value	Gap	LB
200	168.56	10.11%	**167.12**	**8.40%**	153.08
300	230.52	7.24%	**229.4**	**6.75%**	214.96
400	291.36	5.81%	**291.32**	**5.48%**	275.36

that improvements for the different runs of HYBRID are less common after a time limit of 1000 s, which justifies the limit of 1200+1200 s for the experiments in Table 4.

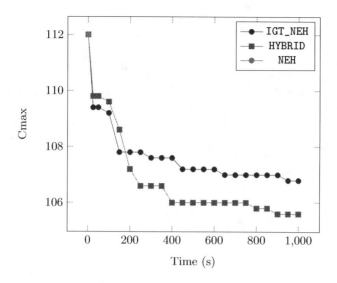

Fig. 2. Comparing average Cmax values for IGT_NEH and HYBRID over time for a single instance with 100 jobs

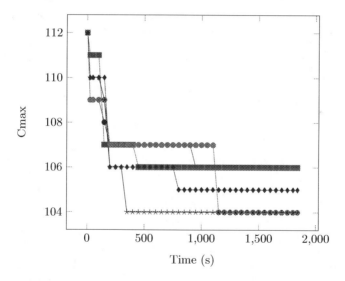

Fig. 3. Comparing Cmax value over time for five runs of HYBRID for a single instance with 100 Jobs

In conclusion, the experiments are show a clear benefit in hybridizing IGT_NEH with a CP-based step, since HYBRID outperforms both approaches. This is probably due to the complementarity of the two approaches, which use different optimization strategies. In fact, the second phase of HYBRID, based on CP, intensifies the search on the last stages of the schedule. This aspect is not taken into account by reference-guided heuristics.

6   Conclusions and Future Work

This paper proposed a novel hybrid heuristic for the HFFTT problem, which combines a metaheuristic with constraint programming. Despite its simplicity, the proposed approach allowed us to improve the results obtained in [1] by means of different CP frameworks. At the same time, the hybrid also outperformed the state-of-the-art metaheuristic for the HFS, adapted to this problem variant.

An interesting research direction is to study how similar hybrid techniques perform on other variants of hybrid flowshop problems, including the HFP, by considering instances in reference datasets. The proposed hybrid approach can be seen as a CP-based decomposition heuristic, where the problem is decomposed with respect to the stages. Other alternative CP-based hybrid schemes may also be considered. One is to decompose the problem with respect to the jobs, fixing the first jobs for all stages to the value in the initial solution, and then rescheduling for the remaining jobs, even for the initial stages. Another alternative is to decompose the problem with respect to a certain time instant, by fixing all the tasks, regardless of stage, starting before, and solving the remaining problem. These and other decompositions will be explored in our future studies.

Acknowledgements. This publication has emanated from research conducted with the financial support of Science Foundation Ireland under Grant Numbers 16/RC/3918 (Confirm) and 12/RC/2289-P2 (Insight), and co-funded under the European Regional Development Fund. The research also received financial support from Johnson and Johnson Research Centre.

References

1. Armstrong, E., Garraffa, M., O'Sullivan, B., Simonis, H.: The hybrid flexible flowshop with transportation times. In: Michel, L.D. (ed.) 27th International Conference on Principles and Practice of Constraint Programming (CP 2021), vol. 210, pp. 16:1–16:18. Schloss Dagstuhl - Leibniz-Zentrum für Informatik, Dagstuhl, Germany (2021). https://doi.org/10.4230/LIPIcs.CP.2021.16
2. Blum, C., Puchinger, J., Raidl, G.R., Roli, A.: Hybrid metaheuristics in combinatorial optimization: A survey. Appl. Soft Comput. **11**(6), 4135–4151 (2011). https://doi.org/10.1016/j.asoc.2011.02.032
3. Della Croce, F., Garraffa, M., Salassa, F., Borean, C., Di Bella, G., Grasso, E.: Heuristic approaches for a domestic energy management system. Comput. Ind. Eng. **109**, 169–178 (2017)
4. Della Croce, F., Grosso, A., Salassa, F.: Minimizing total completion time in the twomachine noidle nowait flow shop problem. Journal of Heuristics (2019). https://doi.org/10.1007/s1073201909430z
5. Fanjul-Peyro, L., Perea, F., Ruiz, R.: Models and matheuristics for the unrelated parallel machine scheduling problem with additional resources. European Journal of Operational Research **260**(2), 482–493 (2017). https://doi.org/10.1016/j.ejor.2017.01.002
6. Kergosien, Y., Mendoza, J.E., T'kindt, V.: Special issue on matheuristics. J. Heuristics **27**, 1–3 (2021)

7. Laborie, P., Rogerie, J., Shaw, P., Vilím, P.: IBM ILOG CP optimizer for scheduling. Constraints **23**(2), 210–250 (2018). https://doi.org/10.1007/s10601-018-9281-x
8. Lin, S.W., Ying, K.C.: Optimization of makespan for no-wait flowshop scheduling problems using efficient matheuristics. Omega, 64, 115–125 (2015)
9. Maniezzo, V., Stützle, T., Voß, S. (eds.) Matheuristics - Hybridizing Metaheuristics and Mathematical Programming, Annals of Information Systems, vol. 10. Springer (2010). https://doi.org/10.1007/978-1-4419-1306-7
10. Naderi, B., Gohari, S., Yazdani, M.: Hybrid flexible flowshop problems: models and solution methods. Appl. Math. Model. **38**(24), 5767–5780 (2014). https://doi.org/10.1016/j.apm.2014.04.012
11. Nawaz, M., Enscore, E.E., Ham, I.: A heuristic algorithm for the m-machine, n-job flow-shop sequencing problem. Omega **11**(1), 91–95 (1983). https://doi.org/10.1016/0305-0483(83)90088-9
12. Rendl, A., Prandtstetter, M., Hiermann, G., Puchinger, J., Raidl, G.: hybrid heuristics for multimodal homecare scheduling. In: Beldiceanu, N., Jussien, N., Pinson, É. (eds.) Integration of AI and OR Techniques in Constraint Programming for Combinatorial Optimization Problems, pp. 339–355. Springer, Berlin Heidelberg (2012). https://doi.org/10.1007/978-3-642-29828-8_22
13. Ruiz, R., Vázquez Rodríguez, J.A.: The hybrid flow shop scheduling problem. European Journal of Operational Research **205**, 1–18 (2010). https://doi.org/10.1016/j.ejor.2009.09.024
14. Tang, T.Y., Beck, J.C.: CP and hybrid models for two-stage batching and scheduling. In: Hebrard, E., Musliu, N. (eds.) Integration of Constraint Programming, Artificial Intelligence, and Operations Research, pp. 431–446. Springer International Publishing, Cham (2020). https://doi.org/10.1007/978-3-030-58942-4_28
15. Öztop, H., Fatih Tasgetiren, M., Eliiyi, D.T., Pan, Q.K.: Metaheuristic algorithms for the hybrid flowshop scheduling problem. Comput. Oper. Res. **111**, 177–196 (2019). https://doi.org/10.1016/j.cor.2019.06.009

A SAT Encoding to Compute Aperiodic Tiling Rhythmic Canons

Gennaro Auricchio[1] , Luca Ferrarini[1,2,3(✉)] , Stefano Gualandi[1] ,
Greta Lanzarotto[1,2,3,4] , and Ludovico Pernazza[1]

[1] Department of Mathematics, University of Pavia, Pavia, Italy
[2] Department of Mathematics and its Applications, University of Milano-Bicocca,
Milan, Italy
l.ferrarini3@campus.unimib.it
[3] INdAM, Rome, Italy
[4] University of Strasbourg, IRMA, Strasbourg, France

Abstract. In Mathematical Music theory, the Aperiodic Tiling Complements Problem consists in finding all the possible aperiodic complements of a given rhythm A. The complexity of this problem depends on the size of the period n of the canon and on the cardinality of the given rhythm A. The current state-of-the-art algorithms can solve instances with n smaller than 180. In this paper we propose an ILP formulation and a SAT Encoding to solve this mathemusical problem, and we use the Maplesat solver to enumerate all the aperiodic complements. We validate our SAT Encoding using several different periods and rhythms and we compute for the first time the complete list of aperiodic tiling complements of standard Vuza rhythms for canons of period $n = \{180, 420, 900\}$.

Keywords: Mathematical models for music · Aperiodic tiling rhythms · SAT encoding · Integer linear programming

1 Introduction

Mathematical Music Theory is the study of Music from a mathematical point of view. Many connections have been discovered, some of which albeit having already a long tradition, are still offering new problems and ideas to researchers, whether they be music composers or computer scientists. The first attempt to produce music through a computational model dates back to 1957, when the authors composed a string quartet, also known as the Illiac Suite, through random number generators and Markov chains [11]. Since then, a plethora of other works have explored how computer science and music can interact: to compose music [20,21], to analyse existing compositions and melodies [7–9], or even to represent human gestures of the music performer [18]. In particular, Constraint Programming has been used to model harmony, counterpoint and other aspects of music (e.g., see [4]), to compose music of various genres as described in the book [3], or to impose musical harmonization constraints in [16].

© Springer Nature Switzerland AG 2022
P. Schaus (Ed.): CPAIOR 2022, LNCS 13292, pp. 14–23, 2022.
https://doi.org/10.1007/978-3-031-08011-1_2

In this paper, we deal with Tiling Rhythmic Canons, that are purely rhythmic contrapuntal compositions. It is well-known that all canons are periodic: for a fixed period n, a tiling rhythmic canon is then represented by a couple of sets $A, B \subseteq \{0, 1, 2, \ldots, n-1\}$ where A defines the sequence of beats (of a constant metre) played by every voice, B the offsets at which the voices start to play, and such that at every beat there is exactly one voice playing. If one of the sets, say A, is given, it is well-known that the problem of finding a *complement* B has in general no unique solution. It is very easy to find tiling canons in which at least one of the set is *periodic*, i.e. it is built repeating a shorter rhythm. From a mathematical point of view, the most interesting canons are therefore those in which both sets are *aperiodic* (the problem can be equivalently rephrased as a research of tessellations of a special kind). To enumerate all aperiodic tiling canons one has to overcome two main hurdles: on one side, the problem lacks the algebraic structure of other ones, such as those involving ring or group theory; on the other side, the combinatorial size of the domain becomes very soon enormous. From a theoretical point of view, starting from the first works in the 1940 s s research has gradually shed some light on the problem; from a more concrete point of view, several heuristics and algorithms that allow to compute tiling complements have been introduced. A complete solution appears however to be still out of reach.

Contributions. The main contributions of this paper are the Integer Linear Programming (ILP) model and the SAT Encoding to solve the Aperiodic Tiling Complements Problem presented in Sect. 3. Using a modern SAT solver we are able to compute the complete list of aperiodic tiling complements of a class of Vuza rhythms for periods $n = \{180, 420, 900\}$.

Outline. The outline of the paper is as follows. Section 2 reviews the main notions on Tiling Rhythmic Canons and defines formally the problem. In Sect. 3, we introduce an ILP model and a SAT Encoding of the Aperiodic Tiling Complements Problem expressing the tiling and the aperiodicity constraints in terms of Boolean variables. Finally, in Sect. 4, we include our computational results to compare the efficiency of the aforementioned ILP model and SAT Encoding with the previous state-of-the-art algorithms.

2 The Aperiodic Tiling Complements Problem

We begin fixing some notation and giving the main definitions. In the following, we conventionally denote the cyclic group of remainder classes modulo n by $\mathbb{Z}_n$ and its elements with the integers $\{0, 1, \ldots, n-1\}$, i.e. identifying each class with its smallest non-negative member.

Definition 1. *Let* $A, B \subseteq \mathbb{Z}_n$. *Let us define the application*

$$\sigma : A \times B \to \mathbb{Z}_n (a, b) \mapsto a + b.$$

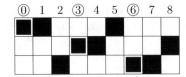

Fig. 1. $A = \{0, 1, 5\}$ is the inner voice; $B = \{0, 3, 6\}$ is the outer voice.

Fig. 2. This time, $B = \{0, 3, 6\}$ is the inner voice, and $A = \{0, 1, 5\}$ is the outer voice.

We set $A + B := Im(\sigma)$; if σ is bijective we say that A and B **are in direct sum,** and we write

$$A \oplus B := Im(\sigma).$$

If $\mathbb{Z}_n = A \oplus B$, we call (A, B) a **tiling rhythmic canon** of period n; A is called the **inner voice** and B the **outer voice** of the canon. In this case, we also say that B is a complement of A, and vice versa.

Remark 1. It is easy to see that the tiling property is invariant under translations, i.e. if A is a tiling complement of some set B, any translate $A + z$ of A is a tiling complement of B, too (and any translate of B is a tiling complement of A). In fact, suppose that $A \oplus B = \mathbb{Z}_n$; for every $k, z \in \mathbb{Z}_n$ by definition there exists one and only one pair $(a, b) \in A \times B$ such that $k - z = a + b$. Consequently, there exists one and only one pair $(a + z, b) \in (A + z) \times B$ such that $k = (a + z) + b$, that is $(A + z) \oplus B = \mathbb{Z}_n$. In view of this, without loss of generality, we shall consider equivalence classes under translation and limit our investigation to rhythms containing 0.

Example 1. Let us consider period $n = 9$ and the two rhythms $A = \{0, 1, 5\} \subset \mathbb{Z}_9$ and $B = \{0, 3, 6\} \subset \mathbb{Z}_9$ in Fig. 1 and Fig. 2. The inner voice is represented with a sequence of black boxes in a row of length n. The outer voice is represented by the starting black box in each row. A and B provide the canon $A \oplus B = \mathbb{Z}_9$, since $\{0, 1, 5\} \oplus \{0, 3, 6\} = \{0, 3, 6, 1, 4, 7, 5, 8, 2\}$, where the last number is obtained by $(5 + 6) \mod 9 = 2$.

Definition 2. A rhythm $A \subseteq \mathbb{Z}_n$ is **periodic (of period z)** if and only if there exists an element $z \in \mathbb{Z}_n$, $z \neq 0$, such that $z + A = A$. In this case, A is also called periodic modulo $z \in \mathbb{Z}_n$. A rhythm $A \subseteq \mathbb{Z}_n$ is **aperiodic** if and only if it is not periodic.

Remark 2. Going back to Example 1, it is easy to note the periodicity $z = 3$ in rhythm $B = \{0, 3, 6\}$: and indeed, $3 + B = B$. The rhythm $A = \{0, 1, 5\}$, instead, is not periodic with respect to any period. Notice that if A is periodic of period z, z must be a strict divisor of the period n of the canon.

Tiling rhythmic canons can be characterised using polynomials. This characterization allows to translate the tiling property into linear constraints and, therefore, to approach the problem through a SAT encoding.

Lemma 1. *Let A be a rhythm in $\mathbb{Z}_n$ and let $p_A(x)$ be the* **characteristic polynomial** *of A, that is, $p_A(x) = \sum_{k \in A} x^k$. Given $B \subseteq \mathbb{Z}_n$ and its characteristic polynomial $p_B(x)$, we have that*

$$p_A(x) \cdot p_B(x) \equiv \sum_{k=0}^{n-1} x^k, \qquad \mod (x^n - 1) \qquad (1)$$

if and only if $p_A(x), p_B(x)$ are polynomials with coefficients in $\{0, 1\}$ and $A \oplus B = \mathbb{Z}_n$.

Example 2. Let us consider the tiling canon introduced in Example 1. In this case, the characteristic polynomial of A is $p_A(x) = x^5 + x + 1$, while the characteristic polynomial of B is $p_B(x) = x^6 + x^3 + 1$. A simple computation shows that $\mod (x^9 - 1)$

$$p_A(x)p_B(x) = x^{11} + x^8 + x^7 + x^6 + x^5 + x^4 + x^3 + x + 1 \equiv \sum_{k=0}^{8} x^k,$$

since $x^{11} = x^9 x^2 \equiv x^2 \mod (x^9 - 1)$.

Definition 3. *A tiling rhythmic canon (A, B) in $\mathbb{Z}_n$ is a* **Vuza canon** *if both A and B are aperiodic.*

Remark 3. Note that a set A is periodic modulo z if and only if it is periodic modulo all the non-trivial multiples of z dividing n. For this reason, when it comes to check whether A is periodic or not, it suffices to check if A is periodic modulo m for every m in the set of maximal divisors of n. We denote by $\mathcal{D}_n$ this set:

$$\mathcal{D}_n := \{ n/p \mid p \text{ is a prime factor of } n \}.$$

We also denote with k_n the cardinality of $\mathcal{D}_n$, so that $n = p_1^{\alpha_1} p_2^{\alpha_2} \ldots p_{k_n}^{\alpha_{k_n}}$ is the unique prime factorization of n, where $\alpha_1, \ldots, \alpha_{k_n} \in \mathbb{N}^+$.

For a complete and exhaustive discussion on tiling problems, we refer the reader to [2]. In this paper, we are interested in the following tiling problem.

Definition 4. *Given a period $n \in \mathbb{N}$ and a rhythm $A \subseteq \mathbb{Z}_n$, the* **Aperiodic Tiling Complements Problem** *consists in finding all its aperiodic complements B i.e., all subsets B of $\mathbb{Z}_n$ such that $A \oplus B = \mathbb{Z}_n$.*

Some problems very similar to the decision of tiling (i.e., the tiling decision problem DIFF in [13]) have been shown to be NP-complete; a strong lower bound for computational complexity of the tiling decision problem is to be expected, too.

3   A SAT Encoding

In this section, we present in parallel an ILP model and a new SAT Encoding for the Aperiodic Tiling Complements Problem that are both used to enumerate all complements of A. We define two sets of constraints: (i) the *tiling constraints* that impose the condition $A \oplus B = \mathbb{Z}_n$, and (ii) the *aperiodicity constraints* that impose that the canon B is aperiodic.

Tiling constraints. Given the period n and the rhythm A, let $\boldsymbol{a} = [a_0, \ldots, a_{n-1}]^{\mathsf{T}}$ be its characteristic (column) vector, that is, $a_i = 1$ if and only if $i \in A$. Using vector $\boldsymbol{a}$ we define the circulant matrix $T \in \{0,1\}^{n \times n}$ of rhythm A, that is, each column of T is the circular shift of the first column, which corresponds to vector $\boldsymbol{a}$. Thus, the matrix T is equal to

$$
T = \begin{bmatrix}
a_0 & a_{n-1} & a_{n-2} & \cdots & a_1 \\
a_1 & a_0 & a_{n-1} & \cdots & a_2 \\
\vdots & \vdots & \vdots & \ddots & \vdots \\
a_{n-1} & a_{n-2} & a_{n-3} & \cdots & a_0
\end{bmatrix}.
$$

We can use the circulant matrix T to impose the tiling conditions as follows. Let us introduce a literal x_i for $i = 0, \ldots, n-1$, that represents the characteristic vector of the tiling rhythm B, that is, $x_i = 1$ if and only if $i \in B$. Note that a literal is equivalent to a 0–1 variable in ILP terminology. Then, the tiling condition can be written with the following linear constraint:

$$
\sum_{i \in \{0, \ldots, n-1\}} T_{ij} x_i = 1, \quad \forall j = 0, \ldots, n-1. \tag{2}
$$

Notice that the set of linear constraints (2) imposes that exactly one variable (literal) in the set $\{x_{n+i-j \bmod n}\}_{j \in A}$ is equal to one. Hence, we encode this condition as an `Exactly-one` constraint, that is, exactly one literal can take the value one. The `Exactly-one` constraint can be expressed as the conjunction of the two constraints `At-least-one` and `At-most-one`, for which standard SAT encoding exist (e.g., see [6,17]). Hence, the tiling constraints (2) are encoded with the following set of clauses depending on $i = 0, \ldots, n-1$:

$$
\bigvee_{j \in A} \left(x_{n-(j-i) \bmod n} \right) \bigwedge_{k,l \in A, k \neq l} \left(\neg x_{n-(k-i) \bmod n} \vee \neg x_{n-(l-i) \bmod n} \right). \tag{3}
$$

Aperiodicity constraints. In view of Definition 2, if there exists a $b \in B$ such that $(d+b) \bmod n \neq b$, then the canon B is not periodic modulo d. Notice that by Remark 3 we need to check this condition only for the values of $d \in \mathcal{D}_n$.

We formulate the aperiodicity constraints introducing auxiliary variables $y_{d,i}, z_{d,i}, u_{d,i} \in \{0,1\}$ for every prime divisor $d \in \mathcal{D}_n$ and for every integer $i = 0, \ldots, d-1$. We set

$$
u_{d,i} = 1 \Leftrightarrow \left(\sum_{k=0}^{n/d-1} x_{i+kd} = \frac{n}{d} \right) \vee \left(\sum_{k=0}^{n/d-1} x_{i+kd} = 0 \right), \tag{4}
$$

for all $d \in \mathcal{D}_n$, $i = 0, \ldots, d-1$, with the condition

$$
\sum_{i=0}^{d-1} u_{d,i} \leq d - 1, \quad \forall d \in \mathcal{D}_n. \tag{5}
$$

Similarly to [5], the constraints (4) can be linearized using standard reformulation techniques as follows:

$$0 \leq \sum_{k=0}^{n/d} x_{i+kd} - \frac{n}{d} y_{d,i} \leq \frac{n}{d} - 1 \qquad \forall d \in \mathcal{D}_n, \ \ i = 0, \ldots, d-1, \qquad (6)$$

$$0 \leq \sum_{k=0}^{n/d} (1 - x_{i+kd}) - \frac{n}{d} z_{d,i} \leq \frac{n}{d} - 1 \qquad \forall d \in \mathcal{D}_n, \ \ i = 0, \ldots, d-1, \qquad (7)$$

$$y_{d,i} + z_{d,i} = u_{d,i} \qquad \forall d \in \mathcal{D}_n, \ \ i = 0, \ldots, d-1. \qquad (8)$$

Notice that when $u_{d,i} = 1$ exactly one of the two incompatible alternatives in the right hand side of (4) is true, while whenever $u_{d,i} = 0$ the two constraints are false. Correspondingly, the constraint (8) imposes that the variables $y_{d,i}$ and $z_{d,i}$ cannot be equal to 1 at the same time. On the other hand, constraint (5) imposes that at least one of the auxiliary variables $u_{d,i}$ be equal to zero.

Next, we encode the previous conditions as a SAT formula. To encode the if and only if clause, we make use of the logical equivalence between $C_1 \Leftrightarrow C_2$ and $(\neg C_1 \vee C_2) \wedge (C_1 \vee \neg C_2)$. The clause C_1 is given directly by the literal $u_{d,i}$. The clause C_2, expressing the right hand side of (4), i.e. the constraint that the variables must be either all true or all false, can be written as

$$C_2 = \left(\bigwedge_{k=0}^{n/d} x_{i+kd} \right) \vee \left(\bigwedge_{k=0}^{n/d} \bar{x}_{i+kd} \right), \qquad \forall d \in \mathcal{D}_n.$$

Then, the linear constraint (5) can be stated as the SAT formula:

$$\neg \left(u_{d,0} \wedge u_{d,1} \wedge \cdots \wedge u_{d(d-1)} \right) = \bigvee_{l=0}^{d-1} \bar{u}_{d,l}, \qquad \forall d \in \mathcal{D}_n.$$

Finally, we express the aperiodicity constraints using

$$\bigwedge_{i=0}^{d-1} [(\neg C_2 \vee u_{d,i}) \wedge (C_2 \vee \bar{u}_{d,i})] \wedge \bigvee_{l=0}^{d-1} \bar{u}_{d,l}, \forall d \in \mathcal{D}_n. \qquad (9)$$

Note that joining (2), (6)–(8) with a constant objective function gives a complete ILP model, which can be solved with a modern ILP solver such as Gurobi to enumerate all possible solutions. At the same time, joining (3) and (9) into a unique CNF formula, we get our complete SAT Encoding of the Aperiodic Tiling Complements Problem. (see Sect. 4 for computational results).

3.1 Existing Solution Approaches

For the computation of all the aperiodic tiling complements of a given rhythm the two most successful approaches already known are the *Fill-Out Procedure* [14] and the *Cutting Sequential Algorithm* [5].

The Fill-Out Procedure. The *Fill-Out Procedure* is the heuristic algorithm introduced in [14]. The key idea behind this algorithm is the following: given a rhythm $A \subseteq \mathbb{Z}_n$ such that $0 \in A$, the algorithm sets $P = \{0\}$ and starts the search for possible expansions of the set P. The expansion is accomplished by adding an element $\alpha \in \mathbb{Z}_n$ to P according to the reverse order induced by a ranking function $r(x, P)$, which counts all the possible ways in which x can be covered through a translation of A. Once every element of $\mathbb{Z}_m \backslash (A \oplus P)$ has been ranked, the algorithm tries to add the element with the lowest rank. Adding a new element defines a new set, namely $\tilde{P} \supset P$, which is again expanded until either it can no longer be expanded or the set becomes a tiling complement. The search ends when all the possibilities have been explored. The algorithm finds also periodic solutions that must removed in post-processing, as well as multiple translations of the same rhythm.

The Cutting Sequential Algorithm (CSA). In [5], the authors formulate the Aperiodic Tiling Complements Problem using an Integer Linear Programming (ILP) model that is based on the polynomial characterization of tiling canons. The ILP model uses auxiliary 0–1 variables to encode the product $p_A(x) \cdot p_B(x)$ which characterizes tiling canons. The aperiodicity constraint is formulated analogously to what done above. The objective function is equal to a constant and has no influence on the solutions found by the model. The ILP model is used within a sequential cutting algorithm that adds a no-good constraint every time a new canon B is found to prevent finding solutions twice. In addition, the sequential algorithm sets a new no-good constraints for every translation of B; hence, in contrast to the *Fill-Out Procedure*, the *CSA Algorithm* needs no post-processing.

4   Computational Results

First, we compare the results obtained using our ILP model and SAT Encoding with the runtimes of the *Fill-Out Procedure* and of the *CSA Algorithm*. We use the canons with periods 72, 108, 120, 144 and 168 that have been completely enumerated by Vuza [19], Fripertinger [10], Amiot [1], Kolountzakis and Matolcsi [14]. Table 1 shows clearly that the two new approaches outperform the state-of-the-art, and in particular, that SAT provides the best solution approach. We then choose some periods n with more complex prime factorizations, such as $n = p^2 q^2 r = 180$, $n = p^2 q r s = 420$, and $n = p^2 q^2 r^2 = 900$. To find aperiodic rhythms A, we apply Vuza's construction [19] with different choices of parameters p_1, p_2, n_1, n_2, n_3. Thus, having n and A as inputs, we search for all the possible aperiodic complements and then we filter out the solutions under translation. Since the post-processing is based on sorting canons, it requires a comparatively small amount of time. We report the results in Table 2: the solution approach based on the SAT Encoding is the clear winner. It is also noteworthy that, from a Music theory perspective, this is the first time that all the tiling complements of the studied rhythms are computed (their number is reported in the last column of the two tables).

Table 1. Aperiodic tiling complements for periods $n \in \{72, 108, 120, 144, 168\}$.

n	$\mathcal{D}_n$	p_1	n_1	p_2	n_2	n_3	runtimes (s)				$\#B$
							FOP	CSA	SAT	ILP	
72	$\{24, 36\}$	2	2	3	3	2	1.59	0.10	<0.01	0.03	6
108	$\{36, 54\}$	2	2	3	3	3	896.06	7.84	0.09	0.19	252
120	$\{24, 40, 60\}$	2	2	5	3	2	24.16	0.27	0.02	0.04	18
		2	2	3	5	2	10.92	0.14	0.01	0.04	20
144	$\{48, 72\}$	4	2	3	3	2	82.53	2.93	0.02	0.11	36
		2	2	3	3	4	>10800.00	>10800.00	11.04	46.96	8640
		2	2	3	3	4	7.13	0.10	<0.01	0.05	6
		2	4	3	3	2	80.04	0.94	0.02	0.08	60
168	$\{24, 56, 84\}$	2	2	7	3	2	461.53	17.61	0.04	0.20	54
		2	2	3	7	2	46.11	0.91	0.02	0.07	42

Table 2. Aperiodic tiling complements for periods $n \in \{180, 420, 900\}$.

n	$\mathcal{D}_n$	p_1	n_1	p_2	n_2	n_3	runtimes (s)		$\#B$
							SAT	ILP	
180	$\{36, 60, 90\}$	2	2	5	3	3	2.57	5.62	2052
		3	3	5	2	2	0.07	0.14	96
		2	2	3	5	3	1.25	2.23	1800
		2	5	3	3	2	0.05	0.16	120
		2	2	3	3	5	8079.07	>10800.00	281232
420	$\{60, 84, 140, 210\}$	7	5	3	2	2	2.13	3.57	720
		5	7	3	2	2	1.52	4.08	672
		7	5	2	3	2	7.73	16.11	3120
		5	7	2	3	2	1.63	4.18	1008
		7	3	5	2	2	4.76	7.45	864
		3	7	5	2	2	12.78	32.19	6720
		7	3	2	5	2	107.83	1186.21	33480
		3	7	2	5	2	0.73	2.36	840
		7	2	5	3	2	11.14	21.19	1872
		2	7	5	3	2	17.31	52.90	10080
		7	2	3	5	2	89.97	691.56	22320
		2	7	3	5	2	1.17	4.13	1120
900	$\{180, 300, 450\}$	2	25	3	3	2	43.60	110.65	15600
		5	10	3	3	2	107.36	741.79	15840
		2	9	5	5	2	958.58	>10800.00	118080
		6	3	5	5	2	5559.76	>10800.00	123840
		3	6	5	5	2	486.39	8290.35	62160

Implementation Details. We have implemented in Python the ILP model and in PySat [12] the SAT Encoding discussed in Sect. 3. We use Gurobi 9.1.1 as ILP solver and Maplesat [15] as SAT solver. The experiments are run on a Dell Workstation with a Intel Xeon W-2155 CPU with 10 physical cores at 3.3GHz and 32 GB of RAM. The source code and the datasets is freely available on GitHub at https://github.com/LucaFerra94/tiling-rhytmic-canons.

Conclusions and Future Work. It is thinkable to devise an algorithm that, for a given n, finds all the pairs (A, B) that give rise to a Vuza canon of period n. This could provide in-depth information on the structure of Vuza canons. Moreover, due to the nature of the problem, it looks like a binary encoder could improve further the performances and allow us to approach even harder instances.

Acknowledgements. The authors thank the unknown referees for their valuable comments and suggestions. This research was partially supported by: Italian Ministry of Education, University and Research (MIUR), Dipartimenti di Eccellenza Program (2018–2022) - Dept. of Mathematics "F. Casorati", University of Pavia; Dept. of Mathematics and its Applications, University of Milano-Bicocca; National Institute of High Mathematics (INdAM) "F. Severi"; Institute for Advanced Mathematical Research (IRMA), University of Strasbourg.

References

1. Amiot, E.: New perspectives on rhythmic canons and the spectral conjecture. J. Math. Music **3**(2), 71–84 (2009)
2. Amiot, E.: Structures, algorithms, and algebraic tools for rhythmic canons. Perspect. New Music **49**(2), 93–142 (2011)
3. Anders, T.: Compositions Created with Constraint Programming. The Oxford Handbook of Algorithmic Music, Oxford University Press, London (2018)
4. Anders, T., Miranda, E.R.: Constraint programming systems for modeling music theories and composition. ACM Comput. Surv. (CSUR) **43**(4), 1–38 (2011)
5. Auricchio, G., Ferrarini, L., Lanzarotto, G.: An integer linear programming model for tilings. arXiv preprint. arXiv:2107.04108 (2021)
6. Bailleux, O., Boufkhad, Y.: Efficient CNF encoding of boolean cardinality constraints. In: Rossi, F. (ed.) CP 2003. LNCS, vol. 2833, pp. 108–122. Springer, Heidelberg (2003). https://doi.org/10.1007/978-3-540-45193-8_8
7. Chemillier, M., Truchet, C.: Two musical csps. In: CP 01 Workshop on Musical Constraints, pp. 1–1 (2001)
8. Courtot, F.: A Constraint-Based Logic Program for Generating Polyphonies. Michigan Publishing, University of Michigan Library, Ann Arbor, MI (1990)
9. Ebcioğlu, K.: An expert system for harmonizing four-part chorales. Comput. Music J. **12**(3), 43–51 (1988)
10. Fripertinger, H., Reich, L., et al.: Remarks on Rhythmical Canons. Citeseer (2005)
11. Hiller Jr, L.A., Isaacson, L.M.: Musical composition with a high speed digital computer. In: Audio Engineering Society Convention 9. Audio Engineering Society (1957)

12. Ignatiev, A., Morgado, A., Marques-Silva, J.: PySAT: a Python toolkit for prototyping with SAT oracles. In: Beyersdorff, O., Wintersteiger, C.M. (eds.) SAT 2018. LNCS, vol. 10929, pp. 428–437. Springer, Cham (2018). https://doi.org/10.1007/978-3-319-94144-8_26
13. Kolountzakis, M.N., Matolcsi, M.: Complex hadamard matrices and the spectral set conjecture. Collectanea Math. Extra **57**, 281–291 (2006)
14. Kolountzakis, M.N., Matolcsi, M.: Algorithms for translational tiling. J. Math. Music **3**(2), 85–97 (2009)
15. Liang, J.H.: Machine learning for SAT solvers. Ph.D. thesis. University of Waterloo, December 2018
16. Pachet, F., Roy, P.: Musical harmonization with constraints: a survey. Constraints **6**(1), 7–19 (2001). https://doi.org/10.1023/A:1009897225381
17. Philipp, T., Steinke, P.: PBLib – a library for encoding pseudo-boolean constraints into CNF. In: Heule, M., Weaver, S. (eds.) SAT 2015. LNCS, vol. 9340, pp. 9–16. Springer, Cham (2015). https://doi.org/10.1007/978-3-319-24318-4_2
18. Radicioni, D.P., Lombardo, V.: A constraint-based approach for annotating music scores with gestural information. Constraints **12**(4), 405–428 (2007). https://doi.org/10.1007/s10601-007-9015-y
19. Vuza, D.T.: sets and regular complementary unending canons (part one, two, three, four). Perspectives of New Music (1991–93)
20. Wiggins, G., Harris, M., Smaill, A.: Representing music for analysis and composition. In: Proceedings of the 2nd IJCAI AI/Music Workshop, pp. 63–71 (1989)
21. Zimmermann, D.: Modelling musical structures. Constraints **6**(1), 53–83 (2001). https://doi.org/10.1023/A:1009801426289

Transferring Information Across Restarts in MIP

Timo Berthold[1]([⊠])(iD), Gregor Hendel[1], and Domenico Salvagnin[2](iD)

[1] Fair Isaac Germany GmbH, Stubenwald -Allee 19, 64625 Bensheim, Germany
{timoberthold,gregorhendel}@fico.com
[2] University of Padova, Via Gradenigo 6/B, 35131 Padova, Italy
salvagni@dei.unipd.it

Abstract. Restarting a solver gives us the chance to learn from things that went good or bad in the search until the restart point. The benefits of restarts are often justified with being able to employ different, better strategies and explore different, more promising parts of the search space. In that light, it is an interesting question to evaluate whether carrying over detected structures and collected statistics across a restart benefits the subsequent search, or even counteracts the anticipated diversification from the previous, unsuccessful search.

In this paper, we will discuss four different types of global information that can potentially be re-used after a restart of a mixed-integer programming (MIP) solver, present technical details of how to carry them through a represolve after a restart, and show how such an information transfer can help to speed up the state-of-the-art commercial MIP solver FICO Xpress by 7% on the instances where a restart is performed.

Keywords: mixed integer programming · restart · global search

1 Introduction

Restarts have been used in SAT solvers for over 20 years [21]. It was quickly picked up by the MIP community in the form of root restarts [2,3] and in other areas of optimization, like global optimization [17]. For an overview on restarting algorithms in the areas of CP, AI, and OR, see [23]. In this paper we consider the solution of *mixed-integer programs (MIPs)*, which are optimization problems of the form

$$x^\star \in \operatorname{argmin}\{c^T x \mid Ax \geq b, x_j \in \mathbb{Z} \text{ for all } j \in \mathcal{I}\},$$

with $A \in \mathbb{R}^{m \times n}$, $b \in \mathbb{R}^m$, $c \in \mathbb{R}^n$, and $\mathcal{I} \subseteq \{1, \dots, n\}$.

For some time, restarts for MIP have primarily revolved around the rule of thumb of restarting the root node processing in the case enough variables have been globally fixed so that another round of full presolving seems beneficial [2]. More recently, restarts of the tree search have gained attention in the MIP community, see, e.g., [6], and most MIP solvers employ them nowadays

P. Schaus (Ed.): CPAIOR 2022, LNCS 13292, pp. 24–33, 2022.
https://doi.org/10.1007/978-3-031-08011-1_3

[11,13,18]. Unlike root restarts, tree restarts are often based on an extrapolation of the remaining time until the tree search is finished [16]. In contrast to SAT solving, aggressive periodic restarts of the tree search have not proven advantageous for MIP in the vast majority of cases for a few reasons:

1. Solving the initial root problem of a MIP is often orders of magnitude more expensive than solving a local node. This makes tree restarts in the early phase of the search way more expensive than their SAT counterpart.
2. While in SAT the set of learned conflicts completely describes the search space already explored, at least until conflict purging kicks in, the same does not apply to the MIP case, where the search strategy is hardly ever a pure DFS and conflict clauses are not always learnt. Thus, a tree restart inevitably leads to some redundant work being performed.
3. MIPs are only very rarely feasibility problems, where a tree restart might just be beneficial because it makes the algorithm find a feasible solution sooner that it would have done without. The optimality proof requires a certain amount of search to be performed no matter what, making restarts in a later phase of the search unattractive, if not harmful.

For those reasons, MIP solvers typically apply very few tree restarts (only one in most cases) and aim at addressing different types of problems with the solver behavior before and after said restarts. The solvers use settings aimed at easy problems in the first phase. Those will typically spare some expensive subroutines and make those problems solve faster. If a restart is conducted, the solver is aware that this problem is relatively hard to solve and will require a certain amount of tree search. Thus, it can adapt and employ more aggressive settings and expensive additional techniques right from the beginning of the subsequent searches. Furthermore, a restart might help in mitigating and even exploiting performance variability effects [19].

Given that one of the primary purposes of tree restarts is to make the search after the restart different from the one before, it is not obvious to which extent information from a given tree search should be carried over to be used in the next search second run, and which type of information is most suitable for such a transfer. The goal of this paper is to address these questions and evaluate the impact of global information transfer, taking the solver restart strategy as a given.

2   Global Information

A main difference between MIP and CP solvers is how constraints interact with each other. In CP solvers, the variables' domain is the central communication object, and each constraint individually contributes to tightening the domain store and guiding the search by propagation. Hence, information is processed very locally, on a constraint-by-constraint level. In MIP solvers, the central object is the LP relaxation, which is solved to optimality, considering all constraints simultaneously.

This is our role model for what we would like to call global information. The data structures that we consider for transfers across restarts have in common that they can connect pieces of information from multiple constraints into a single structure, and the information contained therein is globally valid for the problem at hand. Some of them are gained during the search, like *cutting planes* [25] and *pseudo-costs* [8], and they depend on algorithmic choices of the solver. Hence, rerunning the search with slightly different start conditions would lead to constructing a different object. Others, like the *implication graph* [24], are detected in presolving or during root node processing. These structures might only be partially created for large input problems as the involved detection procedures use certain work limits. Again, a slight change in start conditions would lead to a different result, and we cannot simply recollect the exact same information after a restart.

We will now briefly sketch each type of global information relevant to the present work. For an introduction to how global structures can be used for primal heuristics, see [14].

The Conflict Pool. MIP solvers attempt to create conflicts whenever pruning a node due to infeasibility or cutoff. They are generated by analyzing the sequence of branching decisions and propagations that led to the node being pruned, with the aim of extracting a small subset of those bound changes as an explanation for the infeasibility. From such a conflict, we can derive a conflict clause which is typically formulated as a disjunction of bounds [1]. A conflict clause is a globally valid, not necessarily linear, constraint that states that at least one of the bounds in the conflict must take a different value than in the infeasible subproblem. Note that it is equivalent to state the conflict as a conjunction of the bound changes that led to infeasibility. This is how Xpress stores conflict clauses. MIP solvers typically collect conflicts in a pool, see, e.g., [29] and only keep them for a certain number of nodes. The content of this pool highly depends on the search tree, and how much of it has already been traversed.

Cutting Planes. Throughout the search, and in particular at the root node, MIP solvers generate cutting planes to tighten the LP relaxation, improve the dual bound, and reduce the number of fractional variables. For an overview on cuts in MIP, see [20,27,30]. Globally valid cutting planes, as those separated at the root, can be either directly added to the current LP relaxation or added to a so-called *cut pool*, from which they can be separated again on demand, e.g., at subsequent nodes [25]. Transferring globally valid cuts after restarts can be effective in a twofold way: a) we can warmstart the root cutloop with a set of cuts which have proven useful at the previous root node and b) if we take them into account while represolving they can lead to further reductions.

The Implication Graph. An implication is a relation between two variables that state that a bound change in one variable leads to a bound change in another variable. E.g., for $y \leq ax + b$, it holds that a tightened upper bound on x implies an upper bound on y, which is potentially tighter than the current upper bound.

The set of all such (known) relations for a MIP can be represented as a directed graph, where each arc represents an implication and each node represents a bound change on a variable [24]. Implications might be directly deduced from single constraints but, much more interestingly, we might find them by *probing*, i.e., the process of tentatively changing the bound of a variable and propagating this bound change. Since probing is relatively expensive, it is typically not done on all variables but only on a few promising ones under tight working limits.

Pseudo-costs. On MIP problems with a nonzero objective function, branching decisions that raise the LP objective, and hence the node dual bound in the created children, are essential to prune subproblems and thus create small search trees quickly. The pseudo-costs [8] of a variable x_j summarize the average change of the LP objective observed after branching on x_j. Prioritizing variables based on their recorded pseudo-costs is still a state-of-the-art selection strategy because it provides a good selection quality that is computationally cheap to evaluate. For a recent overview on branching in MIP, see [12]. The main disadvantage of pseudo-costs is that they are uninitialized at the beginning of the search, when the most crucial top-level branching decisions have to be made. The typical remedy is to perform an explicit look-ahead called strong branching [22] for a number of times [5,15], to make an informed branching selection and initialize the pseudo-costs of the tested candidates. In the context of tree restarts, pseudo-costs from the first run can be transferred to have branching statistics available right from the beginning of the subsequent search.

3   Implementation Details

This section highlights the algorithmic choices in FICO Xpress [11] regarding the transfer of global information across restarts. Trivially, we always transfer the primal and dual bounds from a previous run. Further, we preserve the incumbent solution, although it may not remain feasible for the remaining search problem due to additional presolving, in which variables might be fixed to solution values that differ from their value in the incumbent solution, e.g., by dual reductions [4] or even reduced cost fixing.

Mapping Between Presolve Spaces. Whether in the tree or at the root, one major effect of a restart is that the problem gets (re)-presolved another time. To transfer a piece of information across a restart, we need to be able to map a mathematical object involving variables in the presolve space before the restart to the corresponding mathematical object (if any) involving variables in the presolve space after the restart.

We compute a mapping between the two presolve spaces in the following way. Independent of restarts, the solver maintains a stack with the presolve reductions applied so far, as this is needed, among other things, to map presolved solutions back to the original space. When represolving, new reductions are simply appended to the very same stack. To compute a mapping between the two

spaces, we initialize the mapping to the identity and then process the new presolve reductions, in the order in which they were added to the stack, and update
the mapping accordingly, depending on the logic of each individual reduction.
In particular, for each variable in the old presolve space, we distinguish the
following cases:

1. the variable got fixed to some value and hence removed from the problem. In
   this case, we mark the variable as *fixed* in the mapping and store the value
   it was fixed to.
2. the variable was carried through unmodified but has potentially a different
   index in the new presolved space. We simply store the new index of the
   variable.
3. the variable changed meaning (e.g., it changed type) or was eliminated in
   such a way that its value can only be computed a-posteriori during the post-
   solve phase. In this case, we mark the variable as *unmappable*. Any object
   involving an unmappable variable will become unmappable itself and thus
   not transferred across a restart.

Another essential piece of information that we need for a correct transfer is
scaling [26]. However, scaling itself is not considered a presolve reduction in
FICO Xpress, and thus it does not contribute to the stack. Still, we need to
keep track of the scaling factors (w.r.t. the original problem) before and after
represolve to compute the *intermediate* scaling factors between the two presolve
spaces. Note, however, that integer variables are never scaled.

Pseudo-costs. For each variable that is still mappable after a represolve, we con-
dense the entire pseudo-costs from both strong branches and regular branches
on a variable x from the old run into a single pseudo-cost record, in the spirit
of [9]. As a result, we have initialized pseudo-costs available from the beginning
of the search in the new run. They are not considered *reliable* [5] yet, such that
also these variables will eventually be subject to further strong branching evaluations. When recording new branching information during the new run, the
obtained, more recent pseudo-cost information eventually outweighs the transferred information.

The transfer of the variable branching history follows the intuition that the
transferred information is still accurate when the problem does not change too
much during a represolve. Following this logic, we keep track of the relative
change in the number of rows, columns, and matrix nonzero elements after the
restart. If one of these numbers changes by more than 10% (5% in the case
of rows), we discard the collected pseudo-costs and start the search with all
pseudo-costs uninitialized.

Conflicts. Conflicts stored in the conflict pool are processed one by one, using
the mapping and scaling factors computed after represolve. Given a conflict in
conjunctive form, each bound change is mapped to the new space individually,
leading to one of the following cases:

1. the bound change is on an unmappable variable: the conflict will be dropped;
2. the bound change is on a fixed variable: depending on the fixing value, we can either drop the bound change from the conflict (because it is always satisfied) or drop the conflict itself (the condition is never satisfied and thus the conflict would be useless);
3. the bound change is on another variable: simply update to the new index.

If a conflict survives the mapping process, it is finally dealt with as follows:

1. if the conflict is now empty, the problem is infeasible, and we can stop immediately;
2. if the conflict has size one, we turn it into a globally valid bound change;
3. otherwise, the conflict is added to the new conflict pool.

Implication Table. Processing of the implication table is by far and large similar in spirit to the processing of the conflict pool. Each implication in the table is mapped individually, and the outcome depends on a few cases:

1. if either the implying or the implied variable is unmappable, then the implication itself is discarded;
2. if either the implying or the implied variable is fixed, then we can turn the implication into a globally valid bound on the other variable. If both variables got fixed, the implication either directly proves infeasibility or is redundant and can be discarded.
3. if both variables survived, we map the implication by updating the indices of the involved variables and recomputing the coefficients using the intermediate scaling factors.

Cuts. While we could implement the mapping of globally valid cutting planes with the same logic that we used for conflicts and implications, the represolve of cuts follows a completely different logic for historical and technical reasons. In particular:

1. first, cuts are temporarily added to the problem before represolve, as if they were regular constraints in the model;
2. then, we execute the represolve;
3. finally, the (surviving) cuts are removed again from the model and added to the new global cut pool.

There are advantages and disadvantages to this strategy: on the one hand, the presence of cutting planes in the model might hinder some reductions, like column domination; on the other hand, their presence could also lead to a stronger presolved model, as bound tightening can in principle derive tighter bounds exploiting the additional constraints for propagation. In general, cutting planes prevent those reductions that would make the cuts themselves unmappable, which again can be argued both for and against. We did not experiment with alternative approaches, as our current strategy works well in practice: however, that is certainly an interesting direction for future research.

4   Computational Experiments

In this section, we evaluate the transfer of global information across represolves computationally. To this end, we perform two runs with the recently released FICO Xpress 8.13 solver, with the transfer of global information enabled and disabled, respectively. As benchmark set we use a mix of publicly available and customer MIP instances that also serves as one of the main test sets in our daily Xpress MIP development. In order to mitigate the effect of performance variability, we solve three permutations of each of these 619 MIP instances, resulting in a testbed of 1857 instances in total: the unpermuted model and two cyclic permutations [10] characterized by a different initial random seed used for perturbing the rows, columns, and integer variables of the instance.

Both runs are performed on a cluster of 64 identical machines, each equipped with 2 Intel(R) Xeon(R) CPUs E5-2640 v4 @ 2.40 GHz and 64 GB of memory. We set a time limit of 4 h, equaling 14400 s, and allow each job to use 20 parallel threads.

Table 1. Computational results obtained when enabling or disabling the transfer of global information across restarts.

Class	N	no-transfer			transfer			relative	
		Solved	Time	Nodes	Solved	Time	Nodes	Time	Nodes
all	1857	1844	72.97	760	1846	71.03	694	0.97	0.91
[10,14400}	1802	1798	108.22	1039	1800	104.97	952	0.97	0.91
[100,14400}	685	681	409.56	4167	683	390.44	3615	0.95	0.87
affected instances									
all	670	662	113.77	8781	664	105.97	6841	0.93	0.78
[10,14400}	565	561	165.65	17191	563	153.08	13638	0.92	0.79
[100,14400}	338	334	434.21	44505	336	393.87	33397	0.91	0.75

We summarize the results of this experiment in Table 1. The table shows solved instances, time, and nodes for both tested versions for different instance classes. The top part of the table shows the results for subsets of the entire instance bed. The bottom part of the table shows the results only for affected instances, i.e., instances on which the transferred global information affects the solver behavior after the restart. The instance class all in the first row shows the results for the entire testbed. For each class, the number of involved instances is shown in column N. We also present the results for two bracketed subsets [10, 14400} and [100, 14400}. A bracketed subset consists of all instances that could be solved by at least one of the two tested versions, and the slower of the two solves took at least 10 or 100 s or timed out, respectively. The bracketing convention is helpful to filter instances that are solved fast by both versions. Time and node results use a shifted geometric mean [2] with shift values of 10 s and 100 nodes.

The overall results suggest that transferring global information is valuable for the search procedure across the considered subsets of instances, yielding an overall time improvement of 3 % and node improvement of 9 %. The table shows that the number of affected instances amounts to roughly one-third of the test set, which is almost all instances for which FICO Xpress actually performs a restart.

On the affected instances, the performance improvements are naturally more pronounced. Overall, we see a gain of 7% time and 22% nodes. The observed improvement amounts to 9% time and 25% nodes on the most challenging bracket. All time and node improvements are significant according to a Wilcoxon signed rank test [28] with a confidence level of 95%.

Especially the reduction in the solving nodes is a clear indication that the transfer of global information helps the search to make better decisions in the second run. When comparing the relative reduction in time and nodes, we see that the transferred global information comes at a cost. Especially the transferred cutting planes increase the size of the matrix, which increases the time to solve the node LP relaxations in the second run.

When testing the transfer of individual pieces of global information, we found that transferring cutting planes across represolves leads to the most substantial improvements. Transferring cutting planes alone causes a change in the solution path on almost all of the affected instances in Table 1, and is responsible for a speed up of 3 %, across all of the considered brackets. Transferring pseudo-costs only also gives a significant speedup, albeit a bit smaller than the one from cutting planes. A reason for this is that fewer instances (194) are affected by the change: this is partially due to the fact that pseudo-cost transfer will only affect tree restarts, but not root restarts, and partially due to the employed limits on the relative change in the represolved problem. However, if we restrict to the set of instances affected by the change, transferring pseudo-costs is responsible for a speed up of 5 %. As for the two remaining types of global information, namely conflicts and implications, our experiments show a performance-neutral result or slightly detrimental performance impact when transferred across restarts. In particular, transferring the implication graph is performance-neutral to 1% improvement, depending on the bracket you consider. Our interpretation is that many of the implications would be rediscovered in presolving and probing in the second run anyway, so there is not much value. This was also a reason for not yet trying to transfer the clique table, which is similar in nature. Finally, for conflicts, we observed slightly detrimental behavior, which surprised us because conflicts carry global information that cannot be easily, and cheaply, recomputed. Our explanation for this is that the transfer of old conflicts together with the strict limits on the overall conflicts pool size prevents fresh conflicts from being learned. This clearly asks for a more careful re-tuning which is part of future work.

5 Conclusion and Outlook

In this paper we showed that transferring some pieces of global information across a restart can be quite beneficial for the performance of a state of the art MIP solver, both in terms of average runtime and size of the resulting enumeration tree.

As future work we plan to investigate why transferring some of the global structures, namely conflicts and implications, did not result in a measurable performance improvement, and extend the transfer to additional structures, most notably the cliques in the global clique table [7].

References

1. Achterberg, T.: Conflict analysis in mixed integer programming. Discrete Optim. **4**(1), 4–20 (2007)
2. Achterberg, T.: Constraint integer programming. Ph.D. thesis, Technische Universität Berlin (2007)
3. Achterberg, T.: SCIP: solving constraint integer programs. Math. Programm. Comput. **1**(1), 1–41 (2009). https://doi.org/10.1007/s12532-008-0001-1
4. Achterberg, T., Bixby, R.E., Gu, Z., Rothberg, E., Weninger, D.: Presolve reductions in mixed integer programming. Informs J. Comput. **32**(2), 473–506 (2020)
5. Achterberg, T., Koch, T., Martin, A.: Branching rules revisited. Oper. Res. Lett. **33**(1), 42–54 (2005)
6. Anderson, D., Hendel, G., Le Bodic, P., Viernickel, M.: Clairvoyant restarts in branch-and-bound search using online tree-size estimation. Proc. AAAI Conf. Artif. Intell. **33**(01), 1427–1434 (2019)
7. Atamtürk, A., Nemhauser, G.L., Savelsbergh, M.W.: Conflict graphs in solving integer programming problems. Eur. J. Ope. Res. **121**(1), 40–55 (2000)
8. Bénichou, M., Gauthier, J.M., Girodet, P., Hentges, G., Ribière, G., Vincent, O.: Experiments in mixed-integer programming. Math. Programm. **1**, 76–94 (1971)
9. Berthold, T., Feydy, T., Stuckey, P.J.: Rapid learning for binary programs. In: Lodi, A., Milano, M., Toth, P. (eds.) CPAIOR 2010. LNCS, vol. 6140, pp. 51–55. Springer, Heidelberg (2010). https://doi.org/10.1007/978-3-642-13520-0_8
10. Berthold, T., Hendel, G.: Learning to scale mixed-integer programs. Proc. AAAI Conf. Artif. Intell. **35**(5), 3661–3668 (2021)
11. FICO Xpress Optimization (2021). https://www.fico.com/en/products/fico-xpress-optimization
12. Gamrath, G.: Enhanced predictions and structure exploitation in branch-and-bound. Ph.D. thesis, Technical University Berlin (2020)
13. Gamrath, G., et al.: The SCIP Optimization Suite 7.0. Technical report 20–10, ZIB (2020)
14. Gamrath, G., Berthold, T., Heinz, S., Winkler, M.: Structure-driven fix-and-propagate heuristics for mixed integer programming. Math. Programm. Comput. **11**(4), 675–702 (2019). https://doi.org/10.1007/s12532-019-00159-1
15. Hendel, G.: Enhancing MIP branching decisions by using the sample variance of pseudo costs. In: Michel, L. (ed.) CPAIOR 2015. LNCS, vol. 9075, pp. 199–214. Springer, Cham (2015). https://doi.org/10.1007/978-3-319-18008-3_14
16. Hendel, G., Anderson, D., Le Bodic, P., Pfetsch, M.E.: Estimating the size of branch-and-bound trees. Informs J. Comput. **34**(2), 934–952 (2021)

17. Hu, X., Shonkwiler, R., Spruill, M.C.: Random restarts in global optimization. Technical report, Georgia Institute of Technology (2009)
18. ILOG CPLEX Optimization Studio 12.10.0 (2019). https://www.ibm.com/docs/en/icos/12.10.0?topic=v12100-changes-log
19. Lodi, A., Tramontani, A.: Performance variability in mixed-integer programming. In: Theory Driven by Influential Applications, pp. 1–12. INFORMS (2013)
20. Marchand, H., Martin, A., Weismantel, R., Wolsey, L.A.: Cutting planes in integer and mixed integer programming. Discrete Appl. Math. **123**(124), 391–440 (2002)
21. Moskewicz, M.H., Madigan, C.F., Zhao, Y., Zhang, L., Malik, S.: Chaff: engineering an efficient SAT solver. In: Proceedings of the 38th Annual Design Automation Conference, pp. 530–535. DAC 2001. Association for Computing Machinery (2001)
22. Nemhauser, G.L., Wolsey, L.A.: Integer programming and combinatorial optimization Wiley. Chichester. GL Nemhauser, MWP Savelsbergh, GS Sigismondi (1992).Constraint Classification for Mixed Integer Programming Formulations. COAL Bulletin, vol. 20, pp. 8–12 (1988)
23. Pokutta, Sebastian: Restarting algorithms: sometimes there is free lunch. In: Hebrard, Emmanuel, Musliu, Nysret (eds.) CPAIOR 2020. LNCS, vol. 12296, pp. 22–38. Springer, Cham (2020). https://doi.org/10.1007/978-3-030-58942-4_2
24. Savelsbergh, M.W.P.: Preprocessing and probing techniques for mixed integer programming problems. ORSA J. Comput. **6**(4), 445–454 (1994)
25. Savelsbergh, M.W.P., Sigismondi, G.C., Nemhauser, G.L.: Functional description of MINTO, a mixed integer optimizer. Technische Universiteit Eindhoven (1991)
26. Tomlin, J.A.: On scaling linear programming problems. In: Balinski, M.L., Hellerman, E. (eds) Computational practice in mathematical programming. Mathematical Programming Studies vol. 4, pp. 146–166. Springer, Berlin, Heidelberg (1975). https://doi.org/10.1007/BFb0120718
27. Tramontani, A.: Enhanced mixed integer programming techniques and routing problems. Ph.D. thesis, Springer (2011). https://doi.org/10.1007/s10288-010-0140-x
28. Wilcoxon, F.: Individual comparisons by ranking methods. Biometrics Bull. **1**, 80–83 (1945)
29. Witzig, J., Berthold, T., Heinz, S.: Experiments with conflict analysis in mixed integer programming. In: Salvagnin, D., Lombardi, M. (eds.) CPAIOR 2017. LNCS, vol. 10335, pp. 211–220. Springer, Cham (2017). https://doi.org/10.1007/978-3-319-59776-8_17
30. Wolter, K.: Implementation of cutting plane separators for mixed integer programs. Master's thesis, Technische Universität Berlin (2006)

Towards Copeland Optimization in Combinatorial Problems

Sidhant Bhavnani[1] and Alexander Schiendorfer[2]($\boxtimes$)

[1] University of Glasgow, Glasgow, Scotland
2482327b@student.gla.ac.uk
[2] Technische Hochschule Ingolstadt, Ingolstadt, Germany
alexander.schiendorfer@thi.de

Abstract. Traditional approaches to fairness in operations research and social choice, such as the egalitarian/Rawlsian, the utilitarian or the proportional-fair rule implicitly assume that the voters' utility functions are – to a certain degree – comparable. Otherwise, statements such as "maximize the worst-off voter's utility" or "maximize the sum of utilities" are void. But what if the different valuations should truly not be compared or converted into each other? Voting theory only relies on ordinal information and can help to provide democratic rules to define winning solutions. Copeland's method is a well-known generalization of the Condorcet criterion in social choice theory and asks for an outcome that has the best ratio of pairwise majority duel wins to losses. If we simply ask for a feasible solution to a combinatorial problem that maximizes the Copeland score, we are at risk to encounter intractability (due to having to explore *all* solutions) or suffer from (the lack of) irrelevant alternatives. We present first results from optimizing for a Copeland winner to a constraint problem formulated in MiniZinc in a local search fashion based on a changing solution pool. We investigate the effects of diversity constraints on the quality of the estimated Copeland score as well as the gap between the best reported Copeland scores to the actual Copeland scores.

Keywords: Constraint Programming · Social Choice Theory · OR

1 Solving Combinatorial Problems by Voting

Many constraint problems involve the preferences of multiple (human or software) agents – at times even a large number of them in a committee that needs to make joint decisions. Think of designing a new transport network with several bus lines that need to connect stops or agreeing on a travel itinerary subject to time windows of availability of sights. In such cases, each voter $n \in N$ has their own preference ordering $\preceq_n$ over the set of possible outcomes O which

---

This research has been sponsored by DAAD Research Internships in Science and Engineering (RISE).

P. Schaus (Ed.): CPAIOR 2022, LNCS 13292, pp. 34–43, 2022.
https://doi.org/10.1007/978-3-031-08011-1_4

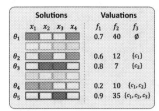 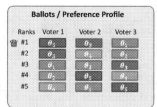 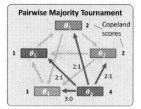

Fig. 1. The problem considered in this paper. Left: Solutions to a combinatorial problem (infeasible ones are grayed out) are valued using classical objective functions by three voters. All use a completely different domain of values – satisfaction degrees in $[0, 1]$, costs to minimize, or sets of violated constraints. Center: The solutions are ranked ordinally according to the voters' valuations. Right: A tournament where an edge indicates which option wins a duel; θ_3 has the highest Copeland score.

is the solution space of a constraint satisfaction problem. Classical approaches in operations research and optimization then turn to utility functions $(f_n)_{n \in N}$ and apply (i) the utilitarian rule [17]: maximize$_x \sum_{n \in N} f_n(x)$, (ii) the egalitarian or Rawlsian rule [24]: maximize$_x \min_{n \in N} f_n(x)$, or (iii) the proportional-fair rule [1,19] which chooses x such that $\sum_{n \in N} \frac{f_n(x') - f_n(x)}{f_n(x)}$ is negative for all other x'. All of these rules assume the utilities to map to a comparable "scale" to be semantically well-defined.

But in many applications, the domains of these valuations need not be comparable (see Fig. 1) – even if they are expressed quantitatively: A (normalized) value of 0.9 for voter i, e.g., is not necessarily "better" than 0.8 for agent $j \neq i$. Instead of trying to convert one agent's "utility units" into another one's, we strive for a purely *ordinal* approach. The field of social choice theory [2], in particular voting theory, offers a mathematical framework to aggregate several individual orderings into a single ordering representing the group through social welfare functions (SWF) or social choice functions (SCF) such as the methods by Condorcet, Copeland, Borda, or majority-based rules. For example, Copeland's method asks to find a candidate that wins the highest number of pairwise majority duels with other candidates (called the Copeland score – see Fig. 1 right). Most often, SWFs and SCFs are considered over small finite sets of outcomes (also called "candidates" due to the heritage of politics). If we aim to apply SCFs/SWFs to a large set of candidates such as the solution space of a combinatorial problem, enumerating them all is out of the question.

In this paper, we present a sampling-based algorithm that sequentially adds new solutions to (and deletes old ones from) a solution pool to maximize an approximate underlying true Copeland score (see Fig. 2). Since this solution pool represents – statistically speaking – a sample of the set of all feasible solutions that would be required to determine the actual Copeland score, during optimization we face a "moving target", i.e., a solution might have a high Copeland score on this particular sample but not overall and the objective value (i.e., Copeland) changes as we explore more solutions. Hence, this paper contributes:

- A formulation of Copeland optimization as a novel modeling method for combinatorial problems involving several voters (see Sect. 2).
- The sampling-based meta-search algorithm with a solver-independent implementation using MiniZinc (see Sect. 3).
- An experimental study that investigates how techniques from constraint programming (diversity maximization [13]) can aid in selecting a high-quality solution pool for reliable Copeland estimates (see Sect. 4).

1.1 Related Work

The strong connections between collective decision making and constraint reasoning have been pointed out by Rossi in 2014 [20], most notably leading to a sequential voting procedure [6, 7] to connect social choice theory and constraint solving. Since then the CP community has reached a higher level of maturity in its constraint solvers and, especially, its modeling languages such as OPL [12], Essence [10], or MiniZinc [18]. Simple Condorcet-voting on top of MiniZinc has been shown to remove bias introduced by weight aggregation in [22]. Of course, pure multi-criteria optimization deals with multiple objectives, most notably via finding Pareto frontiers or minimizing distances to Utopian points [8]. Every stakeholder's preference needs to be encoded as one of the utility functions. Solving for a *large number* of objectives, however, leads to very indecisive optimization criteria [22]. Along those lines, [4] discuss Rawlsian and Utilitarian utility aggregation in a socially desirable sense but work with a common currency that we want to avoid. Copeland voting has been applied in combinatorial optimization [3] but with a focus on hyperparameter/heuristics selection for evolutionary algorithms, not as the objective of the modeled optimization problem itself. Finally, while the distributed constraint optimization (DCOP) community has produced remarkable results in transferring search and propagation algorithms to distributed computing environments [9], there has been little discussion about *what to optimize for* other than a sum of utilities.

2   Foundations: Social Choice and (Soft) CSPs

A constraint satisfaction problem (CSP) is defined by a finite set of variables X with their associated domains of possible values $(D_x)_{x \in X}$ where at least some domain is discrete, and a set of constraints C. A (variable) assignment θ is a map $X \to D$ such that $\theta(x) \in D_x$, for every variable x. It is feasible if $\theta \models c$ holds for every constraint $c \in C$ and then called a *solution*. Finding a solution to a CSP is, in general, NP-complete [21]. A CSP usually becomes a constraint optimization problem (COP) by adding an objective function $f : [X \to D] \to (F, \geq_F)$ where F represents a (at least partially) ordered set, usually $(\mathbb{R}, \geq)$. We seek an assignment θ^* with either maximal or minimal value $f(\theta^*)$. Especially in the context of (algebraic) *soft constraints* that can be abstracted to a c-semiring or (partial) valuation structure taking the role of F, several formalisms other than real-valued objective functions have been motivated and investigated [14, 16].

Examples include *sets* of violated constraints, satisfaction degrees, or valuation tuples in the case of Pareto and lexicographic combinations [11]. If we have multiple objective functions $(f_i)_{i \in N}$ for a set of voters/agents N, we face a multi-objective COP with potentially $|N|$ different valuation domains $(F_i)_{i \in N}$ – then also frequently called utility functions. To combine the several orders over solutions emerging from these objective functions in a purely ordinal way, we turn to social choice theory:

Let $N = \{1, \ldots, n\}$ be a set of voters and O a finite set of outcomes. For every voter $i \in N$, we assume a linear preorder $\succeq_i$ (ties allowed) over O where $o_1 \succeq_i o_2$ indicates that voter i prefers o_1 over o_2. We call $(\succeq_n)_{n \in N}$ a *preference profile*. The pairwise majority relation $\geq_\mu$ over O is defined by:

$$o_1 \geq_\mu o_2 \quad \text{if and only if} \quad |\{i \mid i \in N \wedge o_1 \succeq_i o_2\}| \geq |\{i \mid i \in N \wedge o_2 \succeq_i o_1\}|$$

A social choice function SCF takes a preference profile and returns a set of *winners* $W = SCF((\succeq_n)_{n \in N}) \subseteq O$. It satisfies the Condorcet criterion if $W = \{o_c\}$ for all profiles where o_c is a Condorcet winner, i.e., an option that wins all pairwise duels: $o_c \geq_\mu o, \forall o \neq o_c \in O$. A Condorcet winner needs not exist due to cycles in the pairwise majority relation [2].

There are several extensions to the Condorcet criterion. A rather straightforward extension is the (asymmetric) *Copeland score* $\mathcal{C}$ which awards one point for each won pairwise majority duel and a half point for ties:

$$\mathcal{C}(o) = |\{o' \in O \mid o >_\mu o'\}| + \frac{1}{2}|\{o' \in O \mid o' =_\mu o\}|$$

It is natural to ask for an SCF that picks the Copeland winners as outcomes, i.e., those with the highest Copeland score – then called Copeland's rule [5]. Note that Copeland rewards an outcome for each victory – regardless of the margin.

When we return to our original problem, we set $O = [X \rightarrow D]$, i.e., each solution is a possible outcome to a voting problem and have

$$\theta_1 \succeq_i \theta_2 \Leftrightarrow f_i(\theta_1) \geq_{F_i} f_i(\theta_2)$$

for every voter $i \in N$. This is a well-defined preference profile over solutions – albeit a potentially exponential number of outcomes. Asking for a solution

$$\theta_{\mathcal{C}}^* = \arg\max_\theta \mathcal{C}(\theta)$$

that maximizes the Copeland score is not a classical objective function $f : [X \rightarrow D] \rightarrow F$ which only depends on the variable assignments but must be measured with respect to – in theory – all solutions. In our experiments, we use a normalized variant of $\mathcal{C}$ and call it $\bar{\mathcal{C}}$ with $\bar{\mathcal{C}}(o) = \frac{\mathcal{C}(o)}{|O|}$.

3   Sampling Approach

To solve for $\theta_{\mathcal{C}}^*$ exactly, we would have to explore *all* solutions to the underlying CSP. For many problems, this set is prohibitively large. We, therefore, suggest a sampling-based approach, as shown in Fig. 2.

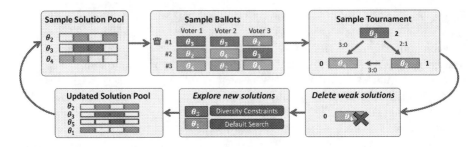

Fig. 2. An iterative approach to approximately searching a Copeland winner. The solution pool is a sample from all solutions (cf. Figure 1) and maintaining a representative sample is essential to reliably estimate Copeland scores.

We start with a CSP formulated in a modeling language – MiniZinc in our case. The valuations $(f_n)_{n \in N}$ are modeled as decision variables that are appropriately constrained[1]. That way, we get a linear preorder $\succeq_n$ (due to possible ties) over solutions for every voter n even though, initially, the feasible solutions are not known.

In every iteration, new solutions are explored and weak solutions are deleted from a sample (called the solution pool) – both ratios are adjustable parameters. It is known [2] that SWFs/SCFs based on pairwise majority such as Copeland may suffer from *irrelevant alternatives* which may change the relative ordering of two options, that is, if $o_1, o_2 \in O$ and $o_1 \in W_{SCF(O)}$ it might be that $o_1 \notin W_{SCF(O \cup O')}$ for some alternatives O' with $O' \cap O = \emptyset$. This can become an issue for the proposed sampling approach: if we (unfortunately) picked a solution pool purely consisting of bad solutions in terms of the underlying Copeland scores, a solution that looks excellent in this pool might be far off the true Copeland winners' scores. Being aware of these theoretical limitations, we hope to mitigate them "in practice" by optimizing the chosen solution pool – ideally to be a representative sample of the whole search space.

To do so, we instruct solvers to maximize the diversity to existing solutions, using a modular model fragment similar to an approach presented in [13]:

```
array[int] of var int: vars; % variables of interest that should be diverse
array[int, index_set(vars)] of int: pool; % old solutions as parameter
array[index_set_1of2(pool), index_set(vars)] of var int: div_abs_dists;

constraint forall(i in index_set_1of2(pool), j in index_set_2of2(pool))(
    div_abs_dists[i, j] = abs(pool[i, j] - vars[j])
);

% Offered objective function as decision variables - can be maximized
var int: diversity_abs = sum(div_abs_dists);
```

This diversity fragment can be tied to any MiniZinc model by equating the `vars` array to the actual variables of interest and maximizing `div_abs_dists`. The next found solution maximizes the sum of absolute distances to the the

---

[1] Our implementation indeed uses integer variables for utilities that need to be maximized but nothing prohibits more general "is-better" predicates.

previously found solutions. Additionally, it has been pointed out that demanding full independence of irrelevant alternatives might be too restrictive in the face of many (substitutable) alternatives [15].

4   Experimental Evaluation

We implemented the proposed sampling approach and evaluated it on several typical combinatorial benchmark problems: *Vehicle routing with time windows, Job-shop scheduling with agents as job-owners, project-to-student assignment,* and *photo placement with friends and enemies.* We selected a range of instances for all models, each with randomly generated preferences that were automatically converted into preference profiles over solutions[2].

We ran our experiments using the Gecode solver [23] to generate all solutions for the CSPs and MiniZinc-Python API to sample the solutions and run our approach. All instances were designed sufficiently small (150–5000 solutions) that we can calculate the underlying Copeland scores exactly to compare them against the sampling approach.

Figure 3 shows an example traversal for one scheduling instance that demonstrates challenges and opportunities the sampling approach faces:

- High-quality solutions in terms of the (normalized) Copeland scores $\bar{\mathcal{C}}$ appear first in the default search traversal of the solver. But this becomes known only after all solutions have been explored! Much of their "Copeland weight" stems from bad solutions that appear very late in the search. When such an "elite group" of solutions is presented to the sampling-approach, bad solutions within the pool might be underestimated with respect to all solutions.
- Conversely, had the search started (due to an unfortunate heuristic) in the bad region, the best among the worst might look very promising although it only wins against bad overall solutions.
- The Copeland scores of the feasible solutions tend to follow a normal distribution but every sample adjacent in the chronological ordering is highly skewed. Estimating the Copeland score based on such a sample might be off.

While this short paper only examines the basic feasibility of the sampling approach to find high-quality Copeland solutions, we emphasize that the distributions we examined showed interesting regularities (Gaussian, Poisson) that should be exploited further.

4.1   Evaluation Metrics

To evaluate our approach, we compute the normalized Copeland Scores on all solutions produced by the CSP and refer to this as the "ground truth" $\bar{\mathcal{C}}^{GT}$. We also measure the sample Copeland scores $\bar{\mathcal{C}}^{SP}$ with respect to the solutions in the sampled pool. The overall winning solution is consequently called $\theta^*_{\bar{\mathcal{C}}_{GT}}$ and

---

[2] Source code: https://github.com/s1db/Local-Search-Copeland-Method.

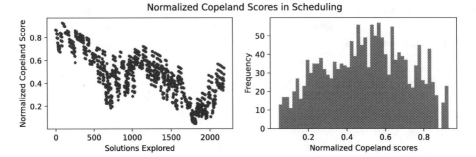

Fig. 3. An example (full) traversal of the search space of a scheduling instance. Normalized Copeland scores are shown (calculated a posteriori) for the solutions in chronological order. The distribution of the scores tends to follow a normal distribution.

the winner with respect to the (final) sample pool is called $\theta^*_{\bar{\mathcal{C}}^{SP}}$. Using this we define the following metrics:

The **Copeland Gap** measures how far off the Copeland score of the selected winner is from the true winner's Copeland score. It measures *quality of the selected solution.*

$$\frac{|\bar{\mathcal{C}}^{GT}(\theta^*_{\bar{\mathcal{C}}^{SP}}) - \bar{\mathcal{C}}^{GT}(\theta^*_{\bar{\mathcal{C}}^{GT}})|}{\bar{\mathcal{C}}^{GT}(\theta^*_{\bar{\mathcal{C}}^{GT}})}$$

The **Copeland Error** measures how far off the Copeland score of the selected winner is from *its own* ground truth Copeland score. It measures *quality of the Copeland score estimate.*

$$\frac{|\bar{\mathcal{C}}^{GT}(\theta^*_{\bar{\mathcal{C}}^{SP}}) - \bar{\mathcal{C}}^{SP}(\theta^*_{\bar{\mathcal{C}}^{SP}})|}{\bar{\mathcal{C}}^{GT}(\theta^*_{\bar{\mathcal{C}}^{SP}})}$$

With those two metrics, we can make judgments about the sampling approach. Does it find solutions close to the true Copeland winner's scores even under a budget (10%–25% of all solutions)? Is the estimated Copeland score reliably close to the true one? It might, e.g., happen that the sampling approach picks a winner that has a good overall Copeland score (low Copeland gap) but it under-/overestimates its true Copeland score (high Copeland error). When inspecting the results in Table 1, we first observe that the sampling approach with default search selects solutions whose Copeland gap is below 35% for all considered instances – often considerably closer. Diversity maximization can aid on some problems more than others. In vehicle routing, e.g., it was able to select the true Copeland winner through a broader exploration and with low Copeland error. However, it can also be misleading: For example, diversity maximization on instance 7 of photo placement picked a solution with a GT Copeland score of 0.579 (best possible was 0.997) but that solution was estimated highly within the pool (0.97) – resulting in high Copeland error and gap. Better management of the solution pool is needed. We suspect that our naive diversity maximization

Table 1. Comparison of the sampling approach with or without diversity maximization on several combinatorial problems. Instance names are consistent with the respository – too large or small instances have been ignored.

Problem	# Solutions	Diversity Maximization		Default Search	
		Gap	Error	Gap	Error
Scheduling 4	2185	0.233	0.252	**0.095**	**0.184**
Scheduling 5	2528	0.134	0.063	**0.047**	**0.043**
Photo Placement 6	720	**0.062**	**0.0339**	0.175	0.176
Photo Placement 7	2520	0.420	0.677	**0.031**	**0.023**
Photo Placement 8	5040	0.234	0.27	**0.095**	**0.026**
Vehicle Routing 3	133	**0**	**0.042**	0.345	0.180
Vehicle Routing 4	133	**0**	**0.043**	0.174	0.227
Project assignment 2	996	0.266	0.291	**0.026**	**0.009**
Project assignment 3	996	**0.083**	**0.11**	0.336	0.485

does not account well for the actual distribution of Copeland scores but produces differently skewed samples – a problem that needs significant future research.

5   Conclusion and Future Work

We presented the first results towards Copeland optimization of combinatorial problems. Certainly, this work asks more questions than it succeeds to answer. Most importantly, the quality of the solution pool in terms of representativity and high-quality solutions is paramount. It would be interesting to assess the sample distributions using statistical methods and apply learning techniques to predict whether a sample winner is going to be strong on the ground truth. This could lead to a proper termination criterion for the sampling approach or give bounds on the sample Copeland scores future solutions need to achieve. This can be propagated as a constraint. Also, the Copeland score (or other pairwise social choice scores) is prone to manipulation via the selection of the sample to exploit irrelevant alternatives. Attacks of that kind should be mitigated. To perform these analyses of relevant problem sizes, we need to take care of computational aspects. Since many similar instances have to be explored, learning solving technologies such as lazy clause generation [25] could be well-suited. Finally, our approach can be easily extended to other SCFs such as Borda or implement more general strategies in weighted tournaments and made accessible for MiniZinc – which we look forward to.

Acknowledgments. We thank Guido Tack and Alexander Knapp for initial discussions that led to the idea of this paper.

References

1. Bertsimas, D., Farias, V.F., Trichakis, N.: The price of fairness. Oper. Res. **59**(1), 17–31 (2011)
2. Brandt, F., Conitzer, V., Endriss, U., Lang, J., Procaccia, A.D.: Handbook of Computational Social Choice. Cambridge University Press, Cambridge (2016)
3. de Carvalho, V.R., Sichman, J.S.: Applying copeland voting to design an agent-based hyper-heuristic. In: Proceedings of the 16th Conference on Autonomous Agents and Multiagent Systems, pp. 972–980 (2017)
4. Chen, V.X., Hooker, J.: Combining leximax fairness and efficiency in a mathematical programming model. Eur. J. Oper. Res. **299**(1), 235–248 (2022)
5. Copeland, A.H.: A reasonable social welfare function. Technical report, mimeo. University of Michigan (1951)
6. Cornelio, C., Pini, M.S., Rossi, F., Venable, K.B.: Multi-agent soft constraint aggregation via sequential voting: theoretical and experimental results. Auton. Agent. Multi-Agent Syst. **33**(1), 159–191 (2019). https://doi.org/10.1007/s10458-018-09400-y
7. Dalla Pozza, G., Rossi, F., Venable, K.B.: Multi-agent soft constraint aggregation: a sequential approach. In: Proceedings 3rd International Conference Agents and Artificial Intelligence ICAART'11, vol. 11 (2010)
8. Ehrgott, M.: Multicriteria Optimization, vol. 491. Springer Science & Business Media, Heidelberg (2005)
9. Fioretto, F., Yeoh, W., Pontelli, E., Ma, Y., Ranade, S.J.: A distributed constraint optimization (DCOP) approach to the economic dispatch with demand response. In: Proceedings 16th International Conference Autonomous Agents and Multiagent Systems (AAMAS 2017), pp. 999–1007. International Foundation for Autonomous Agents and Multiagent Systems (2017)
10. Frisch, A.M., Harvey, W., Jefferson, C., Martínez-Hernández, B., Miguel, I.: Essence: a constraint language for specifying combinatorial problems. Constraints **13**(3), 268–306 (2008). https://doi.org/10.1007/s10601-008-9047-y
11. Gadducci, F., Hölzl, M., Monreale, G.V., Wirsing, M.: Soft constraints for lexicographic orders. In: Castro, F., Gelbukh, A., González, M. (eds.) MICAI 2013. LNCS (LNAI), vol. 8265, pp. 68–79. Springer, Heidelberg (2013). https://doi.org/10.1007/978-3-642-45114-0_6
12. van Hentenryck, P.: The OPL Optimization Programming Language. MIT Press, Cambridge (1999)
13. Ingmar, L., de la Banda, M.G., Stuckey, P.J., Tack, G.: Modelling diversity of solutions. In: Proceedings of the AAAI Conference on Artificial Intelligence, vol. 34, pp. 1528–1535 (2020)
14. Knapp, A., Schiendorfer, A., Reif, W.: Quality over quantity in soft constraints. In: Proceedings 26th International Conference Tools with Artificial Intelligence (ICTAI 2014), pp. 453–460 (2014)
15. McFadden, D., et al.: Conditional Logit Analysis of Qualitative Choice Behavior (1973)
16. Meseguer, P., Rossi, F., Schiex, T.: Soft constraints. In: Rossi, F., van Beek, P., Walsh, T. (eds.) Handbook of Constraint Programming, chap. 9. Elsevier (2006)
17. Moulin, H.: Fair Division and Collective Welfare. MIT press, Cambridge (2004)
18. Nethercote, N., Stuckey, P.J., Becket, R., Brand, S., Duck, G.J., Tack, G.: MiniZinc: towards a standard CP modelling language. In: Bessière, C. (ed.) CP 2007. LNCS, vol. 4741, pp. 529–543. Springer, Heidelberg (2007). https://doi.org/10.1007/978-3-540-74970-7_38

19. Nicosia, G., Pacifici, A., Pferschy, U.: Price of fairness for allocating a bounded resource. Eur. J. Oper. Res. **257**(3), 933–943 (2017)
20. Rossi, F.: Collective decision making: a great opportunity for constraint reasoning. Constraints **19**(2), 186–194 (2013). https://doi.org/10.1007/s10601-013-9153-3
21. Rossi, F., Van Beek, P., Walsh, T.: Handbook of Constraint Programming. Elsevier, Amsterdam (2006)
22. Schiendorfer, A., Reif, W.: Reducing bias in preference aggregation for multiagent soft constraint problems. In: Schiex, T., de Givry, S. (eds.) CP 2019. LNCS, vol. 11802, pp. 510–526. Springer, Cham (2019). https://doi.org/10.1007/978-3-030-30048-7_30
23. Schulte, C., Lagerkvist, M.Z., Tack, G.: Gecode: generic constraint development environment. In: INFORMS Annual Meeting (2006)
24. Sen, A.: Collective Choice and Social Welfare. Harvard University Press, Cambridge (2017)
25. Stuckey, P.J.: Lazy clause generation: combining the power of SAT and CP (and MIP?) solving. In: Lodi, A., Milano, M., Toth, P. (eds.) CPAIOR 2010. LNCS, vol. 6140, pp. 5–9. Springer, Heidelberg (2010). https://doi.org/10.1007/978-3-642-13520-0_3

Coupling Different Integer Encodings for SAT

Hendrik Bierlee[1,2](✉) [iD], Graeme Gange[1] [iD], Guido Tack[1,2] [iD],
Jip J. Dekker[1,2] [iD], and Peter J. Stuckey[1,2] [iD]

[1] Department of Data Science and AI, Monash University, Melbourne, Australia
{hendrik.bierlee,graeme.gange,guido.tack,jip.dekker,
peter.stuckey}@monash.edu
[2] ARC Training Centre in Optimisation Technologies, Integrated Methodologies,
and Applications (OPTIMA), Melbourne, Australia

Abstract. Boolean satisfiability (SAT) solvers have dramatically improved their performance in the last twenty years, enabling them to solve large and complex problems. More recently MaxSAT solvers have appeared that efficiently solve optimisation problems based on SAT. This means that SAT solvers have become a competitive technology for tackling discrete optimisation problems.

A challenge in using SAT solvers for discrete optimisation is the many choices of encoding a problem into SAT. When encoding integer variables appearing in discrete optimisation problems, SAT must choose an encoding for each variable. Typical approaches fix a common encoding for all variables. However, different constraints are much more effective when encoded with a particular encoding choice. This inevitably leads to models where variables have different variable encodings. These models must then be able to *couple* encodings, either by using multiple encoding of single variables and *channelling* between the representations, or by encoding constraints using a mix of representations for the variables involved. In this paper we show how using mixed encodings of integers and coupled encodings of constraints can lead to better (Max)SAT models of discrete optimisation problems.

1 Introduction

Within the last twenty years, Boolean satisfiability (SAT) solving has increased in terms of scalability and performance. More recently optimisation approaches based on SAT, so called MaxSAT technology, have been rapidly developing. SAT and MaxSAT solvers now provide a viable alternative solving technology for many discrete optimisation problems.

Translating discrete optimisation problems to (Max)SAT is a difficult task. Currently, expert SAT modellers build models directly from clauses, or use libraries to encode common constraints such as at-most-one, cardinality or pseudo-Boolean constraints [15,23]. Alternatively, SAT compilers such as Fzn-Tini [13] and Picat-SAT [33], or some modelling languages such as Essence [21], determine a translation of a high-level model.

© Springer Nature Switzerland AG 2022
P. Schaus (Ed.): CPAIOR 2022, LNCS 13292, pp. 44–63, 2022.
https://doi.org/10.1007/978-3-031-08011-1_5

The first fundamental choice a SAT modeller or compiler faces is how an integer decision variable x should be translated to Boolean decisions. At least three possibilities arise: the *direct encoding*, introducing a Boolean that represents $x = d$ for each value d in the domain of x; the *order encoding*, introducing a Boolean that represents $x \geq d$ for each value d in the domain of x; or the *binary encoding*, introducing a Boolean that represents a bit in the (two's complement) binary encoding of x.

Which encoding should be used for each variable in a discrete optimisation problem is a non-trivial question. The answer may depend on the constraints in which the variable appears, and how important they are to solving the overall problem. In the existing approaches, the variable encoding choices made by the SAT compiler or modeller are hard or impossible to change, since the encoding of every constraint depends on it. Some constraints will be more effective if encoded using a particular variable choice, and if these choices are not all the same for the problem of interest we inevitably have to *couple* different integer encodings. This can be managed by encoding an integer variable in two ways and *channelling* the two encodings, or directly encoding constraints that couple different encodings of the variables involved.

In this paper we show that choosing mixed encodings of integer variables in the model can lead to better solving performance. We investigate new possibilities for channelling and coupling encodings of constraints, and also show how *(partial) views* can be used to extend coupling encodings. The experimental results show how the novel channel, coupling, and view constraints enable unique encodings that outperform existing SAT encoding approaches for discrete optimisation problems of interest.

2 Preliminaries

2.1 Constraint Programming

A *constraint satisfaction problem (CSP)* $P = (\mathcal{X}, D_i, \mathcal{C})$ consists of a set of variables $\mathcal{X}$, with each $x \in \mathcal{X}$ restricted to take values from some initial domain $D_i(x)$. For this paper we assume domains are ordered sets of integers, and denote by $lb(x)$ and $ub(x)$ the least and greatest values in $D_i(x)$. We will use interval notation $l..u$ to represent the set of integers $\{l, l+1, .., u\}$. A set of constraints $\mathcal{C}$ expresses relationships between the variables. A *constraint optimisation problem (COP)* is a CSP P together with objective function o, w.l.o.g. to be maximised.

Constraint programming solvers work by propagation over atomic constraints. An *atomic constraint* is (for our purposes) a unary constraint from a given language (which includes the always false constraint *false*). The usual atomic constraints for CP are $x = d$, $x \neq d$, $x \geq d$ and $x \leq d$. A propagator f for constraint C takes a current domain $\mathcal{D}$ given as a set of atomic constraints, and determines a set of new atomic constraints. These are a consequence of the current domain and C, i.e., $a \in f(\mathcal{D})$ implies that $\mathcal{D} \wedge C \to a$.

A propagator f is *propagation complete* for constraint C and a language of atomic constraints $\mathcal{L}$ iff for any $\mathcal{D} \subseteq \mathcal{L}$ and new atom $a \in \mathcal{L}$, $a \notin \mathcal{D}$: if $\mathcal{D} \wedge C \to a$

then $a \in f(\mathcal{D})$. That is, the propagator finds all atomic constraints in $\mathcal{L}$ that are consequences of the constraint and the current domain.

Let $DIRECT(x)$ be the language of atomic constraints $\{false\} \cup \{x = d, x \neq d \mid x \in \mathcal{X}, d \in D_i(x)\}$. A propagator is *domain consistent* if it is propagation complete for $\mathcal{L} = DIRECT(x)$. Let $ORDER(x)$ be the language of atomic constraints $\{false\} \cup \{x \leq d, x \geq d \mid x \in \mathcal{X}, d \in D_i(x)\}$. A propagator is *bounds(Z) consistent* [6] if it is propagation complete for $\mathcal{L} = ORDER(x)$.

CSPs and COPs are typically expressed using a modelling language such as MiniZinc [18]. To illustrate, we show a model for the popular Knight's tour problem in which we are looking for a trajectory of n^2 legal knight moves around an $n \times n$ board. The knight starts and ends at the top-left square (numbered 1), visiting each square exactly once.

```
int: n; % size of board
array[int] of var int: x = [ let
  { % legal knight moves from position r, c
  set of int: neighbours = { n*(r_i-1) + c_i
    | i in 1..8,
      r_i in {r + [-1,-2,-2,-1,1,2,2,1][i]}
          where r_i in 1..n,
      c_i in {c + [-2,-1,1,2,2,1,-1,-2][i]}
          where c_i in 1..n };
  var neighbours: x_r_c;
  } in x_r_c | r,c in 1..n ];
% fix first and last moves (symmetry)
constraint x[1] = n+3;
constraint x[2*n+2] = 1;
include "circuit.mzn";
constraint circuit(x);
```

The variable x[p] gives the next position (in $1..n^2$) to move to from position p. MiniZinc allows us to specify the integer variable domains exactly according to legal knight moves (shown in the figure above). The symmetry breaking constraints reduce the domains of two variables to single, fixed values. The global constraint circuit constrains the x variables to describe a single Hamiltonian circuit traversal of the graph. The constraint may be implemented directly by the solver or defined in terms of basic constraints using the standard library, or a library specialised for a particular solver. For a given target solver, MiniZinc flattens the high-level model into low-level FlatZinc consisting of built-in constraints.

2.2   Boolean Satisfiability

A Boolean Satisfiability (SAT) problem can be seen as a special case of a CSP, where the domain for all variables x is $D_i(x) \in \{0, 1\}$, representing the values *false* (0) and *true* (1). In SAT problems, we usually talk about *literals*, which are either a Boolean variable x or its negation $\neg x$. We extend the negation operation to operate on literals, i.e., $\neg l = \neg x$ if $l = x$ and $\neg l = x$ if $l = \neg x$. We use the notation $l = v$ where l is a literal and $v \in \{0, 1\}$ to encode the appropriate form of the literal, i.e., if $v = 1$ it is equivalent to l and if $v = 0$ it is equivalent to $\neg l$. The notation $l \neq v$ is defined similarly to encode $\neg(l = v)$. A *clause* is a disjunction of literals. In a SAT problem P, the constraints $\mathcal{C}$ are clauses.

MaxSAT problems are a subclass of COP. A MaxSAT problem[1] (P, o) is a SAT problem P and a (usually non-negative) weight o_b associated to each $b \in \mathcal{X}$. The aim is to minimise $\sum_{b \in \mathcal{X}} o_b b$.

The core operation of SAT solvers is *unit propagation*. Given a set of currently true literals ϕ, if there is a clause $C = l_1 \vee \cdots \vee l_n$ where $\neg l_i \in \phi$ for all $1 \leq i \leq n$ then unit propagation detects failure. Similarly, if $\neg l_i \in \phi$ for all $1 \leq i \neq j \leq n$ then unit propagation adds literal l_j to the current partial assignment ϕ.

3   Encoding Integer Variables

In order to use MaxSAT solvers for an arbitrary COP (P, o), we need to encode the variables, constraints, and objective of the COP as a MaxSAT problem. We first examine encoding the variables.

Given an integer x with (possibly non-contiguous) initial domain $D_i(x) = \{d_1, d_2, \ldots, d_m\}$, we will now define three encoding methods: the direct encoding, the order encoding, and the binary encoding. Each encoding method maps x to a set of Boolean *encoding variables*. Additionally, *consistency constraints* on the encoding variables ensure that they correctly represent an integer. We use consistent semantic brackets to name the encoding Booleans, where Boolean $[\![f]\!]$ is true iff the formula f holds.

The *direct* encoding of x introduces m encoding variables $[\![x = v]\!]$, $v \in D_i(x)$. $[\![x = v]\!]$ is true iff x is assigned value v. A propagation complete [19] encoding of the exactly-one *consistency constraint* is posted on the encoding variables,

$$\sum_{d \in D_i(x)} [\![x = d]\!] = 1 \tag{1}$$

to ensure that the separate Booleans faithfully encode an integer (which must take a single value). Unit propagation of Eq. (1) is propagation complete for the constraint $x \in D(x)$ and the language of atomic constraints $DIRECT(x)$.

The *order* encoding of x introduces $m - 1$ encoding variables $[\![x \geq v]\!]$, $v \in \{d_2, \ldots, d_m\}$. $[\![x \geq v]\!]$ is true iff x is assigned a value greater or equal to v. The **decreasing** *consistency constraint*

$$\forall\, 3 \leq i \leq m.\ ([\![x \geq d_i]\!] \rightarrow [\![x \geq d_{i-1}]\!]) \tag{2}$$

enforces that the encoding variables give a consistent view on the integer. Unit propagation on Eq. (2) is propagation complete for constraint $x \in D(x)$ and the

---

[1] The usual definition of MaxSAT makes use of soft clauses with weights but is functionally equivalent to the definition here, and indeed internally many MaxSAT solvers treat the problem in the form we define here.

language of atomic constraints $ORDER(x)$. The order encoding is also known as the ladder, regular, or thermometer encoding.

We extend our semantic brackets notation to make it easier to describe a broader range of bounds constraints as follows: $[\![x \geq d]\!] \equiv 1, d \leq d_1; [\![x \geq d]\!] \equiv [\![x \geq d_{i+1}]\!], d_i < d \leq d_{i+1}; [\![x \geq d]\!] \equiv 0, d > d_m; [\![x > d]\!] \equiv [\![x \geq d+1]\!];$ $[\![x < d]\!] \equiv \neg[\![x \geq d]\!];$ and $[\![x \leq d]\!] \equiv \neg[\![x > d]\!].$

The *binary* encoding of x introduces n encoding variables $[\![\mathsf{bit}(x, k)]\!], k \in 0..n-1$. Then, $[\![\mathsf{bit}(x, k)]\!]$ is true iff the kth most significant bit in the two's complement binary representation of the value assigned to x is true, i.e., in C notation: $\mathsf{bit}(x, k) = (\mathtt{x} \gg \mathtt{k})\ \&\ \mathtt{1}$. We consider two cases: where x is known to be non-negative ($d_1 \geq 0$), and where it may take both positive and negative values. In the first case the number of bits required $n = \lceil log(ub(x) + 1) \rceil)$ is given by the upper bound of x. In the second case the number of required bits $n = max(\lceil log(-lb(x)) \rceil, \lceil log(ub(x) + 1) \rceil) + 1$ is determined by the lower and upper bounds of x. Let $\mathcal{B}(x) = 0..n-1$ and let $\mathcal{R}(x)$ be the set of representable integers for the binary encoding of x. In the first case $\mathcal{R}(x) = 0..2^n - 1$ and in the second $\mathcal{R}(x) = -2^{n-1}..2^{n-1} - 1$.

The *consistency constraint* for the binary encoding enforces that the Boolean representation can only represent values in the initial domain $D_i(x)$. It is the SAT encoding of the constraints (a) $x \geq d_1$ if $d_1 > 0$ or $0 > d_1 > -2^{-n-1}$, (b) $x \leq d_m$ if $d_1 \geq 0 \wedge d_m < 2^{n-1}$ or $d_1 < 0 \wedge d_m < 2^{n-2}$ (c) $x \neq d$ for each $d_1 < d < d_m, d \notin D_i(x)$. Constraints (a) and (b) are encoded using lexicographic decompositions, while (c) can be simply encoded as a clause $\vee_{k \in \mathcal{B}(x)}[\![\mathsf{bit}(x, k)]\!] \neq \mathsf{bit}(d, k)$. For highly sparse domains more efficient approaches are possible for (c). We return to this later. Unfortunately, unit propagation on encodings of these consistency constraints does not achieve propagation completeness on the language of atomic constraints $BINARY(x) = \{false\} \cup \{\mathsf{bit}(x, k), \neg\mathsf{bit}(x, k) \mid x \in \mathcal{X}, k \in \mathcal{B}(x)\}$. The binary encoding is also known as the log(arithmic) encoding. A crucial advantage of the binary encoding is that it is exponentially smaller than the other encodings.

4   Encoding Constraints

Encoding constraints into clauses is a rich area of research. Even for simple constraints defined on Boolean variables such as at-most-one, cardinality, or psuedo-Boolean constraints there are dozens of papers [2,3,19,26,29]. Here we consider encoding constraints over integers. A key consideration is that the encoding of a constraint crucially depends on how the integers it constrains are encoded. We will indicate whether the direct, order or binary encoding is used for a particular integer x as $x{:}\mathbb{D}$, $x{:}\mathbb{O}$ or $x{:}\mathbb{B}$, respectively.

Most encodings of constraints are *uniform*, that is they expect all the integers involved in the constraint to be encoded in the same way (e.g., $x{:}\mathbb{O} \leq y{:}\mathbb{O}$). But different constraints will prefer different integer encodings. The cardinality constraints, such as `global_cardinality`, count how many times particular values occur in an array. Hence, they essentially *prescribe* a direct encoding of integers in the array. Similarly, linear constraints involving large domains essentially blow up in encoding size unless we use binary encoded integers. Many discrete optimisation models will therefore need to deal with different integers being encoded in different ways. Hence, somewhere in the model we must encode constraints with non-uniform encodings of integers (e.g., $x{:}\mathbb{O} \leq y{:}\mathbb{B}$). We call these *coupling* encodings.

There is little existing work on coupling encodings except for channelling constraints, that is, coupling encodings of equality and coupling encodings of linear inequality constraints [27].

4.1 Coupling Equality Constraints

We can couple different integer encodings by allowing a variable x to have multiple encodings, such as $x{:}\mathbb{D}$ and $x{:}\mathbb{O}$, and adding a *channelling* equality constraint between them to our model (e.g., $x{:}\mathbb{D} = x{:}\mathbb{O}$). We now discuss the three coupling encodings of the integer equality constraint $x = y$.

$x{:}\mathbb{D} = y{:}\mathbb{O}$ This is the most common channelling constraint. Its encoding amounts to just defining $[\![x = k]\!]$ in terms of the order variables, for all $k \in D_i(x)$:

$$\forall_{d \in D_i(x)} [\![x = d]\!] \leftrightarrow [\![y \geq d]\!] \wedge [\![y \leq d]\!]. \tag{3}$$

Having introduced this for all d, we no longer need consistency constraints on the direct encoding. It is easy to show [22] that unit propagation for Eq. (2) for y and (3) is propagation complete for $x \in D(x) \wedge x = y \wedge y \in D(y)$ and $\mathcal{L} = DIRECT(x) \cup ORDER(y)$. Hence, we can omit the exactly-one constraint, Eq. (1) on x.

$x{:}\mathbb{D} = y{:}\mathbb{B}$ For this constraint, there is a trade-off between propagation strength and the size of the encoding. This encoding was initially devised in the context of encoding at-most-one [9, 19, 24] constraints, where channelling to the binary representation implicitly prevents two $[\![x = d]\!]$ literals from becoming true:

$$\forall_{d \in D_i(x), k \in \mathcal{B}(y)} [\![x = d]\!] \rightarrow [\![\mathsf{bit}(y, k)]\!] = \mathsf{bit}(d, k). \tag{4}$$

It is extended by Frisch and Peugniez [9] to exactly-one by adding channelling from the binary representation back to the equality literals:

$$\forall_{d \in D_i(x)} \bigwedge_{k \in \mathcal{B}(y)} [\![\mathsf{bit}(y, k)]\!] = \mathsf{bit}(d, k) \rightarrow [\![x = d]\!]. \tag{5}$$

This encoding does not enforce domain consistency between the two representations. For example, consider the case where we have removed all the even values from $D(x)$. In order to achieve domain consistency, we replace this second set of clauses with the modified ones, for each $k \in \mathcal{B}(y), v \in \{0, 1\}$:

$$[\![\mathsf{bit}(y, k)]\!] = v \rightarrow \bigvee_{d \in D_i(x), \mathsf{bit}(d, k) = v} [\![x = d]\!]. \tag{6}$$

This ensures that once all values consistent with $\text{bit}(x, d) = v$ are eliminated, $[\![\text{bit}(x, k)]\!]$ is set to $\neg v$.

As with $x{:}\mathbb{D} = y{:}\mathbb{O}$, the $x{:}\mathbb{D} = y{:}\mathbb{B}$ constraint allows us to drop consistency constraints: both the exactly-one (for direct), and all the binary encoding consistency constraints.

Theorem 1. *Unit propagation on the clauses of Eqs.* (4) *and* (6) *is propagation complete for constraint* $x \in D(x) \wedge x = y \wedge y \in D(y)$ *and* $\mathcal{L} = DIRECT(x) \cup BINARY(y)$. *Hence, it enforces domain consistency on* x. $\qquad\square$

$x{:}\mathbb{O} = y{:}\mathbb{B}$ We are not aware of any previous channelling between order and binary representations. The information that can be exchanged between the order and binary representations is necessarily limited, and can only propagate in two cases. If the lower/upper bound rests on a value d which is inconsistent with some fixed bit $[\![\text{bit}(y, k)]\!]$, we can adjust the bound to the nearest value d' such that $\text{bit}(d', k) = [\![\text{bit}(y, k)]\!]$; and if a bit is constant for all values within the current bounds, we can fix the corresponding variable. A *stable segment* $l..u$ for binary bit k is a range $l..u \subseteq \mathcal{R}(y)$ such that $\text{bit}(d, k)$ is the same for all $d \in l..u$. We can handle both of these cases by posting, for each segment $r_l..r_u \subseteq \mathcal{R}(y)$ which is stable for bit k, the constraint:

$$[\![\text{bit}(y, k)]\!] = \text{bit}(r_l, k) \vee [\![x < r_l]\!] \vee [\![x > r_u]\!]. \tag{7}$$

Example 1. Consider a variable y with $D_i(y) = 0..6$. In the binary representation, the second last bit is fixed for ranges of size 2: $\{\{0, 1\}, \{2, 3\}, \{4, 5\}, \{6\}\}$. The channelling constraints for the second last bit, $k = 1$, are

$$\neg[\![\text{bit}(y, 1)]\!] \vee [\![x \geq 2]\!],$$
$$[\![\text{bit}(y, 1)]\!] \vee [\![x \leq 1]\!] \vee [\![x \geq 4]\!],$$
$$\neg[\![\text{bit}(y, 1)]\!] \vee [\![x \leq 3]\!] \vee [\![x \geq 6]\!], \text{and}$$
$$[\![\text{bit}(y, 1)]\!] \vee [\![x \leq 5]\!].$$

The Order-Binary channelling constraints are propagation complete, and require $O(|\mathcal{B}(y)||D_i(x)|)$ ternary clauses. We can omit the outer bounds consistency constraints on the binary encoding since they are enforced by the channel.

Theorem 2. *Unit propagation on the clauses of Eqs.* (2) *and* (7) *is propagation complete for constraint* $x \in lb(x)..ub(x) \wedge x = y$ *and language* $\mathcal{L} = ORDER(x) \cup BINARY(y)$. *Hence, it enforces the initial bounds on* x. *Note that these equations do not enforce (nor can the encoding variables represent) domain consistency on the domain of* x. $\qquad\square$

4.2   Coupling Inequality Constraints

While the channelling constraints above are propagation complete, introducing a dual representation of a variable and its channelling constraint introduce overhead. When we can avoid this and directly encode a constraint that *couples* two representations we can get more compact models.

Indeed, linear inequality constraints have a simple approach to coupling encodings, which is perhaps folklore. Each encoding can represent the linear term ax as a linear pseudo-Boolean term: direct $a \times d_1 \times [\![x = d_1]\!] + \cdots + a \times d_m \times [\![x = d_m]\!]$, order $a \times d_1 + a \times (d_2 - d_1) \times [\![x \geq d_2]\!] + \cdots + a \times (d_m - d_{m-1}) \times [\![x \geq d_m]\!]$, or binary $a \times [\![\mathsf{bit}(x, 0)]\!] + 2a \times [\![\mathsf{bit}(x, 1)]\!] + \cdots + 2^k a \times [\![\mathsf{bit}(x, k)]\!]$. Hence, an arbitrary linear constraint can be mapped to a linear pseudo-Boolean constraint which is then encoded in any number of ways. Note that we can similarly encode arbitrary integer linear objectives into a pseudo-Boolean objective required for MaxSAT.

Diet-Sugar [27] improves on this simple approach for order+binary coupling[2] of $ax + b_1 y_1 + b_2 y_2 + b_n y_n \leq c$ by essentially encoding the constraints $ad_i [\![x \geq d_i]\!] + b_1 y_1 + b_2 y_2 + b_n y_n \leq c$ for each $d_i \in D_i(x)$ separately, using a BDD to encode each resulting pseudo-Boolean. Although they do not prove it, using the CompletePath encoding (from [1]) of the resulting BDDs will guarantee propagation completeness on the language $ORDER(x) \cup BINARY(y_1) \cup \cdots \cup BINARY(y_n)$.

We consider the simpler inequality coupling constraint $x:\mathbb{O} \leq y:\mathbb{B}$, which we can encode directly as follows.

For each $lb(x) < d \leq ub(x)$ we post the constraint

$$[\![x \geq d]\!] \rightarrow \bigvee_{k \in \mathcal{B}(x), \neg \mathsf{bit}(d-1,k)} [\![\mathsf{bit}(y, k)]\!] \tag{8}$$

The intuition behind this is that the binary representation for d satisfies this because adding any positive amount to $d - 1$ will flip at least one bit from false to true. The encoding consists of $O(|ub(x) - lb(x)|)$ clauses of size $O(|\mathcal{B}(y)|)$. If $D_i(x)$ is not a range then many clauses within domain gaps become redundant. To complete the encoding we need to enforce the initial bounds using the lexicographic encoding of $y \geq lb(x)$ and the unit clause encoding of $x \leq ub(y)$. To extend to ranges with negative numbers we treat the sign bit $[\![\mathsf{bit}(x, n - 1)]\!]$ as if it were negated, but omit details here.

Example 2. Consider encoding $x:\mathbb{O} \leq y:\mathbb{B}$ for variables x and y with $D_i(x) = 1..7$, $D_i(y) = 0..6$. This will result in the following clauses from Eq. (8), shown with the relevant bit representation on the left.

001	$[\![x \geq 2]\!] \rightarrow [\![\mathsf{bit}(y, 2)]\!] \vee [\![\mathsf{bit}(y, 1)]\!]$	
010	$[\![x \geq 3]\!] \rightarrow [\![\mathsf{bit}(y, 2)]\!] \vee$	$[\![\mathsf{bit}(y, 0)]\!]$
011	$[\![x \geq 4]\!] \rightarrow [\![\mathsf{bit}(y, 2)]\!]$	
100	$[\![x \geq 5]\!] \rightarrow$	$[\![\mathsf{bit}(y, 1)]\!] \vee [\![\mathsf{bit}(y, 0)]\!]$
101	$[\![x \geq 6]\!] \rightarrow$	$[\![\mathsf{bit}(y, 1)]\!]$

The encoding also requires $y \geq 1$ encoded as $[\![\mathsf{bit}(y, 0)]\!] \vee [\![\mathsf{bit}(y, 1)]\!] \vee [\![\mathsf{bit}(y, 2)]\!]$ and $x \leq 6$ encoded as $\neg [\![x \geq 7]\!]$. We see that for example $[\![x \geq 4]\!]$, which forces $[\![\mathsf{bit}(y, 2)]\!]$ true, means that it never occurs for later literals. This relies on the order encoding, since $[\![x \geq 5]\!]$ implies $[\![x \geq 4]\!]$.

---

[2] Their approach handles any number of order encoded variables, we restrict to one for simplicity of explanation.

If we suppose instead $D_i(x) = \{1, 2, 6, 7\}$, then we keep only the first, third and last clauses, since the others are subsumed (i.e., $[\![x \geq 3]\!] \equiv [\![x \geq 4]\!] \equiv [\![x \geq 5]\!] \equiv [\![x \geq 6]\!]$). $\hfill \square$

This encoding is propagation complete for $x \leq y$.

Theorem 3. *Unit propagation on the clauses of Eqs. (2) and (8) and clauses for lexicographic encoding of $y \geq lb(x)$ and together with the clause $\neg [\![x \geq ub(y) + 1]\!]$ is propagation complete for $x \leq y$ for the language $\mathcal{L} = ORDER(x) \cup BINARY(y)$.* $\hfill \square$

Interestingly the encoding of Diet-Sugar applied to the constraint $x : \mathbb{O} \leq y : \mathbb{B}$ produces a strict superset of our encoding clauses, adding many redundant clauses.

We can enforce the reverse $x : \mathbb{O} \geq y : \mathbb{B}$ similarly. Apart from the initial bounds constraints $x \geq lb(y)$ and $y \leq ub(x)$ we generate the clauses for $lb(y) < d \leq ub(x)$

$$\neg [\![x \geq d]\!] \rightarrow \bigvee_{k \in \mathcal{B}(y), \text{bit}(d, k)} \neg [\![\text{bit}(y, k)]\!]. \tag{9}$$

We can of course use the conjunction of inequalities, $x : \mathbb{O} \geq y : \mathbb{B} \wedge x : \mathbb{O} \leq y : \mathbb{B}$, to enforce equality $x : \mathbb{O} = y : \mathbb{B}$, but this *inequality channel* does not enforce propagation completeness.

4.3 Coupling Element Constraints

Element constraints are an important component of many CP models. They enable array look-up using a variable index with $y = A[x]$. Our encoding makes use of the fact that the equality constraint $x = d$ for fixed integer d is a conjunction of literals for the direct, order or binary encoding: $[\![x = d]\!]$, $[\![x \geq d]\!] \wedge [\![x \leq d]\!]$, or $\bigwedge_{k \in \mathcal{B}(x)} [\![\text{bit}(x, k)]\!] = \text{bit}(d, k)$, respectively. Consequently, we can define a coupling encoding of element constraint for any encoding choices for y, A and x with for all $d \in D_i(x)$, $(x = d) \rightarrow (A[d] = y)$. In other words, the equality constraint $A[d] = y$ is conditional on the conjunction of literals, $x = d$.

Example 3. Consider the coupling for $y : \mathbb{O} = A : \mathbb{B}[x : \mathbb{D}]$. The clauses take the form

$$\neg [\![x = d]\!] \vee ([\![\text{bit}(A[d], k)]\!] = \text{bit}(r_l, k)) \vee [\![y < r_l]\!] \vee [\![y > r_u]\!]$$

for $d \in D_i(x)$ and stable ranges $r_l .. r_u \subseteq R(y)$ for bit k. $\hfill \square$

When x has the direct encoding and A is fixed, the encoding can be made propagation complete similar to Eq. (6) by adding a backwards clauses for every distinct value u in A: $(y = u) \rightarrow \bigvee_{d \in D_i(x), u = A[d]} [\![x = d]\!]$.

5 Views

Views [25] are a crucial feature in modern CP solvers. For specific constraints, views can simplify propagation construction using an interface to the variable operations. In SAT, whenever an equivalence or negation arises between different expressions, we can represent both with the same Boolean variable. This allows greater scope for views on SAT encoded integers compared to CP integers.

Affine Transformations. The most important view constraint is the affine transformation $y = ax + b$ for fixed integers a and b. For the direct encoding, we have for every $d \in D(x) \cup D(y)$ that $[\![y = ad + b]\!] \equiv [\![x = d]\!]$, so we can use the same Boolean variable to represent both. For the order encoding, we have for non-negative a that $[\![y \geq ad + b]\!] \equiv [\![x \geq d]\!]$ (if a is negative, one $\geq$ flips to $\leq$). For the binary encoding, affine views can be applied in some but not all cases [34].

Minimum/Maximum. Boolean encoding also raises the possibility of *partial views* where only some Boolean variables are reused. In $y = \max(x, m)$ for fixed integer m (and similar for min), after enforcing $y \geq m$ we find a partial view for the order encoding $\bigvee_{d=m}^{ub(y)} [\![y \geq d]\!] \equiv [\![x \geq d]\!]$. For the direct encoding, we have one less equivalence: $\bigvee_{d=m+1}^{ub(y)} [\![y = d]\!] \equiv [\![x = d]\!]$. We require one new Boolean variable for $[\![y = m]\!]$, which is constrained by $\bigvee_{d=lb(x)}^{m} [\![x = d]\!] \rightarrow [\![y = m]\!]$.

Element. Certain compositions of fixed A allow for views. For $y{:}\mathbb{D} = A[x{:}\mathbb{D}]$, every unique value $A[d]$ has binary backwards clause $[\![y = A[d]]\!] \rightarrow [\![x = d]\!]$, so $[\![x = d]\!] \equiv [\![y = A[d]]\!]$. For $y{:}\mathbb{O} = A[x{:}\mathbb{O}]$, we have for all $d \in D_i(x)$ that $[\![x \geq d]\!] \rightarrow [\![y \geq A[d]]\!]$ if $A[d] \leq A[e]$ holds for all $e > d$. Similarly, $[\![x \leq d]\!] \rightarrow [\![y \leq A[d]]\!]$ if $A[c] \leq A[d]$ holds for all $c < d$. If both conditions hold, then $[\![x \geq d]\!] \equiv [\![y \geq A[d]]\!]$. If A is strictly monotone, both conditions hold for all elements and y becomes a total view of x. When the inequalities are flipped, we have negated views $[\![x \geq d]\!] \equiv [\![y < A[d]]\!]$, which are total if A is strictly antitone.

Views allow us to extend our coupling encodings straightforwardly. For example, we can encode $x{:}\mathbb{O} + d \leq y{:}\mathbb{B}$ by using a view to construct $(x + d){:}\mathbb{O}$ and then using the inequality coupling.

6 Experimental Results

To validate the benefit of coupling different integer encodings, we created the MiniZinc-SAT framework, an extension of MiniZinc. Through MiniZinc's *annotations*, the user can declare the integer variable encodings. For example in the knight's tour model, given in Sect. 2.1, we can choose direct encoding for each variable x_r_c in x by adding an annotation as follows.

```
var neighbours: x_r_c :: direct_encoded;
```

Since annotations are first class objects, the user can specify more complex encoding schemes based on, for example, the variable's domain size, its existing encodings, or any other contextual rules. If no encodings are specified, default choices are made. During compilation, MiniZinc-SAT resolves mixed encoding constraints by coupling if possible, or by channelling if necessary. Additionally, views are used whenever the constraint and variable encodings allow it.

Using MiniZinc-SAT, we create various encodings of realistic problems to see if we can improve solver performance. We also compare with Picat-SAT 3.1 [33], a solver that maps MiniZinc to SAT using binary encoding, and Chuffed 0.10.4 [7] a lazy clause generation solver, using lazy direct-order encoding.

Results are given for SAT models and MaxSAT models running Open-WBO 2.1 [16] on a single-core *Intel Xeon 8260* CPU (non-hyperthreaded) with a 10-minute solving timeout and up to 64 GB of RAM. The results are visualised in cactus plots, in which for each configuration the solved instances are sorted by solve time (measured from when the SAT or LCG solver is started). Instances which are not solved (SAT) or proved optimal (MaxSAT) within the time limit are omitted.

Source code, models, instances and run logs are available online[3].

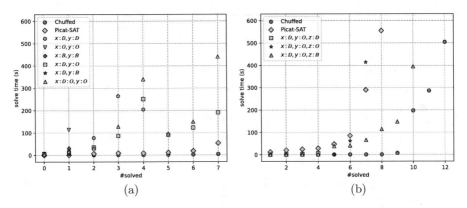

Fig. 1. Cactus plots for solved (a) `knights` instances, and (b) `orienteering` instances.

Knight's Tour. The first example is the `knights` model shown in Sect. 2.1 for instances $n = 8, 10, ..., 22$. The model contains n^2 variables x with relatively sparse domains of up to size 8, but to prevent sub-cycles the `circuit` decomposition introduces another n^2 variables y with contiguous domains of $1..n^2$ that represent the order in which positions are visited. The encodings of x and y are coupled through `element` constraints, extended by a view on the result variable (e.g., $y:\mathbb{O}[x_i:\mathbb{D}] = (y_{i-1} + 1):\mathbb{O}, \forall 1 < i \leq n$). The results for the three uniform and three sensible mixed encoding choices for x and y are shown in Fig. 1a.

Uniform order and binary encodings do not succeed beyond the two easiest instances. This is unsurprising since the `circuit` constraint prescribes $x:\mathbb{D}$. However, since the y variables reason about the order of visits, we see that $y:\mathbb{O}$ is clearly preferred over the uniform approach, $y:\mathbb{D}$. The worst choice is $y:\mathbb{B}$, which has the additional disadvantage that it cannot use the affine transformation view. Creating a redundant order encoding for x variables ($x:\mathbb{D}:\mathbb{O}$) does not seem to make much difference. Chuffed and Picat-SAT both far outperform MiniZinc-SAT, since they have native `circuit` propagator and encoding [32] respectively.

---

[3] https://github.com/hbierlee/cpaior-2022-coupling-sat.

Orienteering. The `orienteering` problem is a COP concerning a complete graph with edge distances $d_{i,j}$ and node rewards. The aim is to find a path from start to finish node that maximises the sum of the rewards of the visited nodes, but which is limited by a linear inequality on the distances of the traversed edges. The principal `subcircuit` constraint is encoded similarly to `circuit` in `knights` by two sets of variables x and y, coupled through `element` constraints. Given the results from `knights`, we will use $x{:}\mathbb{D}$ and $y{:}\mathbb{O}$. However, we will experiment with all possible (non-redundant) encodings on a new set of coupled variables $z_i = d_i[x_i]$ which appear in the linear inequality on the maximum distance. The linear objective can be directly encoded using the $x{:}\mathbb{D}$ variables.

A cactus plot comparing encodings is shown in Fig. 1b. Of the mixed encodings, $z{:}\mathbb{B}$ is clearly the best, since it makes the path length constraint compact. While both Chuffed and Picat-SAT again have native `subcircuit` propagator and encoding, Picat-SAT is now outperformed by the best MiniZinc-SAT encoding. It seems that its `subcircuit` encoding is less effective than the MiniZinc standard decomposition using element coupling.

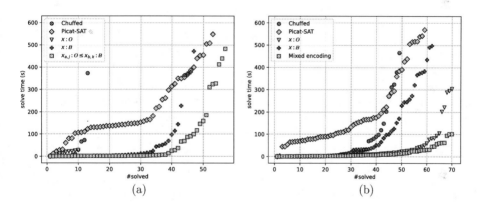

Fig. 2. Cactus plots for solved `jsswet` instances of (a) variant where 40% of jobs are high priority, and of (b) variant with limited average job length.

Job-Shop scheduling with Weighted Earliness/Lateness. Here, we examine a job-shop scheduling problem `jsswet` over n jobs, each with m tasks, with start time variables x. Every task j of job i runs on a different machine for its full duration $d_{i,j}$. Tasks of the same job must finish in sequence ($x_{i,j} + d_{i,j} \leq x_{i,j+1}$), and must not overlap (`disjunctive`) with the tasks of other jobs that run on the same machine. The objective is to minimise the sum of the penalties over all jobs. A job's penalty is its final task's earliness or lateness to its deadline t_i, weighted by its earliness e_i or lateness l_i penalty coefficients, respectively. The order encoding combines affine transformations and max views for the objective: $e_i \times \max(0, t_i - x_{i,m} + d_i){:}\mathbb{O} + l_i \times \max(0, x_{i,m} + d_i - t_i){:}\mathbb{O}$.

In preliminary results not shown here we found that if the schedule's horizon is small, the order encoding works best thanks to its propagation strength. However, when the horizon becomes large this approach will run out of memory due

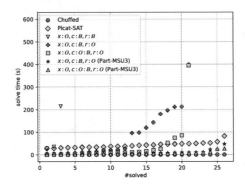

Fig. 3. Cactus plots for solved `table-layout` instances.

to the sheer number of order encoding variables, and only the binary encoding remains viable. So each encoding has benefits and drawbacks. We now consider two variants of the problem.

In the first variant, we consider a large horizon, but designate the first 40% of jobs as *high priority* jobs which must finish within 500 time steps of their deadline, and have much higher penalties. Effectively, this splits the variables into those with small and large domains. We consider encoding all variables with order, or binary, or a mix which uses order for small and binary for large domains. For high priority job-task $x_{a,j}$ and low priority job-task $x_{b,k}$ assigned to the same machine, the `disjunctive` constraint couples the (potentially) mixed encodings via $((x_{a,j} + d_{a,j}):\mathbb{O} \leq x_{b,k}:\mathbb{B}) \vee (x_{b,k}:\mathbb{B} \leq (x_{a,j} - d_{b,k}):\mathbb{O})$. The results in Fig. 2a show that the mixed encoding outperforms the other solvers and uniform encodings.

For the other variant we consider a smaller horizon, which allows a full order encoding of all tasks. However, now the average run time of all jobs is limited by parameter M (2.5 times the sum of all minimal job durations). To effectively constrain this linear inequality $\sum_{i=1}^{n} x_{i,m} - x_{i,1} \leq M$, the mixed encoding adds a redundant binary encoding to the first and last task of each job: $x_{i,1}:\mathbb{O}:\mathbb{B}, x_{i,2}:\mathbb{O}, \ldots, x_{i,m-1}:\mathbb{O}, x_{i,m}:\mathbb{O}:\mathbb{B}$.

The results in Fig. 2b show the mixed encoding convincingly outperforms the other solvers and uniform encodings. The redundant encodings are coupled with $x_{i,1}:\mathbb{O} = x_{i,1}:\mathbb{B}$, but coupling with $x_{i,1}:\mathbb{O} \leq x_{i,1}:\mathbb{B} \wedge x_{i,1}:\mathbb{O} \geq x_{i,1}:\mathbb{B}$ or just $x_{i,1}:\mathbb{O} \leq x_{i,1}:\mathbb{B}$ produces similar results.

Table Layout. Finally, we consider the `table-layout` problem. For a table composed of $n \times m$ cells, our task is to assign a width-height configuration variable $x_{i,j}$ for each cell at row i, column j. The configuration will determine for cell i, j its cell-width $w_{i,j} = W[x_{i,j}]$ and cell-height $h_{i,j} = H[x_{i,j}]$ through `element` constraints. These variables are combined in n row-height variables r_i, where $\bigwedge_{j=1}^{m} r_i \geq h_{i,j}$, and m column-width variables c_j, where $\bigwedge_{i=1}^{m} c_j \geq w_{i,j}$. The objective is then to minimise the table's height $\sum_{i=1}^{n} r_i$, without the table's width $\sum_{j=1}^{m} c_j$ exceeding a given maximum width.

We chose this problem to test two unexplored properties. First, $H_{i,j}$ and $W_{i,j}$ are guaranteed to be respectively monotone and antitone, since the width-height configurations represent sorted, optimal text layouts. Consequently, the element constraint establishes a view between the order encoding variables $[\![x_{i,j} \geq k]\!]$, $[\![h_{i,j} \geq H_{i,j,k}]\!]$ and $[\![w_{i,j} \leq W_{i,j,k}]\!]$. Secondly, $D(h_{i,j})$ is sparse whereas $D(r_i)$ is contiguous. Thus, the coupling $h_{i,j}:\mathbb{O} \leq r_i:\mathbb{B}$ requires far fewer clauses for the domain gaps in $h_{i,j}$, while also skipping a large order encoding of r_i. We compare this very compact encoding against an order encoded objective (using simple binary clauses for $h_{i,j}:\mathbb{O} \leq r_i:\mathbb{O}$). A third approach creates an order encoding for the width ($w_{i,j}:\mathbb{O} \leq c_j:\mathbb{O}$) as well, but still redundantly channels $c_j:\mathbb{O} = c_j:\mathbb{B}$ for the linear inequality.

The results in Sect. 6 show that solver performance suffers greatly if we couple $x:\mathbb{O}$ to a binary encoded objective $r:\mathbb{B}$ rather than using straightforward order encoded objective $r:\mathbb{O}$. The coupling perhaps overcomplicates the objective compared to using binary clauses. Furthermore, channelling rather than coupling to $c:\mathbb{B}$ seems to be marginally better as well. The domain sizes of the `table-layout` instances are too large for the more compact coupling to pay off (in contrast to `jsswet`). For $r:\mathbb{O}$, the pseudo-Boolean objective is unweighted, which means Open-WBO can use its core-guided Part-MSU3 algorithm. This makes the models solve almost instantly. Chuffed does equally well, while Picat-SAT solves all instances but requires more time.

7   Related Work

The choice of variable encoding is a critical decision when encoding CSPs to SAT, and unsurprisingly has seen considerable attention. Most CSP-to-SAT converters fix one of the encodings described in Sect. 3 as their core representation, and convert all variables/constraints based on that encoding. Binary encoding is a popular choice (in two's complement [12] or sign-magnitude [33] variants) despite its relatively weak propagation strength, as it can reliably cope with very large domains. Order encoding, conversely, is adopted by some encoders [17, 28] despite the risk of blow-up due to its effectiveness on primitive arithmetic constraints [30]. This is sometimes paired with a channelled direct encoding for flexibility [5,14]. Not to be confused with our order-binary channelling, some works propose new integer encodings which mix features of the order and binary encoding [20,31].

Some works are concerned with picking the right encoding for the right variable. `Proteus` [14] attempts to predict, for a given instance, which encoding of variables (and constraints) will be most effective, then commits to that encoding for the whole instance. `Satune` [11] selects encodings on a per-variable basis, and optimises the domain representation, based on a training phase for a class of instances. However, it does not resolve the coupling problem, instead requiring connected variables to share a common representation. A more radical approach, adopted by current lazy clause generation solvers [7,8,10] is to implicitly maintain a partial representation of the direct-order encoding, introducing new literals and channelling as necessary.

Diet-Sugar [27] is the only other work that we are aware of that considers encoding non-uniform constraints. It directly introduces coupled order and binary encodings of linear constraints. The SAT translation chooses a single encoding for each variable using a heuristic that leads to an overall small and effective encoding of the constraints. The resulting mixed encodings are shown to yield significant improvement.

General compilers that support MiniZinc have been developed using binary encoding, namely FznTini [12, 13] (two's complement) and Picat-SAT [33] (sign-and-magnitude). Notably, the binary encoding despite theoretically lower propagation strength has continued to prove itself in the Picat-SAT compiler in the yearly MiniZinc competition. It is clear that the black box compilers such as Picat-SAT essentially introduce other forms of integer encoding in some global constraint decompositions [32]. Savile Row [21] similarly converts a high-level model (specified in Essence') into SAT. Savile Row commits to a channelled direct-order encoding for variables, allowing each constraint to choose its preferred uniform encoding, though they apply some transformations [4] to improve the generated SAT encoding.

BEE [17] is an approach to compiling integer models to SAT using only order encoding. It goes beyond our view approach by searching for all Boolean variable equivalences (using a SAT solver) in order to simplify the resulting SAT model. BEE would automatically discover the (partial) views we define on order encoded variables, but none of them are covered by its ad hoc methods.

8   Conclusion

In conclusion, while compilation to SAT is a competitive approach to tackling discrete optimisation problems, efficient encoding of discrete optimisation problems into SAT is challenging. To create the best encodings possible we must allow different representations of integers to be used in the encoding, which means the problem of coupling encodings arises. In this paper we show how we can create coupled models by using channelling equalities, or directly by using coupled encodings of element or inequality constraints. Total and partial views extend the coupling constraints we can encode. We show that coupling encodings can be required for getting the best resulting SAT encoding.

Acknowledgements. This research was partially funded by the Australian Government through the Australian Research Council Industrial Transformation Training Centre in Optimisation Technologies, Integrated Methodologies, and Applications (OPTIMA), Project ID IC200100009.

A Proofs

Theorem 1. *Unit propagation on the clauses of Eqs.* (4) *and* (6) *is propagation complete for constraint* $x \in D(x) \wedge x = y \wedge y \in D(y)$ *and* $\mathcal{L} = DIRECT(x) \cup BINARY(y)$. *Hence, it enforces domain consistency on* x.

Proof. First we show that unit propagation enforces the exactly-one constraint. If we have $[\![x = d]\!]$ and $[\![x = d']\!]$ true simultaneously, then Eq. (4) will force one binary variable in y to take two values, thus failing. If all variables $[\![x = d']\!]$ are set false except one $[\![x = d]\!]$, then Eq. (6) will set the correct bits of y in the remaining solution (via setting the negation to false). The forward direction will then propagate $[\![x = d]\!]$.

Next we show that the unit propagation is propagation complete with respect to the binary variables and missing values in the original domain. Suppose $S \subseteq D(x)$ are the remaining values in the domain. Then $\neg[\![x = d]\!]$ is set for $d \in D_i(x) - S$. Suppose that there is no support for $[\![bit(y, k)]\!]$ in S. Then unit propagation of Eq. (6) will set $\neg[\![bit(y, k)]\!]$.

Finally, we show that unit propagation is propagation complete on the encoding variables. Given the subset of the current assignment literals $L \subseteq \phi$ restricted to the direct encoding variables for x and binary encoding variables of y. Let $S \subset D(x)$ be the set of possible domain values remaining, i.e. $S = \{d \mid d \in D(x), \neg[\![x = d]\!] \notin L\}$. Clearly the direct encoding variables are consistent with this by definition, and if S is a singleton, by the first argument $[\![x = d]\!]$ is set true. We now consider each bit variable $[\![bit(y, k)]\!]$: if $[\![bit(y, k)]\!] \in L$ then by Eq. (4) each value in $d \in S$ has $bit(d, k)$ is true, hence it is supported; similarly if $\neg[\![bit(y, k)]\!] \in L$. Finally, if variable $[\![bit(y, k)]\!]$ does not appear in L, then there must be values in S with both truth values, otherwise one of the equations in Eq. (6) would have propagated.

Theorem 2. *Unit propagation on the clauses of Eqs.* (2) *and* (7) *is propagation complete for constraint* $x \in lb(x)..ub(x) \wedge x = y$ *and language* $\mathcal{L} = ORDER(x) \cup BINARY(y)$. *Hence, it enforces the initial bounds on* x. *Note that these equations do not enforce (nor can the encoding variables represent) domain consistency on the domain of* x.

Proof. Given the subset of the current assignment literals $L \subseteq \phi$ restricted to the order encoding variables for x and binary encoding variables for x, we show there is support for each possible value defined by L. Since each bound or bit is supported or propagated, propagation completeness follows. The order literals in L define a range $l..u$ given by $l = \max\{i \mid [\![x \geq i]\!] \in L\}$ and $u = \min\{i - 1 \mid \neg[\![x \geq i]\!] \in L\}$. The order consistency Eq. (2) enforces that $\neg[\![x \geq i]\!] \in L$ for $i > u$ and $[\![x \geq i]\!] \in L$ for $i < l$. We show l and u are supported. Suppose that $[\![bit(x, b)]\!] \neq bit(l, b) \in L$ for some bit b. Then there is some stable segment for this bit including l, say $r_l..r_u$. Then the clause $[\![bit(x, b)]\!] \vee [\![x < r_l]\!] \vee [\![x > r_u]\!]$ will fire enforcing $[\![x > r_u]\!]$ and hence l cannot be the lower bound. A similar argument applies to the upper bound.

We now show each possible value for each binary encoding variable $[\![\mathrm{bit}(x,b)]\!]$ is supported in the range $l..u$. Suppose to the contrary that w.l.o.g. $[\![\mathrm{bit}(x,b)]\!]$ is not true for any $x \in l..u$. Then there is a stable segment $r_l..r_u$ at least as large as $l..u$ for $\neg[\![\mathrm{bit}(x,b)]\!]$. Then the clause $\neg[\![\mathrm{bit}(x,b)]\!] \vee [\![x < r_l]\!] \vee [\![x > r_u]\!]$ will propagate $\neg[\![\mathrm{bit}(x,b)]\!]$, which then must be in L. The same reasoning holds for the negation.

The channelling also enforces the outer bounds on the binary encoding variables. Clearly $d_1 \leq l \leq u \leq d_m$ so the arguments for support of binary encoding variables automatically take into account the initial bounds.

Theorem 3. *Unit propagation on the clauses of Eqs.* (2) *and* (8) *and clauses for lexicographic encoding of $y \geq lb(x)$ and together with the clause $\neg[\![x \geq ub(y)+1]\!]$ is propagation complete for $x \leq y$ for the language $\mathcal{L} = ORDER(x) \cup BINARY(y)$.*

Proof. Suppose $\mathcal{D} \wedge x \leq y \rightarrow [\![x \leq d]\!]$ we show that unit propagation will enforce this. The initial case given by $y \leq ub(x)$ is enforced by the last clause. Let d be the maximum value y can take given $\mathcal{D}$, then we know that $\neg[\![\mathrm{bit}(y,k)]\!] \in \mathcal{D}$ for all $k \in \mathcal{B}(y)$, $\mathrm{bit}(d,k) = 0$ otherwise y could take a larger value. The clause Eq. (8) for $[\![x \geq d+1]\!]$ has right-hand side $\bigvee_{k\in\mathcal{B}(y),\neg\mathrm{bit}(d,k)}[\![\mathrm{bit}(y,k)]\!]$ and hence propagates $\neg[\![x \geq d+1]\!]$ as required.

Let Y be the set of all possibly values that y can take given the $\mathcal{D}$. Suppose $\mathcal{D} \wedge x \leq y \rightarrow [\![\mathrm{bit}(y,k)]\!]$ for some $k \in \mathcal{B}(y)$. We show that unit propagation will enforce this. If x still sits at its initial lower bound this will be forced by the encoding of $y \geq lb(x)$. Otherwise, this propagation was caused by $[\![x \geq d]\!]$ in the current domain, where $d > \min(Y)$

So clearly (A) $\mathrm{bit}(v,k) = 1$ for all $v \in Y \cap d..\max(y)$. Let (B) $d' = \max\{d'' \mid d'' \in Y, \mathrm{bit}(d'',k) = 0\}$. Clearly such a d' exists otherwise we would already have propagated $[\![\mathrm{bit}(y,k)]\!]$, and, since $d' < d$, $[\![x \geq d'+1]\!]$ is propagated by Eq. (2). We claim the clause 8 for $[\![x \geq d'+1]\!]$ will propagate $[\![\mathrm{bit}(y,k)]\!]$. Suppose to the contrary then there is another bit k' where $\mathrm{bit}(d',k') = 0$ where $\neg[\![\mathrm{bit}(y,k')]\!] \notin \mathcal{D}$. Then $d' + 2^{k'} \in Y$ and either $d' + 2^{k'} \geq d$ contradicting (A), or $d' + 2^{k'} < d$ contradicting (B).

While it is not possible for $\mathcal{D} \wedge x \leq y \rightarrow \neg[\![\mathrm{bit}(y,k)]\!]$ since a lower bound can never force a negative bit (although this can happen subsequently from the clauses for constraint $y \leq ub(y)$).

References

1. Abío, I., Gange, G., Mayer-Eichberger, V., Stuckey, P.J.: On CNF encodings of decision diagrams. In: Quimper, C.-G. (ed.) CPAIOR 2016. LNCS, vol. 9676, pp. 1–17. Springer, Cham (2016). https://doi.org/10.1007/978-3-319-33954-2_1
2. Abío, I., Nieuwenhuis, R., Oliveras, A., Rodríguez-Carbonell, E.: A parametric approach for smaller and better encodings of cardinality constraints. In: Schulte, C. (ed.) CP 2013. LNCS, vol. 8124, pp. 80–96. Springer, Heidelberg (2013). https://doi.org/10.1007/978-3-642-40627-0_9

3. Abío, I., Nieuwenhuis, R., Oliveras, A., Rodríguez-Carbonell, E., Mayer-Eichberger, V.: A new look at BDDs for pseudo-Boolean constraints. J. Artif. Intell. Res. **45**, 443–480 (2012). https://doi.org/10.1613/jair.3653

4. Ansótegui, C., et al.: Automatic detection of at-most-one and exactly-one relations for improved SAT encodings of pseudo-Boolean Constraints. In: Schiex, T., de Givry, S. (eds.) CP 2019. LNCS, vol. 11802, pp. 20–36. Springer, Cham (2019). https://doi.org/10.1007/978-3-030-30048-7_2

5. Barahona, P., Hölldobler, S., Nguyen, V.: Efficient sat-encoding of linear CSP constraints. In: International Symposium on Artificial Intelligence and Mathematics, ISAIM 2014, Fort Lauderdale, FL, USA, 6–8 January 2014 (2014). http://www.cs.uic.edu/pub/Isaim2014/WebPreferences/ISAIM2014_Barahona_etal.pdf

6. Choi, C.W., Harvey, W., Lee, J.H.M., Stuckey, P.J.: Finite domain bounds consistency revisited. In: Sattar, A., Kang, B.-H. (eds.) AI 2006. LNCS (LNAI), vol. 4304, pp. 49–58. Springer, Heidelberg (2006). https://doi.org/10.1007/11941439_9

7. Chu, G.: Improving combinatorial optimization. Ph.D. thesis, University of Melbourne, Australia (2011). http://hdl.handle.net/11343/36679

8. Feydy, T., Stuckey, P.J.: Lazy clause generation reengineered. In: Gent, I.P. (ed.) CP 2009. LNCS, vol. 5732, pp. 352–366. Springer, Heidelberg (2009). https://doi.org/10.1007/978-3-642-04244-7_29

9. Frisch, A.M., Peugniez, T.J.: Solving non-Boolean satisfiability problems with stochastic local search. In: Nebel, B. (ed.) Proceedings of the Seventeenth International Joint Conference on Artificial Intelligence, IJCAI 2001, Seattle, Washington, USA, 4–10 August 2001, pp. 282–290. Morgan Kaufmann (2001)

10. Gange, G., Berg, J., Demirović, E., Stuckey, P.J.: Core-guided and core-boosted search for CP. In: Hebrard, E., Musliu, N. (eds.) CPAIOR 2020. LNCS, vol. 12296, pp. 205–221. Springer, Cham (2020). https://doi.org/10.1007/978-3-030-58942-4_14

11. Gorjiara, H., Xu, G.H., Demsky, B.: Satune: synthesizing efficient SAT encoders. In: Proceedings of the ACM Programming Languages 4(OOPSLA), pp.146:1–146:32 (2020)

12. Huang, J.: Universal Booleanization of constraint models. In: Stuckey, P.J. (ed.) CP 2008. LNCS, vol. 5202, pp. 144–158. Springer, Heidelberg (2008). https://doi.org/10.1007/978-3-540-85958-1_10

13. Huang, J.: Search strategy simulation in constraint booleanization. In: Thirteenth International Conference on the Principles of Knowledge Representation and Reasoning (2012)

14. Hurley, B., Kotthoff, L., Malitsky, Y., O'Sullivan, B.: Proteus: a hierarchical portfolio of solvers and transformations. In: Simonis, H. (ed.) CPAIOR 2014. LNCS, vol. 8451, pp. 301–317. Springer, Cham (2014). https://doi.org/10.1007/978-3-319-07046-9_22

15. Ignatiev, A., Marques-Silva, J., Morgado, A.: A python library for prototyping with sat oracles (2021). https://pypi.org/project/python-sat/

16. Martins, R., Manquinho, V., Lynce, I.: Open-WBO: a modular MaxSAT solver'. In: Sinz, C., Egly, U. (eds.) SAT 2014. LNCS, vol. 8561, pp. 438–445. Springer, Cham (2014). https://doi.org/10.1007/978-3-319-09284-3_33

17. Metodi, A., Codish, M., Stuckey, P.J.: Boolean equi-propagation for concise and efficient SAT encodings of combinatorial problems. J. Artif. Intell. Res. **46**, 303–341 (2013). https://doi.org/10.1613/jair.3809, http://www.jair.org/papers/paper3809.html

18. Nethercote, N., Stuckey, P.J., Becket, R., Brand, S., Duck, G.J., Tack, G.: MiniZinc: towards a standard CP modelling language. In: Bessière, C. (ed.) CP 2007. LNCS, vol. 4741, pp. 529–543. Springer, Heidelberg (2007). https://doi.org/10.1007/978-3-540-74970-7_38

19. Nguyen, V., Mai, S.T.: A new method to encode the at-most-one constraint into SAT. In: Thang, H.Q., et al. (eds.) Proceedings of the Sixth International Symposium on Information and Communication Technology, Hue City, Vietnam, 3–4 December 2015. pp. 46–53. ACM (2015). https://doi.org/10.1145/2833258.2833293

20. Nguyen, V., Velev, M.N., Barahona, P.: Application of hierarchical hybrid encodings to efficient translation of CSPS to SAT. In: 25th IEEE International Conference on Tools with Artificial Intelligence, ICTAI 2013, Herndon, VA, USA, 4–6 November 2013, pp. 1028–1035. IEEE Computer Society (2013). https://doi.org/10.1109/ICTAI.2013.154

21. Nightingale, P., Spracklen, P., Miguel, I.: Automatically improving SAT encoding of constraint problems through common subexpression elimination in Savile row. In: Pesant, G. (ed.) CP 2015. LNCS, vol. 9255, pp. 330–340. Springer, Cham (2015). https://doi.org/10.1007/978-3-319-23219-5_23

22. Ohrimenko, O., Stuckey, P.J., Codish, M.: Propagation via lazy clause generation. Constraints Int. J. **14**(3), 357–391 (2009). https://doi.org/10.1007/s10601-008-9064-x

23. Philipp, T., Steinke, P.: PBLib – A Library for Encoding Pseudo-Boolean Constraints into CNF. In: Heule, M., Weaver, S. (eds.) SAT 2015. LNCS, vol. 9340, pp. 9–16. Springer, Cham (2015). https://doi.org/10.1007/978-3-319-24318-4_2

24. Prestwich, S.: Finding Large Cliques using SAT Local Search, chap. 15, pp. 269–274. John Wiley, Hoboken (2007)

25. Schulte, C., Tack, G.: Views and iterators for generic constraint implementations. In: Hnich, B., Carlsson, M., Fages, F., Rossi, F. (eds.) Recent Advances in Constraints, pp. 118–132. Springer, Berlin Heidelberg, Berlin, Heidelberg (2006). https://doi.org/10.1007/11754602_9

26. Sinz, C.: Towards an optimal CNF encoding of Boolean cardinality constraints. In: van Beek, P. (ed.) CP 2005. LNCS, vol. 3709, pp. 827–831. Springer, Heidelberg (2005). https://doi.org/10.1007/11564751_73

27. Soh, T., Banbara, M., Tamura, N.: A hybrid encoding of CSP to SAT integrating order and log encodings. In: 27th IEEE International Conference on Tools with Artificial Intelligence, ICTAI 2015, Vietri sul Mare, Italy, 9–11 November 2015, pp. 421–428. IEEE Computer Society (2015). https://doi.org/10.1109/ICTAI.2015.70

28. Tamura, N., Banbara, M.: Sugar: a CSP to SAT translator based on order encoding. In: Proceedings of the Second International CSP Solver Competition, pp. 65–69 (2008)

29. Tamura, N., Banbara, M., Soh, T.: Compiling pseudo-boolean constraints to SAT with order encoding. In: 25th IEEE International Conference on Tools with Artificial Intelligence, ICTAI 2013, Herndon, VA, USA, 4–6 November 2013, pp. 1020–1027. IEEE Computer Society (2013). https://doi.org/10.1109/ICTAI.2013.153

30. Tamura, N., Taga, A., Kitagawa, S., Banbara, M.: Compiling finite linear CSP into SAT. Constraints **14**(2), 254–272 (2009)

31. Tanjo, T., Tamura, N., Banbara, M.: A compact and efficient SAT-encoding of finite domain CSP. In: Sakallah, K.A., Simon, L. (eds.) SAT 2011. LNCS, vol. 6695, pp. 375–376. Springer, Heidelberg (2011). https://doi.org/10.1007/978-3-642-21581-0_36

32. Zhou, N.-F.: In pursuit of an efficient SAT encoding for the Hamiltonian cycle problem. In: Simonis, H. (ed.) CP 2020. LNCS, vol. 12333, pp. 585–602. Springer, Cham (2020). https://doi.org/10.1007/978-3-030-58475-7_34
33. Zhou, N.-F., Kjellerstrand, H.: The Picat-SAT compiler. In: Gavanelli, M., Reppy, J. (eds.) PADL 2016. LNCS, vol. 9585, pp. 48–62. Springer, Cham (2016). https://doi.org/10.1007/978-3-319-28228-2_4
34. Zhou, N.-F., Kjellerstrand, H.: Optimizing SAT encodings for arithmetic constraints. In: Beck, J.C. (ed.) CP 2017. LNCS, vol. 10416, pp. 671–686. Springer, Cham (2017). https://doi.org/10.1007/978-3-319-66158-2_43

Model-Based Algorithm Configuration with Adaptive Capping and Prior Distributions

Ignace Bleukx[(✉)], Senne Berden, Lize Coenen, Nicholas Decleyre, and Tias Guns

KU Leuven, Leuven, Belgium
{ignace.bleukx,senne.berden,lize.coenen,
nicholas.decleyre,tias.guns}@kuleuven.be

Abstract. Many advanced solving algorithms for constraint programming problems are highly configurable. The research area of algorithm configuration investigates ways of *automatically* configuring these solvers in the best manner possible. In this paper, we specifically focus on algorithm configuration in which the objective is to decrease the time it takes the solver to find an optimal solution. In this setting, *adaptive capping* is a popular technique which reduces the overall runtime of the search for good configurations by adaptively setting the solver's timeout to the best runtime found so far. Additionally, sequential model-based optimization (SMBO)—in which one iteratively learns a surrogate model that can predict the runtime of unseen configurations—has proven to be a successful paradigm. Unfortunately, adaptive capping and SMBO have thus far remained incompatible, as in adaptive capping, one cannot observe the true runtime of runs that time out, precluding the typical use of SMBO. To marry adaptive capping and SMBO, we instead use SMBO to model the probability that a configuration will *improve* on the best runtime achieved so far, for which we propose several decomposed models. These models also allow defining prior probabilities for each hyperparameter. The experimental results show that our DeCaprio method speeds up hyperparameter search compared to random search and the seminal adaptive capping approach of ParamILS.

Keywords: Algorithm configuration · Adaptive capping · Sequential model-based optimization · Prior distributions

1 Introduction

Constraint solvers are used on a daily basis for solving combinatorial optimization problems such as scheduling, packing and routing. Because of the advanced nature of constraint solvers today, the runtime of a solver on even a single problem instance can vary tremendously depending on its hyperparameter configuration.

Ignace Bleukx and Senne Berden—These authors contributed equally.

P. Schaus (Ed.): CPAIOR 2022, LNCS 13292, pp. 64–73, 2022.
https://doi.org/10.1007/978-3-031-08011-1_6

In *algorithm configuration*, the goal is to find a well-performing hyperparameter configuration automatically. In this field, sequential model-based optimization (SMBO) is a highly successful paradigm [2,9,11]. Applications of SMBO approaches aim to optimize a black-box function that evaluates a configuration by running a solver with that configuration and returning the runtime. These techniques sequentially approximate this function by observing runtimes during the search and updating a probabilistic estimate [2]. This model can then be exploited to direct the search towards good configurations.

However, to obtain these runtimes during the search, one is required to allow the solver to execute until completion, regardless of the quality of the configuration considered. This stands in stark contrast with *adaptive capping*, a technique for optimising runtime in which solver runs using unfruitful configurations are stopped from the moment the runtime surpasses that of the best configuration found so far [3,4,7].

This work is motivated by the observation that with adaptive capping in place, one is especially keen on finding good configurations sooner rather than later, as this allows decreasing the runtime timeout of all subsequent runs; thereby reducing the runtime of the search as a whole.

To marry adaptive capping with SMBO, we step away from the idea of a model that can estimate a configuration's runtime. Instead, our models are related to the probability that a configuration will be better than any of the already considered ones. More specifically, we consider probabilistic models that can be decomposed for each individual hyperparameter independently. They are easy to compute and update, thereby keeping the overhead in between solve calls small, and allowing to scale to an arbitrarily large number of hyperparameters.

One further benefit of this decomposition is that we can now define *prior probabilities* for each hyperparameter. These can overcome the cold start problem by guiding the candidate selection from the very first iteration. To the best of our knowledge, prior information is only used in SMBO within advanced machine learning models, often in the form of transfer learning techniques [14].

As a first step, we focus in this paper on hyperparameter configuration search over a finite grid of discrete hyperparameters, and on finding the best runtime for a single problem instance. Extensions to multiple instances and more are discussed in the conclusion.

Our contributions are summarized as follows:

- We propose a sequential model-based optimization (SMBO) approach with adaptive capping: DeCaprio[1];
- We propose several decomposable probabilistic models that assign a high probability to configurations that are likely to improve on the current best runtime, and that can be updated cheaply using counting-based mechanisms;
- We propose a methodology for computing independent prior distributions for each hyperparameter, based on a heterogeneous dataset of instances; and we discuss how to use these prior distributions in SMBO;

---

[1] DEcomposable adaptive CApping with PRIOrs.

Algorithm 1. generic Sequential Model-Based Optimization, from [2]

---

function SMBO(f, M^0, T, S)
    $\mathcal{H} \leftarrow \emptyset$
    for $t \leftarrow 1$ to T **do**
        $\hat{x} \leftarrow \operatorname{argmax}_x S(x, M^{t-1})$
        Evaluate $f(\hat{x})$ ▷ Expensive step
        $\mathcal{H} \leftarrow \mathcal{H} \cup \{(\hat{x}, f(\hat{x}))\}$
        Fit a new model M^t to $\mathcal{H}$
    return $\mathcal{H}$

– We show that using these decomposed distributions, especially with informed prior probabilities, leads to faster hyperparameter configuration grid search with adaptive capping than both a random grid search baseline and the seminal ParamILS approach.

2   SMBO for Hyperparameter Configuration

Algorithm 1 shows a generic Sequential Model-Based Optimization [2]. It takes as input an unknown (black-box) function f, a model M also called the *surrogate model* that aims to estimate f based on the past evaluations $\mathcal{H}$, a maximum number of iterations T and a scoring function S, also called acquisition function, that uses M^{t-1} to evaluate how promising it is to choose a configuration $\hat{x}$ as the next configuration to run.

SMBO techniques differ in the way they instantiate the four inputs. For algorithm configuration of constraint solvers, the typical choice is to have f be the runtime, objective value or optimality gap [8]; and M a machine learning model that estimates f, such as Gaussian processes [2], tree parzen estimators [2] or random forests [9].

An effective technique from non-SMBO-based algorithm configuration is to do adaptive capping [3,7] of the runtime, with the best runtime known so far. However, in standard SMBO, this would complicate the use of machine learning to build a model M^t of f, as one would no longer observe the real f, but a potentially truncated version of it. So instead, to marry SMBO and adaptive capping, we build a surrogate model M^t that can be used to obtain a configuration $\hat{x}$ that is likely to *improve* on the current best runtime (i.e., where $f(\hat{x}) < min_{(x,f(x)) \in \mathcal{H}} f(x)$), without having to estimate $\hat{x}$'s runtime itself.

For simplicity and efficiency, we keep away from feature-based machine learning methods that can be costly to (re)train and that have many hyperparameters of their own. Instead, we further decompose the surrogate over the individual hyperparameters and use counting-based models that can be updated incrementally as explained in the next section.

Another important difference with generic SMBO is that we assume a finite grid of possible configurations $\mathcal{X}$ that can be exhaustively searched over. We mimic the property of grid search that a configuration is never visited twice, by computing $\operatorname{argmax}_x S(x, M^{t-1})$ over *unseen* configurations only (i.e. configurations not in $\mathcal{H}$).

3   Surrogate Models M and Scoring Functions S

We consider models for which we can compute M^t based on M^{t-1} and whether the just run configuration $\hat{x}$ improved the runtime or not, i.e. whether $f(\hat{x}) < min_{(x,f(x))\in\mathcal{H}}f(x)$ is true.

3.1   Hamming Similarity: Searching Near the Current Best

A first scoring function S is inspired by local search algorithms like ParamILS [7]. In local search, we first explore configurations which are in the direct neighborhood of x^*, the best configuration seen so far. This allows us to define M^t as follows

$$M^t \leftarrow \hat{x} \text{ if } f(\hat{x}) \text{ improved the runtime, else } M^{t-1} \tag{1}$$

The scoring function is then the Hamming similarity between the current best configuration M^t and each of the unseen configurations x. Let H be the set of all possible hyperparameters, and let $value(h, x)$ be the value of hyperparameter $h \in H$ in configuration x. We can then write the Hamming similarity score as

$$S(x, M^t) = \sum_{h\in H} [\![value(h, x) = value(h, M^t)]\!] \tag{2}$$

where $[\![\cdot]\!]$ is the Iverson bracket which converts True/False into 1/0.

3.2   Beta Distribution

In this first probabilistic scheme, we associate a probability $P^t(X = x)$ with every configuration x. Model M^t decomposes $P^t(X)$ into a separate independent Bernoulli distribution $P^t(X_{h,v})$ for every value v of every hyperparameter h, where $X_{h,v}$ is a binary random variable whose value denotes whether hyperparameter h takes value v. The score function then becomes

$$S(x, M^t) = P^t(X = x) = \prod_{h\in H} P^t(X_{h,value(h,x)} = 1) \tag{3}$$

If $X_{h,v} \sim B(p_{h,v})$, we can consider $p_{h,v}$ itself as uncertain—as is common in Bayesian statistics—and use an appropriate prior distribution for $p_{h,v}$, rather than a point probability. For this purpose, we take a beta distribution, as it is a *conjugate* prior for a Bernoulli distribution. This means that when the beta prior is updated with a Bernoulli likelihood, the resulting posterior is again a beta distribution.

A beta distribution is characterized by parameters α and β, which can be thought of as respectively representing a number of successes (i.e. improvements to the current best time) and failures (i.e. timeouts due to adaptive capping).

$$P^t(X_{h,v} = 1) = \mathrm{E}(p_{h,v}) = \frac{\alpha_{h,v}^t}{\alpha_{h,v}^t + \beta_{h,v}^t} \tag{4}$$

So, for every hyperparameter-value combination (h, v), we count the number of successes $\alpha_{h,v}$ and failures $\beta_{h,v}$, which are both initialised to 1. Updating the model M^t with respect to a chosen $\hat{x}$ then amounts to updating these counts based on whether $\hat{x}$ improved the current timeout r^{t-1} or not:

$$\alpha^t_{h,value(h,\hat{x})} \leftarrow \alpha^{t-1}_{h,value(h,\hat{x})} + [\![f(\hat{x}) < r^{t-1}]\!] \tag{5}$$

$$\beta^t_{h,value(h,\hat{x})} \leftarrow \beta^{t-1}_{h,value(h,\hat{x})} + [\![f(\hat{x}) \geq r^{t-1}]\!] \tag{6}$$

3.3   Dirichlet Distribution

A disadvantage of the above approach is that by maintaining a separate beta distribution for every value, we do not utilize the knowledge that in any configuration, every hyperparameter takes only a single value.

In what follows, model M^t instead decomposes $P^t(X)$ into a separate independent categorical distribution $P^t(X_h)$ for every hyperparameter h, where X_h is a random variable that denotes the value of h. The score function then becomes

$$S(x, M^t) = \prod_{h \in H} P^t_h(X_h = value(h, x)) \tag{7}$$

For any hyperparameter h with k distinct values $v_1 \ldots v_k$, the associated categorical distribution is characterized by parameters $p_{h,v_1} \ldots p_{h,v_k}$. We can again consider these uncertain and use a prior distribution. For this purpose, we use a Dirichlet distribution, as it is a conjugate prior for the categorical distribution. A Dirichlet distribution is characterized by parameters $\alpha_{h,v_1} \ldots \alpha_{h,v_k}$, which can again be thought of as success counts.

$$P^t_h(v) = E(p_{h,v}) = \frac{\alpha^t_{h,v}}{\sum_{j=1}^{k} \alpha^t_{h,v_j}} \tag{8}$$

For every hyperparameter-value combination (h, v), we count the number of successes $\alpha_{h,v}$, which are all initialised to 1. When a considered configuration $\hat{x}$ improves the current timeout r^{t-1}, the model M^t is updated by incrementing the appropriate counts:

$$\alpha^t_{h,value(h,\hat{x})} = \begin{cases} \alpha^{t-1}_{h,value(h,\hat{x})} + 1 & \text{if } f(\hat{x}) \text{ improved the runtime} \\ \alpha^{t-1}_{h,value(h,\hat{x})} & \text{otherwise} \end{cases} \tag{9}$$

When $\hat{x}$ does not improve the current timeout, we add a success count for every *other* hyperparameter value, thereby discounting the values of $\hat{x}$.

While this score function can be used on its own, it can also be used as a tie breaker when multiple configurations have the same Hamming score. This was also considered for the beta-distribution-based model, but empirically proved to be less effective.

4   Learning Priors

The previously described approaches all start with a uniform initial distribution and learn which hyperparameters are better during search. However, this leaves them with a *cold start* problem, meaning that the first choices are uninformed. However, in the adaptive capping setting, the first configurations are the most crucial, as any improvement in runtime saves time in all subsequent runs. Hence, we determine the first runtime cap using the default configuration.

We now aim to take this one step further. Our goal is to learn a *prior probability distribution* that plays well with the above models, that is, a prior distribution that is decomposed for each of the hyperparameters independently. We model each hyperparameter h's prior as a Dirichlet distribution, which is characterized by a pseudocount value $\alpha_{h,v}$ for each value v.

To determine an informed prior, we propose to make use of a heterogeneous set of problems K on which the solver can be run. We refer to this set as the *prior set*. To cope with the sensitivity of modern solvers' runtimes [5], and to account for the existence of equivalent optimal configurations, we propose not to base the priors solely on the best configuration for a model (i.e. $x^k = argmin_x f^k(x)$). Instead, to make the priors more robust, we propose to collect all configurations whose runtime is at most 5% worse than the best one. Let us denote this set as O_k for any individual problem $k \in K$:

$$O_k = \{x \in \mathcal{X} \mid f^k(x) \leq 1.05 \cdot \min_{x' \in \mathcal{X}}(f^k(x'))\} \tag{10}$$

This gives us a *set* of best configurations, on which we compute Dirichlet pseudocounts $\hat{\alpha}_{h,v}$ using the relative number of times hyperparameter h's value v was used in O_k, across all k:

$$\hat{\alpha}_{h,v} = \sum_{k \in K} \frac{|\{x \in O_k \mid value(h,x) = v\}|}{|O_k|} \tag{11}$$

The resulting pseudocounts will have a magnitude that depends on the size of K, which could make individual updates during SMBO insignificant if K is large. So, to use these pseudocounts as a Dirichlet prior, we first normalize and scale them as $\alpha_{h,v} = |Values(h)| \cdot \hat{\alpha}_{h,v} / \sum_{v' \in Values(h)} \hat{\alpha}_{h,v'}$, where $Values(h)$ denotes the set of all possible values of hyperparameter h.

5   Experiments

In this section, we empirically answer the following research questions:

Q1 How do the different scoring functions compare in their ability to find good configurations quickly?

Q2 Does the use of an informed prior improve the ability to find good configurations quickly?

Q3 How does DeCaprio perform against ParamILS, a seminal hyperparameter configurator?

5.1   Experimental Setup

All algorithms were implemented in Python3 and run on a Intel Xeon 4214 CPU. We used a collection of 190 heterogeneous CP models[2] modelled in the CPMpy library [6]. They have a runtime[3] ranging from 1 ms to 2 s when solved using the default parameters of OR-Tools' CP-SAT Solver v9.1 [13]. On each of the instances, we configured 9 CP-SAT hyperparameters, totalling to a grid of 13608 configurations. All code, models and data are available in the repository accompanying this paper[4].

For the surrogates that use an informed prior, we use leave-one-out cross validation to split the models into prior and test set. To compare results across different models, we introduce a metric that denotes the relative improvement of the timeout. Here, the default's runtime has value 0, while value 1 denotes the runtime of the optimal configuration. All results are averages over all 190 instances and 10 random seeds.

5.2   Comparison of Models and Surrogates

We want to find good configurations as soon as possible. In Fig. 1, this corresponds to an early rising curve. As baseline we use a uniform random search. From Fig. 1 (left), one can conclude that most surrogates perform more or less equally at the start of the search. However, further into the search, both Hamming and Hamming + Dirichlet tiebreaker surrogates clearly outperform the alternatives. This can also be seen in Table 1. This answers **Q1**.

Using the same metric, we analyze the performance of the surrogates with an informed prior. We introduce two additional baseline algorithms: *Sample prior* and *Sorted prior*. These

Table 1. Rel. improvement on full grid

Iteration	10	10^2	10^3	10^4
Random uniform	0.67	0.87	0.95	**1.00**
Hamming	0.56	0.92	**0.99**	1.00
Beta	0.63	0.87	0.97	**1.00**
Dirichlet	0.64	0.88	0.95	**1.00**
Hamm + D	0.58	0.93	**0.99**	1.00
Sample prior	0.79	0.92	0.98	**1.00**
Sorted prior	0.74	0.87	0.98	**1.00**
Dirichlet prior	0.81	0.87	0.94	**1.00**
Hamm + D prior	**0.82**	**0.94**	**0.99**	**1.00**

respectively sample and take the argmax of the prior probabilities. Neither algorithm updates its model during the search. The results of these baseline algorithms clearly show the importance of a good update rule. Figure 1 (right) and Table 1 show that using an informed prior clearly yields better performance early in the search. This answers **Q2**.

5.3   Comparison with ParamILS

We cannot compare to SMBO techniques like SMAC [10] as they do not support adaptive capping and hence would have unwieldy large total runtimes. ParamILS

---

[2] From Håkan Kjellerstrand's collection: http://www.hakank.org/cpmpy/.

[3] The number of threads was limited to 1 for every solver call.

[4] https://github.com/ML-KULeuven/DeCaprio.

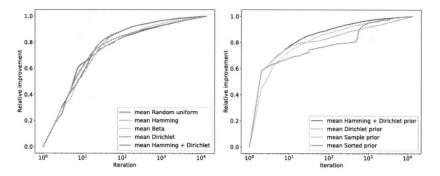

Fig. 1. Relative improvement per iteration. Left: no prior, Right: informed prior

is a seminal algorithm configuration method supporting adaptive capping [7]. Still, its computational overhead is significant on small instances so we have to compare on a smaller grid of 200 configurations. We use the same settings used by the original authors and calculate the relative improvement based on the global best runtime found.

The results of this experiment are summarized in Table 2. Here, we notice that ParamILS performs about equally well as the uninformed Hamming + Dirichlet. When comparing with an *informed* prior, however, our method obtains notably better performance in the first iterations (and hence in the total solve time, not shown).

Table 2. Small grid

Iteration	10	50	100	200
Random uniform	0.71	0.94	0.97	**1.00**
Hamm + D	0.77	**0.98**	**0.99**	1.00
Hamm + D prior	**0.89**	**0.98**	**0.99**	**1.00**
ParamILS	0.76	**0.98**	**0.99**	**1.00**

6   Conclusion and Future Work

In this paper we proposed a scheme for marrying SMBO-based approaches with adaptive capping in the context of algorithm configuration, by using decomposable models and surrogates related to the probability of a runtime improvement.

To make more informed decisions in early iterations we show how to obtain and use Dirichlet *priors* per hyperparameter. Using such prior knowledge showed big improvements over all other methods. Their decomposed nature also means they are easy to extend and combine with new hyperparameters and values, and we envision solver developers to precompute these on their own benchmarks.

In general, algorithm configuration techniques, even SMBO ones, make extensive use of sampling and our light-weight prior-based approach can replace or be combined with current (often random) sampling schemes used, potentially guiding the methods better in their early iterations.

Many directions for future work remain, including evaluating our method on larger datasets with more solvers and models and extending it to configuration over multiple instances, where racing and successive halving [1,12] are orthogonal techniques that can be combined with the ideas presented in this paper.

Acknowledgments. This research was partly funded by the Flemish Government (AI Research Program), the Research Foundation - Flanders (FWO) projects G0G3220N and S007318N and the European Research Council (ERC) under the EU Horizon 2020 research and innovation programme (Grant No 101002802, CHAT-Opt).

A   Adapted SMBO

Algorithm 2. our Sequential Model-Based Optimization

---

function $\text{SMBO}(f, M^0, \mathcal{X}, S, x^{\text{default}}, \hat{T})$

$\quad \hat{x} \leftarrow x^{\text{default}}$

$\quad r \leftarrow f(x^{\text{default}})$

$\quad \mathcal{H} \leftarrow \emptyset$

$\quad T \leftarrow min(|\mathcal{X}|, \hat{T})$

$\quad$ **for** $t \leftarrow 1$ to T **do**

$\quad\quad \hat{x} \leftarrow \text{argmax}_{x \in X \setminus \mathcal{H}} S(x, M^{t-1})$

$\quad\quad$ Evaluate $f(\hat{x})$, cap runtime at r $\quad\quad\quad\quad\quad\quad\quad$ ▷ Expensive step

$\quad\quad$ Fit a new model M^t by updating M^{t-1} with $(\hat{x}, [f(\hat{x}) < r])$

$\quad\quad \mathcal{H} \leftarrow \mathcal{H} \cup \{\hat{x}\}$

$\quad\quad r \leftarrow min(r, f(\hat{x}))$

$\quad$ **return** $\mathcal{H}$

References

1. Anastacio, M., Hoos, H.: Model-based algorithm configuration with default-guided probabilistic sampling. In: Bäck, T. (ed.) PPSN 2020. LNCS, vol. 12269, pp. 95–110. Springer, Cham (2020). https://doi.org/10.1007/978-3-030-58112-1_7
2. Bergstra, J., Bengio, Y.: Algorithms for hyper-parameter optimization. In: In NIPS, pp. 2546–2554 (2011)
3. Cáceres, L.P., López-Ibáñez, M., Hoos, H., Stützle, T.: An experimental study of adaptive capping in irace. In: Battiti, R., Kvasov, D.E., Sergeyev, Y.D. (eds.) LION 2017. LNCS, vol. 10556, pp. 235–250. Springer, Cham (2017). https://doi.org/10.1007/978-3-319-69404-7_17
4. De Souza, M., Ritt, M., López-Ibáñez, M.: Capping methods for the automatic configuration of optimization algorithms. Comput. Oper. Res. **139**, 105615 (2021)
5. Fichte, J.K., Hecher, M., McCreesh, C., Shahab, A.: Complications for computational experiments from modern processors. In: 27th International Conference on Principles and Practice of Constraint Programming (CP 2021). Schloss Dagstuhl-Leibniz-Zentrum für Informatik (2021)
6. Guns, T.: Increasing modeling language convenience with a universal n-dimensional array, cppy as python-embedded example. In: Proceedings of the 18th workshop on Constraint Modelling and Reformulation, Held with CP, vol. 19 (2019)
7. Hutter, F., Hoos, H., Leyton-Brown, K., Stützle, T.: Paramils: an automatic algorithm configuration framework. J. Artif. Intell. Res. (JAIR) **36**, 267–306 (2009)

8. Hutter, F., Hoos, H.H., Leyton-Brown, K.: Automated configuration of mixed integer programming solvers. In: Lodi, A., Milano, M., Toth, P. (eds.) CPAIOR 2010. LNCS, vol. 6140, pp. 186–202. Springer, Heidelberg (2010). https://doi.org/10.1007/978-3-642-13520-0_23

9. Hutter, F., Hoos, H.H., Leyton-Brown, K.: Sequential model-based optimization for general algorithm configuration. In: Coello, C.A.C. (ed.) LION 2011. LNCS, vol. 6683, pp. 507–523. Springer, Heidelberg (2011). https://doi.org/10.1007/978-3-642-25566-3_40

10. Hutter, F., Hoos, H.H., Leyton-Brown, K.: Sequential model-based optimization for general algorithm configuration. In: Coello, C.A.C. (ed.) Learning and Intelligent Optimization, pp. 507–523. Springer, Berlin Heidelberg, Berlin, Heidelberg (2011). https://doi.org/10.1007/978-3-642-25566-3_40

11. Kerschke, P., Hoos, H.H., Neumann, F., Trautmann, H.: Automated algorithm selection: survey and perspectives. Evol. Comput. **27**(1), 3–45 (2019). https://doi.org/10.1162/evco_a_00242

12. López-Ibáñez, M., Dubois-Lacoste, J., Cáceres, L.P., Birattari, M., Stützle, T.: The irace package: iterated racing for automatic algorithm configuration. Oper. Res. Perspect. **3**, 43–58 (2016)

13. Perron, L., Furnon, V.: Or-tools. https://developers.google.com/optimization/

14. Yogatama, D., Mann, G.: Efficient transfer learning method for automatic hyperparameter tuning. In: Artificial Intelligence and Statistics, pp. 1077–1085. PMLR (2014)

Shattering Inequalities for Learning Optimal Decision Trees

Justin J. Boutilier, Carla Michini, and Zachary Zhou[✉]

Department of Industrial and Systems Engineering,
University of Wisconsin-Madison, Madison, WI, USA
{jboutilier,michini,zzhou246}@wisc.edu

Abstract. Recently, mixed-integer programming (MIP) techniques have been applied to learn optimal decision trees. Empirical research has shown that optimal trees typically have better out-of-sample performance than heuristic approaches such as CART. However, the underlying MIP formulations often suffer from slow runtimes, due to weak linear programming (LP) relaxations. In this paper, we first propose a new MIP formulation for learning optimal decision trees with multivariate branching rules and no assumptions on the feature types. Our formulation crucially employs binary variables expressing how each observation is routed throughout the entire tree. We then introduce a new class of valid inequalities for learning optimal multivariate decision trees. Each inequality encodes an inclusion-minimal set of points that cannot be shattered by a multivariate split, and in the context of a MIP formulation, the inequalities are sparse, involving at most the number of features plus two variables. We leverage these valid inequalities within a Benders-like decomposition, where the master problem determines how to route each observation to a leaf node to minimize misclassification error, and the subproblem checks whether, for each branch node of the decision tree, it is possible to construct a multivariate split that realizes the given routing of observations; if not, the subproblem adds at least one of our valid inequalities to the master problem. We demonstrate through numerical experiments that our MIP approach outperforms (in terms of training accuracy, testing accuracy, solution time, and relative gap) two other popular MIP formulations, and is able to improve both in and out-of-sample performance, while remaining competitive in terms of solution time to a wide range of popular approaches from the literature.

Keywords: Decision trees · Mixed-integer programming · Machine learning

1 Introduction

Decision trees are among the most popular techniques for interpretable machine learning [9]. In addition to their use as a standalone method, decision trees form the foundation for several more sophisticated machine learning algorithms such

P. Schaus (Ed.): CPAIOR 2022, LNCS 13292, pp. 74–90, 2022.
https://doi.org/10.1007/978-3-031-08011-1_7

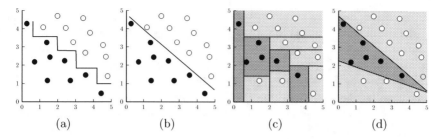

Fig. 1. (a) Eight univariate branching rules are required to correctly separate the black and white observations; (b) only one multivariate branching rule suffices. For a slightly different toy dataset: (c) and (d) show how the feature space is partitioned when using univariate and multivariate branching rules, respectively.

as random forest [8,23]. Although there are many ways to express a decision tree, the majority of the literature, including this paper, focuses on binary trees. In a binary decision tree, each internal node, referred to as a *branch node*, has exactly two children and observations are routed to the left or right child node according to a *branching rule*. Terminal nodes in the tree are referred to as *leaf nodes* and each leaf node is assigned a class k such that any observation routed to that leaf node is classified as belonging to class k. Almost all algorithms (heuristic and exact) that generate decision trees focus on *univariate* branching rules, which check if the value of a single feature exceeds a prescribed threshold. In this work, we instead focus on *multivariate* branching rules, which are separating hyperplanes checking several features at a time. Multivariate branching rules are less easily interpretable, however they provide more flexibility than univariate branching rules, which can only resort to axis-aligned hyperplanes. As a consequence, multivariate branching rules can yield much more compact decision trees. In other words, even if the tests performed at the branching nodes are more complex, the total number of tests needed to achieve a target accuracy can be dramatically smaller; see the toy example in Fig. 1.

Related Work. The problem of learning optimal decision trees is NP-hard [20]. As a result, there exist several famous top-down induction algorithms for learning decision trees such as CART [9] and ID3 [28]. These heuristic methods do not provide any guarantee on the quality of the decision trees computed. More recently, a number of exact approaches have been proposed that typically aim at minimizing the training error and possibly some measure of the tree complexity.

One stream of work uses mixed-integer programming (MIP) to compute optimal decision trees. Motivated by algorithmic advances in integer optimization, Bertsimas and Dunn [7] first formulated the problem of learning an optimal decision tree as a MIP. Their work spurred a series of subsequent papers that propose a variety of MIP tools to model and solve the problem of learning optimal decision trees [1,2,12,16,34,35,37]. One main advantage of MIP approaches is their flexibility: the problem objective can be easily modified to enhance feature selection and/or tree size, and the feasible set can be modified by adding additional constraints of practical interest [1,16]. Moreover, MIP formulations can easily handle multivariate branching rules [7,37]. Typically, MIP formulations contain two key components: 1) a framework to model the routing of observations through the decision tree to leaf nodes, and 2) a framework that properly constructs the tree by devising branching rules at each branch node. Most MIP approaches employ big-M constraints to unify these two components in a single optimization problem [7,34,37], but this modeling technique suffers from the fact that big-M constraints notoriously lead to poor LP relaxations [36]. One notable exception is the work of Aghaei et al. [2], who consider datasets with binary features and formulate the problem of routing observations through a fixed decision tree with univariate branching rules as a max-flow problem. Thanks to these two key assumptions – having binary features and restricting to univariate branching rules – they can avoid using big-M constraints. Moreover, their formulation is amenable to a Benders decomposition, where the master problem is tasked with constructing the decision tree, and the routing of observations to leaf nodes is accomplished by solving a subproblem for each observation that adds optimality cuts to the master. Unfortunately, their flow-based approach does not generalize if we relax either of the two key assumptions.

Another stream of work uses Boolean satisfiability (SAT) [5,22,26,29], constraint programming (CP) [32,33], and dynamic programming [3,4,13,19,25,27] to compute optimal decision trees. These methods address the scalability issues of general MIP methods by using a different range of techniques to explore the search space, such as sub-sampling, branch and bound search, and caching. Some of these algorithms are tailored to the specific structure of decision trees, which is used to speed-up computation. However, 1) all of them assume binary features or use binarization techniques to transform numerical features into binary ones in a preprocessing step, which can dramatically increase the size of the input (causing memory problems) and is not guaranteed to preserve optimality [24]; 2) most of them are designed and/or implemented to work only for binary classification; and 3) all of them are only suited to construct decision trees with univariate branching rules.

Our Contribution. We first propose a new MIP formulation for learning optimal decision trees with multivariate branching rules and no assumptions on the feature types. Our formulation employs only binary variables to (i) express how each observation is routed in the decision tree and (ii) express the objective function as a weighed sum of the training accuracy and the size of the tree. Moreover, we exploit the structure of decision trees and use a geometric interpretation of the optimal decision tree problem to devise a specialized class of valid inequalities, called *shattering* inequalities, which intuitively detect problematic sub-samples of the dataset that cannot be linearly separated. We leverage these inequalities within a Benders-like decomposition [6] to decompose our formulation into a master problem that determines how to route each observation to a leaf node, and a collection of linear programming (LP) feasibility subproblems that certify whether, for each branch node of the decision tree, it is possible to construct a multivariate branching rule that realizes the given routing of observations. If it is not possible to realize the routing, then we add one of our valid inequalities to the master problem as a feasibility cut. Each of our inequalities encodes a minimal set of points that cannot be shattered by a multivariate branching rule and in the context of a MIP formulation, the inequalities are sparse, with at most the number of features plus two variables. Our approach does not require big-M constraints, but generates sparse cuts that capture the combinatorial structure of the problem to strengthen the LP relaxation and decrease training time. Although we use these cuts in a decomposition algorithm for our formulation, they can be directly applied as valid inequalities to other MIP formulations that may not be suited to decomposition (e.g., OCT-H [7] and SVM1-ODT [37]). We demonstrate through numerical experiments that our MIP approach outperforms (in terms of training accuracy, testing accuracy, solution time, and relative gap) two other popular MIP formulations, and is able to improve both in and out-of-sample performance, while remaining competitive in terms of solution time to a wide range of popular approaches from the literature.

2   The Optimal Decision Tree Problem

In this section, we formally introduce the problem setting, our notation, and our formulation.

2.1   The Optimal Decision Tree Problem

We first define our data, which includes a training set of N observations, each of which has p numerical features and belongs to one of K classes. For $n \in \mathbb{N}$, we denote by $[n]$ the set $\{1, \ldots, n\}$. Without loss of generality, we normalize the training set so that all features are scaled to $[0, 1]$. Thus, each observation $i \in [N]$ is a vector $(\mathbf{x}^i, y^i) \in [0, 1]^p \times [K]$.

As noted in Sect. 1, we focus on learning optimal binary decision trees with multivariate branching rules, which we refer to as *multivariate splits*. Each node in the tree is either a branch node or a leaf node. Note that the first node in the tree is colloquially referred to as the root node (even though it is a branch node). The model is built upon a binary tree where every branch node has exactly two children and the leaves are all on the same level. The maximum depth of the decision tree $D \in \mathbb{N}$ is defined as the length of the path from the root to any leaf. We denote the set of branch nodes as $\mathcal{B} = \{1, \ldots, 2^D - 1\}$. Each branch node t corresponds to a multivariate split defined by learned parameters $\mathbf{a}_t \in \mathbb{R}^p$ and $b_t \in \mathbb{R}$. The multivariate split is applied as follows: for each observation $\mathbf{x} \in [0, 1]^p$, if $\mathbf{a}_t^\top \mathbf{x} \le b_t$, then $\mathbf{x}$ is sent to the left child of t, denoted by $2t$; otherwise it is sent to the right child, denoted by $2t + 1$. The key difference between multivariate and univariate splits is that a univariate split allows only one component of $\mathbf{a}_t$ to be non-zero. We denote the set of leaf nodes by $\mathcal{L} = \{2^D, \ldots, 2^{D+1} - 1\}$. Each leaf node t is a terminal node (i.e., it has no children) and is assigned a class $k \in [K]$. All observations routed to leaf t are classified as belonging to class k.

The max depth $D \in \mathbb{N}$ is often used as a hyperparameter to control the complexity and size of the tree. To deter the model from constructing a full binary tree of depth D, we use another hyperparameter $\alpha \geq 0$ that penalizes the number of leaf nodes in our objective function (more on this in Sect. 2.2).

2.2 Problem Formulation

We now present a formulation for learning an optimal decision tree – the training problem – that includes a set of complicating constraints. For each branch node $t \in \mathcal{B}$, we can either define a branching rule establishing whether an incoming observation should be sent to the left or to the right child of t, in which case we say that *node t applies a split*, or we can direct all of the incoming observations to the left child of t. Correspondingly, we introduce a binary variable d_t that is equal to 1 if t applies a split, and to 0 otherwise. The decision variables $\mathbf{d}$ thus define the tree topology. For each $t \in \mathcal{B}$ we have $p + 1$ variables $(\mathbf{a}_t, b_t)$ defining the multivariate split associated with the branch node. If t does *not* apply a split, it is feasible to set these variables to $(\mathbf{0}, 0)$.

For each observation $i \in [N]$ and for each node $t \in \mathcal{B} \cup \mathcal{L}$ of the decision tree, we introduce a binary variable w_{it} that is equal to 1 if observation i is sent to node t, and to 0 otherwise. The decision variables $\mathbf{w}$ thus define how to route the observations from the root node to the leaf nodes.

For each class $k \in [K]$ and leaf node $t \in \mathcal{L}$, we introduce a binary decision variable c_{kt} that is equal to 1 if t is assigned class label k, and to 0 otherwise. Finally, for each observation $i \in [N]$ and leaf node $t \in \mathcal{L}$, we introduce a binary-valued decision variable z_{it} that is equal to 1 if i is sent to leaf t and is correctly classified as y^i, and to 0 otherwise. We will later see that the integrality constraints on $\mathbf{z}$ can be relaxed. Our formulation for the training problem is

$$\underset{\mathbf{c},\mathbf{d},\mathbf{w},\mathbf{z},\mathbf{a},\mathbf{b}}{\text{minimize}} \quad \frac{1}{N} \sum_{i=1}^{N} \left(1 - \sum_{t \in \mathcal{L}} z_{it} \right) + \alpha \sum_{t \in \mathcal{B}} d_t \tag{1a}$$

$$\text{subject to} \quad \sum_{t \in \mathcal{L}} w_{it} = 1 \qquad\qquad\qquad \forall i \in [N], \tag{1b}$$

$$w_{it} = w_{i,2t} + w_{i,2t+1} \qquad\qquad \forall i \in [N],\ t \in \mathcal{B}, \tag{1c}$$

$$w_{i,2t+1} \leq d_t \qquad\qquad\qquad \forall i \in [N],\ t \in \mathcal{B}, \tag{1d}$$

$$\sum_{k=1}^{K} c_{kt} = 1 \qquad\qquad\qquad \forall t \in \mathcal{L}, \tag{1e}$$

$$z_{it} \leq w_{it} \qquad\qquad\qquad \forall i \in [N],\ t \in \mathcal{L}, \tag{1f}$$

$$z_{it} \leq c_{y^i,t} \qquad\qquad\qquad \forall i \in [N],\ t \in \mathcal{L}, \tag{1g}$$

$$c_{kt} \in \{0,1\} \qquad\qquad\qquad \forall k \in [K],\ t \in \mathcal{L}, \tag{1h}$$

$$d_t \in \{0,1\} \qquad\qquad\qquad \forall t \in \mathcal{B}, \tag{1i}$$

$$w_{it} \in \{0,1\} \qquad\qquad\qquad \forall i \in [N],\ t \in \mathcal{B} \cup \mathcal{L}, \tag{1j}$$

$$z_{it} \in \mathbb{R} \qquad\qquad\qquad \forall i \in [N],\ t \in \mathcal{L}, \tag{1k}$$

$$(\mathbf{a}_t, b_t) \in \mathcal{H}_t(\mathbf{w}) \qquad\qquad \forall t \in \mathcal{B}, \tag{1l}$$

where, for each branch node $t \in \mathcal{B}$ and integral $\mathbf{w}$ satisfying (1b)–(1d), the set $\mathcal{H}_t(\mathbf{w})$ is defined as

$$\mathcal{H}_t(\mathbf{w}) = \big\{(\mathbf{a}_t, b_t) \in \mathbb{R}^{p+1} : \mathbf{a}_t^\top \mathbf{x}^i + 1 \le b_t \quad \forall\, i \in [N] : w_{i,2t} = 1, \tag{2}$$

$$\mathbf{a}_t^\top \mathbf{x}^i - 1 \ge b_t \quad \forall\, i \in [N] : w_{i,2t+1} = 1\big\}. \tag{3}$$

Note that, for a fixed $\mathbf{w}$, the set $\mathcal{H}_t(\mathbf{w})$ is a (possibly empty) polyhedron in $\mathbb{R}^{p+1}$.

The objective function (1a) is derived from CART's cost-complexity measure: for fixed $i \in [N]$, the term $1 - \sum_{t \in \mathcal{L}} z_{it}$ is 1 if observation i is misclassified, 0 otherwise, therefore the misclassification rate over the training set is $\frac{1}{N} \sum_{i=1}^{N} \big(1 - \sum_{t \in \mathcal{L}} z_{it}\big)$; the second term weights the number of leaf nodes (to which at least one observation is directed), which is equal to the number of branch nodes (that apply a nontrivial split) plus one.

Constraints (1b) ensure that each observation is mapped to exactly one leaf, while constraints (1c) guarantee that each observation routed to a branch node t is sent to either the left or the right child of t. For a branch node t that does not apply a split, constraints (1d) automatically send any incoming observations to the left child of t. Constraints (1e) assign each leaf node a class in $[K]$. Constraints (1f) and (1g) enforce the condition that if $z_{it} = 1$, then $w_{it} = 1$ and $c_{y^i,t} = 1$ (i.e., observation i is sent to leaf t and is correctly classified as y^i). Note that integrality constraints are not required for the $\mathbf{z}$ variables, since they are implied by the integrality of $\mathbf{w}$ and $\mathbf{c}$, and by the fact that at an optimal solution constraints (1f) and (1g) hold with equality. Complicating constraints (1l) are the only ones involving variables $(\mathbf{a}_t, b_t)$, $t \in \mathcal{B}$, which ensure that the routing defined by $\mathbf{w}$ can be *realized* by multivariate splits. This is possible if and only if for each branch node t we have $\mathcal{H}_t(\mathbf{w}) \ne \varnothing$.

We highlight some technical differences between our formulation and other MIP models in the literature. First, we focus on an alternative means to characterizing the set of feasible routings $\mathbf{w}$. As mentioned in Sect. 1, most MIP formulations link the $\mathbf{w}$ and $(\mathbf{a}_t, b_t)$, $t \in \mathcal{B}$ variables using big-M constraints. However, we have formulated the problem in such a way that these big-M constraints are not necessary. Second, we define $\mathbf{w}$ over all nodes of the decision tree, while previous literature defines $\mathbf{w}$ over only the leaf nodes [7,16,37]. Our primary motivation for defining additional (roughly double) $\mathbf{w}$ variables is that we can exploit these additional variables to create stronger valid inequalities for characterizing the set of feasible routings. A secondary reason for the introduction of $\mathbf{w}$ variables over the branch nodes is that these extra variables give us the option to formulate a model using big-M constraints, but with far fewer of them than existing formulations. For instance, OCT and OCT-H [7] require $2^D N D$ big-M constraints for defining the feasible routings $\mathbf{w}$, whereas we only require $N(2^{D+1} - 2)$. Finally, we penalize the total number of splits used as part of our objective function. Unlike the univariate setting where CART's cost-complexity measure can be directly used as a template, the multivariate setting has no universally accepted objective. For example, OCT-H [7] penalizes the total number of *features* used over all splits in the tree and SVM1-ODT [37] penalizes the ℓ_1 norm of $\mathbf{a}_t$ over all splits in the tree.

3   Shattering Inequalities

In this section, we propose a new class of valid inequalities for (1), called shattering inequalities, which correspond to subsets of observations that cannot be shattered by a multivariate split, and we propose a separation algorithm to generate these inequalities.

Let $\mathcal{C}$ be a family of binary classifiers in $\mathbb{R}^p$. A set of observations is *shattered* by $\mathcal{C}$ if, for any assignment of binary labels to these observations, there exists some classifier in $\mathcal{C}$ that can perfectly separate all the observations. The maximum number of observations that can be shattered by $\mathcal{C}$ is called the *Vapnik-Chervonenkis (VC) dimension* of $\mathcal{C}$ [31].

We now consider the family of binary classifiers $\mathcal{H}$ consisting of the multivariate splits in $\mathbb{R}^p$. Let $\mathcal{I}$ be a collection of subsets $I \subseteq [N]$ of observations such that $\{x^i\}_{i \in I}$ cannot be shattered by $\mathcal{H}$. For each $I \in \mathcal{I}$, denote by $\Lambda(I) \subset \{-1, 1\}^I$ the assignments of binary labels to observations in I so that they cannot be perfectly separated by any multivariate split in $\mathbb{R}^p$. Then, the following inequalities are valid for (1):

$$\sum_{i \in I: \lambda_i = -1} w_{i,2t} + \sum_{i \in I: \lambda_i = +1} w_{i,2t+1} \leq |I| - 1, \quad \forall I \in \mathcal{I}, \ \lambda \in \Lambda(I), \ t \in \mathcal{B}. \quad (4)$$

The shattering inequalities (4) have the form of *packing constraints* [11] and impose the condition that at least one observation in I is *not* routed to the children of t as prescribed by the label assignment λ. We can restrict our attention to the minimal (w.r.t. set inclusion) subsets of $\mathcal{I}$. Indeed, if $I \in \mathcal{I}$ is *not* minimal, then each inequality (4) associated to I is implied by an inequality (4) associated to some $I' \subset I$ in $\mathcal{I}$.

Moreover, if I is a minimal set of observations in $\mathbb{R}^p$ that cannot be shattered by $\mathcal{H}$, then $|I| \leq p+2$. This follows from the fact that the VC dimension of $\mathcal{H}$ is $p+1$. Note that we might still be unable to perfectly split $|I| < p+2$ observations in $\mathbb{R}^p$ if there exists an hyperplane that contains more than p points. For example, for $p = 2$, three points on a line labeled (in sequence) $1, -1, 1$ cannot be perfectly split. As a consequence, the support of inequalities (4) corresponding to minimal sets of observations in $\mathbb{R}^p$ that cannot be shattered by $\mathcal{H}$, is at most $p + 2$. In particular, if $p \ll N$, these inequalities are sparse.

Figure 2 shows an example using a dataset with points $\mathbf{x}^i = (x_1^i, x_2^i) \in \mathbb{R}^2$, where for the first four observations, $\mathbf{x}^1 = (0,0)$, $\mathbf{x}^2 = (0,1)$, $\mathbf{x}^3 = (1,0)$, $\mathbf{x}^4 = (1,1)$; the full dataset may contain many more observations. $I = \{1,2,3,4\}$ is an example of a minimal subset of observations that cannot be shattered by $\mathcal{H}$; no hyperplane is capable of separating $\{\mathbf{x}^1, \mathbf{x}^4\}$ from $\{\mathbf{x}^2, \mathbf{x}^3\}$, however $I \setminus \{i\}$ can be shattered for any $i \in I$. We can derive the shattering inequalities:

$$w_{2,2t} + w_{3,2t} + w_{1,2t+1} + w_{4,2t+1} \leq 3, \quad \forall t \in \mathcal{B} \quad (5)$$

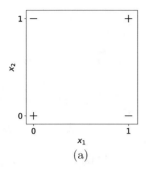
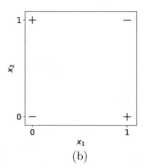

(a)   (b)

Fig. 2. $I = \{1, 2, 3, 4\}$ is a minimal subset of observations that cannot be shattered by $\mathcal{H}$. $\Lambda(I)$ contains exactly two vectors $\lambda = (\lambda_1, \lambda_2, \lambda_3, \lambda_4)$. (a) shows $\lambda = (+1, -1, -1, +1)$, which is used to derive (5); (b) shows $\lambda = (-1, +1, +1, -1)$, which is used to derive (6).

and
$$w_{1,2t} + w_{4,2t} + w_{2,2t+1} + w_{3,2t+1} \le 3, \quad \forall t \in \mathcal{B}. \tag{6}$$

3.1 Decomposition and Separation

We decompose (1) into a master problem and an LP feasibility subproblem. The master problem is obtained from (1) by removing the complicating constraints (1l) and by (possibly) adding some valid inequalities (4). Thus, the master problem is a MIP with decision variables $\mathbf{c}, \mathbf{d}, \mathbf{w}, \mathbf{z}$. The LP feasibility subproblem includes decision variables $(\mathbf{a}_t, b_t)$, $t \in \mathcal{B}$, and verifies that, for a given assignment of $\mathbf{w}$, $\mathcal{H}_t(\mathbf{w})$ is a nonempty polyhedron for all $t \in \mathcal{B}$. Intuitively, the master problem attempts to find an optimal routing of the observations to the leaves defined by $\mathbf{w}$, and the LP feasibility subproblem attempts to find, at each branch node t, a multivariate split $(\mathbf{a}_t, b_t)$ that realizes this routing. If at some branch node there is no multivariate split that is able to route the incoming observations according to $\mathbf{w}$, then our goal is to generate an inequality (4) that is violated by $\mathbf{w}$. This inequality is added to the master problem as a feasibility cut, and the process is repeated until the subproblem becomes feasible, meaning that $\mathbf{w}$ satisfies all inequalities (4).

Next we show how to dynamically generate inequalities (4). As we shall see later, this process can be interpreted as an application of *combinatorial Benders (CB) cuts* [10, 18] to the decomposition of (1) outlined above. Let $(\mathbf{c}, \mathbf{d}, \mathbf{w}, \mathbf{z})$ be a solution to the master problem. For each $t \in \mathcal{B}$, define $I(t) = \{i \in [N] : w_{it} = 1\}$ as the set of observations arriving at node t, and consider the partition of $I(t)$ into $I(2t)$ and $I(2t + 1)$. From (2) we have that $\mathcal{H}_t(\mathbf{w}) \ne \varnothing$ if and only if the observations in $I(2t)$ can be perfectly separated from the observations in $I(2t + 1)$ by a multivariate split, i.e., if and only if the following system of linear inequalities in variables $(\mathbf{a}_t, b_t)$ is feasible:

$$\begin{cases} \mathbf{a}_t^\top \mathbf{x}^i + 1 \le b_t, \ \forall i \in I(2t), \\ \mathbf{a}_t^\top \mathbf{x}^i - 1 \ge b_t, \ \forall i \in I(2t+1). \end{cases} \tag{7}$$

Our goal is to either certify that system (7) is feasible for all $t \in \mathcal{B}$ or, if (7) is infeasible for some $t \in \mathcal{B}$, to return an inclusion-minimal subset of observations $I' \subseteq I(t)$, such that $I' \cap I(2t)$ cannot be perfectly separated from $I' \cap I(2t+1)$ by a multivariate split. Each such subset I' corresponds to an *Irreducible Infeasible Subsystem* (IIS) of the infeasible system (7), which is defined as a subsystem of (7) that would become feasible by discarding one arbitrary inequality. Once we have found an IIS of (7) indexed by $I' \subseteq I(t)$, we add the following cut to the master problem:

$$\sum_{i \in I' \cap I(2t)} w_{i,2t} + \sum_{i \in I' \cap I(2t+1)} w_{i,2t+1} \le |I'| - 1. \tag{8}$$

Inequality (8) is clearly violated by $\mathbf{w}$, and is a shattering inequality (4).

To find an IIS of the infeasible system (7), we construct a dual polyhedron Q_t which is obtained by applying Farkas' lemma to (7):

$$Q_t = \left\{ \mathbf{q} \in \mathbb{R}_+^{I(t)} : \quad \sum_{i \in I(2t)} q_i \mathbf{x}^i = \sum_{i \in I(2t+1)} q_i \mathbf{x}^i, \quad \sum_{i \in I(2t)} q_i = \sum_{i \in I(2t+1)} q_i = 1 \right\}. \tag{9}$$

In fact, there is a one-to-one correspondence between the IISs of (7) and the vertices of Q_t [15]. Specifically, the indices of the inequalities appearing in an IIS of (7) correspond to the support of a vertex of Q_t, and vice versa. We remark that the polyhedron Q_t has a very nice geometric interpretation. The decision variables $\mathbf{q}$ are associated with the observations indexed by $I(t) = I(2t) \cup I(2t+1)$, and they can be interpreted as the coefficients of two convex combinations, one on the observations in $I(2t)$ and the other on the observations in $I(2t+1)$. It is evident from (9) that Q_t is nonempty if and only if there exists a point that is both in the convex hull of $\{x^i\}_{i \in I(2t)}$ and in the convex hull of $\{x^i\}_{i \in I(2t+1)}$.

Based on the above discussion, we can define a separation algorithm to dynamically generate the shattering inequalities (4). This algorithm receives as input a feasible solution $(\mathbf{c}, \mathbf{d}, \mathbf{w}, \mathbf{z})$ of the master problem, and it either establishes that $\mathbf{w}$ satisfies all inequalities (4), or it returns an inequality of family (4) that is violated by $\mathbf{w}$. Precisely, for each $t \in \mathcal{B}$, the algorithm checks the feasibility of the dual polyhedron Q_t. If Q_t is empty for all $t \in \mathcal{B}$, then by Farkas' Lemma system (7) is feasible for all $t \in \mathcal{B}$, thus the LP subproblem is feasible and we can construct an optimal solution to (1) which realizes the routing prescribed by $\mathbf{w}$. If Q_t is nonempty for some $t \in \mathcal{B}$, then by Farkas' Lemma system (7) is infeasible, and from each vertex of Q_t we can construct an IIS of (7) and a corresponding shattering inequality (8) that is violated by $\mathbf{w}$. In practice, it is possible to efficiently generate multiple inequalities (8) by finding multiple vertices of Q_t. One method for finding multiple vertices is to optimize over Q_t

multiple times with different objective functions. For instance, let $\mathbf{f} \in \mathbb{Z}_+^{I(t)}$ be a counter for the number of inequalities (8) that each observation in $I(t)$ has appeared in thus far. One can repeatedly solve $\max\{\mathbf{f}^\top \mathbf{q} : \mathbf{q} \in Q_t\}$, each time updating $\mathbf{f}$ as a cut is added.

The separation algorithm can be implemented via LP with a run time that is polynomial in 2^D and $\text{size}(X)$, where X is the $N \times p$ matrix encoding the features of the observations in the training set and $\text{size}(X)$ is defined as the number of bits required to encode X [30]. Note that there is an exponential dependence with respect to the depth parameter D. However, the number of variables and constraints of our MIP formulation (as well as the other formulations from the literature) are already exponential in D and in practice, we want D to be small so that we can obtain a more interpretable decision tree.

We conclude this section by observing that our formulation (1) and its decomposition can be used to generate the shattering inequalities (8) as CB cuts. CB cuts are a specialization of Hooker's logic-based Benders decomposition [18]. They are formally introduced by Codato and Fischetti [10], who study MIP problems that can be decomposed into a master problem with binary variables, and an LP subproblem whose feasibility depends from the solution of the master problem.

The shattering inequalities (8) can be interpreted as CB cuts by enforcing the complicating constraints (11) through the following logical constraints (note that left children have an even index, while right children have an odd index):

$$\forall i \in [N], \; t \in \mathcal{B} \cup \mathcal{L} \setminus \{1\}, \; w_{it} = 1 \implies \begin{cases} (\mathbf{a}_{t/2})^\top x^i + 1 \leq b_{t/2} & \text{if } t \text{ is even} \\ (\mathbf{a}_{\lfloor t/2 \rfloor})^\top x^i - 1 \geq b_{\lfloor t/2 \rfloor} & \text{if } t \text{ is odd.} \end{cases}$$

After solving the master problem, if the inequality system given by the activated logical constraints (i.e., those where $w_{it} = 1$) is infeasible, the IISs of the system can be used to derive CB cuts. A key observation is that, in our setting, each IIS involves only the components of $\mathbf{w}$ that pertain to a specific branch node $t \in \mathcal{B}$. As a result, we can separately consider the inequality systems (7) associated with each individual branch node $t \in \mathcal{B}$.

4 Experiments

In this section, we provide two sets of numerical experiments to benchmark two implementations of our approach (S-OCT-FULL and S-OCT-BEND) with four approaches from the literature: OCT and OCT-H [7], MIP models for learning optimal univariate and multivariate trees respectively, implemented by Interpretable AI [21]; FlowOCT (solved using Benders decomposition) [2], MIP models for learning univariate trees on datasets with binary features; and DL8.5 [3], an itemset mining-based approach that uses branch-and-bound and caching.

4.1 Experimental Setup

We applied all decision tree implementations to the following fifteen commonly used datasets obtained from the UCI Machine Learning Repository [14]: (A)

Balance Scale, (B) Banknote Authentication, (C) Blood Transfusion, (D) Breast Cancer, (E) Climate Model Crashes, (F) Congressional Voting Records, (G) Glass Identification, (H) Hayes-Roth, (I) Image Segmentation, (J) Ionosphere, (K) Iris, (L) Parkinsons, (M) Soybean (Small), (N) Tic-Tac-Toe Endgame, and (O) Wine. Since FlowOCT and DL8.5 require binary features, we perform an additional *bucketization* step for the numerical datasets [25]; for every feature j, we sort the observations according to this feature, find consecutive observations with different class labels (and different values for feature j), and define a binary feature that has value 1 if and only if x_j is less than the average of the two adjacent feature values. Aside from this peculiarity, we perform the standard one-hot encoding for categorical features and normalize numerical features to the $[0, 1]$ interval.

We partitioned each dataset so that 75% of the observations are used for training and 25% for testing. We tuned the complexity hyperparameter α for S-OCT and FlowOCT[1] by partitioning the training set so that two-thirds is truly used for training and one-third is used for validation. We searched over five α values: $\{0.00001, 0.0001, 0.001, 0.01, 0.1\}$. Interpretable AI automatically tunes α for OCT and OCT-H. There is no α for DL8.5.

Our experiments were programmed using Python 3.8.10 and all optimization problems were solved using Gurobi 9.5 [17] on a machine with a 3.00 GHz 6-core Intel Core i5-8500 processor and 16 GB RAM. A 10-minute time limit was imposed for all optimization problems. Our code can be found at https://github.com/zachzhou777/S-OCT.

Direct MIP Comparison. Our entire approach is grounded in formulation (1), which crucially uses the binary variables **w** to model how observations are routed throughout the decision tree. A straightforward way to turn our formulation (1) into a MIP formulation is to use big-M constraints to model constraints (1l) (we test the implementation without big-M constraints in the next section). We call the corresponding MIP formulation S-OCT-FULL. Our first goal is to test the strength of S-OCT-FULL against two other MIP formulations, namely OCT-H and FlowOCT. Note that OCT-H uses multivariate branching rules, while FlowOCT is tailored to datasets with binary features and uses univariate branching rules. When dealing with datasets having numerical features, we apply the preprocessing step described in Sect. 4.1 before applying FlowOCT. Note that since (in this experiment) we focus on comparing the strength of the MIP formulation rather than its efficacy as practical machine learning method, we do not employ warm starts or early stopping (as we do in the comprehensive experiments).

Comprehensive Comparison. In the second set of experiments, we compare the practical performance of our approach against a wider range of other

---

[1] We modify the FlowOCT objective to 1) minimize error rate plus a regularization term and 2) use α rather than $\lambda \in [0, 1)$ for regularization, to be consistent with other models.

methods, both within and outside of MIP. Besides S-OCT-FULL, we also consider the decomposition approach that uses shattering inequalities described in Sect. 3.1, which we call S-OCT-BEND. We compare S-OCT-FULL and S-OCT-BEND against OCT and OCT-H (as implemented by Interpretable AI [21]), FlowOCT, and DL8.5. Our goal is to assess the performance of these models, even given limited training time. Oftentimes, MIP-based decision trees are able to produce near-optimal solutions within seconds, and additional solving time produces marginal improvements (if any). Therefore, for the MIP models (aside from OCT and OCT-H, as Interpretable AI's implementations of these are a black box), we terminate the solve if 60 s pass without a new incumbent solution being found. We also feed the MIP models a warm start; for FlowOCT, we feed CART's solution, and for S-OCT, we feed the results of a top-down greedy induction where an S-OCT stump (a tree with depth 1) is applied at each branch node so as to maximize training accuracy (similar to [7]).

4.2 Results

Direct MIP Comparison. Figure 3 displays boxplots of the training accuracy, testing accuracy, solution time, and the relative gap for S-OCT-FULL, OCT-H, and FlowOCT at depth 2 and 3. The average ± standard deviation training (testing) accuracy was 0.938 ± 0.122 (0.870 ± 0.136) for S-OCT-FULL, 0.912 ± 0.146 (0.836 ± 0.145) for OCT-H, and 0.833 ± 0.139 (0.811 ± 0.137) for FlowOCT. The average solution time (relative gap) was 158.7 ± 255.1 (0.237 ± 0.408) for S-OCT-FULL, 194.5 ± 258.8 (0.289 ± 0.453) for OCT-H, and 312.5 ± 269.1 (0.463 ± 0.366) for FlowOCT. Across all 30 instances (15 for each depth), S-OCT-FULL timed out for seven instances, OCT-H timed out for eight instances, and FlowOCT timed out for eighteen instances. Overall, we find that S-OCT-FULL achieved the highest average training and testing accuracy across all 15 datasets with the fastest average solution time and the smallest average relative gap.

Comprehensive Comparison. Table 1 provides a detailed comparison of the numerical results for each dataset and for six different model implementations. We first compare the results between S-OCT-FULL and S-OCT-BEND to highlight the improvement in solution time offered by our benders implementation. Both models achieved similar training accuracy (0.960 ± 0.088 for S-OCT-FULL and 0.951 ± 0.105 for S-OCT-BEND), while S-OCT-FULL achieved a higher testing accuracy (0.901 ± 0.115 vs. 0.876 ± 0.145) and S-OCT-BEND achieved a faster solution time (44.6 ± 10.7 vs. 29.2 ± 9.9). These differences are likely due to the early stopping criteria used as part of our S-OCT-BEND implementation, which allows us to achieve similar testing performance with shorter solution times as compared to S-OCT-FULL. However, even in instances where both S-OCT-FULL and S-OCT-BEND solved to optimality, the in-sample and out-of-sample accuracy may differ due to the existence of multiple optimal solutions. Note that these models optimize a combination of in-sample accuracy and

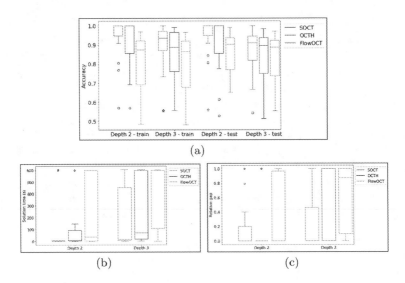

Fig. 3. Boxplots of the training accuracy, testing accuracy, solution time, and the relative gap for S-OCT, OCT-H, and FlowOCT at depth 2 and 3.

a regularization term on the number of branch nodes being used; hence even the in-sample accuracy may differ between S-OCT-FULL and S-OCT-BEND.

Next, we compare the in-sample accuracy achieved by all models. We find that S-OCT-FULL performed best (0.960 ± 0.088), followed by S-OCT-BEND (0.951 ± 0.105) and OCT-H (0.948 ± 0.086). The remaining models all performed significantly worse: OCT (0.882 ± 0.118), DL8.5 (0.883 ± 0.121), and FlowOCT (0.859 ± 0.140). We observe that the multivariate models (S-OCT-FULL, S-OCT-BEND, and OCT-H) significantly outperform the univariate models (DL8.5, FlowOCT, OCT). Intuitively, this makes sense because univariate models require more depth to achieve comparably complex branching rules (see Fig. 1).

We find similar results for out-of-sample accuracy, where S-OCT-FULL performed best (0.901 ± 0.115), followed closely by OCT-H (0.885 ± 0.115) and S-OCT-BEND (0.876 ± 0.145). The remaining models achieve an average testing accuracy of 0.849 ± 0.125 (OCT), 0.846 ±0.126 (DL8.5), and 0.821 ± 0.155 (FlowOCT). Finally, we compare solutions times between all approaches. The average solution time was 44.6 ± 56.3 for S-OCT-FULL, 29.2 ± 30.0 for S-OCT-BEND, 0.7 ± 1.7 for OCT, 15.0 ± 18.6 for OCT-H, 18.0 ± 46.1 for DL8.5, and 57.4 ± 38.6 for FlowOCT. Although our models had the second and third slowest solutions times, they are able to find the (provably) optimal solution in 9/15 instances.

Figure 4 displays line plots of the training accuracy, testing accuracy, and solution time across all 15 datasets (sorted for clarity) and for all six models at depth four. The line plots visualize the results in Table 1 for depth four; we see that both S-OCT-FULL and S-OCT-BEND achieve the highest training accuracy for all datasets and the highest testing accuracy for 12/15 datasets. Our models are comparable in terms of solution time for 9/15 datasets; for the remaining seven datasets, our models are slightly slower.

5 Conclusion

We proposed a new MIP formulation for the optimal decision tree problem. Our approach directly deals with numerical features and leverages the higher modeling power of multivariate branching rules. We also introduced a new class of valid inequalities and an exact decomposition approach that uses these inequalities as feasibility cuts. These inequalities exploit the structure of decision trees and express the geometrical properties of the dataset at hand. We demonstrate through numerical experiments that our MIP approach outperforms (in terms of training accuracy, testing accuracy, solution time, and relative gap) two other popular MIP formulations, and is able to improve both in and out-of-sample performance, while remaining competitive in terms of solution time to a wide range of popular approaches from the literature. Finally, we note that our formulation and the shattering inequalities (4) are general and can be extended to any binary classifier used to implement the branching rules. When the branching rules are implemented via multivariate splits, the separation of the shattering inequalities can be performed efficiently. However, the separation may become more challenging if we consider more complex classifiers.

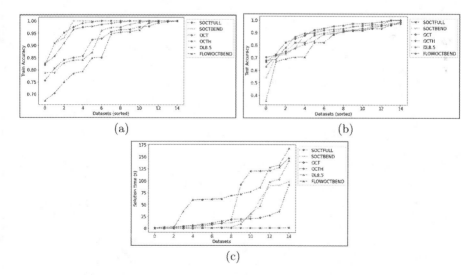

Fig. 4. Line plots across all 15 datasets (sorted for clarity) of the training accuracy, testing accuracy, and solution time for all six models at depth four.

Table 1. Detailed summary of comprehensive comparisons.

	(A)	(B)	(C)	(D)	(E)	(F)	(G)	(H)	(I)	(J)	(K)	(L)	(M)	(N)	(O)
								Dataset							
Observations (N)	468	1029	561	426	405	326	160	132	210	263	112	146	35	718	133
Features (p)	20	4	4	30	18	48	9	15	19	34	4	22	72	27	13
Buckets	N/A	1303	105	3499	1149	N/A	487	N/A	1657	1779	35	892	N/A	N/A	456
Classes (K)	3	2	2	2	2	2	6	3	7	2	3	2	4	2	3
Results for $D = 2$															
							In-sample accuracy (%)								
SOCTFULL	**100.0**	**100.0**	80.57	**100.0**	99.26	**100.0**	70.0	**90.91**	**57.14**	**100.0**	**100.0**	**100.0**	**100.0**	**100.0**	**100.0**
SOCTBEND	**100.0**	**100.0**	80.57	**100.0**	**100.0**	**100.0**	74.38	**90.91**	**57.14**	**100.0**	**100.0**	97.26	**100.0**	**100.0**	**100.0**
OCT	68.59	93.0	79.5	96.01	93.33	95.4	69.38	60.61	**57.14**	90.87	96.43	91.1	**100.0**	65.74	93.98
OCTH	99.79	99.51	**82.17**	97.18	98.02	**100.0**	73.12	90.15	**57.14**	95.82	96.43	97.26	97.14	**100.0**	96.99
DL8.5	68.59	93.0	79.68	96.01	93.83	96.01	69.38	60.61	**57.14**	91.25	96.43	91.1	**100.0**	72.14	96.99
FLOWOCT	68.59	91.55	78.07	94.13	91.11	95.4	64.38	60.61	42.86	90.87	96.43	91.1	**100.0**	72.14	96.99
							Out-of-sample accuracy (%)								
SOCTFULL	**98.73**	**100.0**	73.26	95.1	**95.56**	**97.25**	**59.26**	78.57	56.14	90.91	**97.37**	79.59	**100.0**	**95.83**	**97.78**
SOCTBEND	**98.73**	**100.0**	73.26	95.1	93.33	**97.25**	**59.26**	**82.14**	55.57	88.64	**97.37**	77.55	**100.0**	**95.83**	**97.78**
OCT	66.88	91.84	70.59	94.41	94.07	96.33	55.56	67.86	**56.57**	**92.05**	89.47	89.8	**100.0**	64.17	84.44
OCTH	98.09	98.54	76.5	**96.5**	94.81	89.91	53.7	78.57	**56.57**	**92.05**	89.47	**93.88**	91.67	95.42	95.56
DL8.5	66.88	91.55	69.52	94.41	94.07	95.41	55.56	75.0	56.52	90.91	89.47	**93.88**	**100.0**	65.83	95.56
FLOWOCT	66.88	91.55	70.59	93.01	92.59	96.33	55.56	67.86	42.71	**92.05**	89.47	91.84	**100.0**	65.83	95.56
							Computational time (s)								
SOCTFULL	1.27	6.36	108.15	4.55	55.49	0.81	149.88	8.68	1.03	109.04	0.2	33.28	0.21	10.69	0.28
SOCTBEND	0.64	5.48	106.57	3.71	11.39	0.43	98.43	20.65	19.34	30.55	0.14	137.43	0.16	9.69	0.18
OCT	8.48	**0.76**	0.16	**0.54**	**0.29**	0.3	**0.09**	0.07	**0.19**	**0.33**	0.03	**0.17**	7.83	0.33	**0.09**
OCTH	25.0	10.3	3.01	15.81	10.9	19.06	2.72	2.39	4.38	14.69	0.39	4.64	3.62	50.55	2.49
DL8.5	**0.0**	4.5	**0.03**	35.93	3.56	**0.01**	0.59	**0.0**	8.27	6.94	**0.0**	1.64	**0.0**	**0.01**	0.33
FLOWOCT	1.07	65.23	60.13	65.97	64.45	2.75	60.41	0.47	60.72	72.15	0.23	93.48	0.51	61.5	61.9
Results for $D = 3$															
							In-sample accuracy (%)								
SOCTFULL	**100.0**	**100.0**	81.46	**100.0**	**100.0**	**100.0**	**90.62**	**90.91**	**98.57**	**100.0**	**100.0**	95.21	**100.0**	**100.0**	**100.0**
SOCTBEND	**100.0**	**100.0**	81.46	**100.0**	99.26	**100.0**	85.0	**90.91**	57.14	**100.0**	**100.0**	**100.0**	**100.0**	75.63	99.25
OCT	73.08	97.76	80.57	98.36	95.56	95.4	76.25	72.73	84.76	93.54	98.21	98.63	**100.0**	75.63	99.25
OCTH	**100.0**	98.54	**82.53**	99.3	99.01	**100.0**	83.12	89.39	93.33	98.1	96.43	**100.0**	**100.0**	99.44	97.74
DL8.5	74.15	93.68	81.64	96.48	93.58	97.85	80.0	74.24	65.24	92.02	**100.0**	**100.0**	**100.0**	78.69	**100.0**
FLOWOCT	73.72	93.97	80.75	95.54	96.3	97.24	74.38	74.24	57.14	92.4	**100.0**	95.89	**100.0**	76.6	99.25
							Out-of-sample accuracy (%)								
SOCTFULL	**98.73**	**100.0**	73.26	95.1	93.33	**97.25**	**70.37**	85.71	**93.0**	89.77	**97.37**	87.76	**100.0**	**95.83**	**97.78**
SOCTBEND	**98.73**	**100.0**	73.26	95.1	**95.56**	**97.25**	53.7	71.43	56.0	88.64	**97.37**	81.63	**100.0**	**95.83**	**97.78**
OCT	68.79	96.5	74.87	**95.8**	93.33	96.33	64.81	75.0	81.43	92.05	**97.37**	**97.96**	91.67	72.92	**97.78**
OCTH	95.54	**100.0**	77.54	95.1	94.07	92.66	61.11	82.14	83.57	**96.59**	89.47	**97.96**	91.67	**96.25**	86.67
DL8.5	71.97	92.42	72.73	90.91	90.37	93.58	68.52	**89.29**	63.71	89.77	**97.37**	95.92	**100.0**	73.75	93.33
FLOWOCT	71.97	93.0	75.4	91.61	91.85	93.58	62.96	**89.29**	56.67	92.05	**97.37**	93.88	91.67	71.67	91.11
							Computational time (s)								
SOCTFULL	2.28	6.96	97.09	6.3	13.52	1.68	182.6	51.54	148.03	88.04	0.43	109.17	0.32	12.57	0.48
SOCTBEND	0.98	5.87	94.63	3.89	82.26	0.56	89.06	60.94	60.54	20.63	0.2	20.55	0.17	9.91	0.27
OCT	0.5	0.65	**0.19**	1.0	**0.51**	0.46	**0.16**	0.1	**0.32**	**0.73**	0.04	**0.26**	0.05	0.69	**0.14**
OCTH	21.18	10.97	4.07	23.19	16.58	19.12	4.37	3.03	10.66	24.26	0.49	7.56	2.97	77.92	2.99
DL8.5	**0.02**	**0.01**	2.69	2.19	1.35	**0.13**	244.93	**0.01**	0.53	2.67	**0.01**	54.99	**0.0**	**0.06**	3.07
FLOWOCT	77.7	88.52	60.61	109.1	62.81	63.15	61.36	20.59	75.47	66.38	4.08	87.22	0.01	60.79	60.86
Results for $D = 4$															
							In-sample accuracy (%)								
SOCTFULL	**100.0**	**100.0**	82.0	**100.0**	**100.0**	**100.0**	97.5	**90.91**	99.05	99.62	**100.0**	95.21	**100.0**	**100.0**	**100.0**
SOCTBEND	**100.0**	**100.0**	82.0	**100.0**	**100.0**	**100.0**	93.12	**90.91**	**100.0**	**100.0**	**100.0**	**100.0**	**100.0**	**100.0**	**100.0**
OCT	75.64	99.71	80.57	97.89	95.56	97.55	85.62	84.09	92.38	92.78	96.43	96.58	**100.0**	84.96	99.25
OCTH	**100.0**	**100.0**	82.0	97.42	99.01	**100.0**	85.62	**90.91**	99.52	98.48	96.43	**100.0**	**100.0**	**100.0**	97.74
DL8.5	78.85	97.47	**83.96**	98.59	96.05	99.39	78.75	84.09	82.86	97.34	**100.0**	**100.0**	**100.0**	87.19	**100.0**
FLOWOCT	75.0	96.31	78.43	99.3	95.56	95.4	79.38	67.42	70.48	94.68	**100.0**	84.93	**100.0**	85.1	**100.0**
							Out-of-sample accuracy (%)								
SOCTFULL	**98.73**	**100.0**	72.73	95.1	93.33	**97.25**	66.67	82.14	**92.1**	85.23	**97.37**	87.76	**100.0**	**95.83**	**97.78**
SOCTBEND	**98.73**	**100.0**	72.73	95.1	93.33	**97.25**	53.7	71.43	90.24	88.64	**97.37**	77.55	**100.0**	**95.83**	**97.78**
OCT	71.97	98.83	74.87	95.8	93.33	93.58	**70.37**	82.14	87.9	92.05	89.47	95.92	91.67	82.08	91.11
OCTH	95.54	**100.0**	78.07	**96.5**	**94.81**	92.66	68.52	67.86	89.52	93.18	89.47	91.84	**100.0**	**96.67**	86.67
DL8.5	73.89	95.34	72.19	91.61	88.89	90.83	62.96	**89.29**	77.48	92.05	**97.37**	**97.96**	83.33	80.83	95.56
FLOWOCT	70.7	93.29	70.59	90.91	93.33	96.33	66.67	35.71	69.05	88.64	**97.37**	91.84	91.67	82.08	82.22
							Computational time (s)								
SOCTFULL	6.05	10.74	92.25	8.37	15.42	4.43	128.31	120.29	147.42	120.64	0.54	121.07	0.55	18.08	0.94
SOCTBEND	1.53	6.63	90.82	4.29	11.81	0.83	89.9	60.61	98.44	15.82	0.32	25.06	0.19	11.12	0.36
OCT	0.72	0.85	**0.29**	1.17	0.6	**0.63**	0.22	0.15	**0.42**	**0.71**	0.05	0.31	0.06	1.44	**0.18**
OCTH	27.4	10.22	4.79	20.29	21.87	19.23	6.11	3.49	19.57	34.92	0.49	8.72	2.63	91.83	3.02
DL8.5	**0.15**	**0.05**	142.38	97.92	30.4	2.07	0.96	**0.03**	9.28	103.14	**0.01**	47.48	**0.0**	**0.72**	2.44
FLOWOCT	128.38	86.22	168.05	77.36	61.6	60.54	71.92	133.35	62.13	68.93	0.03	35.45	0.01	60.25	0.37

References

1. Aghaei, S., Azizi, M.J., Vayanos, P.: Learning optimal and fair decision trees for non-discriminative decision-making (2019)
2. Aghaei, S., Gomez, A., Vayanos, P.: Learning optimal classification trees: strong max-flow formulations (2020)
3. Aglin, G., Nijssen, S., Schaus, P.: Learning optimal decision trees using caching branch-and-bound search. Proc. AAAI Conf. Artif. Intell. **34**(04), 3146–3153 (2020)
4. Aglin, G., Nijssen, S., Schaus, P.: Pydl8.5: a library for learning optimal decision trees. In: Bessiere, C. (ed.) Proceedings of the Twenty-Ninth International Joint Conference on Artificial Intelligence, IJCAI-20, pp. 5222–5224. International Joint Conferences on Artificial Intelligence Organization, demos (2020)
5. Avellaneda, F.: Efficient inference of optimal decision trees. Proc. AAAI Conf. Artif. Intell. **34**(04), 3195–3202 (2020)
6. Benders, J.F.: Partitioning procedures for solving mixed-variables programming problems. Numer. Math. **4**(1), 238–252 (1962)
7. Bertsimas, D., Dunn, J.: Optimal classification trees. Mach. Learn. **106**(7), 1039–1082 (2017). https://doi.org/10.1007/s10994-017-5633-9
8. Breiman, L.: Random forests. Mach. Learn. **45**(1), 5–32 (2001). https://doi.org/10.1023/A:1010933404324
9. Breiman, L., Friedman, J., Stone, C.J., Olshen, R.A.: Classification and Regression Trees. CRC Press, Boca Raton (1984)
10. Codato, G., Fischetti, M.: Combinatorial benders' cuts for mixed-integer linear programming. Oper. Res. **54**(4), 756–766 (2006)
11. Cornuéjols, G.: Combinatorial optimization: packing and covering. CBMS-NSF Regional Conference Series in Applied Mathematics, Society for Industrial and Applied Mathematics (2001)
12. Dash, S., Günlük, O., Wei, D.: Boolean decision rules via column generation (2020)
13. Demirović, et al.: Murtree: optimal classification trees via dynamic programming and search (2021)
14. Dua, D., Graff, C.: UCI machine learning repository (2017). http://archive.ics.uci.edu/ml
15. Gleeson, J., Ryan, J.: Identifying minimally infeasible subsystems of inequalities. INFORMS J. Comput. **2**, 61–63 (1990)
16. Gunluk, O., Kalagnanam, J., Li, M., Menickelly, M., Scheinberg, K.: Optimal generalized decision trees via integer programming (2019)
17. Gurobi Optimization, L.: Gurobi optimizer reference manual (2021). http://www.gurobi.com
18. Hooker, J., Ottosson, G.: Logic-based benders decomposition. Math. Prog. 96 (2001)
19. Hu, H., Siala, M., Hebrard, E., Huguet, M.J.: Learning optimal decision trees with maxsat and its integration in adaboost. In: Bessiere, C. (ed.) Proceedings of the Twenty-Ninth International Joint Conference on Artificial Intelligence, IJCAI-20, pp. 1170–1176. International Joint Conferences on Artificial Intelligence Organization (2020)
20. Hyafil, L., Rivest, R.L.: Constructing optimal binary decision trees is np-complete. Inf. Process. Lett. **5**(1), 15–17 (1976)
21. Interpretable AI, L.: Interpretable ai documentation (2021). https://www.interpretable.ai

22. Janota, M., Morgado, A.: SAT-based encodings for optimal decision trees with explicit paths. In: Pulina, L., Seidl, M. (eds.) SAT 2020. LNCS, vol. 12178, pp. 501–518. Springer, Cham (2020). https://doi.org/10.1007/978-3-030-51825-7_35
23. Liaw, A., Wiener, M., et al.: Classification and regression by randomforest. R News **2**(3), 18–22 (2002)
24. Lin, J., Zhong, C., Hu, D., Rudin, C., Seltzer, M.: Generalized and scalable optimal sparse decision trees. In: International Conference on Machine Learning, pp. 6150–6160. PMLR (2020)
25. Lin, J.J., Zhong, C., Hu, D., Rudin, C., Seltzer, M.I.: Generalized and scalable optimal sparse decision trees. In: ICML (2020)
26. Narodytska, N., Ignatiev, A., Pereira, F., Marques-Silva, J.: Learning optimal decision trees with sat. In: Proceedings of the Twenty-Seventh International Joint Conference on Artificial Intelligence, IJCAI-18, pp. 1362–1368. International Joint Conferences on Artificial Intelligence Organization (2018)
27. Nijssen, S., Fromont, E.: Mining optimal decision trees from itemset lattices. In: Proceedings of the 13th ACM SIGKDD International Conference on Knowledge Discovery and Data Mining, pp. 530–539. KDD 2007, Association for Computing Machinery, New York (2007)
28. Quinlan, J.R.: Induction of decision trees. Mach. Learn. **1**(1), 81–106 (1986)
29. Schidler, A., Szeider, S.: Sat-based decision tree learning for large data sets. Proc. AAAI Conf. Artif. Intell. **35**(5), 3904–3912 (2021)
30. Schrijver, A.: Theory of Linear and Integer Programming. Wiley, Chichester (1986)
31. Vapnik, V.: Statistical Learning Theory. Wiley, New York (1998)
32. Verhaeghe, H., Nijssen, S., Pesant, G., Quimper, C.-G., Schaus, P.: Learning optimal decision trees using constraint programming. Constraints **25**(3), 226–250 (2020). https://doi.org/10.1007/s10601-020-09312-3
33. Verhaeghe, H., Nijssen, S., Pesant, G., Quimper, C.G., Schaus, P.: Learning optimal decision trees using constraint programming (extended abstract). In: Bessiere, C. (ed.) Proceedings of the Twenty-Ninth International Joint Conference on Artificial Intelligence, IJCAI-20, pp. 4765–4769. International Joint Conferences on Artificial Intelligence Organization (2020)
34. Verwer, S., Zhang, Y.: Learning optimal classification trees using a binary linear program formulation. In: Proceedings of the Thirty-Third AAAI Conference on Artificial Intelligence (AAAI-19), pp. 1625–1632. 27 Jan 2019—01 Feb 2019. AAAI Press (2019)
35. Verwer, S., Zhang, Y.: Learning decision trees with flexible constraints and objectives using integer optimization. In: Salvagnin, D., Lombardi, M. (eds.) Integration of AI and OR Techniques in Constraint Programming, pp. 94–103. Springer International Publishing, Cham (2017). https://doi.org/10.1007/978-3-319-59776-8_8
36. Wolsey, L.: Integer Programming. Wiley Series in Discrete Mathematics and Optimization, Wiley, Hoboken (1998)
37. Zhu, H., Murali, P., Phan, D.T., Nguyen, L.M., Kalagnanam, J.: A scalable mip-based method for learning optimal multivariate decision trees. In: Larochelle, H., Ranzato, M., Hadsell, R., Balcan, M., Lin, H. (eds.) Advances in Neural Information Processing Systems 33: Annual Conference on Neural Information Processing Systems 2020, NeurIPS 2020 6–12 December 2020, virtual (2020)

Learning Pseudo-Backdoors for Mixed Integer Programs

Aaron Ferber[1]([✉]) [iD], Jialin Song[2] [iD], Bistra Dilkina[1] [iD], and Yisong Yue[3] [iD]

[1] University of Southern California, Los Angeles, USA
{aferber,dilkina}@usc.edu
[2] NVIDIA, Santa Clara, USA
jialins@nvidia.com
[3] California Institute of Technology, Pasadena, USA
yyue@caltech.edu

Abstract. We propose a machine learning approach for quickly solving Mixed Integer Programs (MIPs) by learning to prioritize sets of branching variables at the root node which result in faster solution times, which we call pseudo-backdoors. Learning-based approaches have seen success in combinatorial optimization by flexibly leveraging common structures in a given distribution of problems. Our approach takes inspiration from the concept of strong backdoors, which are small sets of variables such that only branching on these variables yields an optimal integral solution and a proof of optimality. Our notion of pseudo-backdoors corresponds to a small set of variables such that prioritizing branching on them when possible leads to faster solve time. A key advantage of pseudo-backdoors over strong backdoors is that they retain the solver's optimality guarantees and are amenable to data-driven identification. Our proposed method learns to estimate the relative solver speed of a candidate pseudo-backdoor and determine whether or not to use it. This pipeline can be used to identify high-quality pseudo-backdoors on unseen MIP instances for a given MIP distribution. We evaluate our method on five problem distributions and find that our approach can efficiently identify high-quality pseudo-backdoors. In addition, we compare our learned approach against Gurobi, a state-of-the-art MIP solver, demonstrating that our method can be used to improve solver performance.

1 Introduction

Mixed integer programs (MIPs) are flexible combinatorial optimization problems [8], which consist of setting variables to optimize a linear objective function subject to both linear and integrality constraints. MIPs can be solved via the branch-and-bound algorithm [30], which explores the solution space using tree search by branching on decision variables and pruning provably suboptimal subtrees. Many heuristic components of MIP solvers impact runtime such as node selection, primal heuristics, cut generation, and variable selection [1]. Recent work has focused on data-driven machine learning approaches for variable selection to dynamically recommend which variable to branch on throughout the search process, based on features computed at each search node [4,16,25,34]. We examine

© Springer Nature Switzerland AG 2022
P. Schaus (Ed.): CPAIOR 2022, LNCS 13292, pp. 91–102, 2022.
https://doi.org/10.1007/978-3-031-08011-1_8

a different problem of predicting high-priority variables at the top of the branch-and-bound tree, leveraging the concept of backdoors, which represent variables that are core to solving a given combinatorial problem. Backdoors are introduced in [40] for solving Constraint Satisfaction Problems (CSPs), with "weak backdoors" being a small set of variables that satisfy the following property: there exists an assignment to this subset of variables such that the remaining unassigned CSP subproblem can be solved in polynomial time. In a "strong backdoor" any setting of the backdoor variables leads to a polynomially-solvable subproblem. Later, backdoors are generalized to combinatorial optimization and MIPs [10,11], where a strong backdoor is a subset of integer variables such that only branching on them yields a provably optimal solution, making the problem complexity dependent on backdoor size rather than problem size. While backdoors have theoretical and practical limitations in CSP and SAT [23,24,36], recent work has shown that prioritizing "backdoor" sets of branching variables can speed up MIP solving [15,27]. Prioritizing "backdoor" sets of branching variables at the start of tree search means that they take precedence for branching when possible. As a result, if the MIP is not solved by only branching on "backdoor" variables, the solver can finish by using non-backdoor variables. Thus, a set of prioritized variables resulting in fast runtimes may not strictly be a strong or weak backdoor and to clarify this distinction, we refer to a fast-solving set of integer variables as a *pseudo-backdoor*.

Our goal in finding pseudo-backdoors is to quickly solve MIPs. In a MIP we need to find real values for n decision variables $x \in \mathbb{R}^n$, maximizing a linear objective function $c^T x$, subject to m linear constraints $Ax \leq b$, with a subset $\mathcal{I} \subseteq [n]$ of variables required to be integral $x_i \in \mathbb{Z} \forall i \in \mathcal{I}$. Formally we solve $\min\{c^T x | Ax \leq b, x_j \in \mathbb{Z} \forall i \in \mathcal{I}\}$. Given a MIP $P = (c, A, b, \mathcal{I})$, we aim to find a pseudo-backdoor $\mathcal{B} \subseteq \mathcal{I}$ which quickly solves the MIP. We consider a distributional MIP solving setting where we train a pseudo-backdoor identifier on training instances to quickly solve unseen MIPs.

We propose a data-driven approach to identifying pseudo-backdoors for MIP distributions by learning a scoring model to select a pseudo-backdoor with fast runtime and a classification model to predict whether the selected pseudo-backdoor will beat the default solver. If suggested, we instruct the MIP solver to branch on pseudo-backdoor variables before other variables. We represent MIPs as featurized bipartite graphs with variable and constraint nodes as in [16], using graph attention networks [39] with global attention pooling [32] to learn both models. We evaluate empirically on neural network verification [17], facility location [9], and the generalized independent set problem (gisp) [7], showing that our models identify pseudo-backdoors that quickly solve MIPs compared to Gurobi.

2   Related Work

Backdoors in SAT & MIP: Backdoors were introduced to represent the core hardness in Constraint Satisfaction Problems (CSP) [40], where the authors analyzed CSP time complexity assuming various sized backdoors exist. [10] extend the definition of backdoors to general combinatorial optimization problems. Both works study the practical existence of backdoors, finding small backdoors for realistic problems. [28] propose collecting highly-branched variables for identifying backdoors in SAT. In MIP, [12] study fracture backdoors which are variables whose removal would result in a natural MIP decomposition. Additionally, [15] identify MIP pseudo-backdoors by solving a set covering problem, and observe that prioritizing branching on pseudo-backdoors can quickly solve some MIPLIB [29] instances. Recent work uses Monte Carlo Tree Search to sample pseudo-backdoors for a given MIP [27]. These approaches identify pseudo-backdoors from scratch in individual MIP instances rather than MIP distributions, and can't operate on new instances without re-running the entire procedure. Our method is the first data-driven attempt to predict pseudo-backdoors using labeled data which frontloads the computational overhead of finding fast-solving pseudo-backdoors on a training set for deployment on unseen instances without rerunning pseudo-backdoor search.

Learning in Combinatorial Optimization: Machine learning has shown promise improving combinatorial optimization. In exact MIP solving, machine learning operates in various heuristic components like selecting variables at search nodes [4,16,25], running primal heuristics [26], or performing neural diving [34]. These algorithms operate inside the solver, and can complement our approach which treats the solver as a black box, whether learning-augmented or not. Several approaches improve runtime by learning performant solver configurations [2,21,22]. Our approach leverages the structural connection between the MIP formulation and a set of hyperparameters, the variable branching priorities, to improve solve time rather than predicting global hyperparamters.

3   Learning Pseudo-Backdoors

We use two learned models: a scorer that ranks subsets of integer variables according to their runtime, and a classifier that predicts whether the best-scoring subset will improve solve time compared to running a default solver. The classifier ensures that even when it is hard to sample pseudo-backdoors we can still algorithmically run the default solver.

Figure 1 illustrates our method's deployment. Given an unseen MIP, we randomly sample fixed-sized subsets of integer variables according to their LP fractionality as in [10], scoring the sampled subsets and predicting the best-scoring subset to be a pseudo-backdoor. The classifier then predicts whether the identified pseudo-backdoor is faster than the default solver. If so, we assigning higher branching priorities to variables in the pseudo-backdoor and solve; otherwise, we use the default solver. The scorer $S(P, \mathcal{B}; \theta_S)$ has neural network parameters θ_S, using the MIP P and candidate subset $\mathcal{B}$ to score $\mathcal{B}$ based on runtime. The classifier $C(P, \mathcal{B}; \theta_C)$ has neural network parameters θ_C taking the same input $(P, \mathcal{B})$ to predict whether $\mathcal{B}$ will outperform the default solver.

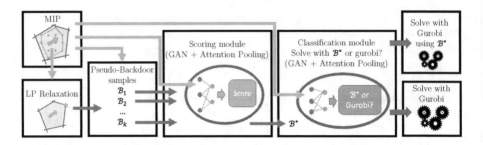

Fig. 1. The pseudo-backdoor pipeline for a single MIP instance consisting of a scorer $S(P, \mathcal{B}; \theta_S)$ and classifier $C(P, \mathcal{B}; \theta_C)$. k pseudo-backdoor sets of decision variables $\mathcal{B}_1, \ldots, \mathcal{B}_k$ are sampled according to LP fractionality. These candidate sets are ranked by the scoring module $S(P, \mathcal{B}; \theta_S)$ to predict the best pseudo-backdoor $\mathcal{B}^*$. The classification module determines whether or not to run the solver using $\mathcal{B}^*$ based on the predicted pseudo-backdoor success probability $C(P, \mathcal{B}^*; \theta_C)$.

3.1   MIP and Backdoor Data Representation

We want to ensure that our architecture can operate on variable-sized MIP instances, incorporate pseudo-backdoor information $\mathcal{B}$, and yield predictions invariant to unimportant changes in MIP formulations such as variable or constraint permutations. Thus, we represent the MIP as a bipartite graph as in [16]. The bipartite graph representation ensures the MIP encoding is invariant to variable and constraint permutation. Additionally, we can use a variety of predictive models designed for variable-sized graphs, enabling deployment on problems with varying numbers of variables and constraints. The MIP is represented as a featurized bipartite graph $P = (\mathcal{G}, \mathbf{C}, \mathbf{E}, \mathbf{V})$, with graph $\mathcal{G}$ containing variable nodes and constraint nodes. Additionally, there is an edge $(i, j) \in \mathcal{E}$ between a variable node i and a constraint node j iff the variable i appears in constraint j with nonzero coefficient i.e. $A_{i,j} \neq 0$. Constraint, variable, and edge features are represented as matrices $\mathbf{C} \in \mathcal{R}^{m \times c}, \mathbf{E} \in \mathcal{R}^{|\mathcal{E}| \times e}, \mathbf{V} \in \mathcal{R}^{n \times v}$, and are described in Table 1. Pseudo-backdoor sets $\mathcal{B}$ are represented with a binary variable feature which is 1 for variables in $\mathcal{B}$ and 0 otherwise. Encoding the input $(P, \mathcal{B})$ as a featurized bipartite graph allows us to leverage state-of-the-art techniques in making predictions on variable-sized graphs.

3.2   Neural Network Architecture

Given the graph encoding of the MIP and candidate backdoor set, we can apply a variety of graph prediction methods such as those in the pytorch geometric library [14]. These approaches generate high-level node embeddings via message passing, where a node's embedding is an aggregation of those of it's neighbors. Previous work used the Graph Convolutional Network (GCN) [16] to predict branching scores on individual variable nodes at a given branch and bound node. However, we were unable to train a GCN model to predict either scores or success probabilities. Alternatively, we found that Graph Attention Network (GAT)

Table 1. Variable, constraint, and edge features that encode a MIP instance.

Category	Name (count)	Description
V Vars	type (3)	Variable type (continuous, binary, integer)
	obj_coef	The decision variable's objective coefficient
	has_lb	Does the variable have a lower bound?
	has_ub	Does the variable have an upper bound?
	root_lp_at_lb	Variable at it's lower bound in the root LP?
	root_lp_at_ub	Variable at it's upper bound in the root LP?
	root_lp_frac	Is the variable fractional in the root LP?
	lp_basis (4)	Variable root LP status (basic, lower, upper, superbasic)
E Edges	coef	Constraint coefficient
C Constrs	obj_cos_sim	Cosine similarity of objective and constraint
	bias	Constant bias
	root_lp_tight	Constraint tightness in the root LP
	root_lp_dual	Dual value In the root LP
	sense (3)	Constraint direction ($\leq, \geq, =$)

worked in our setting [39]. Here a node's high-level embedding is a weighted aggregation of the embeddings of its neighbors and edges, where the weights are themselves predicted via a network layer as done in attention-based models [38], thus allowing the model to attend to variable or constraint nodes which are most useful for generating the next iteration's embedding. Formally, the GAT performs the message passing iteration in Eq. 1 to compute embeddings x'_i for graph node i based on the neighbors of $i, \mathcal{N}(i)$ with attention weights $\alpha_{i,j}$ using neural network weights θ. The attention weights are given by Eq. 2 using learnable parameters a on node embeddings x and edge features $\mathbf{E}$.

$$x'_i = \alpha_{i,i}\theta x_i + \sum_{j \in \mathcal{N}(i)} \alpha_{i,j}\theta x_j \tag{1}$$

$$\alpha_{i,j} = \frac{\exp\left(\text{LeakyReLU}\left(a^T\left[\theta x_i \| \theta x_j \| \theta^e \mathbf{E}_{i,j}\right]\right)\right)}{\sum_{k \in \mathcal{N}(i) \cup \{i\}} \exp\left(\text{LeakyReLU}\left(a^T\left[\theta x_i \| \theta x_k \| \theta^e \mathbf{E}_{i,j}\right]\right)\right)} \tag{2}$$

Importantly, the attention weights $\alpha_{i,j}$ sum to 1 for source node i and are non-negative so the embeddings x'_i are a weighted average of adjacent node embeddings. At the first message passing iteration, the node features $\mathbf{V}, \mathbf{C}$ are used. After several iterations of message passing, we aggregate the node embeddings into a global feature vector representing the MIP using global attention pooling [32], which similarly computes a weighted average of node embeddings with weights given by neural network layer. This fixed-length graph representation vector is then passed to a leaky-relu activated MLP to produce a single scalar output representing the score for the scoring module or logit for the classification module. Obtaining node embeddings followed by aggregation across the entire MIP enables us to use the same network architecture for MIPs of various sizes.

3.3 Learning the Scorer Model

We train the scorer to rank candidate backdoors by solver runtime when prioritizing pseudo-backdoor variables, which we label by setting high branching priority and solving to optimality. For a MIP P and a pair of candidate sets $\mathcal{B}_1, \mathcal{B}_2$, the model estimates scores $s_1 = S(P, \mathcal{B}_1; \theta_S)$, $s_2 = S(P, \mathcal{B}_2; \theta_S)$ to match ranking label y which is -1 if solving with $\mathcal{B}_1$ is faster than with $\mathcal{B}_2$, and 1 otherwise. We train with the margin ranking loss on all pairs of pseudo-backdoors for a given MIP, $L(s_1, s_2, y) = \max(0, -y(s_1 - s_2) + m)$ [37] for margin value $m = 0.1$. Learning to rank has been pioneered in data-driven information retrieval systems [6,33] and in our use case, it is sufficient that we *rank* pseudo-backdoors correctly for each MIP rather than model the intricacies of runtime. At deployment time, we sample several candidate backdoors and score them, taking the best-scoring pseudo-backdoor for deployment. This best-scoring pseudo-backdoor can then be deployed on its own by prioritizing the variables in the pseudo-backdoor, representing our SCORER model. We initially experimented with predicting normalized runtime as well as the probability that the candidate backdoor outperforms Gurobi, but found that these losses weren't tuning the model toward isolating high-quality pseudo-backdoors for a given MIP as top-scorers. Ultimately, the approach of learning to rank closely aligned our model's loss with the overall deployment problem of identifying the fastest-solving pseudo-backdoor.

3.4 Learning the Classifier Model

For a given MIP instance, it may be difficult to sample high-quality pseudo-backdoors. As a result, we learn a classifier to determine whether to use the best-scoring pseudo-backdoor or the default MIP solver. The classifier has the same architecture as the scoring model, taking in the MIP and candidate subset encoding $(P, \mathcal{B})$, and outputting an estimated probability that the pseudo-backdoor outperforms the standard MIP solver. We train the model to minimize a binary cross-entropy loss between the model outputs and labels indicating whether the scorer's best-scoring pseudo-backdoor beat the default solver. At test time we use a threshold of 0.5 to determine whether to use the best-scoring pseudo-backdoor or the default solver. SCORER+CLS denotes the combined scorer and classifier identifying the best pseudo-backdoor and whether it beats Gurobi.

4 Experiment Results

4.1 Problem Domains

Many real-world setting require solving a homogeneous family of problems, where instances share similar structures, differing slightly in the problem size or numerical coefficients. We evaluate the two components of our method on problem instances drawn from three domains[1]. We refrained from evaluating our methodology on heterogeneous MIP datasets like MIPLIB [29] as we found difficulty sampling fast-solving pseudo-backdoors to train with for a variety of realistic MIP distributions, and we target homogeneous settings where latent problem information that would normally require domain knowledge to extract may help automatically identify high-quality pseudo-backdoors.

Neural Network Verification: Neural network verification determines a fixed network's robustness to input data perturbation. Previous work [17] formulates a MIP that bounds the perturbation required to fool a neural network. Our MIP instances are a random subset from [34] which considers verification of MNIST images against a small convolutional neural network.

Facility Location: Capacitated Facility Location asks which facilities to open and how to route limited facility supply to satisfy customer demand. The goal is to minimize facility setup cost and per-unit cost transporting between facilities and customers. As in [16], we generate MIPs according to [9] having 100 facilities supplying 200 (easy) and 400 (hard) customers.

Generalized Independent Set Problem: The Generalized Independent Set Problem (GISP) is a graph optimization problem proposed for forestry management [20]. The input consists of a graph, a subset of removable edges with deletion cost, and revenues for each vertex upon selection. The goal is to select a set of vertices and removable edges that maximize the net profit of total vertex revenues minus total edge costs such that the selected vertices are independent in the modified subgraph. We randomly generate GISP instances from two hardness settings with nodes between 150 (easy) and 175 (hard), with node reward $r = 100$ and edge removal cost $c = 1$. Erdős-Rényi graphs [13] are generated with edge probability $p = 0.3$, and are made removable with probability $\alpha = 0.25$.

4.2   Data Generation and Model Evaluation

Each setting uses 300 MIPs: 100 $\mathcal{D}_s$ for training the scorer, 100 $\mathcal{D}_c$ for training the classifier, and 100 $\mathcal{D}_t$ for testing. To train our two models, we first generate candidate pseudo-backdoors for each instance by randomly sampling $p\%$ of integer variables proportional to their fractionality in the LP relaxation as in [10]. We weight variables by their absolute distance to the nearest integer, with more fractional decision variables having higher probability. Note that this sampling method is done only at the root node rather than fractionality branching during branch-and-bound which is shown to perform poorly [1]. On the contrary, in [10] the authors successfully found backdoors using LP fractionality sampling. From initial experiments sampling variables of size $1\%, 3\%, 5\%$ on NNVerify and GISP easy, we found that sampling 1% of the decision variables yielded high-quality pseudo-backdoors on training data. As a result, our models are trained using

---

[1] **Other domains:** We initially collected data on two hardness settings of set cover [3], combinatorial auctions [31], and maximum independent set [5] but found that the best sampled backdoors underperformed Gurobi, meaning that no matter how accurate our models, they could never outperform standard Gurobi.

pseudo-backdoors with $p = 1\%$ of the decision variables, presuming that smaller pseudo-backdoors yield faster runtimes due to the more limited search trees. For each MIP, we label performance on 50 random candidate backdoors, yielding a dataset of 15,000 pseudo-backdoors. We find that the best sampled backdoors yield average runtime improvements of 44%, 30%, 16%, and 37% for NNVerify, Facilities easy and hard, and GISP easy and hard respectively, meaning that correctly identifying pseudo-backdoors has the potential for large runtime improvements. This improvement firstly indicates that high-quality pseudo-backdoors exist in our datasets, and secondly that there is potential benefit for an algorithm that correctly identifies these pseudo-backdoors.

We evaluate the pipeline in Fig. 1 on the test set $\mathcal{D}_t$, by sampling subsets using the LP relaxation, predicting the best pseudo-backdoor with the scorer, and finally determining whether or not to use the selected pseudo-backdoor with the classifier. The backdoor set is deployed by setting Gurobi's Branch-Priority to be 2 for variables in the pseudo-backdoor and 1 for other variables. As noted in Gurobi documentation [18] the BranchPriority is the primary criterion for selecting fractional branching variables, with higher-priority variables always taking precedence over lower-priority variables with ties being broken using internal variable selection criteria. We consider the runtime of the full pipeline including solving the LP relaxation, LP-based sampling, batched inference on sampled sets to obtain a pseudo-backdoor, classification of the predicted pseudo-backdoor, and solving the MIP. As Gurobi doesn't allow access the initial LP relaxation, we need to re-solve the LP from scratch incurring additional runtime which would not be present with deeper solver integration. We solve MIPs with single-threaded Gurobi 9.1 [19] (avoiding conflicts between parallel solves) on five 32-core machines with Intel 2.1 GHz cpus and 264 GB of memory. Neural Networks are trained with Pytorch [35] on 4 GTX 1080 Ti GPUs.

4.3 Main Results

We present results on 100 test MIPs $\mathcal{D}_t$ in Table 2. We evaluate the standalone score model (SCORER), which uses the candidate backdoor with best predicted score, and the full scorer and classification pipeline (SCORER+CLS). The table presents win/tie/loss over Gurobi and runtime in seconds comprised of solving the root LP, sampling, and model inference. When SCORER+CLS suggests using Gurobi we record a "tie" to give insight into model predictions even though it is a slight loss in runtime. Overall, batched model inference takes 50ms on average to compute scores for all 50 candidate backdoors and predict success probability on the best-scoring pseudo-backdoor. The root LP solve times average 0.02, 0.08, 1.2, and 1.9 s for GISP, nnverify, facilities easy, and facilities hard, incurring runtime which could be avoided with tighter solver integration.

As shown in Table 2, SCORER quickly solves many MIPs, with lower average runtime than Gurobi on facilities easy (6%) and GISP hard (9%). Additionally, SCORER outperforms Gurobi at different percentiles for NNVerify while losing on average. Importantly, the win/tie/loss demonstrates that SCORER can often identify quality pseudo-backdoors. In NNVerify, facilities hard, and GISP easy,

Table 2. Runtime (secs) of standard Gurobi (grb), the score model (SCORER), and the joint score and classify model (SCORER+CLS). We report mean and standard deviation, 25th, 50th, and 75th percentiles. Finally, we report Win/Tie/Loss vs Gurobi.

dataset	solver	mean	stdev	25 pct	median	75 pct	W/T/L vs grb
nnverify	grb	6.5	7.9	2.9	4.3	6.1	0 / 100 / 0
nnverify	scorer	7.0	12.8	**2.5**	3.5	5.1	68 / 0 / 31
nnverify	scorer+cls	**5.6**	**7.4**	2.6	**3.3**	**5.0**	62 / 18 / 20
facilities easy	grb	27.4	21.6	14.0	22.1	32.7	0 / 100 / 0
facilities easy	scorer	24.9	**18.2**	12.9	**19.7**	32.1	65 / 0 / 35
facilities easy	scorer+cls	**22.8**	18.8	**11.7**	22.7	**29.3**	55 / 33 / 12
facilities hard	grb	46.9	31.6	26.2	37.1	55.2	0 / 100 / 0
facilities hard	scorer	56.0	42.5	29.3	42.5	69.1	28 / 0 / 72
facilities hard	scorer+cls	**44.5**	**30.9**	**25.2**	**35.5**	**52.7**	25 / 74 / 1
gisp easy	grb	611	**182**	488	580	681	0 / 100 / 0
gisp easy	scorer	960	755	515	649	915	41 / 0 / 59
gisp easy	scorer+cls	**601**	247	**481**	**568**	**663**	24 / 70 / 6
gisp hard	grb	2533	939	1840	2521	2976	0 / 100 / 0
gisp hard	scorer	2373	**855**	1721	2262	2926	47 / 0 / 53
gisp hard	scorer+cls	**2326**	**855**	**1654**	**2215**	**2866**	41 / 39 / 20

the increased average runtime is partially explained by the higher standard deviation. Additionally, in facilities hard, SCORER suffers 72 losses to Gurobi.

SCORER+CLS is faster than Gurobi across all distributions on average and at most quantiles. The SCORER+CLS pipeline outperforms Gurobi on mean solve time by 14%, 17%, 5%, 2% and 8% on NNVerify, facilities easy, facilities hard, GISP easy and GISP hard respectively. In terms of win/tie/loss, SCORER+CLS retains most of SCORER's wins and only incorrectly selects the pseudo-backdoor in 20 of 100 instances. Furthermore, SCORER+CLS's variability is reduced in NNVerify, Facilities hard, and GISP easy to be comparable with Gurobi. Overall, SCORER identifies performant pseudo-backdoors, and SCORER+CLS leverages the pseudo-backdoors for consistently faster solving by deciding when to run Gurobi.

5   Conclusion

We present a method for learning pseudo-backdoors for faster MIP solving. Inspired by previous work showing that small sets of variables can sometimes improve MIP solving, our method is comprised of two parts: an initial scorer which ranks candidate pseudo-backdoors according to how fast they solve a given MIP, and a classifier to predict whether the selected pseudo-backdoor will improve over a standard solver. We ablate these models on several realistic MIP distributions, demonstrating that fast-solving pseudo-backdoors sometimes exist for the examined MIP distributions, the scorer can identify pseudo-backdoors, and the classifier can determine when these pseudo-backdoors speed up MIP solving.

Acknowledgments. AF and BD were partially supported by NSF AI Institute for Advances in Optimization Award #2112533 and Qualcomm. JS and YY were partially supported by NSF #1763108, Raytheon, Beyond Limits, and JPL.

References

1. Achterberg, T., Wunderling, R.: Mixed integer programming: analyzing 12 years of progress. In: Jünger, M., Reinelt, G. (eds) Facets of Combinatorial Optimization, pp. 449–481. Springer, Berlin (2013). https://doi.org/10.1007/978-3-642-38189-8_18
2. Ansótegui, C., Malitsky, Y., Samulowitz, H., Sellmann, M., Tierney, K.: Model-based genetic algorithms for algorithm configuration. In: International Conference on Artificial Intelligence, pp. 733–739 (2015)
3. Balas, E., Ho, A.: Set covering algorithms using cutting planes, heuristics, and subgradient optimization: a computational study. In: Padberg, M.W. (eds) Combinatorial Optimization, pp. 37–60. Springer, Berlin (1980). https://doi.org/10.1007/BFb0120886
4. Balcan, M.F., Dick, T., Sandholm, T., Vitercik, E.: Learning to branch. In: International Conference on Machine Learning (ICML) (2018)
5. Bergman, D., Cire, A.A., Hoeve, W.J., Hooker, J.: Decision Diagrams for Optimization, 1st edn. Springer Publishing Company Inc., Cham (2016). https://doi.org/10.1007/978-3-319-42849-9
6. Burges, C., et al.: Learning to rank using gradient descent. In: Proceedings of the 22nd International Conference on Machine learning, pp. 89–96 (2005)
7. Colombi, M., Mansini, R., Savelsbergh, M.: The generalized independent set problem: polyhedral analysis and solution approaches. Eur. J. Oper. Res. **260**(1), 41–55 (2017)
8. Conforti, M., Cornuéjols, G., Zambelli, G., et al.: Integer Programming. Springer, Cham (2014). https://doi.org/10.1007/978-3-319-11008-0
9. Cornuéjols, G., Sridharan, R., Thizy, J.M.: A comparison of heuristics and relaxations for the capacitated plant location problem. Eur. J. Oper. Res. **50**(3), 280–297 (1991)
10. Dilkina, B., Gomes, C.P., Malitsky, Y., Sabharwal, A., Sellmann, M.: Backdoors to combinatorial optimization: feasibility and optimality. In: van Hoeve, W.-J., Hooker, J.N. (eds.) CPAIOR 2009. LNCS, vol. 5547, pp. 56–70. Springer, Heidelberg (2009). https://doi.org/10.1007/978-3-642-01929-6_6
11. Dilkina, B., Gomes, C.P., Sabharwal, A.: Backdoors in the context of learning. In: Kullmann, O. (ed.) SAT 2009. LNCS, vol. 5584, pp. 73–79. Springer, Heidelberg (2009). https://doi.org/10.1007/978-3-642-02777-2_9
12. Dvořák, P., Eiben, E., Ganian, R., Knop, D., Ordyniak, S.: Solving integer linear programs with a small number of global variables and constraints. In: International Joint Conference on Artificial Intelligence (IJCAI), pp. 607–613 (2017)
13. Erdős, P., Rényi, A.: On the evolution of random graphs. Publ. Math. Inst. Hung. Acad. Sci **5**(1), 17–60 (1960)

14. Fey, M., Lenssen, J.E.: Fast graph representation learning with PyTorch geometric. In: ICLR Workshop on Representation Learning on Graphs and Manifolds (2019)
15. Fischetti, M., Monaci, M.: Backdoor branching. In: Günlük, O., Woeginger, G.J. (eds.) IPCO 2011. LNCS, vol. 6655, pp. 183–191. Springer, Heidelberg (2011). https://doi.org/10.1007/978-3-642-20807-2_15
16. Gasse, M., Chételat, D., Ferroni, N., Charlin, L., Lodi, A.: Exact combinatorial optimization with graph convolutional neural networks. In: Advances in Neural Information Processing Systems (NeurIPS) (2019)
17. Gowal, S., et al.: On the effectiveness of interval bound propagation for training verifiably robust models (2018). arXiv preprint http://arxiv.org/abs/1810.12715
18. Gurobi Optimization, L.: Gurobi optimizer reference manual (2021). www.gurobi.com/documentation/9.1/refman/branchpriority.html
19. Gurobi Optimization, L.: Gurobi optimizer reference manual (2021). http://www.gurobi.com
20. Hochbaum, D.S., Pathria, A.: Forest harvesting and minimum cuts: a new approach to handling spatial constraints. For. Sci. **43**(4), 544–554 (1997)
21. Hoos, H.H.: Automated algorithm configuration and parameter tuning. In: Hamadi, Y., Monfroy, E., Saubion, F. (eds.) Auton. Search, pp. 37–71. Springer, Heidelberg (2011). https://doi.org/10.1007/978-3-642-21434-9_3
22. Hutter, F., Hoos, H.H., Leyton-Brown, K., Stützle, T.: Paramils: an automatic algorithm configuration framework. J. Artif. Intell. Res. **36**, 267–306 (2009)
23. Järvisalo, M., Junttila, T.: Limitations of restricted branching in clause learning. In: Bessière, C. (ed.) CP 2007. LNCS, vol. 4741, pp. 348–363. Springer, Heidelberg (2007). https://doi.org/10.1007/978-3-540-74970-7_26
24. Järvisalo, M., Niemelä, I.: The effect of structural branching on the efficiency of clause learning sat solving: an experimental study. J. Algorithms **63**(1–3), 90–113 (2008)
25. Khalil, E., Le Bodic, P., Song, L., Nemhauser, G., Dilkina, B.: Learning to branch in mixed integer programming. In: AAAI Conference on Artificial Intelligence (2016)
26. Khalil, E.B., Dilkina, B., Nemhauser, G., Ahmed, S., Shao, Y.: Learning to run heuristics in tree search. In: International Joint Conference on Artificial Intelligence (IJCAI) (2017)
27. Khalil, E.B., Vaezipoor, P., Dilkina, B.: Finding backdoors to integer programs: a monte carlo tree search framework (2022)
28. Kilby, P., Slaney, J., Thiébaux, S., Walsh, T.: Backbones and backdoors in satisfiability. In: AAAI, pp. 1368–1373 (2005)
29. Koch, T., et al.: Miplib 2010. Math. Program. Comput. **3**(2), 103–163 (2011). https://doi.org/10.1007/s12532-011-0025-9
30. Land, A.H., Doig, A.G.: An automatic method for solving discrete programming problems. In: Jünger, M. (ed.) 50 Years of Integer Programming 1958-2008, pp. 105–132. Springer, Heidelberg (2010). https://doi.org/10.1007/978-3-540-68279-0_5
31. Leyton-Brown, K., Pearson, M., Shoham, Y.: Towards a universal test suite for combinatorial auction algorithms. In: ACM Conference on Electronic Commerce, pp. 66–76 (2000)
32. Li, Y., Zemel, R., Brockschmidt, M., Tarlow, D.: Gated graph sequence neural networks. In: ICLR (2016)
33. Liu, T.Y., et al.: Learning to rank for information retrieval. Found. Trends® Inf. Retrieval **3**(3), 225–331 (2009)
34. Nair, V. et al.: Solving mixed integer programs using neural networks (2020)

35. Paszke, A., et al.: An imperative style, high-performance deep learning library. In: Wallach, H., Larochelle, H., Beygelzimer, A., d' Alché-Buc, F., Fox, E., Garnett, R. (eds.) Advances in Neural Information Processing Systems 32, pp. 8024–8035. Curran Associates, Inc. (2019). http://papers.neurips.cc/paper/9015-pytorch-an-imperative-style-high-performance-deep-learning-library.pdf
36. Semenov, A., Zaikin, O., Otpuschennikov, I., Kochemazov, S., Ignatiev, A.: On cryptographic attacks using backdoors for SAT. In: AAAI Conference on Artificial Intelligence (2018)
37. Tsochantaridis, I., Joachims, T., Hofmann, T., Altun, Y., Singer, Y.: Large margin methods for structured and interdependent output variables. J. Mach. Learn. Res. **6**(9) (2005)
38. Vaswani, A., et al.: Attention is all you need. Advances in neural information processing systems, vol. 30 (2017)
39. Veličković, P., Cucurull, G., Casanova, A., Romero, A., Liò, P., Bengio, Y.: Graph attention networks. In: International Conference on Learning Representations (ICLR) (2018). https://openreview.net/forum?id=rJXMpikCZ, accepted as poster
40. Williams, R., Gomes, C.P., Selman, B.: Backdoors to typical case complexity. In: International Joint Conference on Artificial Intelligence (IJCAI), vol. 3, pp. 1173–1178 (2003)

Leveraging Integer Linear Programming to Learn Optimal Fair Rule Lists

Ulrich Aïvodji[1], Julien Ferry[2(✉)], Sébastien Gambs[3], Marie-José Huguet[2], and Mohamed Siala[2]

[1] École de Technologie Supérieure, Montréal, Canada
[2] LAAS-CNRS, Université de Toulouse, CNRS, INSA, Toulouse, France
jferry@laas.fr
[3] Université du Québec à Montréal, Montréal, Canada

Abstract. Fairness and interpretability are fundamental requirements for the development of responsible machine learning. However, learning optimal interpretable models under fairness constraints has been identified as a major challenge. In this paper, we investigate and improve on a state-of-the-art exact learning algorithm, called CORELS, which learns rule lists that are certifiably optimal in terms of accuracy and sparsity. Statistical fairness metrics have been integrated incrementally into CORELS in the literature. This paper demonstrates the limitations of such an approach for exploring the search space efficiently before proposing an Integer Linear Programming method, leveraging accuracy, sparsity and fairness jointly for better pruning. Our thorough experiments show clear benefits of our approach regarding the exploration of the search space.

Keywords: Fairness · Interpretability · Rule lists · Machine learning

1 Introduction

The combination of the availability of large datasets as well as algorithmic and computational progress has led to a significant increase in the performance of machine learning models. Despite their usefulness for numerous applications, the use of such models also raises several issues when their outcome impacts individuals' lives (*e.g.*, credit scoring or scholarships granting). Fairness and interpretability are key properties for the development of trustworthy machine learning and have become legal requirements defined in legislative texts [14].

The interpretability of a machine learning model is defined in [10] as "the ability to explain or to present in understandable terms to a human". This definition is quite general, and its precise instantiation depends on the task at hand, the context considered and the target of the explanation. Several methods have been proposed to explain machine learning models' predictions, which can be categorized into two main families. On one side, *black-box explanations* [15]

Julien Ferry—First author.

P. Schaus (Ed.): CPAIOR 2022, LNCS 13292, pp. 103–119, 2022.
https://doi.org/10.1007/978-3-031-08011-1_9

can be useful in non-sensitive contexts to provide *a posteriori* explanations of a black-box, but can be manipulated [24]. On the other side, *transparent-box design* [21] aims at building inherently interpretable models (*e.g.*, rule-based or tree-based models of reasonable size) [13,21].

Fairness is a central requirement for high-stake decision systems. Indeed, learning algorithms try to extract useful correlations from the training data but real-world datasets may include negative biases that should not be captured (*e.g.*, historical discrimination). Several fairness notions have been proposed to address this issue [6,7,26]. Among them, statistical fairness metrics ensure that a given statistical measure has similar values between groups as determined by the value of a sensitive feature. They are widely used as they can implement legal requirements and are easily quantifiable. Several approaches to fair learning have emerged in the literature, categorized into three main families. *Preprocessing* techniques [20] directly modify the training data to remove undesirable correlations so that any classifier trained on this data does not learn such correlations. *Postprocessing* approaches [16] modify the outputs of a previously trained classifier to meet some fairness criteria. Finally, *algorithmic modification* techniques [27] directly incorporate the fairness requirements into the learning algorithm and output a model satisfying a given fairness definition. In this paper, we focus on statistical fairness metrics using algorithmic modification approaches, which usually offer the best trade-offs between accuracy and fairness [6].

While many heuristic approaches for learning have been proposed, exact approaches offer a considerable advantage as a lack of optimality can have societal implications [4]. For instance, CORELS [3,4] produces rule lists that are certifiably optimal in terms of accuracy and sparsity. It relies on a branch-and-bound algorithm leveraging several dedicated bounds to prune the search space efficiently. FairCORELS [1,2] is a bi-objective extension of CORELS handling both statistical fairness and accuracy. FairCORELS consists in an ϵ-constraint method that leverages CORELS' original search tree and bounds for the accuracy objective and considers the fairness objective as a constraint. However, handling such constraints modifies the set of acceptable solutions, which makes the exploration considerably harder. Indeed, learning optimal interpretable machine learning models under constraints (*e.g.*, fairness constraints) has been identified as one of the main technical challenges towards interpretable machine learning [25].

In this paper, we address this issue and propose a method that harnesses the fairness constraints to efficiently prune the search space and optionally guide exploration. More precisely, we argue that CORELS' original bounds are not sufficient to efficiently explore the search space in this bi-objective setup. To address this, we design Integer Linear Programming (ILP) models combining both accuracy and fairness requirements for well-known statistical fairness metrics. These models are incorporated into FairCORELS through effective pruning mechanisms and can also be used to guide the exploration towards fair and accurate rule lists. Our large experimental study using three datasets with various fairness measures and requirements demonstrates clear benefits of the proposed approaches in terms of search exploration, memory consumption and learning quality.

The outline of the paper is as follows. First, we provide the relevant background and notations in Sect. 2. Then in Sect. 3, after describing the fair learning algorithm used, we discuss the theoretical claims motivating the necessity of efficient pruning. Afterwards, in Sect. 4, we propose pruning approaches based on ILP models, before evaluating empirically their efficiency and quality in Sect. 5 through a large experimental study. Finally, we conclude in Sect. 6.

2 Technical Background and Notations

In this section, we introduce the necessary background as well as the different notations used throughout the paper.

2.1 Rule Lists and Associated Notations

In supervised machine learning, the purpose of a classification problem is to learn a classifier function that maps as accurately as possible an input space to an output space. We use $\mathcal{F} = \{f_1, \ldots, f_G\}$ to denote a set of G binary features, all of them take their value in $\{0,1\}$. The training data, denoted by $\mathcal{E} = \{e_1, \ldots, e_M\}$, is a set of M examples. The examples in $\mathcal{E}$ are partitioned into $\mathcal{E}^+$ and $\mathcal{E}^-$, which correspond respectively to positive examples and negative ones. Precisely, an example $e_j \in \mathcal{E}$ is represented as a 2-tuple (x_j, y_j), in which $x_j \in \{0,1\}^G$ denotes the value vector for all binary features associated with the example and $y_j \in \{0,1\}$ is the label indicating its class. We have $e_j \in \mathcal{E}^+$ if $y_j = 1$ and $e_j \in \mathcal{E}^-$ if $y_j = 0$.

We consider classifiers that are expressed as *rule lists* [23], which are formed by an ordered list of *if-then* rules, followed by a default prediction. More precisely, a *rule list* is a tuple $d = (\delta_d, q_0)$ in which $\delta_d = (r_1, r_2, \ldots, r_k)$ is d's *prefix*, and $q_0 \in \{0,1\}$ is a *default prediction*. A prefix is an ordered list of k distinct association rules $r_i = a_i \to q_i$. Each rule r_i is composed of an *antecedent* a_i and a *consequent* $q_i \in \{0,1\}$. Each antecedent a_i is a Boolean assertion over $\mathcal{F}$ evaluating either to true or false for each possible input $x \in \{0,1\}^G$. If a_i evaluates to true for example e_j, we say that rule r_i *captures* e_j. Similarly, if at least one of the rules in δ_d captures e_j, we say that prefix δ_d captures example e_j. Rule list 1.1 predicts whether a given individual has a [low] or [high] salary. Its prefix is composed of five rules, and its default decision is [low].

Rule list 1.1. found by `FairCORELS` on the Adult Income dataset.

```
if    [occupation:Blue-Collar]    then  [low]
else if  [occupation:Service]     then  [low]
else if  [capital gain: > 0]      then  [high]
else if  [not(workclass:Government)]  then  [low]
else if  [education:Masters/Doctorate]  then  [high]
else  [low]
```

Using a rule list $d = (\delta_d, q_0)$ to classify an example e is straightforward as rules in δ_d are applied sequentially. If e is not captured by prefix δ_d, then the default prediction q_0 is returned. Finally, remark that rule list $((), q_0)$ is well defined, and simply consists of a default prediction (hence representing a constant classifier).

Table 1. Summary of four statistical fairness metrics widely used in the literature.

Metric	Statistical Measure	Mathematical Formulation
Statistical Parity (SP)	Probability of Positive Prediction	$\left\| \dfrac{TP^c_{\mathcal{E},p} + FP^c_{\mathcal{E},p}}{|\mathcal{E}^p|} - \dfrac{TP^c_{\mathcal{E},u} + FP^c_{\mathcal{E},u}}{|\mathcal{E}^u|} \right\| \le \epsilon$
Predictive Equality (PE)	False Positive Rate	$\left\| \dfrac{FP^c_{\mathcal{E},p}}{|\mathcal{E}^p \cap \mathcal{E}^-|} - \dfrac{FP^c_{\mathcal{E},u}}{|\mathcal{E}^u \cap \mathcal{E}^-|} \right\| \le \epsilon$
Equal Opportunity (EOpp)	False Negative Rate	$\left\| \dfrac{FN^c_{\mathcal{E},p}}{|\mathcal{E}^p \cap \mathcal{E}^+|} - \dfrac{FN^c_{\mathcal{E},u}}{|\mathcal{E}^u \cap \mathcal{E}^+|} \right\| \le \epsilon$
Equalized Odds (EO)	PE and EOpp	Conjunction of PE and EOpp

2.2 Statistical Fairness

The rationale of statistical fairness notions is to ensure that a given statistical measure has similar values between several protected groups, defined by the value(s) of some sensitive feature(s) of $\mathcal{F}$. The underlying principle is that such sensitive features (*e.g.*, race, gender, ...) should not influence predictions. While the exact formulation of such metrics would enforce equality for the given measure over the protected groups, a common relaxation consists of bounding the difference. Depending on the particular value being equalized across groups, several metrics have been proposed in the literature. In this paper, we consider the four most commonly used metrics: Statistical Parity [12] (SP), Predictive Equality [9] (PE), Equal Opportunity [16] (EOpp) and Equalized Odds [16] (EO).

Let $\mathcal{E}$ denote a training set and c a classifier. Throughout the paper, we assume that $\mathcal{E}$ is partitioned into two groups: a protected group $\mathcal{E}^p$ and an unprotected group $\mathcal{E}^u$ (this partition depends on the value of the sensitive feature(s)). Let also $\epsilon \in [0, 1]$ denote the unfairness tolerance (*i.e.*, the maximum acceptable value for the unfairness measure). Thus, the fairness requirement gets harder as ϵ gets smaller. For a classifier c, among a group $\mathcal{E}^h$, with $h \in \{p, u\}$, we denote by $TP^c_{\mathcal{E},h}$ the number of true positives, $TN^c_{\mathcal{E},h}$ the number of true negatives, $FP^c_{\mathcal{E},h}$ the number of false positives and $FN^c_{\mathcal{E},h}$ the number of false negatives. Table 1 gives the definition of the four metrics considered.

3 CORELS and FairCORELS

CORELS [4] is a state-of-the-art supervised learning algorithm that outputs a certifiably optimal rule list minimizing the following objective function on a given training dataset $\mathcal{E}$:

$$\mathsf{obj}(d, \mathcal{E}) = \mathsf{misc}(d, \mathcal{E}) + \lambda \cdot K_d, \tag{1}$$

in which $\mathsf{misc}(d, \mathcal{E}) \in [0, 1]$ denotes the training classification error of the rule list d, K_d is the length of d (*i.e.*, number of association rules in d) and λ is a regularization hyper-parameter for sparsity. CORELS is a branch-and-bound algorithm, representing the search space of rule lists $\mathcal{R}$ as a prefix tree. Each node is a prefix in this tree, and each child node is an extension of its parent, obtained by adding exactly one rule at the end of the parent's prefix. Finally, the root node corresponds to the empty prefix. Each node is a possible solution (*i.e.*, rule list), obtained by adding a default decision (based on majority prediction) to the prefix associated with this node. While this search space corresponds to an exhaustive enumeration of the candidate solutions, CORELS leverages several bounds to prune it efficiently. Thanks to these bounds, along with several smart data structures, CORELS is able to find optimal solutions with a reasonable amount of time and memory. The set of antecedents A is pre-mined and given as input to the algorithm. While CORELS is agnostic to the rule mining procedure used as preprocessing, an overview of existing techniques can be found in [8].

FairCORELS [1,2] is a bi-objective extension of CORELS jointly addressing accuracy and statistical fairness, integrating several metrics from the literature. Formally, given a statistical fairness notion, whose violation by a rule list d on dataset $\mathcal{E}$ is quantified by an unfairness function $\mathsf{unf}(d,\mathcal{E})$ and a maximum acceptable violation ϵ, FairCORELS solves the following optimization problem:

$$\underset{d\in\mathcal{R}}{\arg\min}\quad \mathsf{obj}(d,\mathcal{E}) \tag{2}$$
$$\text{such that } \mathsf{unf}(d,\mathcal{E}) \leq \epsilon$$

FairCORELS is presented in Algorithm 1. In this algorithm, d^c denotes the current best solution and z^c is its objective value. Moreover, a priority queue Q of prefixes is used to store its exploration frontier. The priority queue ordering defines the exploration heuristic. The function $\mathsf{b}(\delta,\mathcal{E})$ (coming from the CORELS algorithm) gives an objective lower bound for any rule list built upon prefix δ on the dataset $\mathcal{E}$. At each iteration of the main loop, a prefix δ is removed from the priority queue (Line 4). When the lower bound of δ is less than the current best objective value (Line 5), two operations are considered. First, the rule list d formed by prefix δ along with a default prediction is accepted as a new best solution if it improves the current best objective value while respecting the unfairness tolerance (Line 9). Second, extensions of δ using the antecedents not involved in δ's rules are added to the queue (Line 12).

The constrained optimization formulation of the fair learning problem used in FairCORELS allows for the construction of different trade-offs between accuracy and fairness using a simple ϵ-constraint method [22]. However, the fairness constraints modify the set of acceptable solutions and the resulting search space is considerably harder to work with. Indeed, CORELS' original bounds are less efficient as the fairness constraint gets stronger. In addition, some data structures used by CORELS to speed up the exploration are no longer usable. For instance, a prefix permutation map that reduces considerably the running time and the memory consumption [3,4] does not apply anymore. This symmetry-aware map ensures that only the best permutation of each set of rules containing the same antecedents is kept. However, it cannot be used within FairCORELS without sacrificing optimality. Indeed, a given permutation may allow for better objective function values than others but may not lead to solutions meeting the fairness

Algorithm 1. FairCORELS

Input: Training data $\mathcal{E}$ with set of pre-mined antecedents A; unfairness tolerance ϵ; initial best known rule list d^0 such that $\mathrm{unf}(d^0, \mathcal{E}) \leq \epsilon$

Output: (d^*, z^*) in which d^* is a rule list with the minimum objective function value z^* such that $\mathrm{unf}(d^*, \mathcal{E}) \leq \epsilon$

1: $(d^c, z^c) \leftarrow (d^0, \mathrm{obj}(d^0, \mathcal{E}))$
2: $Q \leftarrow queue(())$ ▷ Initially the queue contains the empty prefix ()
3: **while** Q not empty **do**   ▷ Stop when the queue is empty
4:     $\delta \leftarrow Q.pop()$
5:     **if** $\mathrm{b}(\delta, \mathcal{E}) < z^c$ **then**
6:         $d \leftarrow (\delta, q_0)$ ▷ Set default prediction q_0 to minimize training error
7:         $z \leftarrow \mathrm{obj}(d, \mathcal{E})$
8:         **if** $z < z^c$ **and** $\mathrm{unf}(d, \mathcal{E}) \leq \epsilon$ **then**
9:             $(d^c, z^c) \leftarrow (d, z)$ ▷ Update best rule list and objective
10:            **for** a in $A \setminus \{a_i \mid \exists r_i \in \delta, r_i = a_i \rightarrow q_i\}$ **do** ▷ Antecedent a not involved in δ
11:                $r \leftarrow (a \rightarrow q)$ ▷ Set a's consequent q to minimize training error
12:                $Q.push(\delta \cup r)$ ▷ Enqueue extension of δ with r
13: $(d^*, z^*) \leftarrow (d^c, z^c)$

---

requirement. In this situation, one could miss solutions that exhibit lower objective function values and meet the fairness requirement. Since we are interested in preserving the guarantee of optimality, we cannot use such a data structure. However, we note that a weaker permutation map can be designed and used without losing the guarantee of optimality (we precisely do that later in Sect. 5.3). Overall, both observations motivate the need for a new pruning approach, leveraging both the objective function value and the fairness constraint to efficiently explore FairCORELS' search space.

4   The Proposed Pruning Approach

This section presents our proposition to prune the search space by reasoning about the number of well-classified examples and fairness. The main idea is to discard prefixes that cannot improve the current objective while satisfying the fairness requirement before being treated. To realize this, one has to guarantee that for any prefix discarded, none of its extensions can satisfy both requirements, which is the purpose of Sect. 4.1. Afterwards, Sect. 4.2 exploits this property in the presentation of our proposition.

4.1   A Sufficient Condition to Reject Prefixes

Let $\mathcal{E}$ be a training set and d be a rule list. We use $W_{\mathcal{E}}^d$ to denote the number of examples of dataset $\mathcal{E}$ well classified by d:

$$W_{\mathcal{E}}^d = TP_{\mathcal{E},p}^d + TP_{\mathcal{E},u}^d + TN_{\mathcal{E},p}^d + TN_{\mathcal{E},u}^d \tag{3}$$

$$= TP_{\mathcal{E},p}^d + TP_{\mathcal{E},u}^d + |\mathcal{E}^p \cap \mathcal{E}^-| - FP_{\mathcal{E},p}^d + |\mathcal{E}^u \cap \mathcal{E}^-| - FP_{\mathcal{E},u}^d \tag{4}$$

We slightly extend the notation introduced in Sect. 2. For a prefix δ, among a group $\mathcal{E}^h$ with $h \in \{p, u\}$, we denote by $TP_{\mathcal{E},h}^{\delta}$ (respectively $TN_{\mathcal{E},h}^{\delta}$, $FP_{\mathcal{E},h}^{\delta}$ and $FN_{\mathcal{E},h}^{\delta}$) the number of true positives (respectively true negatives, false positives and false negatives) among the examples of $\mathcal{E}$ captured by δ. Similarly, we define $W_{\mathcal{E}}^{\delta}$ as the number of examples well classified by δ, among the examples of $\mathcal{E}$ that δ captures. Clearly, $W_{\mathcal{E}}^{\delta} = TP_{\mathcal{E},p}^{\delta} + TP_{\mathcal{E},u}^{\delta} + TN_{\mathcal{E},p}^{\delta} + TN_{\mathcal{E},u}^{\delta}$.

We define $\sigma(\delta)$ to be the set of all rule lists whose prefixes start with δ: $\sigma(\delta) = \{(\delta_d, q_0) \mid \delta_d \text{ starts with } \delta\}$. Formally, we say that δ_d starts with δ (a prefix of length K) if and only if the K first rules of δ_d are precisely those of δ, appearing in the same order.

Consider $d = (\delta_d, q_0)$ such that $d \in \sigma(\delta)$. On the one hand, some examples of $\mathcal{E}$ cannot be captured by δ. On the other hand, all examples of $\mathcal{E}$ captured by δ are captured by δ_d and have the same prediction as with δ.

Proposition 1. *Given a prefix δ, a rule list $d \in \sigma(\delta)$ and $h \in \{p, u\}$, we have:*

$$TP_{\mathcal{E},h}^{\delta} \leq TP_{\mathcal{E},h}^{d} \leq |\mathcal{E}^h \cap \mathcal{E}^+| - FN_{\mathcal{E},h}^{\delta}$$
$$FP_{\mathcal{E},h}^{\delta} \leq FP_{\mathcal{E},h}^{d} \leq |\mathcal{E}^h \cap \mathcal{E}^-| - TN_{\mathcal{E},h}^{\delta}$$

Proof. The lower bounds are an immediate consequence of the fact that all examples captured by δ are captured by d's prefix and have the same predictions that in δ. Concerning the upper bounds, we show the proof for the first inequality as the second can be proven using a similar argument. Define T as the set of examples in $\mathcal{E}^h \cap \mathcal{E}^+$ that are not determined by δ. When constructing d from δ, the maximum possible augmentation of true positives within protected group h is to predict all the examples correctly in T. The size of the set containing true positives of δ and T is equal to $|\mathcal{E}^h \cap \mathcal{E}^+| - FN_{\mathcal{E},h}^{\delta}$. Hence the upper bound. $\square$

As a consequence of Proposition 1, $W_{\mathcal{E}}^{d} \geq W_{\mathcal{E}}^{\delta}$. We now define four integer decision variables that are used in our Integer Linear Programming (ILP) models. These variables are used to model the confusion matrix of any rule list whose prefix starts with δ as well as to define constraints modelling accuracy and fairness requirements over such matrix.

$$x^{TP_{\mathcal{E},p}} \in [TP_{\mathcal{E},p}^{\delta}, |\mathcal{E}^p \cap \mathcal{E}^+| - FN_{\mathcal{E},p}^{\delta}], \quad x^{TP_{\mathcal{E},u}} \in [TP_{\mathcal{E},u}^{\delta}, |\mathcal{E}^u \cap \mathcal{E}^+| - FN_{\mathcal{E},u}^{\delta}],$$
$$x^{FP_{\mathcal{E},p}} \in [FP_{\mathcal{E},p}^{\delta}, |\mathcal{E}^p \cap \mathcal{E}^-| - TN_{\mathcal{E},p}^{\delta}], \quad x^{FP_{\mathcal{E},u}} \in [FP_{\mathcal{E},u}^{\delta}, |\mathcal{E}^u \cap \mathcal{E}^-| - TN_{\mathcal{E},u}^{\delta}].$$

Consider the following constraint in which L and U are two integers such that $0 \leq L \leq U \leq |\mathcal{E}|$:

$$L \leq x^{TP_{\mathcal{E},p}} + x^{TP_{\mathcal{E},u}} + |\mathcal{E}^p \cap \mathcal{E}^-| - x^{FP_{\mathcal{E},p}} + |\mathcal{E}^u \cap \mathcal{E}^-| - x^{FP_{\mathcal{E},u}} \leq U. \quad (5)$$

We define $ILP(\delta, \mathcal{E}, L, U)$ to be the ILP model defined by the four variables $x^{TP_{\mathcal{E},p}}, x^{FP_{\mathcal{E},p}}, x^{TP_{\mathcal{E},u}}, x^{FP_{\mathcal{E},u}}$ and Constraint (5).

Proposition 2. *Given a prefix δ and $0 \leq L \leq U \leq |\mathcal{E}|$, if $ILP(\delta, \mathcal{E}, L, U)$ is unsatisfiable then we have:*

$$\nexists d \in \sigma(\delta) \mid L \leq W_{\mathcal{E}}^{d} \leq U$$

Proof. Assume that there exists some $d \in \sigma(\delta)$ such that $L \leq W_{\mathcal{E}}^d \leq U$. Then, $x^{TP_{\mathcal{E},p}} = TP_{\mathcal{E},p}^d$, $x^{TP_{\mathcal{E},u}} = TP_{\mathcal{E},u}^d$, $x^{FP_{\mathcal{E},p}} = FP_{\mathcal{E},p}^d$ and $x^{FP_{\mathcal{E},u}} = FP_{\mathcal{E},u}^d$ is a solution to $ILP(\delta, \mathcal{E}, L, U)$. Indeed, Constraint (5) is satisfied by hypothesis, and the bounds of the four variables are respected due to Proposition 1 and the fact that d is an extension of δ. Finally, if $\exists d \in \sigma(\delta) \mid L \leq W_{\mathcal{E}}^d \leq U$, then $ILP(\delta, \mathcal{E}, L, U)$ is satisfiable, which completes the proof by contrapositive. ☐

In the following paragraph, we show how the $ILP(\delta, \mathcal{E}, L, U)$ model can be extended to include the different considered statistical fairness metrics (defined in Table 1). For the sake of conciseness, we detail the procedure for the Statistical Parity metric and provide the key elements for the three other metrics. Note that propositions similar to Proposition 3 can be adapted and proved for the three other metrics, following the same reasoning.

Integrating Statistical Parity. We introduce a constant $C_1 = \epsilon \times |\mathcal{E}^p| \times |\mathcal{E}^u|$ and the following constraint:

$$- C_1 \leq |\mathcal{E}^u| \times (x^{TP_{\mathcal{E},p}} + x^{FP_{\mathcal{E},p}}) - |\mathcal{E}^p| \times (x^{TP_{\mathcal{E},u}} + x^{FP_{\mathcal{E},u}}) \leq C_1. \tag{6}$$

Let $ILP_{SP}(\delta, \mathcal{E}, L, U, \epsilon)$ be the Integer Linear Programming model defined by the four variables $x^{TP_{\mathcal{E},p}}, x^{FP_{\mathcal{E},p}}, x^{TP_{\mathcal{E},u}}, x^{FP_{\mathcal{E},u}}$ and Constraints (5) and (6).

Proposition 3. *Given a prefix δ, an unfairness tolerance $\epsilon \in [0, 1]$, and $0 \leq L \leq U \leq |\mathcal{E}|$, if $ILP_{SP}(\delta, \mathcal{E}, L, U, \epsilon)$ is unsatisfiable then we have:*

$$\nexists d \in \sigma(\delta) \mid L \leq W_{\mathcal{E}}^d \leq U \text{ and } \mathsf{unf}_{SP}(d, \mathcal{E}) \leq \epsilon$$

Proof. Assume that there exists some $d \in \sigma(\delta)$ such that $L \leq W_{\mathcal{E}}^d \leq U$ and $\mathsf{unf}_{SP}(d, \mathcal{E}) \leq \epsilon$. First, observe that Constraint (6) is equivalent to the mathematical formulation of the Statistical Parity condition defined in Table 1. Indeed, $\mathsf{unf}_{SP}(d, \mathcal{E}) \leq \epsilon$ if and only if $-C_1 \leq |\mathcal{E}^u| \times (TP_{\mathcal{E},p}^d + FP_{\mathcal{E},p}^d) - |\mathcal{E}^p| \times (TP_{\mathcal{E},u}^d + FP_{\mathcal{E},u}^d) \leq C_1$. Then, $x^{TP_{\mathcal{E},p}} = TP_{\mathcal{E},p}^d$, $x^{TP_{\mathcal{E},u}} = TP_{\mathcal{E},u}^d$, $x^{FP_{\mathcal{E},p}} = FP_{\mathcal{E},p}^d$ and $x^{FP_{\mathcal{E},u}} = FP_{\mathcal{E},u}^d$ is a solution to $ILP_{SP}(\delta, \mathcal{E}, L, U, \epsilon)$. Finally, if $\exists d \in \sigma(\delta) \mid L \leq W_{\mathcal{E}}^d \leq U$ and $\mathsf{unf}_{SP}(d, \mathcal{E}) \leq \epsilon$, then $ILP_{SP}(\delta, \mathcal{E}, L, U, \epsilon)$ is satisfiable, which completes the proof by contrapositive. ☐

Integrating Other Statistical Fairness Metrics. Consider a prefix δ, an unfairness tolerance $\epsilon \in [0, 1]$ and $0 \leq L \leq U \leq |\mathcal{E}|$. We define the following useful constants $C_2 = \epsilon \times |\mathcal{E}^u \cap \mathcal{E}^-| \times |\mathcal{E}^p \cap \mathcal{E}^-|$, and $C_3 = \epsilon \times |\mathcal{E}^p \cap \mathcal{E}^+| \times |\mathcal{E}^u \cap \mathcal{E}^+|$.

Predictive Equality. Consider the following constraint:

$$- C_2 \leq |\mathcal{E}^u \cap \mathcal{E}^-| \times x^{FP_{\mathcal{E},p}} - |\mathcal{E}^p \cap \mathcal{E}^-| \times x^{FP_{\mathcal{E},u}} \leq C_2. \tag{7}$$

Let $ILP_{PE}(\delta, \mathcal{E}, L, U, \epsilon)$ be the ILP model defined by the four variables $x^{TP_{\mathcal{E},p}}, x^{FP_{\mathcal{E},p}}, x^{TP_{\mathcal{E},u}}, x^{FP_{\mathcal{E},u}}$ and Constraints (5) and (7). If $ILP_{PE}(\delta, \mathcal{E}, L, U, \epsilon)$ is unsatisfiable, then: $\nexists d \in \sigma(\delta) \mid L \leq W_{\mathcal{E}}^d \leq U$ and $\mathsf{unf}_{PE}(d, \mathcal{E}) \leq \epsilon$.

Equal Opportunity. Consider the following constraint:

$$-C_3 \leq |\mathcal{E}^p \cap \mathcal{E}^+| \times x^{TP_{\mathcal{E},u}} - |\mathcal{E}^u \cap \mathcal{E}^+| \times x^{TP_{\mathcal{E},p}} \leq C_3. \tag{8}$$

Let $ILP_{EOpp}(\delta, \mathcal{E}, L, U, \epsilon)$ be the ILP model defined by the four variables $x^{TP_{\mathcal{E},p}}, x^{FP_{\mathcal{E},p}}, x^{TP_{\mathcal{E},u}}, x^{FP_{\mathcal{E},u}}$ and Constraints (5) and (8). If $ILP_{EOpp}(\delta, \mathcal{E}, L, U, \epsilon)$ is unsatisfiable, then: $\nexists d \in \sigma(\delta) \mid L \leq W_{\mathcal{E}}^d \leq U$ and $\mathsf{unf}_{EOpp}(d, \mathcal{E}) \leq \epsilon$.

Equalized Odds. Since the Equalized Odds metric is the conjunction of Equal Opportunity and Predictive Equality, we simply use the conjunction of Constraints (7) and (8) to integrate it.

Let $ILP_{EO}(\delta, \mathcal{E}, L, U, \epsilon)$ be the ILP model defined by the four variables $x^{TP_{\mathcal{E},p}}, x^{FP_{\mathcal{E},p}}, x^{TP_{\mathcal{E},u}}, x^{FP_{\mathcal{E},u}}$ and Constraints (5), (7) and (8). If $ILP_{EO}(\delta, \mathcal{E}, L, U, \epsilon)$ is unsatisfiable then: $\nexists d \in \sigma(\delta) \mid L \leq W_{\mathcal{E}}^d \leq U$ and $\mathsf{unf}_{EO}(d, \mathcal{E}) \leq \epsilon$.

4.2 Integration Within `FairCORELS`

We have proposed a sufficient condition to reject prefixes that do not respect a given fairness metric within a requirement of well-classified examples. One can use this property to reject prefixes before being they are treated in the main loop of `FairCORELS`. This pruning idea can be integrated using two approaches.

The first one called the *eager* approach, checks the sufficient condition before adding an extension of a prefix to the priority queue (before Line 12 with $\delta \cup r$ being the prefix given in the ILP). The second approach called the *lazy* approach, checks the sufficient condition when a prefix is removed from the priority queue and passed the branch and bound lower bound test at Line 5 with δ being the prefix tested. If the corresponding ILP (called with valid bounds) is unsatisfiable, then the prefix δ being tested can safely be discarded since no rule list whose prefix starts with δ can satisfy the conjunction of fairness and well-classified examples requirements. The difference between the two approaches can be seen as the trade-off between memory consumption and computational time. Indeed, given the same inputs and exploration strategies, the *eager* approach consumes less memory than the *lazy* approach as it prunes prefixes before adding them to the queue. However, it requires more calls to the ILP solver.

Finally, we also consider using the ILP models to guide exploration. To realize this, we add an objective to the previously defined ILP, maximizing $x^{TP_{\mathcal{E},p}} - x^{FP_{\mathcal{E},p}} + x^{TP_{\mathcal{E},u}} - x^{FP_{\mathcal{E},u}}$. The ILP is then called as in the *eager* approach, just before adding an extension of a prefix to the priority queue (before Line 12). Whenever it is unsatisfiable, the corresponding prefix is pruned. However, when it is satisfiable, we additionally get the best accuracy reachable (*e.g.*, a lower bound on the objective function value) while also meeting the fairness constraint and improving the objective function. We use this value to order the priority queue Q and define the *ILP-Guided* search heuristic. Intuitively, it guides the exploration towards the prefixes whose fairness may conflict least with accuracy (those with highest ILP objective function).

When building the ILP models, we use tight lower and upper bounds on the number of well-classified examples, whose computations are detailed hereafter.

Lower Bound Computation. Let $L(k, d, \mathcal{E}) = |\mathcal{E}| \cdot (1 - (\text{misc}(d, \mathcal{E}) + \lambda \cdot (K_d - k)))$.

Proposition 4. *Consider a rule list d_2. A rule list $d_1 = (\delta_{d_1}, q_0)$ has better objective value on $\mathcal{E}$ than d_2 if and only if $W_{\mathcal{E}}^{d_1} > L(|\delta_{d_1}|, d_2, \mathcal{E})$, in which $|\delta_{d_1}|$ is the length of d_1's prefix.*

Proof. $\text{obj}(d_1, \mathcal{E}) < \text{obj}(d_2, \mathcal{E}) \iff \text{misc}(d_1, \mathcal{E}) + \lambda \cdot K_{d_1} < \text{misc}(d_2, \mathcal{E}) + \lambda \cdot K_{d_2} \iff |\mathcal{E}| \cdot (1 - \text{misc}(d_1, \mathcal{E})) > |\mathcal{E}| \cdot (1 - (\text{misc}(d_2, \mathcal{E}) + \lambda \cdot (K_{d_2} - |\delta_{d_1}|))) \iff W_{\mathcal{E}}^{d_1} > L(|\delta_{d_1}|, d_2, \mathcal{E})$ □

Consider the prefix δ and the current best solution d^c of the main loop. Let $d = (\delta_d, q_0) \in \sigma(\delta)$. Using Proposition 4, we have d has a better objective value than d^c if and only if $W_{\mathcal{E}}^d > L(|\delta_d|, d^c, \mathcal{E}) \geq L(|\delta|, d^c, \mathcal{E})$ because $|\delta_d| \geq |\delta|$. Therefore $L(|\delta|, d^c, \mathcal{E})$ is a valid lower bound for the ILP, ensuring that rule list d improves over the current best objective value.

Upper Bound Computation. We leverage two observations to compute a tight value $U(\delta, \mathcal{E})$ such that $\forall d \in \sigma(\delta), W_{\mathcal{E}}^d \leq U(\delta, \mathcal{E})$. First, the examples captured and misclassified by δ will always be misclassified for any $d \in \sigma(\delta)$. Second, among the examples not captured by δ, some may conflict (*i.e.*, have the same features vector associated with different labels) and can never be simultaneously predicted correctly. This computation corresponds to the Equivalent Points Bound of `CORELS` (described in details in Sect. 3.14 of [4]).

5   Experimental Study

The purpose of this section is two-fold. First, after describing our experimental setup, we show the efficiency of the proposed pruning approaches using two biased datasets and the four considered fairness metrics of Table 1. Afterwards, we demonstrate the scalability of our method as well as its complementarity with a new prefix permutation map, using a larger real-world dataset.

5.1   Experimental Protocol

We implement and solve the ILP models in C++ using the `ILOG CPLEX 20.10` solver[1], with an efficient memoisation mechanism. Sensitive features are used for measuring and mitigating unfairness but are not used in the model's construction in order to prevent disparate treatment [27]. For each dataset, we generate 100 different training sets by randomly selecting 90% of the dataset's instances, with reported values being averaged over the 100 instances. Test values are measured

---

[1] Source code of this enhanced version of the `FairCORELS` Python package is available on https://github.com/ferryjul/fairCORELSV2. The use of the CPLEX solver is possible but not mandatory, as our released code also embeds an open-source solver (whose configuration has been tuned to handle our pruning problem efficiently). This solver is Mistral-2.0 [17,19], in its version used for the Minizinc Challenge 2020.

on the remaining 10% instances for each random split. All experiments are run on a computing grid over a set of homogeneous nodes using Intel Xeon E5-2683 v4 Broadwell @ 2.1GHz CPU.

We use three exploration heuristics: a best-first search *ILP-Guided*, a best-first search guided by `CORELS`'s objective and a Breadth-First-Search (BFS). The former inherently comes with an *eager* pruning. For the latter two, we compare the *original* `FairCORELS` (no ILP pruning), as well as *lazy* and *eager* integrations of our pruning approach. Then, we evaluate the seven exploration settings. However, results for the three best-first searches guided by `CORELS`'s objective are omitted because they consistently provided worst performances (considering all evaluated criteria) than the BFS with equivalent pruning integration. This can be explained by the fact that this approach guides exploration towards accurate solutions first, which conflicts with fairness in practice.

5.2 Evaluation of the Proposed ILP-Based Pruning Approaches

To empirically assess the effectiveness of our proposed pruning on `FairCORELS`, we perform experiments for the four metrics of Table 1 using two well-known classification tasks of the literature with several fairness requirements. The first task consists in predicting which individuals from the COMPAS dataset [5] will re-offend within two years. We consider race (African-American/Caucasian) as the sensitive feature. Features are binarized using one-hot encoding for categorical ones and quantiles (with 5 bins) for numerical ones. Rules are generated as single features without minimum support. The resulting preprocessed dataset contains 18 rules and 6150 examples.

The second task consists in predicting whether individuals from the German Credit dataset [11] have a good or bad credit score. We consider age (low/high) as the sensitive feature, with both groups separated by the median value. Features are binarized using one-hot encoding for categorical ones and quantiles (2 bins) for numerical ones. Rules are generated as single features with minimum support of 0.25 or conjunctions of two features with minimum support of 0.5. Gender-related features were excluded. The resulting preprocessed dataset contains 49 rules and 1000 examples. For experiments on the COMPAS (respectively German Credit) dataset, the maximum running time is set to 20 min (respectively 40 min). For each experiment, the maximum memory use is fixed to 4 Gb. Due to the limited space available, we detail our evaluation for the Statistical Parity metric. Results for all other metrics show similar trends.

Figure 1(a) displays the proportion of instances solved to optimality as a function of the fairness requirement (which gets harder as $1-\epsilon$ increases) to illustrate the joint action of `CORELS`' bounds and the proposed ILP-based pruning. For low fairness requirements, all evaluated methods reach optimality, thanks to the action of `CORELS`' bounds. However, these bounds are less effective for strong fairness requirements, and without the ILP pruning, optimality can hardly be reached. Conversely, the higher the value of $1-\epsilon$, the larger the pruning of the search space. Hence, optimality is reached most of the time when performing

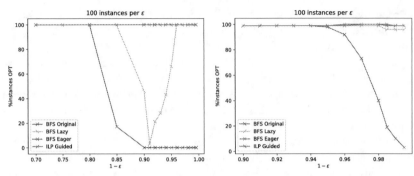

(a) Proportion of instances solved to optimality as a function of $1 - \epsilon$.

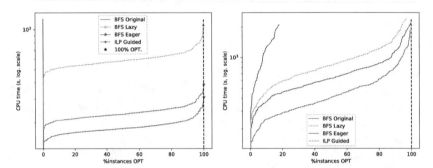

(b) CPU time as a function of the proportion of instances solved to optimality.

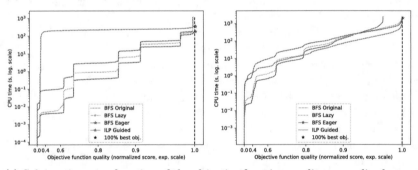

(c) Solving time as a function of the objective function quality normalized score.

Fig. 1. Experimental results (left: COMPAS, right: German Credit).

an eager pruning (*eager* BFS or *ILP-Guided*). This joint effect is particularly visible with the *lazy* BFS approach on the COMPAS dataset.

Figures 1(b) and 1(c) are generated using high fairness requirements (unfairness tolerances ranging between 0.005 and 0.02). Figure 1(b) presents the solving time as a function of the proportion of instances solved to optimality (lower is better). It shows a clear dominance of the proposed pruning approaches. For COMPAS, the *original* FairCORELS does not prove optimality to any of the

Table 2. Learning quality evaluation ($\epsilon \in [0.005, 0.05]$): Proportion of instances for which each method led to the best train (resp. test) accuracy, and average violation of the fairness constraint at test time.

Dataset	UNF	BFS Original			BFS Lazy			BFS Eager			ILP Guided		
		Train Acc	Test Acc	Unf viol.	Train Acc	Test Acc	Unf viol.	Train Acc	Test Acc	Unf viol.	Train Acc	Test Acc	Unf viol.
COMPAS dataset	SP	.951	.971	**.009**	1	.98	**.009**	1	**.981**	.009	1	.98	**.009**
	PE	.927	.956	**.033**	1	**.977**	.034	1	**.977**	.034	1	**.977**	.034
	EOpp	.941	.961	**.03**	1	.98	.031	1	**.983**	.031	1	**.983**	.031
	EO	.897	.934	**.035**	.997	.974	.036	1	**.976**	.036	1	.974	.036
German Credit dataset	SP	.567	**.799**	**.045**	.994	.77	**.045**	**.999**	.783	**.045**	.996	.779	**.045**
	PE	.967	.914	.138	1	.914	**.137**	1	.914	.138	.997	**.927**	.138
	EOpp	.683	.816	.056	.99	.799	.055	1	.806	.055	.991	**.829**	**.054**
	EO	.52	.759	**.158**	.979	.751	.161	.997	.741	.16	1	**.771**	.159

instances, whereas all pruning methodologies prove optimality to all instances. For German Credit, similar trends are observed. Overall, the *eager* approach appears more suitable to prove optimality, as it keeps the size of the queue as small as possible. For experiments with German Credit, the *ILP-Guided* approach effectively speeds up convergence and proof of optimality by guiding exploration towards fair and accurate solutions. This is not the case when using COMPAS, but the approach is still able to reach the best solutions, thanks to the performed pruning. Figure 1(c) shows the learning time as a function of the objective function quality (normalized objective score proposed in [18]). The proposed pruning allows finding better solutions within the time and memory limits after a slow start. Indeed, the pruning slows the beginning of the exploration, but pays off, given enough time, by effectively limiting the growth of the priority queue. The *lazy* approach is faster than the *eager* one at the beginning of the exploration. However, this trend is inverted given sufficient time. Again, the *ILP-Guided* approach speeds up convergence on German Credit, but worsens it on COMPAS.

Finally, the reported results illustrate the efficiency of the proposed pruning approaches to speed up the exploration of the prefix tree. The *lazy* approach less slows exploration at the beginning, but the *eager* approach gives better results given sufficient time. The *ILP-Guided* strategy showed an ability to speed up convergence, but its performances depend on the problem at hand.

Test results are reported in Table 2, and suggest that building optimal models does not result in worsening accuracy nor fairness generalization.

5.3 Scalability and Complementarity with the Permutation Map

As discussed in Sect. 3, a prefix permutation map speeds up the CORELS algorithm by leveraging symmetries but cannot be used within FairCORELS without compromising optimality. We modify it to enforce a weaker symmetry-breaking mechanism while maintaining the guarantee of optimality. More precisely, the

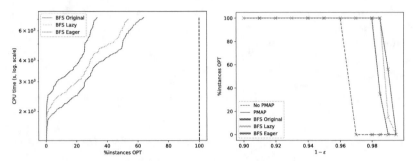

(a) Left: CPU time as a function of the proportion of instances solved to optimality. Right: proportion of instances solved to optimality as a function of $1 - \epsilon$.

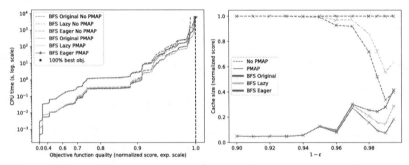

(b) Left: CPU time as a function of the objective function score. Right: relative cache size as a function of $1 - \epsilon$.

Fig. 2. Results of our experiments on the Adult Income dataset.

proposed new prefix permutation map (PMAP) considers that two prefixes of equal length are equivalent if and only if they have exactly the same confusion matrix and their rules imply the same antecedents. It pushes a new prefix to the priority queue Q (Line 12) only if Q contains no equivalent prefix.

To evaluate the scalability of our pruning approaches, we consider Adult Income [11], a larger dataset that gathers records of individuals from the 1994 U.S. census. We consider the task of predicting whether an individual earns more than $50,000\$$ per year, with gender (male/female) being the sensitive attribute. Categorical attributes are one-hot encoded and numerical ones are discretized using quantiles (3 bins). The resulting dataset contains $48,842$ examples and 47 rules (attributes or their negation), with a minimum support of 0.05. We consider only the Statistical Parity metric, as the three others do not conflict strongly with accuracy in this setting as observed in Fig. 1(a) of [1]. Experiments are performed with and without the new PMAP. The maximum running time is set to two hours, with a maximum memory use of 8 Gb. Results for the *ILP-Guided* approach are excluded as they show no clear improvement over the *eager* pruning, suggesting that the guidance was not beneficial overall.

Table 3. Learning quality evaluation (Adult Income dataset, $\epsilon \in [0.005, 0.1]$)

ϵ	Map Type	BFS Original			BFS Lazy			BFS Eager		
		Train Acc	Test Acc	Test Unf viol.	Train Acc	Test Acc	Test Unf viol.	Train Acc	Test Acc	Test Unf viol.
All	No PMAP	.938	.942	**-.004**	.963	.966	**-.004**	.964	.967	**-.004**
	PMAP	.966	.97	**-.004**	.998	.987	**-.004**	1	**.989**	**-.004**
< 0.02	No PMAP	.815	.835	**.0**	.89	.907	.001	.892	.91	.001
	PMAP	.897	.91	.001	.993	.96	.001	1	**.968**	.001

Results are summarized in Fig. 2. The left plot of Fig. 2(a) shows the proportion of instances solved to optimality, for $\epsilon \in [0.005, 0.02]$. For these strong fairness requirements, the approaches not using the new PMAP were never able to prove optimality (as can be seen in the right plot) and are not represented. The complementarity with our pruning approach is particularly visible, with the methods using both the PMAP and the ILP pruning having the best performances, both in terms of objective function quality (Fig. 2(b), left plot) and proof of optimality. This is also observed in terms of memory use in Fig. 2(b) (right plot). Indeed, the PMAP considerably reduces the size of the queue, leveraging the prefix tree symmetries. However, its effect is weakened for strong fairness constraints. The use of the ILP pruning mitigates this trend and for very strong fairness requirements, the *eager* pruning alone proposes lower memory consumption than the PMAP alone, to reach the same solutions. Finally, learning quality results are provided in Table 3 and confirm these observations. More precisely, they consistently show that the approaches improving train accuracy also improve test accuracy, without impacting fairness violation.

6 Conclusion

We propose effective ILP models leveraging accuracy and fairness jointly to prune the search space of `FairCORELS`. Our large experimental study shows clear benefits of our approach to speed-up the learning algorithm on well-known datasets from the literature. This gain is illustrated on three dimensions: achieving better training objective function values (without loss of the learning quality), using less memory footprint (*i.e.*, reduced cache size) and certifying optimality in limited amounts of time and memory. Combined with a proposed simple data structure, the ILP pruning approaches allow the learning of optimal rule lists under fairness constraints for datasets of realistic size.

Thanks to the declarative nature of our pruning approach, our framework is flexible and can simultaneously handle multiple fairness criteria for any number of sensitive groups. Indeed, each group's confusion matrix is modelled using two variables in our ILP. Considering more than two groups would require declaring additional variables, along with desired constraints using these variables.

Overall, our work illustrates the fact that statistical fairness and accuracy, when considered jointly, can be leveraged to reduce the scope of acceptable solutions efficiently. In the future, it would be interesting to pursue this line of work by considering other learning algorithms and machine learning requirements.

Guiding the exploration by leveraging on the ILP models (as attempted with the *ILP-Guided* approach) also seems to be a promising direction.

References

1. Aïvodji, U., Ferry, J., Gambs, S., Huguet, M.J., Siala, M.: Learning fair rule lists (2019). arXiv preprint http://arxiv.org/abs/1909.03977
2. Aïvodji, U., Ferry, J., Gambs, S., Huguet, M.J., Siala, M.: Faircorels, an open-source library for learning fair rule lists. In: Proceedings of the 30th ACM International Conference on Information & Knowledge Management, pp. 4665–4669. CIKM 2021, Association for Computing Machinery, New York (2021). https://doi.org/10.1145/3459637.3481965
3. Angelino, E., Larus-Stone, N., Alabi, D., Seltzer, M., Rudin, C.: Learning certifiably optimal rule lists. In: Proceedings of the 23rd ACM SIGKDD International Conference on Knowledge Discovery and Data Mining, pp. 35–44. KDD 2017, Association for Computing Machinery (2017). https://doi.org/10.1145/3097983.3098047
4. Angelino, E., Larus-Stone, N., Alabi, D., Seltzer, M., Rudin, C.: Learning certifiably optimal rule lists for categorical data. J. Mach. Learn. Res. 18(234), 1–78 (2018). http://jmlr.org/papers/v18/17-716.html
5. Angwin, J., Larson, J., Mattu, S., Kirchner, L.: Machine bias: there's software used across the country to predict future criminals and it's biased against blacks. propublica (2016). ProPublica, 23 May 2016
6. Barocas, S., Hardt, M., Narayanan, A.: Fairness and machine learning (2019). http://www.fairmlbook.org
7. Caton, S., Haas, C.: Fairness in machine learning: a survey (2020). arXiv preprint http://arxiv.org/abs/2010.04053
8. Chikalov, I., et al.: Logical analysis of data: theory, methodology and applications, pp. 147–192. Springer, Berlin Heidelberg (2013). https://doi.org/10.1007/978-3-642-28667-4_3
9. Chouldechova, A.: Fair prediction with disparate impact: a study of bias in recidivism prediction instruments. Big Data 5(2), 153–163 (2017). https://doi.org/10.1089/big.2016.0047
10. Doshi-Velez, F., Kim, B.: Towards a rigorous science of interpretable machine learning (2017). arXiv preprint http://arxiv.org/abs/1702.08608
11. Dua, D., Graff, C.: UCI machine learning repository (2017). http://archive.ics.uci.edu/ml
12. Dwork, C., Hardt, M., Pitassi, T., Reingold, O., Zemel, R.: Fairness through awareness. In: Proceedings of the 3rd Innovations in Theoretical Computer Science Conference, pp. 214–226. ITCS 2012, Association for Computing Machinery, New York (2012). https://doi.org/10.1145/2090236.2090255
13. Freitas, A.A.: Comprehensible classification models: a position paper. SIGKDD Explor. Newsl. 15(1), 1–10 (2014). https://doi.org/10.1145/2594473.2594475
14. Goodman, B., Flaxman, S.: European union regulations on algorithmic decision-making and a "right to explanation". AI Mag. 38(3), 50–57 (2017)

15. Guidotti, R., Monreale, A., Ruggieri, S., Turini, F., Giannotti, F., Pedreschi, D.: A survey of methods for explaining black box models. ACM Comput. Surv. **51**(5), 1–42 (2018). https://doi.org/10.1145/3236009
16. Hardt, M., Price, E., Price, E., Srebro, N.: Equality of opportunity in supervised learning. In: Advances in Neural Information Processing Systems, vol. 29. Curran Associates, Inc. (2016). https://proceedings.neurips.cc/paper/2016/file/9d2682367c3935defcb1f9e247a97c0d-Paper.pdf
17. Hebrard, E.: Mistral, a constraint satisfaction library. Proc. Third Int. CSP Solver Competition **3**(3), 31–39 (2008)
18. Hebrard, E., Siala, M.: Explanation-based weighted degree. In: Salvagnin, D., Lombardi, M. (eds.) CPAIOR 2017. LNCS, vol. 10335, pp. 167–175. Springer, Cham (2017). https://doi.org/10.1007/978-3-319-59776-8_13
19. Hebrard, E., Siala, M.: Solver engine (2017). https://www.cril.univ-artois.fr/CompetitionXCSP17/files/Mistral.pdf
20. Kamiran, F., Calders, T.: Data preprocessing techniques for classification without discrimination. Knowl. Inf. Syst. **33**(1), 1–33 (2012). https://doi.org/10.1007/s10115-011-0463-8
21. Lipton, Z.C.: The mythos of model interpretability: in machine learning, the concept of interpretability is both important and slippery. Queue **16**(3), 31–57 (2018). https://doi.org/10.1145/3236386.3241340
22. Miettinen, K.: Nonlinear Multiobjective Optimization, International Series in Operations Research & Management Science, vol. 12. Springer, Boston (2012). https://doi.org/10.1007/978-1-4615-5563-6
23. Rivest, R.L.: Learning decision lists. Mach. Learn. **2**(3), 229–246 (1987). https://doi.org/10.1007/BF00058680
24. Rudin, C.: Stop explaining black box machine learning models for high stakes decisions and use interpretable models instead. Nat. Mach. Intell. **1**(5), 206–215 (2019). https://doi.org/10.1038/s42256-019-0048-x
25. Rudin, C., Chen, C., Chen, Z., Huang, H., Semenova, L., Zhong, C.: Interpretable machine learning: fundamental principles and 10 grand challenges (2021). arXiv preprint http://arxiv.org/abs/2103.1125
26. Verma, S., Rubin, J.: Fairness definitions explained. In: Proceedings of the International Workshop on Software Fairness, pp. 1–7. FairWare 2018, Association for Computing Machinery, New York (2018). https://doi.org/10.1145/3194770.3194776
27. Zafar, M.B., Valera, I., Gomez Rodriguez, M., Gummadi, K.P.: Fairness beyond disparate treatment & disparate impact: learning classification without disparate mistreatment. In: Proceedings of the 26th International Conference on World Wide Web, pp. 1171–1180. WWW 2017, International World Wide Web Conferences Steering Committee, Republic and Canton of Geneva, CHE (2017). https://doi.org/10.1145/3038912.3052660

Solving the Extended Job Shop Scheduling Problem with AGVs – Classical and Quantum Approaches

Marc Geitz[1], Cristian Grozea[2]([⊠]) [iD], Wolfgang Steigerwald[1], Robin Stöhr[1], and Armin Wolf[2][iD]

[1] Telekom Innovation Laboratories, Berlin, Germany
{marc.geitz,wolfgang.steigerwald}@telekom.de
[2] Fraunhofer FOKUS, Berlin, Germany
{cristian.grozea,armin.wolf}@fokus.fraunhofer.de

Abstract. In this article we approach an extended Job Shop Scheduling Problem (JSSP). The goal is to create an optimized duty roster for a set of workpieces to be processed in a flexibly organized workshop, where the workpieces are transported by one or more Autonomous Ground Vehicles (AGV), that are included in the planning.

We are approaching this extended, more complex variant of JSSP (still NP-complete) using Constraint Programming (CP) and Quantum Annealing (QA) as competing methods.

We present and discuss: a) the results of our classical solution based on CP modeling and b) the results with modeling as quadratic unconstrained binary optimisation (QUBO) solved with hybrid quantum annealers from D-Wave, as well as with tabu search on current CPUs.

The insight we get from these experiments is that solving QUBO models might lead to solutions where some immediate improvement is achievable through straight-forward, polynomial time postprocessing. Furthermore, QUBO proves to be suitable as an approachable modelling alternative to the expert CP modelling, as it was possible to obtain for medium sized problems similar results, but requiring more computing power. While we show that our CP approach scales now better with increased problem size than the hybrid Quantum Annealing, the number of qubits available for direct QA is increasing as well and might eventually change the winning method.

Keywords: Constraint Programming · Job Shop Scheduling · Quadratic Unconstrained Boolean Optimization Problem · Quantum Annealing · Quantum Computing · Sequence-Dependent Setup-Times

The presented work was funded by the German Federal Ministry for Economic Affairs and Climate Action within the project "PlanQK" (BMWi, funding number 01MK20005) [7].

P. Schaus (Ed.): CPAIOR 2022, LNCS 13292, pp. 120–137, 2022.
https://doi.org/10.1007/978-3-031-08011-1_10

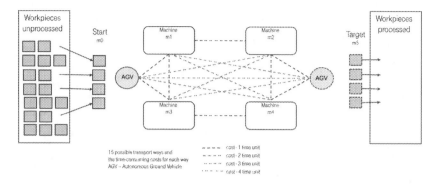

Fig. 1. Sample use case scenario: 4 machines, one AGV

1 Introduction

1.1 Problem Motivation and Description

The calculation of duty rosters is, since many years, a special subject in the departments of Operations Research. JSSP is NP-hard, which means that with each additional node in the manufacturing sequence the solution space expands exponentially. Based on a project about campus networks combined with edge computing at an OSRAM factory [17] we introduce and solve here an extension where one or more Autonomous Ground Vehicles (AGVs) transport the work pieces between the machines. The individualized transport of the work pieces on the shop floor within the paradigms of Industry 4.0 increases the complexity tremendously. For this reason we derive here a use case including several realistic aspects (specialized machine functions, heterogeneous processing times and number of processing steps).

2 Use Case Scenario

For the sample scenario Set1 (depicted in Fig. 1) we consider four machines with different tool settings where an Autonomous Ground Vehicle (AGV) is used to transport the workpieces on the shopfloor. The scenario can be up-scaled with more machines and several AGVs. We solve in the experiments described below problems with up to 10 work pieces, 2 AGVs and 100 tasks.

2.1 Definitions and Process Requirements

- Each machine has a specific functionality.
- Each workpiece must be processed in a predefined sequence of steps (i.e. processing tasks), named "job". Multiple processing tasks requiring the same machine inside a job are allowed.
- Each machine has unlimited storage for workpieces at input and output.

- The individual processing tasks and transportation tasks are non-preemptive.
- At most one workpiece can be processed on a machine at any time.
- For each processing task the processing sequence and the duration is given.
- Every processing task must be executed exactly once.
- Every processing task inside a job can only start after the previous one.
- The transportation of the workpieces is performed by one of the AGVs.
- The AGVs and all the workpieces are positioned at the beginning on the start point ("Start") and at the end on the target point ("Target"). There can be several start and target points.
- Each AGV can carry at most one workpiece on each transportation task between two nodes: from "Start" to a machine, from a machine to another machine, and from a machine to "Target".
- The required transit time between every two nodes must be incorporated during the planning and optimization of the duty roster.
- The transit times between every two nodes are known in advance.
- The total processing time should be minimized. The total processing time is the time consumed for all parts being manufactured during their processing tasks, according to the duty roster, including transport.

A sample scenario is presented schematically in Fig. 1.

3   The Extended Job Shop Scheduling Problem with AGVs – Definition

In this section the considered extended JSSP with AGVs problem is formally defined. In detail the fixed parameters of this constrained optimization problem, the decision variables to be determined, the constraints to be satisfied and the objective to be optimized are specified.

3.1   Parameters

There are *machines* $m = 0, \ldots, M$ where some machines represent *start locations* like machine $m = 0$ in Fig. 1 and some machines represent *target locations* like machine $m = M$ in Fig. 1. For any two machines m and m' the *transition duration* $d(m, m')$ from machine m to machine m' is given. For any machines m, m' and m'' it holds that

$$d(m, m) = 0, \quad d(m, m') \geq 0 \quad \text{and} \quad d(m, m'') \leq d(m, m') + d(m', m'') \quad (1)$$

These conditions are in general satisfied in real world scenarios, as the transition duration are directly proportional with according distances (for constant speed) and the distance is a metric.

There are *jobs* $j = 1, \ldots, J$, one for each work piece j. Each job j has $F_j + 1$ *processing tasks*: $t_{j,0}, t_{j,2} \ldots, t_{j,2 \cdot F_j}$. By convention we use *even indices* to identify them as *processing* tasks, because there are *transportation* tasks – with odd indices – between them (see below). Each processing task $t_{j,2 \cdot i}$ is processed on machine $m_{j,i}$ where for each job the first machine $m_{j,0}$ is its start location and the last machine m_{j,F_j} is its target location. For each processing task $t_{j,2i}$ its *processing time* on machine $m_{j,i}$ is $p_{j,i}$. In particular, for each job j the processing time at its start location and its target location is zero: $p_{j,0} = p_{j,F_j} = 0$.

Further, each job j has of F_j *transportation tasks*: $t_{j,1}, t_{j,3}, \ldots, t_{j,2\cdot(F_j-1)+1}$. By convention we use *odd indices* to identify them as *transportation* tasks. For each transportation task $t_{j,2\cdot i+1}$ its *processing time* is the *travelling time* $d(m_{j,i}, m_{j,i+1})$ from machine $m_{j,i}$ to machine $m_{j,i+1}$.

Further there are *AGVs* $v = M + 1, \ldots, M + V$ transporting work pieces between machines. For each AGV v there is a *(dummy) transportation task* $t_{v,1}$ performed by this AGV. Its zero processing time is the zero *travelling time* $d(m_v, m_v)$ at its start location, i.e. at machine m_v, where the AGV v is located initially. Its start time $s(t_{v,1}) = 0$ by definition.

3.2   Variables

For each processing or transportation task t its *start time* $s(t) \in \{0, \ldots, T\}$ has to be determined where T is the *scheduling horizon*[1] and further for each transportation task $t_{j,2\cdot i+1}$ the transporting AGV $v(j, i) \in \{M + 1, \ldots, M + V\}$ must be determined, too.

3.3   Constraints

For each job $j = 1, \ldots, J$ it must hold that the the corresponding tasks – either for work or transport – are in linear order:

$$s(t_{j,0}) + p_{j,0} \le s(t_{j,1}) \wedge s(t_{j,1}) + d(m_{j,0}, m_{j,1}) \le s(t_{j,2})$$
$$\wedge \ldots \wedge s(t_{j,2\cdot F_j-1}) + d(m_{j,F_j-1}, m_{j,F_j}) \le s(t_{j,2\cdot F_j}). \tag{2}$$

For each machine $m = 0, \ldots, M$ and at each time $\tau \in \{0, \ldots, T\}$ it must hold that at most one processing task is performed by this machine:

$$\sum_{\substack{j \in \{1,\ldots,J\} \wedge i \in \{0,\ldots,M\} \wedge \\ m_{j,i} = m \wedge s(t_{j,2\cdot i}) \le \tau < s(t_{j,2\cdot i}) + p_{j,i}}} 1 \le 1, \tag{3}$$

i.e. the processing tasks on each machine will be processed sequentially.

For each AGV $v = M + 1, \ldots, M + V$ and at each time $\tau \in \{0, \ldots, T\}$ it must hold that at most one transportation task is performed by this AGV:

$$\sum_{\substack{v \in \{M+1,\ldots,M+V\} \wedge j \in \{1,\ldots,J\} \wedge i \in \{0,\ldots,M\} \wedge \\ v(j,i) = v \wedge s(t_{j,2\cdot i+1}) \le \tau < s(t_{j,2\cdot i+1}) + d(m_{j,i}, m_{j,i+1})}} 1 \le 1, \tag{4}$$

---

[1] An upper bound of the scheduling horizon is the sum of all processing times and all transition durations.

i.e. the transportation tasks will be processed sequentially by each AGV. Further, there must be time for empty drives between the destinations and departures of successive transportation tasks:

For any two successive transportation tasks $t_{j,2\cdot i+1}$ and $t_{j',2\cdot i'+1}$ performed by the same AGV it must hold that there is enough time for an *empty drive* between the machines, i.e. if there is an AGV $v \in \{M+1,\ldots,M+V\}$ such that

$$v(j,i) = v = v(i',j') \wedge s(t_{j,2\cdot i+1}) + d(m_{j,i},m_{j,i+1}) \leq s(t_{j',2\cdot i'+1}), \qquad (5)$$

then it must hold that

$$s(t_{j,2\cdot i+1}) + d(m_{j,i},m_{j,i+1}) + d(m_{j,i+1},m_{j',i'}) \leq s(t_{j',2\cdot i'+1}). \qquad (6)$$

3.4 Objective

The objective is to determine the variables defined in Sect. 3.2 such that the constraints in Sect. 3.3 are satisfied and the *make-span*, i.e. the latest start time (and thus end time by definition) of the last tasks of all jobs at the target location and thus the overall completion time of all jobs

$$\max_{j=1,\ldots,J} s(t_{j,2M}) \quad \text{is minimized.} \qquad (7)$$

4 Related Work

Our extension to JSSP adds additional transportation tasks to be scheduled on *alternative exclusive resources* (namely AGVs) with *sequence-dependent setup / transition times*. All these scheduling problems are NP-hard [6].

In optimization problems, like extended JSSP with AGVs, we are trying to find the best solution from all feasible solutions. We can convert those problems into energy minimization problems and use the effects of quantum physics to find the solution of the lowest potential energy [9]. In quantum annealing we have, instead of classical bits, so-called *qubits*, which on top of being in a state of 0 or 1, those can also be in a superposition, which corresponds to both states at the same time.

Quantum Annealing (QA) assumes a system in full superposition being in its lowest energy state. The system is then slowly (adiabatically) conveyed to an energy function describing the optimization problem - the adiabatic theorem [30] assures that the system will remain on its lowest energy state during the transition. At the end of the annealing process, the system has turned from a fully superimposed quantum state to a classical state, where the qubits can easily be read out.

The basic JSSP has been solved using the technique of simulated annealing [8,35,36]. A quantum annealing solution has first been introduced in [25] where the JSSP has been modelled as cost or energy function to be minimized (QUBO, see below). A basic implementation for a D-Wave quantum annealer is presented in [10,18]. The modelling or implementation of a more complicated JSSP problem, especially with the introduction of AGVs or robots to transport goods between machines on the shopfloor has not been done previously, as far as we could find.

A *quadratic unconstrained binary optimization problem* (QUBO) is a minimization problem of this form:

$$\min_{x\in\{0,1\}^N} \sum_i Q_{i,i}x_i + \sum_{i<j} Q_{i,j}x_i x_j \tag{8}$$

where $x = (x_i)_{i=1...N}$ are binary variables and the real-valued matrix Q of size $N \times N$ fully defines the QUBO instance. The matrix Q is by convention upper diagonal. Thanks to the similarity of the QUBO formula to the Ising Model [11], QUBO matrices can be transformed trivially into Ising Hamiltonian matrices and thus solutions for the QUBO problems can be produced by quantum annealers like those of D-Wave. The constraints of the problems to be modelled as QUBOs are integrated into the objective function as additional terms that achieve their minimum only when the respective constraints are satisfied.

A quantum computer will compute the minimum of the QUBO function, hence solve the optimization problem with all constraints fulfilled.

Constraint Programming (CP) deals with discrete decision and optimization problems combining pruning algorithms known from Operations Research (OR) reducing the search space and heuristic tree search algorithms, known from Artificial Intelligence (AI) in order to explore the search space. The advantage of CP is that each decision during search will be used to prune branches in the search tree. Using specialized OR pruning algorithms, CP is an appropriate approach for solving resource scheduling problems [3] like extended JSSP with AGVs.

Compared to local search approaches, e.g., *Simulated Annealing* or *Quantum Annealing*, CP offers several advantages:

- The solutions found (schedules) always satisfy the specified constraints, which is not always the case in local search, where the violation of the constraints is only minimized.
- Whereas both approaches can find good or even optimal solutions (depending on the error surface, in the case of the local search), in the case of CP, the optimality of the best solutions can be proven with complete search.

In CP there exist highly efficient methods for reducing the finite value ranges of the decision variables (start times, durations, end times, possible resources etc.) of activities on exclusive resources. Some of those are: *overload checking, forbidden regions, not-first/not-last detection, detectable precedences* and *edge finding* [4,27,31]. Some recent version of those pruning methods and related approaches support *(sequence-dependent) setup/transition times* [1,2,14,24,26, 28] and pruning for *optional activities* [5,29,32]. Optional activities (that can be optionally scheduled, but must not necessarily) are the basis for certain modelling approaches for scheduling activities on *alternatively exclusive resources* [34]. Such activities are in our case the transport tasks, that can be scheduled on alternative available AGVs. However these activities change the locations of the AGV which has to be respected: Transition times occur between two consecutive transport tasks when the destination of the first transport task differs from the origin of the second one. In those cases the used AGV has to travel (empty) from one location to another.

Optimization in CP is generally done by *Branch-and-Bound* (B&B), i.e. through an iterative search for ever better solutions. The optimization follows either a *monotonous* or a *dichotomous* bounding strategy combined with a mostly problem specific one, generally full depth search [16]. This combined strategy (B&B with full depth search) guarantees that the best solutions are found and their optimality is proven. Furthermore, dichotomous bounding allows to quantify the quality of already found, intermediate solutions.

Recent advances in CP follow a hybrid approach that combines search space pruning with the satisfiability problem solving (SAT Solving) for classical propositional logic [23]. There, clauses are generated during constraint processing (propagation) and further processed by SAT Solving [13]. Promising results show that this hybrid approach is ideally suitable for solving scheduling problems in a classical way [19–22].

5   Methods

5.1   Solving the Extended JSSP with AGVs Using Constraint Programming

We use the Constraint Programming library firstCS [33] to model and solve the extended JSSP with AGVs. We handle the linear order of the tasks in the jobs (cf. Eq. (2)) either by order constraints between consecutive tasks stating that the start time plus the duration, i.e. the end time of the predecessor task is less than or equal to the start time of the successor task or more globally while using an implementation of *global difference constraints* for the same order conditions [12]. The firstCS library offers objects and algorithms adopted from [27,29,31] to handle task scheduling constraints on *unary*, *single* resp. *exclusive* resources. We use these objects and algorithms to schedule the processing tasks on the machines satisfying the constraints defined in Eq. (3) and also to schedule the transportation tasks sequentially on one AGV (cf. Eq. (4)). We handle sequence-dependent transition times (cf. Eq. (5) and (6)) either on the modeling level (see below) or alternatively, we handle sequential scheduling of transportation tasks with sequence-dependent transition times on AGVs while using a rather new approach presented in [24]. Even though this approach supports *optional activities* and thus can be generalized to schedule tasks alternatively on several resources, we currently use this approach only to schedule transportation tasks on one AGV. For extended JSSP with several AGVs we adapted the approach presented in [34] to deal with such *alternative exclusive resources* (cf. Eq. (4)).

On the modeling level we handle sequence-dependent transition times on AGVs alternatively to the approach presented in [24]. In this case, we assume that an AGV must perform the transportation tasks in $\mathcal{T} = \{t_1, \ldots, t_n\}$ between some locations in $\mathcal{L} = \{L_0, \ldots, L_m\}$, i.e. of the machines of the extended JSSP with AGVs. If the destination and the departure locations between two consecutive transportation tasks are different, then some additional transitions (empty drives) have to be performed. Further, let $d(P, Q)$ be the *travel time* between any two locations $P, Q \in \mathcal{L}$ (of machines). Then, it is assumed that the Conditions (1) are satisfied, i.e.

$$d(L, L) = 0, d(P, Q) \geq 0 \text{ and } d(P, Q) \leq d(P, R) + d(R, Q) \text{for } L, P, Q, R \in \mathcal{L}. \tag{9}$$

Each transportation task $t_i \in \mathcal{T}$ is annotated by its *departure location* $P_i \in \mathcal{L}$ (of a machine) where the AGV picks up a workpiece and its *destination location* $Q_i \in \mathcal{L}$ (of a machine) where the AGV releases a workpiece, i.e. let $t_i = t_i(P_i, Q_i)$. Further, let $s_i = s_i(P_i)$ be *the (variable) start time* of the transportation task $t_i(P_i, Q_i)$ at location P_i and $e_i = e_i(Q_i)$ be *the (variable) end time* of the transportation task $t_i(P_i, Q_i)$ at location Q_i. Further let $d_i = d(P_i, Q_i)$ be the *duration* of task $t_i(P_i, Q_i)$ determined by the travel time $d(P_i, Q_i)$ between both locations (between the machines) such that

$$s_i(P_i) + d(P_i, Q_i) = e_i(Q_i) \text{ holds for } i = 1, \ldots, n. \tag{10}$$

The transportation tasks in $\mathcal{T}$ of an AGV must be scheduled in *linear order*; i.e. $\mathcal{T} = \{t_1, \ldots, t_n\}$ must be ordered in such a way that

$$e_i(Q_i) + d(Q_i, P_{i+1}) \leq s_{i+1}(P_{i+1}) \text{ holds for } i = 1, \ldots, n - 1. \tag{11}$$

There, $d(Q_i, P_{i+1})$ is the *transition time* for the empty drive between the destination location Q_i of the i-th transportation task and the departure location P_{i+1} of the successive $(i + 1)$-th transportation task. In order to satisfy these transition times for empty drives on an AGV, i.e. an exclusively available resource we adapted the approach presented in [32] where additional exclusively available resources – one for each location – are introduced as well as additional transportation tasks: For each "real" transportation task $t_i(P_i, Q_i)$ and each location L we consider an additional "virtual" transportation tasks t_i^L to be scheduled on the "virtual" resource L having start and end times

$$s_i^L = s_i(P_i) + d(P_i, L) \text{ and } e_i^L = e_i(Q_i) + d(Q_i, L).$$

It matters that each "virtual" task is well-defined: the duration of each "virtual" task t_i^L namely $e_i^L - s_i^L = e_i(Q_i) + d(Q_i, L) - s_i(P_i) - d(P_i, L) = s_i(P_i) + d(P_i, Q_i) + d(Q_i, L) - s_i(P_i) - d(P_i, L) = d(P_i, Q_i) + d(Q_i, L) - d(P_i, L)$ is always non-negative due to the triangle inequality (cf.Conditions (9)). The important point is that we proved for an AGV with more than two transportation tasks that

$$e_i(Q_i) + d(Q_i, P_j) \leq s_j(P_j) \iff e_i^L \leq s_j^L \tag{12}$$

holds for any two different transportation tasks $t_i, t_j \in \mathcal{T}, t_i \neq t_j$ and for each location $L \in \mathcal{L}$.[2] This means that scheduling the "virtual" tasks t_i^L in the same

---

[2] Details on the proof of this equivalence can be found in [32].

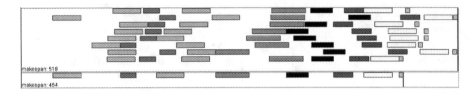

Fig. 2. The interrelationship between "real" transportation tasks (bottom line) with empty drives in between and their according "virtual" tasks (having the same color in the upper part with one line, i.e. exclusive resource, for each location) scheduled in the same linear order.

linear order on each "virtual" resource $L \in \mathcal{L}$ results in a schedule where the "real" transportation tasks are scheduled in the same order respecting the necessary transition times for empty drives between different locations as shown in Fig. 2. Consequently, we force such a scheduling with the same linear order of the "real" and the "virtual" tasks when using this approach.

For several alternative AGVs, *optional* tasks (a.k.a. *optional activities*) are used, such that pruning related to these tasks will be performed when a task is assigned to one of the alternative AGVs. The presented constraint models and the according filtering algorithms for pruning the search space are used together with a branch-and-bound tree-search based on *domain reduction* [15] for minimizing the make-span (cf. Eq. (7)).

5.2   Solving the Extended JSSP with AGVs on a Quantum Annealer

We model our problem as a *quadratic unconstrained binary optimization* problem (*QUBO*, cf. (8)).

Variable Definitions and Notations: let T be the planning horizon and I be the number of processing tasks. A safe, but usually overestimating choice for the planning horizon is the sum of the processing time of all tasks. We use here lower guesses that we increase until we reach one that produces admissible solutions (i.e. with all constraints satisfied). Starting with the variable structure used in [25], we index our variable array x by processing task index i and running time $0 \le t < T$: $x_{i,t} = 1$ if and only if processing task O_i starts at time t, where O_i denotes the processing task number i. The processing tasks are ordered in a lexicographical way, starting from the first processing task of the first job, going to the last processing task of the last job. Additionally we denote the processing time of the processing task i with p_i, and with k_n the index of the last processing task of the job j_n. Let $W = (w_{a,b})$ be the point-to-point transition duration matrix, where $w_{a,b}$ stores the time the AGV needs from machine a to machine b.

Transport Tasks: we model the start point and the end point as two additional machines and add to each job dummy processing tasks of null duration at the beginning and end. Therefore we can assume the workpieces are at start time at the (pseudo)machine where they are first needed. To include the AGV movements into the model, we introduce transport tasks in addition to the standard processing tasks. Between every two successive standard processing tasks of a job, we add B transport tasks, where B denotes the number of AGVs. The individual transport tasks are denoted with ξ. All B transport tasks between two standard processing tasks are combined in a set that we name "the transport task group". One of these transport tasks groups, denoted with Ψ, will have the structure $\Psi_x = (\xi_{x0}, \xi_{x1}, ...)$. From each transport tasks group, a single transport task must be picked, to transport the piece. Individual transport tasks are defined as a tuple $\xi_x = (b, w_{m_s m_e}, m_s, m_e)$ where b is the AGV executing this operation, m_s is the start machine the workpiece needs to be picked up from and m_e is the end machine the piece needs to be delivered to. Those tasks are appended at the end of the standard processing tasks in our indexing. Additionally we define Ω_i as the predecessor transport task of the standard processing task i. This contains the transport tasks of all AGVs, that come directly before i. For those indices i for which O_i is the first task of a job, Ω_i is an empty set, as that task is dummy and only has the role to ensure the transport following it starts in the start point. We define Φ_i as the standard processing task, previous to the transport task i. I_m is defined to be the set of all processing tasks executed on machine m. I_b is defined to be the set of all transport tasks executed by the AGV b. I_Ψ is the set of all transport tasks in the group Ψ.

We need to account for the following **constraints:**

h1 *Every processing task must run exactly once.* With a planning horizon T, our variables for the processing task O_i would consist of $x_{i,0}, x_{i,1}, \ldots, x_{i,T-1}$. Here we need to enforce, that exactly one of these variables is 1, and the rest are 0. That means we want to increase the penalty, when none, or more than one of them is 1. We can do that by iterating over the set of all tasks I, and for each $i \in I$ we subtract 1 from the sum off all $t \in T$, and square the result. This way the penalty term for this constraint is 0 if exactly one time slot is assigned for each task, and higher than that otherwise, which will enforce exactly one time slot being picked for each task: $h_1(\bar{x}) = \sum_{i=0}^{I-1} \left(\sum_{t=0}^{T-1} x_{i,t} - 1 \right)^2$.

h2 *Every machine can only work on one task at a time*:
$h_2(\bar{x}) = \sum_{m=0}^{M-1} \sum_{(i,t,i',t') \in R_m} x_{i,t} x_{i',t'}$, where R_m is the set that contains all indices of task i with its corresponding starting time t and a second task i' with its time t', both to be scheduled on the same machine, for which scheduling them as such leads to an overlap: $R_m = \{(i, t, i', t') : (i, i') \in I_m \times I_m, i \neq i', 0 \leq t, t' \leq T, 0 \leq t' - t < p_i\}$

h3 *The processing tasks of a job cannot overlap.* For this constraint, we need to iterate over all jobs n, and pick the tasks i and execution start times t and t', for which the sum of the processing time of i with its execution start time t is larger than the execution start time of the next processing task t', and therefore violates the constraint: $h_3(\bar{x}) = \sum_{n=0}^{N-1} \sum_{t,t';i,i+1 \in I_m;t+p_i>t'} x_{i,t} x_{i+1,t'}$.

The **objective function** is built by introducing a set of "upper bound" variables U_t such that exactly one of them is true. Together they model a virtual

M. Geitz et al.

finite domain variable V with the domain $0 \ldots T$, such that $V = t$ if and only if $U_t = 1$, i.e. $\sum_{t=0}^{T} U_t = 1$.

h4 *Exactly one of the variables U_t is set.* Enforcing this is done by adding the following quadratic constraint to the target function: $h_4(\bar{U}) = (\sum_{t=0}^{T} U_t - 1)^2$.

h5 *No planned task ends after the upper bound.* The index t where $U_t = 1$ is understood to be an upper bound of the whole schedule, therefore we impose that no i, t_1, t_2 exist such that $x_{i,t_1} = 1$, $U_{t_2} = 1$ and $t_1 + p_i > t_2$ (no planed task ends after the upper bound). Enforcing this condition can be done by adding this penalty term to the target function $h_5(\bar{x}, \bar{U}) = \sum_{i,t_1,t_2 ; t_1 + p_i > t_2} x_{i,t_1} U_{t_2}$.

With those, we can introduce the core of the objective function $o(\bar{U}) = \sum_{t=0}^{N} t U_t$. Its meaning is fairly straightforward: by minimizing $o(\bar{U})$ one gets a lowest upper bound that is also minimum amongst all consistent solutions (solutions satisfying all constraints) thus the solution that guarantees the shortest make-span.

h6 *For each transport between two processing tasks , exactly one transport task must be picked.* Here we also make sure that for all tasks belonging to the same transport tasks group and for all times, there is only one variable $x_{i,t}$, that will have the value 1, which means one and exactly one of the AGVs is selected to perform that transport: $h_6(\bar{x}) = \sum_{\substack{j=0 \\ \Omega_j \neq \emptyset}}^{I-1} \left(\sum_{i \in \Omega_j} \sum_{t=0}^{T-1} x_{i,t} - 1 \right)^2$.

h7 *AGVs must have enough time to switch to the next machine.* Here we iterate over all transport tasks of each AGV, denoted as I_b. For all possible tasks of these AGVs, we make sure that the starting time of the task $i' \in I_b$, is not before the task $i \in I_b$ is finished, plus the time the AGV needs to switch machines. In here, $me(i)$ defines the machine end of transport task i or the fourth element in the tuple. $ms(i')$ defines the machine start of the transport task i', or the third element in the tuple:

$$h_7(\bar{x}) = \sum_{b=0}^{B-1} \sum_{\substack{i,i' \in I_b \\ i \neq i'}} \sum_{\substack{t,t' \\ t' > t}} x_{i,t} x_{i',t'} max \left(t + p_i + w_{me(i)ms(i')} - t', 0 \right).$$

h8 *A processing task can only be scheduled at time t, if the corresponding transport task is finished before or at t.* Using our previously defined *Predecessor Transport Task*, we iterate through all processing tasks and their corresponding transport tasks, and apply a penalty if the transport task i' is finished after the time t of the processing task: $h_8(\bar{x}) = \sum_{i=0}^{I-1} \sum_{i' \in \Omega_i} \sum_{t,t'} x_{i,t} x_{i',t'} max \left(t' + p_{i'} - t, 0 \right)$.

h9 *A transport task can only be picked at time t, if the corresponding processing task is finished before or at t.* This will be the inverse of $h8$. Using the *Predecessor Processing Task*, we iterate through all transport tasks and their corresponding processing tasks and apply a penalty if the the previous processing task starting at time t' with its individual processing time, is not finished before or at t: $h_9(\bar{x}) = \sum_{j=0}^{I-1} \sum_{i \in \Omega_j} \sum_{t,t'} x_{i,t} x_{\Phi_i,t'} max \left(t' + p_{\Phi_i} - t, 0 \right)$.

Final QUBO. We add the constraints weighted with α (chosen to satisfy $\alpha > T$) to the core of the objective function, to get the final target function for the

Table 1. Empirical results based on the `firstCS` implementation (MS=Make-Span)

Id	problem size	GD	B&B	model-based trans. times			algorithm-based trans. times		
				MS	runtime	proof of opt.	MS	runtime	proof of opt.
Set1	4x4x24x1	−	M	65	3064 sec.	—	82	2156 sec.	—
		−	D	58	20313 sec.	20531 sec.	63	503 sec.	—
		+	M	58	151 sec.	164 sec.	58	112 sec.	131 sec.
		+	D	58	34 sec.	54 sec.	58	36 sec.	68 sec.
Set2	4x8x24x1	−	M	57	1636 sec.	1638 sec.	57	2509 sec.	2532 sec.
		−	D	57	167 sec.	167 sec.	57	723 sec.	723 sec.
		+	M	57	37 sec.	38 sec.	57	9 sec.	9 sec.
		+	D	57	7 sec.	7 sec.	57	2 sec.	2 sec.
Set3	4x4x24x1	−	M	90	3527 sec.	—	108	264 sec.	—
		−	D	58	620 sec.	748 sec.	62	879 sec.	—
		+	M	58	131 sec.	149 sec.	58	347 sec.	603 sec.
		+	D	58	57 sec.	75 sec.	58	335 sec.	710 sec.
Set4	4x4x21x1	−	M	57	3299 sec.	—	59	57 sec.	—
		−	D	58	1954 sec.	—	81	0.165 sec.	—
		+	M	48	38 sec.	41 sec.	48	36 sec.	57 sec.
		+	D	48	15 sec.	18 sec.	48	22 sec.	39 sec.
Set5	4x4x21x1	−	M	57	3304 sec.	—	59	55 sec.	—
		−	D	58	1847 sec.	—	81	0.092 sec.	—
		+	M	48	46 sec.	50 sec.	48	25 sec.	40 sec.
		+	D	48	17 sec.	21 sec.	48	20 sec.	35 sec.
Set6	4x4x24x2	−	M	39	192 sec.	1629 sec.		—	
		−	D	39	134 sec.	1599 sec.		—	
		+	M	39	162 sec.	1695 sec.		—	
		+	D	39	155 sec.	1802 sec.		—	

Id	problem size	GD	B&B	MS	runtime	Id	problem size	GD	B&B	MS	runtime
Set7	10x10x120x1	−	M	412	1263 sec.	Set8	10x10x120x2	−	M	224	238 sec.
		−	D	467	167 sec.			−	D	241	84 sec.
		+	M	412	2274 sec.			+	M	224	1049 sec.
		+	D	467	331 sec.			+	D	241	236 sec.

extended JSSP with AGVs: $H_T(\bar{x}, \bar{U}) = \alpha \sum_{i=1..9} h_i + o(\bar{U})$. This is now an unconstrained minimization problem in binary variables, as the constraints have been integrated into the objective function – a QUBO problem.

6 Results

6.1 Experimental Results Using the Constraint Programming Approach

We performed runtime experiments with CP-based solutions of the extended JSSP with AGVs (cf. Sec. 5.1). The experiments were performed on a Intel(R) Xeon(R) CPU E5-2695 v4 @ 2.10 GHz (in single core mode) running Ubuntu 20.04.2 and using OpenJDK 1.8.0_302. The results of the experiments are shown

Table 2. Empirical comparison of the results on various quantum, quantum inspired and classical platforms. The quantum approaches processed for each problem the same QUBO matrix. In parentheses the make-span before eliminating the gaps, when available. All times are time budgets, except for the CPU-based constraints based optimisation (CBO) with `firstCS`, where they are measured durations.

Problem	Method	Time	Make-span
Set1	QC D-Wave Leap Hybrid v2 (5000 qubits)	2 min.	58 (62)
	Quantum-inspired Fujitsu DAU (8192 sim qubits)	15 min.	59 (62)
	CPU Tabu Search (single core)	20 min.	58 (63)
	CPU-based CBO `firstCS` (single core)	34 sec.	58
Set2	QC D-Wave Leap Hybrid v2 (5000 qubits)	1 min.	57 (62)
	CPU Tabu Search (32 cores)	120 min.	57
	CPU-based CBO `firstCS` (single core)	2 sec.	57
Set3	QC D-Wave Leap Hybrid v2 (5000 qubits)	5 min.	59 (60)
	CPU Tabu Search (32 cores)	60 min.	58
	CPU-based CBO `firstCS` (single core)	57 sec.	58
Set4	QC D-Wave Leap Hybrid v2 (5000 qubits)	5 min.	53 (57)
	CPU Tabu Search (single core)	10 sec.	48 (59)
	CPU-based CBO `firstCS` (single core)	15 sec.	48
Set5	QC D-Wave Leap Hybrid v2 (5000 qubits)	5 min.	53 (58)
	CPU Tabu Search (32 cores)	60 min.	48
	CPU-based CBO `firstCS` (single core)	17 sec.	48
Set6	QC D-Wave Leap Hybrid v2 (5000 qubits)	60 min.	42 (58)
	Quantum-inspired Fujitsu DAU (8192 sim. qubits)	15 min	43 (53)
	CPU Tabu Search (32 cores)	120 min.	39
	CPU-based CBO `firstCS` (single core)	134 sec.	39
Set7 (95403 qubits)	QC D-Wave Leap Hybrid v2 (5000 qubits)	Failure:	QUBO too dense
	CPU-based CBO `firstCS` (single core)	21 min.	412

in Table 1. The first column contains the problem identifiers, the second column the problem sizes consisting of #machines x #jobs x #tasks x #AGVs.[3] Column 3 shows whether the *global difference constraints* (GD) are used (+) or not (−). Column 4 shows whether dichotomous (D) or monotonous (M) branch-and-bound (B&B) was performed for minimizing the make-spans shown in columns 5 and 8. Optimal make-spans are in bold. Make-spans in italic are the best ones found within one hour search. Columns 6 and 9 shows the runtimes to find the schedules with the presented make-spans. We made experiments based on the modelling approach (columns 5–7) presented in Sect. 5.1 to deal with transition

---

[3] The data is available at https://github.com/cgrozea/Data4ExtJSSAGV.

times between transportation tasks and based on the alternative algorithmic approach presented in [24] (columns 8–10). Please note, that the algorithmic approach is only applied to problems with one AGV, an extension for multiple AGVs is not yet available. Columns 7 and 10 shows the total runtimes for proving the optimality of found best solutions, i.e., including time for the exhaustive search for even better solutions.

The applied B&B optimization is based on different kinds of problem-specific, depth-first search: For the rather small, single AGV problem instances (Set1–Set5) the transport tasks on the AGV are sequentially ordered first, then the start times of all tasks are determined by successively splitting the domains of the start time variables in "earlier" and "later" subsets starting with the smallest domain. For problem instances with two AGVs (Set6 and Set8) the transport tasks are assigned to the AGVs beforehand. For the larger problem instances (Set7 and Set8) the ordering decisions on transport tasks on the AGVs during search is omitted. Not any solution was found using algorithmic approach for Set 7 within one hour, thus we presented for Set7 and Set8 only the make-spans and runtimes for the modelling approach. Further, B&B optimization starts with two trivial bounds of the make-span: the sum of all transport durations between the machines divided by the number of AGVs as an initial lower bound and the sum of all task durations plus the longest travel time between all machines as an initial upper bound.

6.2   Experimental Results Using the Quantum Approach

We have run several experiments, their results are summarized in Table 2. Solving each problem has been attempted on D-Wave Leap Hybrid v2 (which included running subproblems on the D-Wave quantum annealer) and with Tabu search (using the qbsolv library on CPU). For two of the problems we had temporarily access to the quantum-inspired Fujitsu digital annealer as well. A special note about the way we proceeded with the tabu search: we first attempted to solve the problem on a single core, when this did not produce good results we first increased the time limit, and when this failed to led to improvement we ran multiple instances in parallel and kept the best run – in each case we report the best result achieved. The problem Set7 leads to a QUBO matrix of 95403 qubits that is too dense for D-Wave Leap Hybrid v2 and requires too much RAM to attempt to solve it with qbsolv. The problem Set8 is even larger, thus it was not attempted with QUBO-based methods. We examined multiple solution plots with the different approaches. In the case of optimality, the solutions are very similar, mostly they are only differing in the start times/end time of the drives that allow such variations.

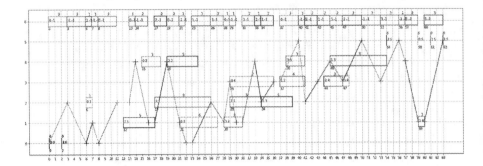

Fig. 3. Set1 - A tabu search solution with make-span 63 containing gaps, e.g. at time unit 11. It can be compacted to a solution wit make-span 58 (optimal) by simply eliminating the gaps. A color per workpiece, special machines 0=start, 5=target, 6=AGV.

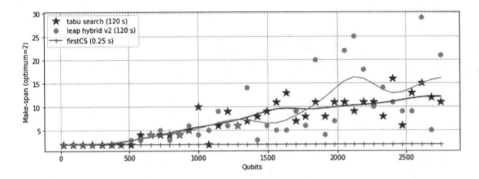

Fig. 4. Comparison of the solution quality for the same running time budget. The smooth versions were computed using the scipy function *gaussian_filter1d* with $\sigma = 2$.

7   Discussion and Conclusion

An interesting behavior of all optimizers applied on the QUBO matrices (thus not of the CP) is the presence of obviously unnecessary *gaps* in the produced solutions. We have found out that by simply eliminating those gaps by doing any action as soon as the prerequisites are fulfilled, one obtains clearly better solutions, sometimes even the optimal ones. An example is given in Fig. 3.

Quantum solvers are capable of finding similar results to those produced by the classical methods for small problem sizes. These solutions prove the QUBO models working but currently there is limited benefit for real life use cases. We used a simple problem to test the scalability: a single machine, a single AGV, two jobs, each with two tasks (durations: 0, 1), both to be executed on the single machine. As we increase the required number of qubits and qubit couplings by increasing the horizon, we observe a degrading of the quantum results versus the classical results (Figure 4). The winner in this comparison is the CP-based approach, which delivers constantly within 250 ms the optimum make-span value.

To conclude, in this work we were able to model an extended JSSP using CP and as QUBO that is automatically converted to Ising Hamiltonian for adiabatic quantum computers or simulators. The model has been extended to incorporate the transport of workpieces through multiple AGVs. The QUBO problems have been minimized on several solvers, including Tabu Search, and adiabatic quantum computers. As a contrast, the original problem (not converted to QUBO) has been solved also with CP-based optimisation. The problem sizes that we are currently able to address are below business relevance and can be executed on classical computers with similar performance. It is uncertain right now, when the quantum computers will bring advantages for practical decision or optimization applications. Once that happens, it is likely that quantum computers will replace classical computers, on specialized problems like optimization problems, that have often exponential time complexity.

References

1. Artigues, C., Belmokhtar, S., Feillet, D.: A New Exact Solution Algorithm for the Job Shop Problem with Sequence-Dependent Setup Times. In: Régin, J.C., Rueher, M. (eds.) Integration of AI and OR Techniques in Constraint Programming for Combinatorial Optimization Problems First International Conference, CPAIOR 2004, Nice, France, April 20–22, 2004. Proceedings. Lecture Notes in Computer Science, vol. 3011, pp. 37–49. Springer-Verlag (2004). https://doi.org/10.1007/978-3-540-24664-0_3, http://citeseerx.ist.psu.edu/viewdoc/download?doi=10.1.1.460.8724&rep=rep1&type=pdf
2. Artigues, C., Feillet, D.: A branch and bound method for the job-shop problem with sequence-dependent setup times. Ann. Oper. Res. **159**(1), 135–159 (2008). https://doi.org/10.1007/s10479-007-0283-0
3. Baptiste, P., le Pape, C., Nuijten, W.: Constraint-Based Scheduling: Applying Constraint Programming to Scheduling Problems. No. 39 in international series in operations research & management science. Kluwer Academic Publishers (2001)
4. Baptiste, P., Pape, C.L.: Edge-finding constraint propagation algorithms for disjunctive and cumulative scheduling. In: Proceedings 15th Workshop of the U.K. Planning Special Interest Group (1996)
5. Barták, R., Čepek, O.: Incremental propagation rules for a precedence graph with optional activities and time windows. Transactions of the Institute of Measurement and Control **32**(1), 73–96 (2010). https://doi.org/10.1177/0142331208100099
6. Brucker, P. (ed.): Scheduling Algorithms. Springer, Berlin, Heidelberg (2007). https://doi.org/10.1007/978-3-540-69516-5
7. Bundesministerium für Wirtschaft und Energie: Plattform und Ökosystem für Quantenunterstützte Künstliche Intelligenz. https://planqk.de
8. Chakraborty, S., Bhowmik, S.: Job shop scheduling using simulated annealing. In: Proceedings of the First International Conference on Computation and Communication Advancement. pp. 69–73. McGrawHill Publication, JIS College of Engineering, Kalyani, India, January 2013
9. D-Wave Systems Inc.: Introduction to quantum annealing. https://docs.dwavesys.com/docs/latest/c_gs_2.html. Accessed 4 May 2020
10. D-Wave Systems Inc.: Job shop scheduling. https://github.com/dwave-examples/job-shop-scheduling. Accessed 4 May 2020

11. D-Wave Systems Inc.: Problem formulations: Ising and qubo. https://docs. dwavesys.com/docs/latest/c_gs_3.html. Accessed 4 May 2020

12. Feydy, T., Schutt, A., Stuckey, P.J.: Global difference constraint propagation for finite domain solvers. In: Proceedings of the 10th International ACM SIGPLAN Symposium on Principles and Practice of Declarative Programming - PPDP '08. p. 226. ACM Press, Valencia, Spain (2008). DOI: https://doi.org/10.1145/1389449. 1389478

13. Feydy, T., Stuckey, P.J.: Lazy clause generation reengineered. In: Gent, I.P. (ed.) CP 2009. LNCS, vol. 5732, pp. 352–366. Springer, Heidelberg (2009). https://doi. org/10.1007/978-3-642-04244-7_29

14. Focacci, F., Laborie, P., Nuijten, W.: Solving scheduling problems with setup times and alternative resources. In: Proceedings of the AIPS-2000, p. 10. AAAI (2000)

15. Goltz, H.-J.: Reducing domains for search in CLP(FD) and its application to job-shop scheduling. In: Montanari, U., Rossi, F. (eds.) CP 1995. LNCS, vol. 976, pp. 549–562. Springer, Heidelberg (1995). https://doi.org/10.1007/3-540-60299-2_33

16. Hofstedt, P., Wolf, A.: Einführung in die Constraint-Programmierung: Grundlagen, Methoden, Sprachen, Anwendungen. eXamen. press, Springer-Verlag, Berlin Heidelberg (2007). https://doi.org/10.1007/978-3-540-68194-6

17. Lambrecht, J., Steffens, E.J., Geitz, M., Vick, A., Funk, E., Steigerwald, W.: Cognitive edge for factory: a case study on campus networks enabling smart intralogistics. In: 2019 24th IEEE International Conference on Emerging Technologies and Factory Automation (ETFA). pp. 1325–1328, September 2019. https://doi.org/10. 1109/ETFA.2019.8869394

18. Lobe, E.: Lösen von QUBO-Problemen auf einem Adiabatischen Quanten-Annealer. GOR Workshop. 19 May 2017

19. Schutt, A., Feydy, T., Stuckey, P.J.: Explaining time-table-edge-finding propagation for the cumulative resource constraint. In: Gomes, C., Sellmann, M. (eds.) CPAIOR 2013. LNCS, vol. 7874, pp. 234–250. Springer, Heidelberg (2013). https:// doi.org/10.1007/978-3-642-38171-3_16

20. Schutt, A., Feydy, T., Stuckey, P.J., Wallace, M.G.: Explaining the cumulative propagator. Constraints 16(3), 250–282 (2011). https://doi.org/10.1007/s10601-010-9103-2

21. Schutt, A., Feydy, T., Stuckey, P.J., Wallace, M.G.: Solving RCPSP/max by lazy clause generation. J. Sched. 16(3), 273–289 (2013). https://doi.org/10.1007/ s10951-012-0285-x

22. Schutt, A., Stuckey, P.J.: Explaining producer/consumer constraints. In: Rueher, M. (ed.) CP 2016. LNCS, vol. 9892, pp. 438–454. Springer, Cham (2016). https:// doi.org/10.1007/978-3-319-44953-1_28

23. Stuckey, P.J.: Lazy clause generation: combining the power of SAT and CP (and MIP?) solving. In: Lodi, A., Milano, M., Toth, P. (eds.) CPAIOR 2010. LNCS, vol. 6140, pp. 5–9. Springer, Heidelberg (2010). https://doi.org/10.1007/978-3-642-13520-0_3

24. Van Cauwelaert, S., Dejemeppe, C., Schaus, P.: An efficient filtering algorithm for the unary resource constraint with transition times and optional activities. J. Sched. 23(4), 431–449 (2020). https://doi.org/10.1007/s10951-019-00632-8

25. Venturelli, D., Marchand, D.J.J., Rojo, G.: Quantum annealing implementation of job-shop scheduling. October 2016. http://arxiv.org/abs/1506.08479, comment: p. 15, 6 figure, Presented at Constraint Satisfaction Techniques for Planning and Scheduling (COPLAS) Workshop of the 26th International Conference on Automated Planning and Scheduling (2016)

26. Vilím, P.: Batch processing with sequence dependent setup times: new results. In: Proceedings of the 4th Workshop of Constraint Programming for Decision and Control, CPDC'02, p. 6. Gliwice, Poland (2002)

27. Vilím, P.: $O(nlogn)$ filtering algorithms for unary resource constraint. In: Régin, J.-C., Rueher, M. (eds.) CPAIOR 2004. LNCS, vol. 3011, pp. 335–347. Springer, Heidelberg (2004). https://doi.org/10.1007/978-3-540-24664-0_23

28. Vilím, P.: Batch processing with sequence dependent setup times. In: Van Hentenryck, P. (ed.) CP 2002. LNCS, vol. 2470, pp. 764–764. Springer, Heidelberg (2002). https://doi.org/10.1007/3-540-46135-3_62

29. Vilím, P., Barták, R., Čepek, O.: Unary resource constraint with optional activities. In: Wallace, M. (ed.) CP 2004. LNCS, vol. 3258, pp. 62–76. Springer, Heidelberg (2004). https://doi.org/10.1007/978-3-540-30201-8_8

30. Wittek, P.: Quantum Machine Learning: What Quantum Computing Means to Data Mining. Elsevier, Amsterdam (2014)

31. Wolf, A.: Pruning while sweeping over task intervals. In: Rossi, F. (ed.) CP 2003. LNCS, vol. 2833, pp. 739–753. Springer, Heidelberg (2003). https://doi.org/10.1007/978-3-540-45193-8_50

32. Wolf, A.: Constraint-based task scheduling with sequence dependent setup times, time windows and breaks. In: Fischer, S., Maehle, E., Reischuk, R. (eds.). In: Proceedings of the Informatik 2009: Im Focus Das Leben, Beiträge Der 39. Jahrestagung Der Gesellschaft Für Informatik e.V. (GI), 28.9.-2.10.2009, Lübeck. Lecture Notes in Informatics (LNI) - Proceedings Series of the Gesellschaft Für Informatik (GI), vol. 154, pp. 3205–3219. Gesellschaft für Informatik e.V. (2009)

33. Wolf, A.: firstCS—New Aspects on Combining Constraint Programming with Object-Orientation in Java. KI - Künstliche Intelligenz **26**(1), 55–60 (2012). https://doi.org/10.1007/s13218-011-0161-4

34. Wolf, A., Schlenker, H.: Realising the alternative resources constraint. In: Seipel, D., Hanus, M., Geske, U., Bartenstein, O. (eds.) INAP/WLP -2004. LNCS (LNAI), vol. 3392, pp. 185–199. Springer, Heidelberg (2005). https://doi.org/10.1007/11415763_12

35. Yamada T., N.R.: Job-shop scheduling by simulated annealing combined with deterministic local search. In: Osman, I.H., Kelly, J.P. (eds.) Meta-Heuristics, pp. 237–248. Springer, Boston, MA (1996). https://doi.org/10.1007/978-1-4613-1361-8_15

36. Zhang, R.: A simulated annealing-based heuristic algorithm for job shop scheduling to minimize lateness. International Journal of Advanced Robotic Systems (2013). https://doi.org/10.5772/55956

Stochastic Decision Diagrams

J. N. Hooker[(✉)]

Carnegie Mellon University, Pittsburgh, USA
jh38@andrew.cmu.edu

Abstract. We introduce stochastic decision diagrams (SDDs) as a generalization of deterministic decision diagrams, which in recent years have been used to solve a variety of discrete optimization and constraint satisfaction problems. SDDs allow one to extend the relaxation techniques of deterministic diagrams to stochastic dynamic programming problems in which optimal controls are state-dependent. In particular, we develop sufficient conditions under which node merger operations applied during top-down compilation of the SDD yield a valid relaxed SDD whose size can be limited as desired. The relaxed SDD provides bounds on the optimal value that can be used to evaluate the quality of solutions obtained heuristically or to accelerate the search for an optimal solution. This results in a general and completely novel method for obtaining optimization bounds for stochastic dynamic programming, and the only method that can be applied to the original state space. We report computational experience on stochastic maximum clique (equivalently, maximum independent set) problem instances.

Keywords: stochastic decision diagrams · stochastic dynamic programming · relaxed decision diagrams · maximum clique problem

1   Introduction

Decision diagrams have proved to be a useful tool for solving a variety of discrete optimization and constraint programming problems [1,5,8,9,14,16–20,23,28,31]. One of their distinctive features is that they are naturally suited for dynamic programming (DP) formulations and provide an alternative to traditional state space enumeration as a means to solving them. They do so by creating a novel framework for branch-and-bound search [8] and a discrete relaxation method for calculating bounds [7]. Yet many, if not most, useful DP models are stochastic [29]. This suggests that it may be beneficial to extend the concept of a decision diagram to the stochastic case by introducing probabilities. Such is the goal of the present paper.

Stochastic decision diagrams (SDDs) can represent a very general class of discrete stochastic DP models with state-dependent transitions and controls. This is because the state transition graph of the DP model can be interpreted as an SDD. Despite this, the concept of an SDD is very different from that of a state

© Springer Nature Switzerland AG 2022
P. Schaus (Ed.): CPAIOR 2022, LNCS 13292, pp. 138–154, 2022.
https://doi.org/10.1007/978-3-031-08011-1_11

transition graph, partly because the nodes of the SDD need not correspond to states. In addition, a DP model can sometimes be radically simplified by reconceiving the transition graph as a decision diagram (without state information) and applying well-known reduction techniques to the diagram [24]. Perhaps more importantly, an SDD-based perspective opens the door to relaxation techniques that have been developed for decision diagrams, such as node merger and node splitting. Decision-diagram-based relaxation is attractive in that it does not presuppose linearity or convexity, and it allows one to obtain bounds of any desired quality by controlling the size of the relaxed diagram. It has already been widely applied to deterministic models, as recounted in [9].

We therefore focus on how to build relaxed SDDs, using node merger in particular. We develop sufficient conditions under which nodes can be merged during top-down compilation of an SDD so as to yield a valid relaxed SDD. The relaxed SDD can then provide bounds on the optimal value of the original problem. A complicating element of this analysis is that an optimal solution of a stochastic DP is not simply an assignment of values to variables, but a *policy* for selecting an optimal control in any state one happens to reach. Nonetheless, the analysis yields a general and completely novel method for bounding stochastic dynamic programming models. Such bounds are essential for judging the quality of a heuristic solution, and they can be very helpful for accelerating the search for an exact solution by excluding unpromising regions of the search space.

We assess the quality of SDD-based bounds experimentally using a stochastic version of the maximum clique problem, which is equivalent to the maximum independent set (maximum stable set) problem defined on the complementary graph. We select this problem because decision diagrams have already been evaluated as a bounding mechanism for the deterministic case, using the well-known DIMACS instance set [7]. It was found that decision diagrams supply tighter bounds, in less time, than the full cutting plane resources of a commercial integer programming solver. This suggests that bounds from an SDD may likewise be useful for the stochastic case, even while they cannot be directly compared with an integer programming model because no such practical model exists. As a test bed, we use a variety of random and DIMACS instances. We find that SDD-based bounds degrade rather modestly as one reduces the size of the relaxed SDD and consequently the time invested in building it, even when the time is reduced to a few seconds. These results suggest that SDDs can provide useful bounds of continuously adjustable quality.

2   Related Work

Decision diagrams were introduced as an optimization method in [21,22]. The idea of a relaxed diagram first appears in [1] as a technique for enhancing propagation in constraint programming, by means of both node splitting and node merger. Relaxed diagrams were first used to obtain optimization bounds in [7,10], and much subsequent work is described in [9]. Sufficient conditions under which node merger creates a relaxed diagram are presented in [25]. They are simpler than the conditions developed below for SDDs because there is no need to accommodate solutions in the form of policies. Relaxed decision diagrams are combined with Lagrangian methods in [6,14,26].

Dynamic programming, both deterministic and stochastic, is credited to Bellman [3,4]. Various DP techniques are described in [12,13]. Due to the astronomical size of state spaces often encountered in stochastic DP, practitioners generally resort to approximate DP, which estimates the cost-to-go, rather than attempting to establish valid bounds [12,29]. Connections between decision diagrams and deterministic DP, including nonserial DP [11], are discussed in [24].

When DP bounds are desired, they are traditionally obtained by state space relaxation, which approximates the original state space with a smaller, computationally feasible space [2,15,27,30]. It differs in fundamental ways from relaxation based on decision diagrams. A relaxed decision diagram uses the same variables and state space as the original problem, rather than mapping the problem into a smaller space. This allows the relaxed diagram to provide a branching framework for an exact branch-and-bound method [8]. Furthermore, the relaxation is constructed dynamically during compilation rather than specified *a priori* and is thereby potentially more sensitive to problem structure. It can be tightened by filtering techniques and Lagrangian methods. Finally, the relaxed diagram can be sized to provide a bound of any desired quality.

3   Stochastic Decision Diagrams

A decision diagram can be defined as a directed, acyclic multigraph in which the nodes are partitioned into *layers*. Each arc of the graph is directed from a node in layer i to a node in layer $i + 1$ for some $i \in \{1, \ldots, n\}$. We let L_i represent the set of nodes in layer i. Layers 1 and $n + 1$ contain a single node, namely the root r and the terminus t, respectively. Each layer i is associated with a variable x_i with finite domain $\mathcal{X}_i$. The arcs leaving any node in layer i have *labels* in $\mathcal{X}_i$, representing possible values of x_i (*controls*) at that node.

Conventional decision diagrams are *deterministic*, meaning that the control at a given node determines a transition to particular node on the next layer. A path from r to t defines a sequence of controls $x = (x_1, \ldots, x_n)$ as indicated by the arc labels on the path. A decision diagram is *weighted* if there is a length (cost) associated with each arc.

Any discrete optimization problem of the form $\min_{x \in X} \{\sum_{i=1}^{n} c_i(x_i)\}$ with finite-domain variables $x_1, \ldots, x_n$ and feasible set X can be represented by a weighted decision diagram.[1] The diagram is constructed so that its r–t paths correspond to feasible solutions $x \in \mathcal{X}$, and the length of any r–t path is the cost of the corresponding solution. The optimal value of the problem is the length of a shortest r–t path.

A *stochastic decision diagram* (SDD) associates probabilities as well as labels and costs with the arcs. An SDD is significantly different from a deterministic diagram in that a given control x_i at a node u does not, in general, determine

---

[1] Decision diagrams can also represent problems with nonseparable cost functions $c(x)$ as described in [24]. However, this generalization is not relevant here.

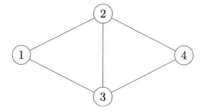

Fig. 1. Graph for a small maximum clique problem.

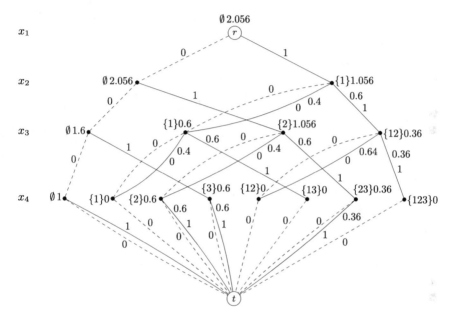

Fig. 2. SDD for a small stochastic maximum clique problem. Transition probabilities less than 1 are shown on the arcs. Zero-cost solid arcs from layer 4 to the terminus are not shown. States and maximum expected costs-to-go are indicated at nodes.

a transition to a particular node in the next layer. Rather, control x_i can lead to any of several possible *outcomes* ω, each with a probability $p_{i\omega}(u, x_i)$ and cost $c_{i\omega}(u, x_i)$. We suppose, without loss of generality, that the same set Ω_i of outcomes is available at each node in layer i, although many of the outcomes may have probability zero. Each outcome ω at node u with positive probability gives rise to an arc (u, u') that leads to a node $u' = \phi_{i\omega}(u, x_i)$.

Due to the probabilistic nature of transitions, a sequence of controls $\boldsymbol{x} = (x_1, \ldots, x_n)$ does not correspond to a single r–t path. Furthermore, the control exercised in layer i may depend on which node is reached in that layer. We therefore define a *policy* $\boldsymbol{\pi}$ rather than a vector $\boldsymbol{x}$ of controls for SDDs, where $\pi_i(u)$ is the specified control at node $u \in L_i$. The policy is *feasible* if $\pi_i(u) \in \mathcal{X}_i(u)$ for all $u \in L_i$ and all i, where $\mathcal{X}_i(u)$ is the set of permissible controls at node u.

As an example, consider a stochastic maximum clique problem associated with the graph of Fig. 1. Each edge of the graph appears with probability 0.6. The objective is to find a clique of maximum expected size, without knowing in advance which edges are present. The problem is represented by the SDD in Fig. 2. There are two controls x_i in each layer i, with $x_i = 1$ indicating a decision to include vertex i in the clique (solid arcs), and $x_i = 0$ a decision to exclude vertex i (dashed arcs). There are two solid arcs leaving most nodes, because there are two possible outcomes when $x_i = 1$: vertex i can be added to the clique ($\omega = 1$), or it cannot be added ($\omega = 0$). The former occurs when all the edges from vertex i to vertices currently in the clique exist. This outcome has probability 0.6^m, where m is the size of the clique, while outcome $\omega = 0$ has probability $1 - 0.6^m$. The resulting probabilities are indicated next to the arcs in Fig. 2 when they are less than 1. For readability, zero-cost solid arcs from layer 4 to the terminus (those corresponding to $\omega = 0$) are not shown in the figure.

Arc costs are also indicated in Fig. 2, where arcs with cost 1 correspond to outcome $\omega = 1$ and those with cost 0 to outcome $\omega = 0$. We seek a policy that maximizes expected cost; i.e., maximizes the expected size of the resulting clique. A policy π consists of a decision $\pi_i(u)$ at each node $u \in L_i$ as to whether vertex i should included in the clique. Thus one can consider the outcome of previous controls when deciding whether to include a given vertex.

The expected cost $c(D, \pi)$ of a policy π on a stochastic diagram D can be computed with the recursion

$$h_i(u, \pi) = \sum_{\omega \in \Omega_i} p_{i\omega}\big(u, \pi_i(u)\big) \big[c_{i\omega}\big(u, \pi_i(u)\big) + h_{i+1}\big(\phi_{i\omega}(u, \pi_i(u)), \pi\big)\big]$$

for $i = n, n - 1, \ldots, 1$, where $h_{n+1}(t, \pi) = 0$ and $c(D, \pi) = h_1(r, \pi)$. A policy π^* is optimal if it minimizes $c(D, \pi)$ over all feasible policies π. The minimum expected cost can be computed with the recursion

$$h_i(u) = \min_{x_i \in \mathcal{X}_i(u)} \left\{ \sum_{\omega \in \Omega_i} p_{i\omega}(u, x_i)\big[c_{i\omega}(u, x_i) + h_{i+1}\big(\phi_{i\omega}(u, x_i)\big)\big] \right\} \qquad (1)$$

for $i = n, n - 1, \ldots, 1$, where $h_{n+1}(t) = 0$. Here, $h_i(u)$ is the optimal expected *cost-to-go* at node u in layer i. The overall optimal cost is $c(D) = h_1(r)$. An optimal policy π^* is obtained by setting $\pi_i^*(u)$ to an optimizing value of x_i in (1) for each i and each $u \in L_i$.

Optimal expected costs-to-go for the example are shown at the nodes in Fig. 2 (along with the corresponding states, to be discussed in the next section). They are computed by maximizing rather than minimizing in (1), since we seek a maximum clique. The maximum expected cost for this problem instance is 2.056. This means that if an optimal choice is made at every node, the resulting clique will have size 2.056 on the average. Note that at the root node, either choice is optimal. At both nodes in layer 2, the optimal choice is to add vertex 2, and so forth.

4   Stochastic Dynamic Programming

Stochastic decision diagrams can represent a very general class of stochastic dynamic programming (DP) problems in which costs, probabilities and controls are state-dependent. Let $p_{i\omega}(S_i, x_i)$ be the probability of outcome $\omega \in \Omega_i$ in state S_i under control x_i, and $c_{i\omega}(S_i, x_i)$ the cost of that outcome.[2] Outcome ω effects a transition from state S_i to state $\phi_{i\omega}(S_i, x_i)$. A solution of the problem is a policy $\pi = (\pi_1, \ldots \pi_n)$ that maps each possible state S_i to a control $\pi_i(S_i)$. A policy π is feasible if $\pi_i(S_i)$ is an available control for every state S_i; that is, $\pi_i(S_i) \in \mathcal{X}_i(S_i)$. The expected cost of policy π is $h_1(S_1, \pi)$, where

$$h_i(S_i, \pi) = \sum_{\omega \in \Omega_i} p_{i\omega}(S_i, \pi_i(S_i)) \left[c_{i\omega}(S_i, \pi_i(S_i)) + h_{i+1}(\phi_{i\omega}(S_i, \pi_i(S_i)), \pi) \right] \quad (2)$$

for $i = n, n-1, \ldots, 1$. There is a single initial state S_1 and a single terminal state S_{n+1}, with $h_{n+1}(S_{n+1}, \pi) = 0$. We refer to $h_i(S_i, \pi)$ as the *expected cost-to-go* of state S_i under policy π. A policy π^* is optimal if it minimizes $h_1(S_1, \pi)$ over all feasible policies π. An optimal solution can be found with the recursion

$$h_i(S_i) = \min_{x_i \in \mathcal{X}_i(S_i)} \left\{ \sum_{\omega \in \Omega_i} p_{i\omega}(S_i, x_i) \left[c_{i\omega}(S_i, x_i) + h_{i+1}(\phi_{i\omega}(S_i, x_i)) \right] \right\} \quad (3)$$

with $h_{n+1}(S_{n+1}) = 0$ and $h_1(S_1)$ equal to the optimal expected cost. An optimal policy can be obtained by setting $\pi_i(S_i)$ to an optimizing value of x_i in (3) for each state S_i.

An SDD corresponding to model (3) contains a node in layer i for every state S_i that can be reached with positive probability under some policy. We let $S_i(u)$ refer to the state associated with node $u \in L_i$. For each node $u \in L_i$ and each control $x_i \in \mathcal{X}(S_i(u))$, there is an outgoing arc to a node associated with state $\phi_{i\omega}(S_i(u), x_i)$ for every outcome ω with positive probability in state S_i. Each of these arcs has label x_i, probability $p_{i\omega}(S_i, x_i)$, and cost $c_{i\omega}(S_i, x_i)$. Several arcs may connect the same two nodes, but they must have different labels. An optimal policy can be computed in the SDD using (1).

In the DP model (3) for the stochastic maximum clique problem, the min operator is max, and each state S_i is the set S_i of vertices currently in the clique. The state corresponding to each node of the SDD is shown in Fig. 2. The transitions and probabilities are as follows:

$$p_{i\omega}(S_i, x_i) = \begin{cases} p_i(S_i), & \text{if } (x_i, \omega) = (1,1) \\ 1 - p_i(S_i), & \text{if } (x_i, \omega) = (1,0) \\ \omega, & \text{if } x_i = 0 \end{cases} \quad (4)$$

$$\phi_{i\omega}(S_i, x_i) = \begin{cases} S_i \cup \{i\}, & \text{if } (x_i, \omega) = (1,1) \\ S_i, & \text{if } \omega = 0 \end{cases} \quad (5)$$

$$c_{i\omega}(S_i, x_i) = \omega \text{ for } x_i = 0, 1 \text{ and all } S_i \quad (6)$$

where $p_i(S_i) = \prod_{j \in S_i} p_{ij}$ and p_{ij} is the probability of edge (i, j).

---

[2] Following convention in the DP literature, the subscript i in S_i does not index the state but indicates that control x_i is applied in state S_i.

As a second example, we consider a stochastic job sequencing problem with time windows. Each job j has release time r_j and due date d_j. Its processing time is a random variable, and each possible processing time $t_{j\omega}$ corresponds to an outcome $\omega \in \Omega_i$. The set Ω_i can in principle be infinite, but for the purpose of building an SDD, we suppose it is finite. We also suppose that the outcome probabilities are state-independent, so that the probability of a processing time $t_{j\omega}$ is $p_{j\omega}$ regardless of which jobs have been scheduled so far. The objective is to minimize total tardiness, where the tardiness of a job j that starts at time s_j is $(s_j + t_{j\omega} - d_j)^+$, and where we use the notation α^+ for $\max\{0, \alpha\}$. The control x_i in stage i is which job will be ith in the sequence.

In the DP model (3) for the problem, each state $\boldsymbol{S}_i$ is a tuple (S_i, f_i), where S_i is the set of jobs scheduled so far, and f_i is the finish time of the last job scheduled. The recursion is defined by

$$p_{i\omega}\big((S_i, f_i), x_i\big) = p_{x_i\omega} \tag{7}$$

$$\phi_{i\omega}\big((S_i, f_i), x_i\big) = \big(S_i \cup \{x_i\},\ \max\{r_{x_i}, f_i\} + t_{x_i\omega}\big) \tag{8}$$

$$c_{i\omega}\big((S_i, f_i), x_i\big) = \big(\max\{r_{x_i}, f_i\} + t_{x_i\omega} - d_{x_i}\big)^+ \tag{9}$$

The nodes of the an SDD representing the problem correspond to the states (S_i, f_i). Each arc with label x_i leaving state (S_i, f_i) is generated by an outcome ω with positive probability $p_{i_i\omega}$ and has cost $c_{i\omega}((S_i, f_i), x_i)$.

5   Relaxed SDDs

In the deterministic case, a decision diagram $\bar{D}$ relaxes a diagram D when every r–t path of D occurs in $\bar{D}$ with equal or smaller cost. Since policies rather than paths represent solutions in the stochastic case, it is convenient to define a relaxed SDD in terms of its optimal expected cost, rather than the cost of individual paths. This, of course, ensures that a relaxed SDD provides a valid bound on the optimal value of the original SDD. Thus, if two diagrams D and $\bar{D}$ have the same number of layers, and the same possible controls and outcomes, then $\bar{D}$ *relaxes* D if the minimum expected cost $c(D)$ of D is at least the minimum expected cost $c(\bar{D})$ of $\bar{D}$.

Under suitable conditions, a relaxed SDD can be obtained by node merger during top-down compilation, as in the deterministic case. That is, when a given layer i is created from the previous layer, some of the nodes in layer i are merged, so as to reduce the size of the diagram. A larger number of mergers yield a smaller diagram but, in general, a weaker relaxation. When two nodes $u, u' \in L_i$ are merged, the state associated with the resulting node is obtained by applying a merger operation $\boldsymbol{S}_i(u) \oplus \boldsymbol{S}_i(u')$ to the states associated with the merged nodes. Layer $i + 1$ is then created on the basis of the resulting states in layer i.

In the maximum clique problem, we can use the merger operation $S_i \oplus S_i' = S_i \cap S_i'$. We will see in the next section that this yields a valid relaxed SDD. Figure 3 displays the relaxed SDD that results from merging two particular nodes in the maximum clique SDD of Fig. 2, namely the nodes in layer 3 with states

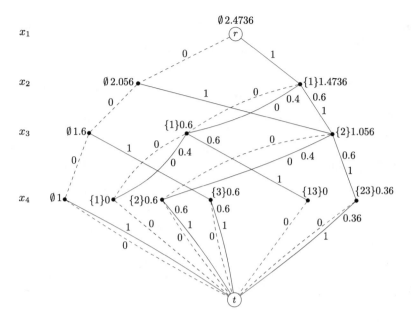

Fig. 3. Relaxed SDD that results from the merger of two nodes in the SDD of Fig 2.

$\{2\}$ and $\{1, 2\}$. The maximum expected cost of the relaxed SDD is 2.4736, which is an upper bound on the optimal expected cost of 2.056 of the original problem.

In the job sequencing problem, we can use the merger operation

$$(S_i, f_i) \oplus (S'_i, f'_i) = \left(S_i \cap S'_i, \min\{f_i, f'_i\}\right) \tag{10}$$

We likewise show in the next section that this creates a valid relaxed SDD.

6   Conditions for Node Merger

We now develop sufficient conditions under which a merger operation yields a relaxed SDD. We again suppose that a concept of relaxation is defined for states. We assume that relaxation has the property that state $\bar{S}_i$ relaxes state S_i only if

$$p_{i\omega}(S_i, x_i)c_{i\omega}(S_i, x_i) \geq p_{i\omega}(\bar{S}_i, x_i)c_{i\omega}(\bar{S}_i, x_i) \tag{11}$$

for any control $x_i \in \mathcal{X}_i$ and any outcome $\omega \in \Omega_i$, and for $i = 1, \ldots, n$. Let the expected immediate cost of a state S_i under control x_i be

$$c_i(S_i, x_i) = \sum_{\omega \in \Omega_i} p_{i\omega}(S_i, x_i)c_{i\omega}(S_i, x_i)$$

Thus, we have from (11)

Lemma 1. *Relaxing a state does not increase its expected immediate cost.*

We will show that node merger yields a valid relaxed SDD if the merger operation $\oplus$ and transition function ϕ satisfy the following conditions for $i = 1, \ldots, n$:

(C1) For any two states S_i and S'_i, state $S_i \oplus S'_i$ relaxes both S_i and S'_i.

(C2) If state $\bar{S}_i$ relaxes state S_i, then $\phi_{i\omega}(\bar{S}_i, x_i)$ relaxes $\phi_{i\omega}(S_i, x_i)$ for any $\omega \in \Omega_i$ and any $x_i \in \mathcal{X}_i$.

(C3) If state $\bar{S}_i$ relaxes state S_i, then given any control $x_i \in \mathcal{X}_i$ and any set of $\{\eta_\omega \mid \omega \in \Omega_i\}$ of numbers, there is a control $\bar{x}_i \in \mathcal{X}_i$ such that

$$\sum_{\omega \in \Omega_i} p_{i\omega}(S_i, x_i)\big(c_{i\omega}(S_i, x_i) + \eta_\omega\big) \geq \sum_{\omega \in \Omega_i} p_{i\omega}(\bar{S}_i, \bar{x}_i)\big(c_{i\omega}(\bar{S}_i, \bar{x}_i) + \eta_\omega\big) \quad (12)$$

The argument to follow is simplified if we suppose, without loss of generality, that a given diagram and its relaxation are *fully articulated*. This means that the diagrams contain nodes representing all possible states, even those that cannot be reached with positive probability. This device will allow us to merge nodes by changing only the arc probabilities, with no change in the structure of the diagram. Thus each layer L_{i+1} of a fully articulated diagram contains all states S_{i+1} such that $S_{i+1} = \phi_{i\omega}(S_i(u), x_i)$ for some node $u \in L_i$, outcome $\omega \in \Omega_i$, and control $x_i \in \mathcal{X}_i$, even if the probability $p_{i\omega}(S_i, x_i)$ of reaching S_{i+1} from u is zero for all ω, x_i.

When two nodes $u, u' \in L_{i+1}$ are merged to form node $\hat{u}$, we leave all nodes in place but change only the probabilities on arcs from nodes in layer L_i to u, u', and $\hat{u}$. That is, we transfer the probability on any arc (v, u) to an arc $(v, \hat{u})$ with the same label, and similarly for any arc (v, u'). Thus, if $u = \phi_{i\omega}(v, x_i)$ and $\hat{u} = \phi_{i\hat{\omega}}(v, x_i)$, we set $p_{i\hat{\omega}}(v, x_i) = p_{i\omega}(v, x_i)$ and then set $p_{i\omega}(v, x_i)$ to zero, and similarly for any arc (v, u'). The following lemma is key.

Lemma 2. *If conditions (C2) and (C3) are satisfied, and state $\bar{S}_k$ relaxes state S_k, then the optimal cost to go of S_k is at least the optimal cost to go of $\bar{S}_k$.*

Proof. It suffices to show by induction that

$$h_i(S_i) \geq h_i(\bar{S}_i) \quad (13)$$

for $i = n, n-1, \ldots, k$. Claim (13) is true for $i = n$ by virtue of Lemma 1, because $h_n(S_n, x_n) = c_n(S_n, x_n)$ for any control $x_n \in \mathcal{X}_n$. We now suppose (13) is true for $i+1$ and show it is true for i. It suffices to show that for any control $x_i \in \mathcal{X}_i$, there is a control $\bar{x}_i \in \mathcal{X}_i$ for which

$$\sum_{\omega \in \Omega_i} p_{i\omega}(S_i, x_i)\big[c_{ij}(S_i, x_i) + h_{i+1}(\phi_{i\omega}(S_i, x_i))\big]$$
$$\geq \sum_{\omega \in \Omega_i} p_{i\omega}(\bar{S}_i, \bar{x}_i)\big[c_{i\omega}(\bar{S}_i, \bar{x}_i) + h_{i+1}(\phi_{i\omega}(\bar{S}_i, \bar{x}_i))\big] \quad (14)$$

We have from condition (C2) that $\phi_{i\omega}(\bar{S}_i, \bar{x}_i)$ relaxes $\phi_{i\omega}(S_i, x_i)$ for all $\omega \in \Omega_i$. This and the induction hypothesis imply $h_{i+1}(\phi_{i\omega}(S_i, x_i)) \geq h_{i+1}(\phi_{i\omega}(\bar{S}_i, \bar{x}_i))$ for $\omega \in \Omega_i$. Thus to show (14) it suffices to show

$$\sum_{\omega \in \Omega_i} p_{i\omega}(\boldsymbol{S}_i, x_i)\big[c_{i\omega}(\boldsymbol{S}_i, x_i) + h_{i+1}(\phi_{i\omega}(\boldsymbol{S}_i, x_i))\big]$$

$$\geq \sum_{\omega \in \Omega_i} p_{i\omega}(\bar{\boldsymbol{S}}_i, \bar{x}_i)\big[c_{i\omega}(\bar{\boldsymbol{S}}_i, \bar{x}_i) + h_{i+1}(\phi_{i\omega}(\boldsymbol{S}_i, x_i))\big] \tag{15}$$

But (15) follows from condition (C3) by setting $\eta_\omega = h_{i+1}(\phi_{i\omega}(\boldsymbol{S}_i, x_i))$ for each $\omega \in \Omega_i$. $\qquad\square$

Theorem 1. *If conditions* (C1)–(C3) *are satisfied, the merger of nodes during compilation of diagram D results in a relaxation of D.*

Proof. Suppose without loss of generality that D is fully articulated, and let $\bar{D}$ be the fully articulated diagram that results from top-down compilation with node merger. Also let $\bar{D}_k$ be the compiled diagram that results when nodes are merged only in levels $1, \ldots, k$. It suffices to show the following inductively for $k = 1, \ldots, n$:

$$c(D) \geq c(\bar{D}_k) \tag{16}$$

This is trivially true for $k = 1$ because $D_1 = \bar{D}_1$. We therefore suppose (16) is true for k and show it is true for $k + 1$. Due to the induction hypothesis, it suffices to show

$$c(\bar{D}_k) \geq c(\bar{D}_{k+1}) \tag{17}$$

Let $\boldsymbol{\pi}^k$ be an optimal policy for $\bar{D}_k$, so that $c(\bar{D}_k) = h_1(\boldsymbol{S}_1, \boldsymbol{\pi}^k)$. We will suppose that only two nodes u, u' in layer $k+1$ of $\bar{D}_k$ are merged to obtain a node $\hat{u}$, since the argument can be repeated for additional mergers. The merger transfers the transition probabilities on arcs to nodes u and u' to the arcs to $\hat{u}$. In addition, by condition (C1), state $\boldsymbol{S}_{k+1}(\hat{u})$ is a relaxation of both $\boldsymbol{S}_{k+1}(u)$ and $\boldsymbol{S}_{k+1}(u')$. Thus, by Lemma 2, the optimal cost to go at $\hat{u}$ is no larger than the optimal cost to go at u and u'. That is,

$$\begin{aligned} h_{k+1}\big(\boldsymbol{S}_{k+1}(\hat{u})\big) &\leq h_{k+1}\big(\boldsymbol{S}_{k+1}(u)\big) \\ h_{k+1}\big(\boldsymbol{S}_{k+1}(\hat{u})\big) &\leq h_{k+1}\big(\boldsymbol{S}_{k+1}(u')\big) \end{aligned} \tag{18}$$

Now due to (18) and the redistribution of transition probabilities into level $k+1$ of $\bar{D}_{k+1}$, the cost-to-go $h_k(v, \boldsymbol{\pi}^k)$ at any node v in layer k under policy $\boldsymbol{\pi}^k$ is no less in $\bar{D}_k$ than in $\bar{D}_{k+1}$. This means that the optimal cost $h(\bar{D}_k)$ of $\bar{D}_k$ is no less than the expected cost $h_1(\boldsymbol{S}_1, \boldsymbol{\pi})$ of $\bar{D}_{k+1}$ under policy $\boldsymbol{\pi}^k$, and is therefore no less than the optimal cost of $\bar{D}_{k+1}$. This establishes (17), as desired. $\quad\square$

We can now verify that the merger operations used earlier in the maximum claim and job sequencing problems yield valid relaxed SDDs.

Corollary 1. *The merger operation $S_i \oplus S_i' = S_i \cap S_i'$ results in a relaxed SDD for the maximum clique problem.*

Proof. In this problem, a state $\boldsymbol{S}_i$ is the set S_i of vertices already selected. We will say that $\bar{S}_i$ relaxes S_i when $\bar{S}_i \subseteq S_i$. We first observe that the definitional requirement (11) for a relaxation is satisfied, by substituting the values given in (4)–(6) into (11) for the four cases $(x_i, \omega) = (1, 1), (1, 0), (0.1), (0.0)$.

Now, due to Theorem 1, it suffices to show that conditions (C1)–(C3) are satisfied. Condition (C1) is obviously satisfied, since $S_i \cap S_i' \subseteq S_i$ and $S_i \cap S_i' \subseteq S_i'$. To show (C2), we suppose that $\bar{S}_i \subseteq \bar{S}_i$. Then $\phi_{i0}(\bar{S}_i) = \bar{S}_i \subseteq S_i = \phi_{i0}(S_i)$ and $\phi_{i1}(\bar{S}_i) = \bar{S}_i \cup \{i\} \subseteq S_i \cup \{i\} = \phi_{i1}(S_i)$. Thus, $\phi_{i\omega}(\bar{S}_i)$ relaxes $\phi_{i\omega}(S_i)$ for $\omega = 0, 1$.

We now show (C3) with the sense of the inequality reversed, since we are maximizing rather than minimizing. We note that

$$
\sum_{\omega \in \Omega_i} p_{i\omega}(\boldsymbol{S}_i, x_i)\big(c_{i\omega}(\boldsymbol{S}_i, x_i) + \eta_\omega\big) =
\begin{cases}
\eta_0, & \text{if } x_i = 0 \\
\big(1 - p_i(S_i)\big)\eta_0 + p_i(S_i)(1 + \eta_1), & \text{if } x_i = 1
\end{cases}
$$

$$
\sum_{\omega \in \Omega_i} p_{i\omega}(\bar{\boldsymbol{S}}_i, \bar{x}_i)\big(c_{i\omega}(\bar{\boldsymbol{S}}_i, \bar{x}_i) + \eta_j\big) =
\begin{cases}
\eta_0, & \text{if } \bar{x}_i = 0 \\
\big(1 - p_i(\bar{S}_i)\big)\eta_0 + p_i(\bar{S}_i)(1 + \eta_1), & \text{if } \bar{x}_i = 1
\end{cases}
$$

If $x_i = 0$, we select the control $\bar{x}_i = 0$, and (12) becomes $\eta_0 \geq \eta_0$, which is obviously satisfied. If $x_i = 1$, we consider two cases. We first suppose that $\eta_0 \geq 1 + \eta_1$, in which case we select $\bar{x}_i = 0$. Then (12) becomes

$$
\big(1 - p_i(S_i)\big)\eta_0 + p_i(S_i)(1 + \eta_1) \leq \eta_0
$$

which simplifies to $\eta_0 \leq 1 + \eta_1$ or $\eta_0 \leq \eta_0$, and (12) is therefore satisfied. We now suppose $\eta_0 < 1 + \eta_1$, in which case we select $\bar{x}_i = 1$. Then (12) becomes

$$
\big(1 - p_i(S_i)\big)\eta_0 + p_i(S_i)(1 + \eta_1) \leq \big(1 - p_i(\bar{S}_i)\big)\eta_0 + p_i(\bar{S}_i)(1 + \eta_1)
$$

Since $p_i(\bar{S}_i) \geq p_i(S_i)$, this simplifies to $\eta_0 \leq 1 + \eta_1$ or $0 \leq 0$, and (12) is again satisfied.   □

The conditions for a valid relaxation simplify when transition probabilities are state-independent. In this case, it suffices to satisfy (C1) and (C2).

Corollary 2. *If conditions* (C1) *and* (C2) *are satisfied, and transition probabilities are state independent, then merger of nodes during compilation results in a relaxed SDD.*

Proof. Due to Theorem 1, it suffices to show that condition (C3) is satisfied as well as (C1) and (C2). In fact, we can show that for any control $x_i \in \mathcal{X}_i$ and any set $\{\eta_\omega \mid \omega \in \Omega_i\}$, (12) holds for $\bar{x}_i = x_i$. Since transition probabilities are state independent, we have $p_{i\omega}(\boldsymbol{S}_i, x_i) = p_{i\omega}(\bar{\boldsymbol{S}}_i, x_i)$, and (12) becomes

$$
\sum_{\omega \in \Omega_i} p_{i\omega}(\boldsymbol{S}_i, x_i)c_{i\omega}(\boldsymbol{S}_i, x_i) + p_{i\omega}(\boldsymbol{S}_i, x_i)\eta_\omega \geq \sum_{\omega \in \Omega_i} p_{i\omega}(\bar{\boldsymbol{S}}_i, x_i)c_{i\omega}(\bar{\boldsymbol{S}}_i, x_i) + p_{i\omega}(\boldsymbol{S}_i, x_i)\eta_\omega
$$

This follows immediately from (11), and (C3) is therefore satisfied.   □

Corollary 3. *The merger operation* (10) *results in a relaxed SDD for the stochastic job sequencing problem.*

Proof. In this problem, we say that a state $(\bar{S}_i, \bar{f}_i)$ relaxes a state (S_i, f_i) when $\bar{S}_i \subseteq S_i$ and $\bar{f}_i \leq f_i$. To show that this relaxation satisfies (11), it suffices to show

$$c_{i\omega}\big((S_i, f_i), x_i\big) \geq c_{i\omega}\big((\bar{S}_i, \bar{f}_i), x_i\big)$$

for any x_i and ω, because the transition probabilities are state-independent. But due to (9), this becomes

$$\big(\max\{r_{x_i}, f_i\} + t_{x_i\omega} - d_{x_i}\big)^+ \geq \big(\max\{r_{x_i}, \bar{f}_i\} + t_{x_i\omega} - d_{x_i}\big)^+$$

which holds because $f_i \geq \bar{f}_i$. Conditions (C1) and (C2) can be checked in a similar manner, and the claim follows due to Corollary 2.   □

7 Computational Evaluation

The primary challenge in computational evaluation of SDD bounds is finding nontrivial instances that can be solved to optimality, so that the tightness of bounds can be assessed. Unfortunately, the DIMACS instances of the maximum clique problem are much harder to solve as a stochastic problem, and nearly all are intractable. We were able to solve only five instances, three of which are too trivial for evaluation of bounds, and one of which ran for almost 24 h. We therefore generated a number of random instances that are calibrated in size to be both tractable and nontrivial. Since state-space enumeration is the only currently available method for optimal solution, we solved the instances to optimality by generating an exact SDD, which is essentially state-space enumeration with an intelligent variable ordering heuristic. We performed tests on both random and DIMACS instances, the latter mostly without comparison with optimal values.

We generated graphs for random instances by selecting each possible edge with a specified probability (which determines the average density of the resulting graphs). We then randomized the graphs by assigning each edge (i, j) probability $p_{ij} = 0.5 + 17ij \bmod(50)/100$. Thus, all probabilities were drawn from the interval $[0.5, 1]$. Probabilities were assigned to edges in DIMACS instances in the same manner.

The size of the relaxed SDDs is controlled by placing an upper bound on the width (the maximum number of node in a layer). The SDDs are compiled using the same heuristics as in [7]. In particular, the least attractive nodes on a given layer are merged until the desired width is obtained, where attractiveness is measured by the length of the longest path from the root in the deterministic problem. The rationale for this is that unattractive nodes are less likely to be part of an optimal solution and can therefore be merged without affecting the value of that solution. We do not check whether the state resulting from a merger already occurs at a node in the layer, because such a check is quite time consuming, although this can result in slightly weaker bounds.

Figure 4 displays the results for the random instances. Each plot shows the bounds that result from three random instances of different sizes. The bounds are plotted against the time invested in compiling relaxed SDDs of various widths

Table 1. DIMACS Instances Tested

Instance	Vertices	Density	Instance	Vertices	Density
brock200_1	200	0.7417	hamming6-2	64	0.8906
cfat200-1	200	0.0767	johnson8-4-4	70	0.7571
cfat500-1	500	0.0357	keller4	171	0.6453
c125.9	125	0.8913	p_hat300-1	300	0.2430
DSJC500_5	500	0.5010	san200_0.7_1	200	0.6965
gen200_p0.9_44	200	0.8955	sanr_0.7	200	0.6934

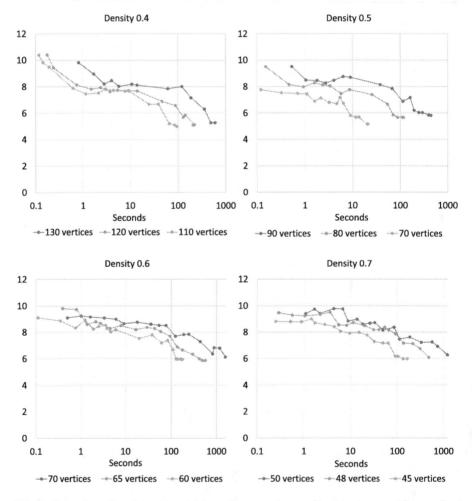

Fig. 4. Bound vs. time investment for random maximum clique instances. The smallest (rightmost) bound shown for each instance is the optimal value.

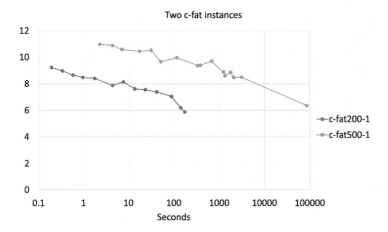

Fig. 5. Bound vs. time investment for two DIMACS instances solved to optimality.

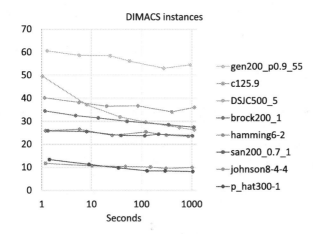

Fig. 6. Bound vs. time investment for various DIMACS instances, none of which are solved to optimality.

and computing the optimal expected clique size for each. Each point results from setting the maximum width to a certain value. The bound is plotted against the resulting computation time rather than the max width, because this better indicates the cost of obtaining the bound.[3] Nearly all of the computational cost is incurred by generating and merging nodes. The right-most data point for each instance represents an optimal solution. The plots indicate that the quality of the bound degrades rather modestly as the time investment ranges over about three orders of magnitude. The relationship is not entirely monotone, as would

---

[3] Detailed computational results, including specified max widths, are posted at public.tepper.cmu.edu/jnh/CPAIOR2022data.pdf.

be expected because the quality of the bound depends on the chance effects of the merger heuristic.

Tables 1 lists the DIMACS instances tested. Figure 5 shows bound-vs.-time plots for the two nontrivial DIMACS instances solved to optimality. The bound degradation is again gradual, and it is almost the same over a comparable time span for the large, difficult instance as for the small, easy instance. Figure 6 contains plots for various additional DIMACS instances, none of which could be solved to optimality. With one exception, they show a similar pattern of gradual and roughly logarithmic bound improvement as the time investment increases.

8   Conclusion

We introduced a concept of stochastic decision diagrams (SDDs) and showed how they can represent discrete stochastic dynamic programming (DP) problems. In particular, we indicated how state-dependent policies can be understood in a decision diagram context. Due to the difficulty and importance of deriving valid bounds for DP models, we focused on extending relaxation technology that has been developed for deterministic decision diagrams to the stochastic case, so that these bounds can be available for DP applications. We established sufficient conditions under which a node merger operation applied during top-down compilation yields valid relaxed SDDs. These provide optimization bounds whose quality can be adjusted at will by controlling the size of the relaxed SDD. As corollaries, we showed that simple node merger rules for stochastic maximum clique and job sequencing problems satisfy these conditions.

To test the performance of SDD bounding in practice, we computed bounds for random and benchmark instances of the stochastic maximum clique problem. We found that the bound quality degrades only gradually as the SDD size (and the time investment) is reduced. The time-quality relationship is roughly logarithmic in most cases, which allows one to estimate how much the bound would improve with further time investment after computing bounds for a few sample SDD widths.

A number of research directions can be pursued at this point, but two are particularly interesting. One is to extend to stochastic diagrams the decision-diagram-based branch-and-bound procedure introduced in [8]. This would provide a novel alternative to traditional state-space enumeration as an exact solution method for stochastic DP problems of moderate size.

A second direction is to investigate SDD-derived bounds on costs-to-go in approximate stochastic DP. The cost-to-go at any state in any stage of the DP recursion can be bounded by building an SDD that begins with that state at its root node. The solution of the DP problem would still be calculated on the basis of estimated costs-to-go, as in traditional approximate DP. Then one would compute the expected cost of this same policy using SDD-based bounds on the cost-to-go rather than estimates of the cost-to-go. This would provide a bound on the quality of the policy obtained by approximate DP.

References

1. Andersen, H.R., Hadzic, T., Hooker, J.N., Tiedemann, P.: A constraint store based on multivalued decision diagrams. In: Bessière, C. (ed.) CP 2007. LNCS, vol. 4741, pp. 118–132. Springer, Heidelberg (2007). https://doi.org/10.1007/978-3-540-74970-7_11

2. Baldacci, R., Mingozzi, A., Roberti, R.: New state-space relaxations for solving the traveling salesman problem with time windows. INFORMS J. Comput. **24**(3), 356–371 (2012)

3. Bellman, R.: The theory of dynamic programming. Bull. Am. Math. Soc. **60**, 503–516 (1954)

4. Bellman, R.: Dynamic Programming. Priceton University Press, Princeton, NJ (1957)

5. Bergman, D., Ciré, A.A.: Discrete nonlinear optimization by state-space decompositions. Manag. Sci. **64**, 4700–4720 (2018)

6. Bergman, D., Cire, A.A., van Hoeve, W.-J.: Improved constraint propagation via lagrangian decomposition. In: Pesant, G. (ed.) CP 2015. LNCS, vol. 9255, pp. 30–38. Springer, Cham (2015). https://doi.org/10.1007/978-3-319-23219-5_3

7. Bergman, D., Ciré, A.A., van Hoeve, W.J., Hooker, J.N.: Optimization bounds from binary decision diagrams. INFORMS J. Comput. **26**, 253–268 (2013)

8. Bergman, D., Ciré, A.A., van Hoeve, W.J., Hooker, J.N.: Discrete optimization with binary decision diagrams. INFORMS J. Comput. **28**, 47–66 (2014)

9. Bergman, D., Cire, A.A., van Hoeve, W.-J., Hooker, J.: MDD-based constraint programming. In: Decision Diagrams for Optimization. AIFTA, pp. 157–181. Springer, Cham (2016). https://doi.org/10.1007/978-3-319-42849-9_9

10. Bergman, D., van Hoeve, W.-J., Hooker, J.N.: Manipulating MDD relaxations for combinatorial optimization. In: Achterberg, T., Beck, J.C. (eds.) CPAIOR 2011. LNCS, vol. 6697, pp. 20–35. Springer, Heidelberg (2011). https://doi.org/10.1007/978-3-642-21311-3_5

11. Bertele, U., Brioschi, F.: Nonserial Dynamic Programming. Academic Press, New York (1972)

12. Bertsekas, D.P.: Dynamic programming and optimal control: approximate dynamic programming, vol. 2, 4th edn. Athena Scientific, Nashua, NH (2012)

13. Bertsekas, D.P.: Dynamic Programming and Optimal Control, vol. 1, 4th edn. Athena Scientific, Nashua, NH (2017)

14. Castro, M.P., Ciré, A.A., Beck, J.C.: An MDD-based Lagrangian approach to the multicommodity pickup-and-delivery TSP. INFORMS J. Comput. **32**, 263–278 (2020)

15. Christofides, N., Mingozzi, A., Toth, P.: State-space relaxation procedures for the computation of bounds to routing problems. Networks **11**(2), 145–164 (1981)

16. Ciré, A.A., van Hoeve, W.J.: Multivalued decision diagrams for sequencing problems. Oper. Res. **61**, 1411–1428 (2013)

17. Ciré, A.A., van Hoeve, W.J.: MDD propagation for disjunctive scheduling. In: Proceedings of the Twenty-Second International Conference on Automated Planning and Scheduling (ICAPS) (2012). AAAI Press

18. Gentzel, R., Michel, L., van Hoeve, W.-J.: HADDOCK: A language and architecture for decision diagram compilation. In: Simonis, H. (ed.) CP 2020. LNCS, vol. 12333, pp. 531–547. Springer, Cham (2020). https://doi.org/10.1007/978-3-030-58475-7_31

19. González, J.E., Ciré, A.A., Lodi, A., Rousseau, L.M.: BDD-based optimization for the quadratic stable set problem. Discrete Optim. **44**, 100610 (2020)
20. González, J.E., Cire, A.A., Lodi, A., Rousseau, L.-M.: Integrated integer programming and decision diagram search tree with an application to the maximum independent set problem. Constraints **25**(1), 23–46 (2020). https://doi.org/10.1007/s10601-019-09306-w
21. Hadžić, T., Hooker, J.N.: Discrete global optimization with binary decision diagrams. In: GICOLAG 2006. Vienna, Austria, December 2006
22. Hadžić, T., Hooker, J.N.: Cost-bounded binary decision diagrams for 0-1 programming. In: Van Hentenryck, P., Wolsey, L. (eds.) CPAIOR 2007. LNCS, vol. 4510, pp. 84–98. Springer, Heidelberg (2007). https://doi.org/10.1007/978-3-540-72397-4_7
23. Hoda, S., van Hoeve, W.-J., Hooker, J.N.: A systematic approach to MDD-based constraint programming. In: Cohen, D. (ed.) CP 2010. LNCS, vol. 6308, pp. 266–280. Springer, Heidelberg (2010). https://doi.org/10.1007/978-3-642-15396-9_23
24. Hooker, J.N.: Decision diagrams and dynamic programming. In: Gomes, C., Sellmann, M. (eds.) CPAIOR 2013. LNCS, vol. 7874, pp. 94–110. Springer, Heidelberg (2013). https://doi.org/10.1007/978-3-642-38171-3_7
25. Hooker, J.N.: Job sequencing bounds from decision diagrams. In: Beck, J.C. (ed.) CP 2017. LNCS, vol. 10416, pp. 565–578. Springer, Cham (2017). https://doi.org/10.1007/978-3-319-66158-2_36
26. Hooker, J.N.: Improved job sequencing bounds from decision diagrams. In: Schiex, T., de Givry, S. (eds.) CP 2019. LNCS, vol. 11802, pp. 268–283. Springer, Cham (2019). https://doi.org/10.1007/978-3-030-30048-7_16
27. Mingozzi, A.: State space relaxation and search strategies in dynamic programming. In: Koenig, S., Holte, R.C. (eds.) SARA 2002. LNCS (LNAI), vol. 2371, pp. 51–51. Springer, Heidelberg (2002). https://doi.org/10.1007/3-540-45622-8_4
28. O'Neil, R.J., Hoffman, K.: Decision diagrams for solving traveling salesman problems with pickup and delivery in real time. Oper. Res. Lett. **47**, 197–201 (2019)
29. Powell, W.B.: Approximate Dynamic Programming: Solving the Curses of Dimensionality, 2nd edn. Wiley-Interscience, Hoboken (2011)
30. Righini, G., Salani, M.: New dynamic programming algorithms for the resource constrained shortest path problem. Networks **51**, 155–170 (2008)
31. Tjandraatmadja, C., van Hoeve, W.-J.: Incorporating bounds from decision diagrams into integer programming. Math. Program. Comput. **13**(2), 225–256 (2020). https://doi.org/10.1007/s12532-020-00191-6

Improving the Robustness of EPS
to Solve the TSP

Nicolas Isoart and Jean-Charles Régin[✉]

Université Côte d'Azur - I3S - CNRS, 2000, route des Lucioles - Les Algorithmes -
Euclide B, BP 121, 06903 Sophia Antipolis Cedex, France
{nicolas.isoart,jean-charles.regin}@univ-cotedazur.fr

Abstract. Embarrassingly Parallel Search (EPS) parallelizes the search
for solutions in CP by decomposing the initial problem into a huge number of sub-problems that are consistent with propagation. Then, each
waiting worker takes a sub-problem and solves it. The process is repeated
until all the sub-problems have been solved. EPS is based on the idea
that if there are many sub-problems to solve then the solving times of the
workers will be balanced even if the solving times of the sub-problems
are not. This approach gives rather good results for solving the Traveling Salesman Problem (TSP). Unfortunately, for some instances, sub-
problems with extremely different solving times appear, for example one
requiring a huge part of the total solving time. In this case the load
balancing is poor. We show that a general increase in the number of
sub-problems does not solve this imbalance. We present a method that
identifies the presence of difficult sub-problems during the solving process and decompose them again. This method keeps the advantages of
EPS: the communication is very reduced (the workers do not communicate with each other) and it is independent of the solver. Experimental
results for the TSP show a good improvement of load balancing and a
better scaling with hundred of cores.

1 Introduction

The Traveling Salesman Problem (TSP) consists in searching for a single cycle
covering a graph such that the sum of the cost of the edges of the cycle is minimized. This problem has been widely studied as there is a huge number of direct
applications such as routing problems, school bus problems, *etc.* and indirect
applications such as scheduling where cities are tasks and arcs are transition
times. There are very efficient methods to find near optimal solutions such as
LKH algorithm [8,20]. Therefore, when we solve the problem in parallel, we can
consider TSP as a satisfaction problem rather than an optimization problem.

In constraint programming there are two main approaches for solving a problem in parallel: the search space splitting method (i.e., the work-stealing approach) and problem decomposition (i.e., embarrassingly parallel search).

The work-stealing method dynamically splits the search space during the
solving [1,4]. When a worker has finished exploring a sub-problem, it asks other

P. Schaus (Ed.): CPAIOR 2022, LNCS 13292, pp. 155–172, 2022.
https://doi.org/10.1007/978-3-031-08011-1_12

workers for another sub-problem. If another worker agrees to the demand, then it dynamically splits its current sub-problem into two disjoint sub-problems and sends one sub-problem to the starving worker. The starving worker "steals" some work from the busy one. Several implementation of work stealing approach have been designed [22,27]. The most recent ones, like Bobpp tries to be as independent as possible from the solver [3,16].

The Embarrassingly Parallel Search (EPS) [21,23,25] is a more recent method. It statically decomposes the initial problem into a huge number of sub-problems that are consistent with propagation (*i.e.* running the propagation mechanism on them does not detect any inconsistency). Then, each waiting worker takes a sub-problem and solves it until all the sub-problems have been solved. The assignment of the sub-problems to workers is dynamic, and there is no communication between the workers. EPS is based on the idea that if there is a large number of sub-problems to solve then the solving times of the workers will be balanced even if the solving times of the sub-problems are not. In other words, load balancing should be automatically obtained in a statistical sense. EPS seems to be more robust and to outperform the work-stealing approach.

As EPS gives very good results on many problems, it was legitimate to try it to solve the TSP [12]. In order to use it, some modifications of the EPS decomposition mechanism have been required for two reasons. First, the model of the TSP in CP contains a set-variable with the mandatory edges as lower bound and optional edges as upper bound. Decomposing with a set-variable is not as trivial as a classical Cartesian product, because the order must be carefully handled while enumerating. Second, the most efficient strategy is LCFirst: it selects one node from the graph according to a heuristic and keeps branching on the node until there are no more candidates around it, no matter if we backtrack or not. This strategy is a depth-first process unlike the decomposition mechanism of EPS which is a depth-bounded. The first change due to the set-variable is minor. For the second issue the method *Bound-Backtrack-and-Dive* has been introduced [12]. It proceeds in two steps. A sequential solving of the problem with a bounded number of backtracks is run in order to extract key information from LCFirst. Then, the EPS decomposition uses that information rather than LCFirst. The experimental results shown that almost a linear gain on the number of cores is obtained and that *Bound-Backtrack-and-Dive* may considerably reduce the number of backtracks performed for some problems. It is important to note that without this method it is possible to observe a degradation of performance according to the number of cores.

The authors give experimental results with a classical laptop nowadays, *i.e.* with four cores with hyperthreading. The results are quite good. For many problems, the same kind of results are obtained when increasing the number of cores. Unfortunately, this is not true for all problems and often not the case for very difficult problems. It is precisely for those problems that we would like to be able to use the combined power of many computational cores.

The objective of this article is to propose a method to detect this kind of difficult problems and to improve their solving in parallel. In other words, we propose to slightly modify EPS in order to improve its robustness. EPS has many qualities, such as its simplicity, its independence from the solvers used and search strategies, the absence of communication between workers and weak communication between the master and the workers. It is important to respect these advantages and not to make any changes that could jeopardize them.

The study of EPS, as defined in [12], with a hundred cores for the solving of the TSP showed rather unexpected results. Three major issues were observed:

1. *Unstable decomposition.* The decomposition is no longer stable in the sense that decomposing more may lead to a strong degradation of the performance. This happens with the TSP because the best strategy is a depth first strategy whereas the decomposition mechanism can be seen has a breadth first process. In other words, in sequential solving the CP model uses the previous calculations, whereas in parallel solving, the decomposition makes the sub-problems independent and they are solved in a different order than sequential solving would have done. This means that the decomposition can interact unfavorably with the increase in the number of cores. In other words, increasing the number of cores can lead to no improvement. For instance, problem pcb442 of the TSPLib requires 91,137 s to be solved sequentially whereas EPS (with #sppw $= 30$) requires 10,645 s with 24 workers, 7,458 s with 48 workers, 12,808 s with 72 workers and 11,025 s with 96 workers.

2. *Non-monotonic decomposition.* The decomposition is difficult to comprehend because sometimes decomposing more leads to more problems and sometimes it leads to fewer problems [12]. In addition, the succession of the two phenomena can be observed during the same decomposition.

3. *Extremely heterogeneous sub-problems.* A very small number of sub-problems can take a very large part of the overall solving time. For instance, the decomposition of the problem gr431 of the TSPLIB [26] into 300 sub-problems, leads to 5 sub-problems which take 50% of the overall solving time. The consequence is that at the end of the solving, only 5 workers remain active, independently of the number of the available workers.

In order to remedy these issues, we propose to reduce the number of sub-problems that are considered during the decomposition and to re-decompose the remaining unsolved sub-problems if needed, and to repeat this process while the whole problem is unsolved. Indeed, a re-decomposition allows a better redistribution of the work and so a better load balancing.

However, this approach triggers new questions:

– When should we re-decompose? Under which conditions? If we wait too long then the risk is that very few workers, or even one, will be left to solve a problem while all the others have no more work to do.
– When a worker is aborted, it has already performed some computations, how can we avoid redoing the same computations that has just been done?
– How should we re-decompose? What is the right number of sub-problems per worker to consider for a decomposition? Should it evolve?

We will answer all these questions and show how to define the right parameters in order to obtain a worthwhile re-decomposition.

The article is organized as follows. First, we recall the CP model of the TSP and we introduce EPS. Then, we discuss some problems arising when a hundred cores are used. Next, we present a new method based on re-decompositions when needed. At last, we give some experiments showing the advantages of our approach and we conclude.

2   Preliminaries

2.1   TSP Model in CP

One of the most efficient models in constraint programming for solving the TSP is based on the Weighted Circuit Constraint (WCC) associated with the k-cutset constraint and the mandatory Hamiltonian path constraint coupled with a branch-and-bound procedures [9]. The WCC is composed of a Lagrangian relaxation of a minimum 1-tree [6,7,11] (*i.e.* a special minimum spanning tree) plus two minimum cost edges and degree constraints. The k-cutsets constraint find all sizes 2 and 3 cutsets in the graph and imposes that an even number of edges for each cutset is mandatory [10,14]. The mandatory Hamiltonian path constraint stops the search when a non optimal hamiltonian subpath of the tour is found [13]. A single undirected graph variable where all nodes are mandatory is used to represent the graph. The branch-and-bound considers only the edges, it consists in making a binary search where a left branch is an edge assignment and a right branch is an edge removal.

In addition, this TSP model uses a search strategy integrating a graph interpretation of Last Conflict heuristics [5,18], named LCFirst [2]. This search strategy selects one node from the graph according to a heuristic and keeps branching on the node until there are no more candidates around it, no matter if we backtrack or not. One of the most efficient heuristic and the one used by default is minDeltaDegree. It selects the edges for which the sum of the endpoint degrees in the upper bound minus the sum of the endpoint degrees in the set-variable lower bound is minimal. Hence, this search strategy learns from previous branching choices and tends to keep those that previously caused a failure. Results in [2] show that LCFirst clearly outperforms all other search strategies, it is the only one that really exploits the graph structure. So far, LCFirst beats all other search strategies by one or more orders of magnitude.

2.2   EPS

Embarrassingly Parallel Search (EPS) decomposes the initial problem into a huge number of sub-problems that are consistent with propagation. Then, each waiting worker takes a sub-problem and solves it until all sub-problems have been solved. It is usually considered that a good number of sub-problems per worker is between 30 to 300 [21,23,25].

The generation of q sub-problems is not straightforward because the number of sub-problems consistent with the propagation may not be related to the Cartesian product of some domains. A simple algorithm could be to perform a Breadth First Search (BFS) in the search tree until the desired number of sub-problems consistent with the propagation is reached. Unfortunately, it is not easy to perform a BFS efficiently mainly because BFS is not an incremental algorithm like Depth-First Search (DFS). Therefore, EPS uses a process resembling an iterative deepening depth-first search [15]: we consider a set $Y \subseteq X$ of variables: we only assign the variables of Y and we stop the search when they are all assigned. In other words, we never try to assign a variable that is not in Y. This process is repeated until all assignments of Y consistent with the propagation has been found. Each branch of a search tree computed by this search defines an assignment (*i.e.* a sub-problem). To generate q sub-problems, we repeat the previous method by adding variables to Y if necessary, until the number of sub-problems is greater than or equal to q.

In some cases, frequently encountered with the TSP, the increase of variables in Y leads to a reduction in the number of generated sub-problems. This phenomenon is named: non-monotonic decomposition. If suddenly many branches fail in the search tree, it may be possible that more problems have been removed than generated. For special cases such as this one, it is preferable to stop the decomposition because the number of sub-problems generated can start to oscillate without really progressing globally. Thus, a stopping criterion other than the number of sub-problems generated has been defined [12]: if there are two successive additions of variables in Y for a decrease of the number of generated sub-problems is observed, then the decomposition is stopped.

When we consider a set $Y \subseteq X$ of variables, we pay attention to the set-variables. A classical variable is instantiated by a single value, whereas for a set-variable we will determine how many of its values should be instantiated at most. For instance, a set variable can be instantiated with at most 1, 2 or 3 values. Its cardinality defines only the maximum because we search for partial assignments. In general, all but one set-variable of Y will be potentially instantiated with their maximum possible values (*i.e.* the maximum cardinality).

Note that the decomposition is also performed in parallel.

As the best search strategy for the TSP is LCFirst which is depth-first based whereas the decomposition is depth-bounded, in previous work [12] we introduced the method *Bound-Backtrack-and-Dive* that partially solve this issue. It proceeds in two steps. First, a sequential solving of the problem with a bounded number of backtracks is run in order to extract key information from LCFirst. Then, the EPS decomposition uses that information rather than LCFirst. The experimental results shown a strong improvement with this method.

3   Performance with a Hundred Cores

The modified version of EPS performed well with the TSP model when there are 4 cores [12]. As it is usual in parallelism, one has the right to wonder if the scaling we observe for a few cores can be verified in practice with about a hundred cores. EPS has already been modified for improving its scaling while used on data centers [24].

When using hundred cores, we observe the three issues mentioned in Introduction: unstable decomposition, non-monotonic decomposition and extremely heterogeneous sub-problems. As we already mentioned, the non-monotonic decomposition problem is solved by introducing a stopping criterion other than the number of generated sub-problems. We will now focus on the other two issues.

In order to remedy the unstable decomposition issue we suggest to consider it carefully and to avoid decomposing too much. That is instead of trying to decompose in a lot of sub-problems, we suggest to consider fewer sub-problems. The risk of reducing the number of sub-problems is that it can be difficult to ensure a good load balancing between the workers. However, when there are a hundred workers it is less important to have 2, 5 or 10 workers that are not active than when you have only 4 workers. On the other hand, the second problem we have to solve is that of extremely heterogeneous sub-problems which therefore lead *de facto* to bad load balancing.

In order to remedy the bad load balancing caused by the presence of extremely heterogeneous sub-problems, we have no choice but to re-decompose these sub-problems into many other sub-problems which are more homogeneous. In other words, we have to be prepared to do several decomposition steps.

All this must be done without disturbing the functioning of EPS for problems that do not show these behaviors and that are very well solved by EPS even with a hundred cores. Therefore, systematically performing several decomposition steps worsens the results. One must identify if there are some extremely heterogeneous sub-problems and in this case prepare to restart a decomposition.

This also must be done while keeping the advantages of EPS: a very reduced communication (the workers do not communicate with each other) and an independence from the solver used.

4   Re-decomposition

First, we try to estimate under which conditions it could be worthwhile to re-decompose the unsolved sub-problems. Then, we try to avoid redoing the same work when the solving of a sub-problem is interrupted in order to re-decompose.

The challenge can be resumed as follows: if we wait too long for performing a decomposition then the risk is that very few workers, or even one, will be left to solve a problem while all the others have no more work to do. However, re-decomposing for better distribution means losing part of the previous computations already done and requires a certain minimum time that may not be profitable.

Notation 1 (known values) and Notation 2 (unknown values) describe some information about the search.

Notation 1

- w: total number of workers.
- a: number of active workers (i.e. workers which are currently solving a subproblem).
- c_R: wall clock solving time of the set of the unsolved sub-problems R. Precisely, this is the solving time already done for the remaining sub-problems.

Notation 2

- t_R: wall clock total solving time for the set of remaining sub-problems R. It is the sum of the solving times of the sub-problems not yet solved.
- d_R: wall clock time needed to decompose the set of remaining sub-problems R. We also called it the time to re-decompose.
- rt_R: wall clock time needed to solve the set of remaining sub-problems R after a re-decomposition.

We immediately have the following property:

Proposition 1. *The minimum remaining computation time without re-decomposition is:* $\frac{t_R - c_R}{a}$

This is a lower bound of the real value because it assumes that the remaining time for the set of the remaining sub-problems is perfectly distributed among the workers if $a = w$ and if $a < w$ then it means that the remaining computation time for each sub-problem is the same.

Proposition 2. *The minimum remaining computation time for re-decomposing and solving the remaining sub-problems is:* $\frac{rt_R}{w} + d_R$

From these two properties we propose a new property:

Proposition 3. *Performing a re-decomposition may become worthwhile if*

$$\frac{rt_R}{w} + d_R < \frac{t_R - c_R}{a} \tag{1}$$

If after a decomposition all the computations performed by a stopped worker are lost, then it means that $c_R = 0$ and $rt_R = t_R$. On the other hand, if all the computations previously made are not redone, then we have $rt_R = t_R - c_R$. Unfortunately, it is very difficult to obtain this result because EPS is independent from the solver and from the search. Nevertheless we can expect to have $rt_R < t_R$.

Simplifying this inequality is not simple because we are faced with several unknown variables. We do not know precisely t_R, nor rt_R, nor d_R and we have no guarantee that the solving process will be perfectly homogeneous.

Therefore, we propose to make some assumptions:

- We will consider that rt_R can be rewritten as $t_R - qc_R$ with $0 \le q \le 1$ and q corresponds to the proportion of computations already made that we can avoid redoing. For practical reasons, we suggest to simply estimate the value of q from the previous calculations. Let us consider a sub-problem $p \in R$, we look for the value q_p. While solving p, if the search tree proved that k values can be safely removed from the initial domain of some variables, then we know that we were trying a $(k+1)^{th}$ value when solving. Avoiding reconsidering the kth values allow to avoid redoing $q_p = k/(k+1)$ from the calculations previously made. However, we do not know where we were in the solving of the $(k+1)^{th}$ branch. On average, we can consider that we were halfway in the solving of this branch. It means that we will avoid redoing $q_p = k/(k+0.5)$ from the previous calculations. Thus, we suggest to take as q the average of the values of q_p for each sub-problem p.

- We also observed that the decomposition time does not really change between decompositions. We will therefore consider that it is constant $d_R = d$. In practice, we will consider that a re-decomposition will take the time of the previous decomposition.
- Since no solving time is lost when $a = w$, we only consider a re-decomposition when $a < w$.

We can rewrite the previous property:

Proposition 4. *Performing a re-decomposition may become worthwhile if*

$$t_R > \frac{(w - aq)c_R + wad}{w - a} \tag{2}$$

Proof. With Notation 1, Eq. 1 can be rewritten as $\frac{t_R - qc_R}{w} + d < \frac{t_R - c_R}{a}$ which is equivalent to $(w - a)t_R > (w - aq)c_R + wad$. Since $a < w$ we have $(w - a) > 0$ and the property holds. □

The main issue is that we do not know the value of t_R. However, for a given value of a we can check whether Eq. 2 is satisfied or not and stop the computation when it is satisfied. First of all we know that $t_R \geq c_R$. Next, during a wall clock period of time T of computations performed by a active workers the total solving time computed is aT and so we have $t_R \geq aT + c_R$. Thus if $aT + c_R > \frac{(w - aq)c_R + wad}{w - a}$ then we know that Eq. 2 is satisfied. So we can compute the value of T for which Eq. 2 is satisfied. This value will become the maximum timeout we accept without performing a new decomposition.

$$aT + c_R > \frac{(w - aq)c_R + wad}{w - a} \Leftrightarrow aT > \frac{(w - aq)c_R + wad}{w - a} - c_R$$
$$\Leftrightarrow aT > \frac{(a - aq)c_R + wad}{w - a} \Leftrightarrow aT > \frac{a(1 - q)c_R + wad}{w - a}$$
$$\Leftrightarrow T > \frac{(1 - q)c_R + wd}{w - a}$$

Proposition 5. *Assume that after T units of computation no new sub-problem is solved. It is worthwhile to re-decompose if*

$$T > \frac{(1 - q)c_R + wd}{w - a} \tag{3}$$

Proof. If after $T_1 < T$ units of time we solve 1 task s_1 then we can compute the new inequality. We decrement the number of active workers and work with $R_1 = R - \{s_1\}$ and $c_{R_1} = c_R + aT_1 - t(s_1)$.

Equation 3 suggests that we should wait a certain amount of time to be sure that we should decompose. However, waiting that time may have a cost that leads to poorer results. Thus, we propose to use T as an upper bound of the waiting time before re-decomposing.

Indeed, on the one hand, we do not know how to avoid redoing all the calculations performed (we cannot avoid c_r, but only qc_r) and, on the other hand, we do not calculate with the power of all the workers, which means that we lose a certain amount if the decision is to decompose again. However, if we re-decompose too early when we should not have, we will also lose time. Therefore, instead of directly using the formula in Eq. 3, it is more interesting to introduce a decision factor α, with $0 \leq \alpha \leq 1$. The idea behind this value corresponds to a kind of risk sharing. We have the final property:

Proposition 6. *Assume that after T units of computation no sub-problem is solved. The current active workers are stopped and a decomposition is performed if*

$$T > \alpha \frac{(1-q)c_R + wd}{w - a} \tag{4}$$

This leads to Algorithm 1. Equation 4 will evolve as sub-problems are solved.

Algorithm 1: CHECKDECOMPOSITION algorithm

CHECKDECOMPOSITION(d, α, q): boolean
> // The sub-problems are currently run in parallel ;
> Wait until it remains only $a = w - 1$ active workers ;
> $stopTime \leftarrow$ WALLCLOCKTIME() ;
> $R \leftarrow$ set of remaining sub-problems ;
> $C \leftarrow$ Sum of current solving times of the remaining sub-problems ;
> $timeout \leftarrow \alpha \frac{(1-q)C+wd}{w-a}$;
> **for each** *sub-problem* $p \in R$ *solved before* $timeout$ **do**
>> // the timeout evolves according to the solved sub-problems ;
>> $R \leftarrow R - \{p\}$;
>> $T \leftarrow$ WALLCLOCKTIME() $- stopTime$;
>> $stopTime \leftarrow$ WALLCLOCKTIME() ;
>> $C \leftarrow C + aT - t(p)$; // $t(p)$ is the solving time of the sub-problem p ;
>> $a \leftarrow a - 1$;
>> $timeout \leftarrow \min(timeout - T, \alpha \frac{(1-q)C+wd}{w-a})$;
>
> **if** *all sub-problems are solved* **then** return false;
> Abort all the remaining active workers ;
> return true; // A re-decomposition will be performed ;

Moreover, we need to make sure that each inequality holds. Indeed, when a problem p of R is solved, we compute a new inequality of Eq. 4 for $R - \{p\}$. However, the time between the last task solved and p can be huge. Therefore, the new timeout must verify the old inequality (i.e. $timeout - T$) and the new computed inequality. Thus, we take $\min(timeout - T, \alpha \frac{(1-q)c_R+wd}{w-a})$ as the new timeout. Note that Function CHECKDECOMPOSITION is run in the main loop of the EPS master which manages the workers and the sub-problems. If it returns true then a re-decomposition is performed.

4.1   How to Avoid Redoing Calculations?

When a worker is aborted, it has already performed some work. How can we avoid redoing the same work that has just been done?

In order to avoid redoing part of the previously performed calculations, we could memorize the boundary of the search already made. Nogoods recording methods [17, 19] can be used for managing the savings and the restarts. However, we introduce a simple and general method that requires little intervention in the solver and practically no more transmission of information. In addition, it allows to estimate the proportion of computations that we can avoid redoing. Like the nogoods recording methods, we only assume that the search can be described by a set of decisions and rejections of decisions. Conceptually, the search tree can be viewed as a binary tree where a node has two children: one applies the decision and the other rejects the decision (usually by imposing the opposite). When a sub-tree resulting from a decision is completely finished then we know that the exploration of that part is completely finished and we do not need to do it again. Thus, we propose to no longer consider the nodes linked to the root of which only the refutation of the decision remains. This is easy to express because it simply corresponds to the suppression of values in variable domains. In the case of TSP, it corresponds to a set of edges that should no longer be used for the tour. In other words, we simply define a new sub problem whose domains are included in the initial one and we transmit it to the EPS master.

4.2   Discussion

Instead of an overall decomposition, it is also possible to study the behavior of EPS at the level of each of the sub-problems. Thus, one can try to define criteria to try to know whether a sub-problem should or should not be re-decomposed and then proceed to a specific individual re-decomposition for some sub-problems. We tried this method, but it is not advantageous and raises many questions: when should we look to see if the sub-problems are particularly difficult? Should they be stopped immediately? Moreover, this method leads to a succession of decompositions (one per sub-problem) which quickly becomes expensive and does not bring any particular gain. In fact, EPS always tries to consider the solving of sub-problems as global and there seems to be little interest in questioning this point of view. In addition, only the master can use such a criteria because we search for sub-problems that are more difficult than others and only the master can have this information. A worker has no information about the other workers. This is why it is preferable to let the master manage the solving of the different workers, even if it means interrupting some of them.

5   Experiments

We experimentally show that EPSrd, the re-decomposition method we propose, allows obtaining better solving times and more robustness than the classical EPS method (denoted by EPS).

The algorithms have been implemented in Java 11 in a locally developed CP solver. The experiments were performed on 4 Clear Linux machines, each using two Intel Xeon E5-2696v2 and 64 GB of RAM. Thus the experiments use 96 identical cores. The reference instances are from the TSPLIB [26], a library of reference graphs for the TSP and the set of instances is based on the set given by Fages et al. [2] that can be seen as state-of-the-art instances. Some more difficult instances have been added. The name of each instance is suffixed by its number of nodes. All given times are wall clock time (usually in seconds).

5.1   Satisfaction vs Optimization

Table 1. LKH bound when k-opt is limited to 3-opt.

	optimal bound	LKH bound	LKH time (s)	seq. time (s)
a280	2378	2378	0.48	8.1
ali535	202339	202339	3.19	13013.9
d198	15780	15780	0.7	7
d493	35002	35002	1.2	4989.9
gil262	2378	2378	0.48	2042.2
gr229	134602	134602	0.28	41.2
gr431	171414	171428	1.62	520.6
gr666	294358	294361	4.05	14719.6
kroA200	29368	29368	0.19	63.7
kroB150	26130	26130	0.22	8.7
kroB200	29437	29437	0.14	15.1
lin318	42029	42029	0.53	6.4
pcb442	50778	50778	0.8	91137.7
pr136	96772	96772	0.11	8.9
pr264	49135	49135	0.17	3.8
pr299	48191	48191	0.66	1595.5
rat195	2323	2323	0.33	17.9
rd400	15281	15281	0.53	6721
si175	21407	21407	0.78	100.7
tsp225	3916	3916	0.27	23.5

As shown in Table 1, the famous LKH heuristic [8, 20] allows to find very quickly a solution whose value is very close to the optimal one (they are the same except for gr431 and gr666). Thus, for the purpose of parallelization, the TSP problem can be seen more as a satisfaction problem than as an optimization problem.

5.2   Comparison for a Given Number of Sub-problems

We search for the best configuration for EPS and EPSrd. In Table 2, we consider the number of sub-problems per workers (#sppw). We observe that #sppw = 100 for EPS and #sppw = 10 for EPSrd (with $\alpha = 0.1$) are the best configurations. We observe that most of the time EPSrd improves the results of EPS. EPS is based on the idea that for having a good load balancing we need to have a certain number of sub-problems. The experiments support this idea, even if sometimes decomposing too much may be time consuming. For instance, pcb442 is solved in 22,800 s with #sppw = 10, 11,025 s with #sppw = 30, 7,686 s with #sppw = 50 and 5,168.8 s with #sppw = 100. However, if we carefully look at the results, we can refine this principle. We observe that when the load balancing is not an issue for a problem (such as gr229 where #sppw = 10 is the best), then a small value of #sppw is perfectly fine. This means that we need to have a greater value only when the load balancing is an issue. However, we have no information about the load balancing before the solving. Conversely, EPSrd tries to identify during the solving whether the load balancing is an issue or not. If it detects such an issue, then it performs a re-decomposition. Thus, EPSrd has advantage to start with a low #sppw. Note that this approach can be seen as a dynamic increase of the #sspw value. We observe that the mean solving times of EPSrd with #sppw = 10 is 2.3 times faster than with other #sspw values. In addition, the problems with a huge load balancing issue such as pcb442 are well solved with EPSrd (5,166.8 s for EPS and 159.5 s for EPSrd). Finally, we observe that the best configuration for EPSrd in this table is 5.3 times faster than the best one for EPS.

Table 2. Comparison between EPS and EPSrd for different values of #sppw.

Instances	EPS(s)				EPSrd(s)				EPS/EPSrd			
	#sppw				#sppw				#sppw			
	10	30	50	100	10	30	50	100	10	30	50	100
a280	1.8	9.8	17.4	1.7	1.6	1.6	18.3	1.6	1.1	6.2	0.9	1.0
ali535	3,436.0	4,045.6	3,945.4	4,056.3	463.9	1,108.1	560.5	617.5	7.4	3.7	7.0	6.6
d198	12.2	14.6	15.2	14.5	13.3	12.9	12.6	13.5	0.9	1.1	1.2	1.1
d493	4,188.8	4,426.0	3,810.6	2,247.2	1,084.5	1,449.3	2,259.1	3,708.3	3.9	3.1	1.7	0.6
gil262	75.7	79.1	68.9	113.6	30.6	88.5	75.5	106.9	2.5	0.9	0.9	1.1
gr229	12.4	20.3	19.7	41.4	16.2	30.0	35.1	41.3	0.8	0.7	0.6	1.0
gr431	77.6	82.2	105.3	169.4	82.6	91.8	106.3	168.3	0.9	0.9	1.0	1.0
gr666	4,588.5	1,653.5	1,749.0	1,555.2	926.0	998.3	1,122.8	1,075.8	5.0	1.7	1.6	1.4
kroA200	6.4	6.2	7.5	6.1	7.3	7.1	7.4	6.4	0.9	0.9	1.0	0.9
kroB150	2.5	2.3	2.6	2.5	2.6	2.6	2.4	2.5	1.0	0.9	1.1	1.0
kroB200	6.2	5.2	4.4	4.4	6.6	5.6	4.7	4.5	0.9	0.9	0.9	1.0
lin318	3.6	3.9	3.4	3.6	3.4	3.3	3.3	3.7	1.1	1.2	1.0	1.0
pcb442	22,800.1	11,025.4	7,686.4	5,166.8	159.5	289.7	408.9	331.0	143.0	38.1	18.8	15.6
pr136	1.4	1.6	1.8	1.7	1.7	1.6	1.7	1.6	0.8	1.0	1.1	1.1
pr264	11.9	13.8	13.9	12.8	16.5	13.3	13.1	13.8	0.7	1.0	1.1	0.9
pr299	1,345.2	2,183.9	3,807.3	3,160.6	211.3	220.8	303.4	396.8	6.4	9.9	12.6	8.0
rat195	5.9	5.1	5.1	4.8	5.7	4.9	4.4	4.3	1.0	1.0	1.1	1.1
rd400	706.2	496.2	477.7	560.9	135.9	215.9	244.8	816.1	5.2	2.3	2.0	0.7
si175	4.6	6.9	12.2	15.0	5.0	6.7	10.6	13.5	0.9	1.0	1.2	1.1
tsp225	10.5	9.0	8.3	8.8	9.6	8.2	8.1	8.0	1.1	1.1	1.0	1.1
mean	1,864.9	1,204.5	1,088.1	857.4	159.2	228.0	260.1	366.8				

5.3   The α Value

In Table 3, we consider the α parameter for EPSrd with #sppw = 10. We note s.d. the standard deviation of the solving times. We notice that the importance of the alpha value relies on the number of re-decompositions. For instance, gr229 is re-decomposed between 0 and 1 time for each alpha value. Then, all the solving times are close to each other, and the standard deviation is quite low (2.2). Conversely, the standard deviation for gr666 is equal to 54.7 and we have between 5 and 11 re-decompositions for each alpha value. We also notice that, the more the number of re-decompositions is high the more the alpha value is important. Nevertheless, the alpha value does not have an impact as important as #sppw. Finally, considering the mean, $\alpha = 0.3$ is the best alpha value whereas considering the geometric mean $\alpha = \{0.1, 0.5, 0.7\}$ are the best alpha values.

Table 3. Impact of the α value on the solving times in seconds.

	α					#re-decomp.	s.d.
	0.1	0.3	0.5	0.7	0.9		
a280	1.6	5.2	1.6	1.5	1.5	0	1.6
ali535	463.9	415.2	467.9	471.1	500.0	4 to 7	30.6
d198	13.3	14.5	16.0	12.9	13.2	0 to 2	1.3
d493	1,084.5	1,033.7	1,135.9	1,048.8	1,101.6	6 to 11	41.0
gil262	30.6	26.9	30.0	34.0	31.4	1 to 3	2.6
gr229	16.2	16.4	12.8	11.8	12.6	0 to 1	2.2
gr431	82.6	95.1	86.1	87.2	94.7	1 to 4	5.5
gr666	926.0	822.5	786.9	807.9	810.8	5 to 11	54.7
kroA200	7.3	7.2	7.0	6.8	6.7	0	0.3
kroB150	2.6	2.7	2.7	2.6	2.8	0	0.1
kroB200	6.6	5.6	4.9	6.2	6.7	0	0.8
lin318	3.4	3.3	4.0	3.2	3.5	0	0.3
pcb442	159.5	210.6	243.1	270.4	314.2	5 to 8	58.7
pr136	1.7	1.4	1.8	1.5	1.6	0 to 1	0.2
pr264	16.5	12.0	11.6	11.7	12.9	0 to 1	2.0
pr299	211.3	237.4	236.3	248.3	317.4	8 to 14	40.0
rat195	5.7	5.9	5.0	5.9	6.3	0	0.5
rd400	135.9	132.5	134.0	135.7	140.4	4 to 6	3.0
si175	5.0	5.2	5.5	6.2	4.9	0 to 1	0.5
tsp225	9.6	9.7	10.5	10.4	10.2	0	0.4
mean	159.2	153.1	160.2	159.2	169.7		6.0
geo mean	24.5	25.5	24.5	24.5	25.5		

5.4 Computations that Have Already Been Made

By avoiding to redo computations already done after a re-decomposition we gain between a factor 2 (for pcb442) and a few percent (rd400). We generally gain between 10% and 20%.

5.5 Solving Evolution

In Table 4, we study the behavior of EPS and EPSrd when the best configuration is set. That is #sppw = 100 for EPS and #sppw = 10 and alpha = 0.1 for EPSrd. This experiment shows some drawbacks of EPS for some problems.

For EPS, we give some wall clock times (in s): the decomposition time, the wall clock time of the solving time when all workers are active, the wall clock time of the solving time when some (or most) workers are inactive and the wall clock time to solve the problem (total). Note that the wall clock time is the sum of "decomp.", "all workers are active" and "not all workers are active" times. Next, we give some information about EPSrd: the decomposition time, the wall clock time performed between the end of the first decomposition and the start of the first re-decomposition, the wall clock time of the solving time when all workers are active, the wall clock time of the solving time when some (or most) workers are inactive and the wall clock time to solve the problem (total). We can see that EPSrd obtains a better load balancing. For instance, with EPS gr666 spends 304.8 s when all the workers are active and 1,101 s when some workers are not active. Conversely, with EPSrd gr666 spends 401.6 s when all the workers are active and 142.9 s when some workers are not active. Then, when the load balancing is going to be bad, EPSrd quickly performs a first re-decomposition avoiding spending a lot of time with some inactive workers.

Table 4. Solving evolution of EPS and EPSrd. "W.A." stands for "workers are active"

Instances	EPS time(s)				EPSrd time(s)				
	decomp.	all W.A.	not all W.A.	total	decomp.	before first re-decomp.	all W.A.	not all W.A.	total
a280	1.3	0.0	0.3	1.7	1.3	0.3	0.0	0.3	1.6
ali535	30.1	3.7	4,022.5	4,056.3	141.6	9.9	238.9	83.3	463.9
d198	9.3	1.0	4.2	14.5	9.4	3.4	2.4	1.5	13.3
d493	135.7	8.6	2,102.9	2,247.2	178.5	16.6	801.4	104.6	1,084.5
gil262	105.4	6.7	1.5	113.6	17.8	3.8	6.5	6.3	30.6
gr229	20.7	19.2	1.4	41.4	10.8	4.4	3.9	1.6	16.2
gr431	146.4	5.9	17.1	169.4	63.8	8.8	6.8	12.0	82.6
gr666	149.4	304.8	1,101.0	1,555.2	381.5	56.8	401.6	142.9	926.0
kroA200	5.6	0.3	0.1	6.1	6.7	0.6	0.5	0.1	7.3
kroB150	2.3	0.0	0.2	2.5	2.3	0.2	0.1	0.2	2.6
kroB200	4.2	0.2	0.1	4.4	6.1	0.5	0.2	0.3	6.6
lin318	3.4	0.0	0.3	3.6	3.3	0.1	0.0	0.1	3.4
pcb442	80.9	58.5	5,027.4	5,166.8	51.2	18.2	67.8	40.4	159.5
pr136	1.5	0.1	0.2	1.7	1.0	0.7	0.1	0.6	1.7
pr264	10.4	1.1	1.4	12.8	12.0	4.0	2.8	1.7	16.5
pr299	112.9	5.8	3,041.9	3,160.6	80.6	5.4	85.3	45.4	211.3
rat195	4.6	0.1	0.1	4.8	5.5	0.2	0.1	0.2	5.7
rd400	241.2	25.0	294.6	560.9	60.4	17.0	50.5	24.9	135.9
si175	12.7	0.7	1.7	15.0	3.3	1.6	0.9	0.8	5.0
tsp225	8.0	0.2	0.5	8.8	9.0	0.7	0.2	0.4	9.6

5.6 Overall Results

Table 5. Comparison between EPS and EPSrd.

Instances	Sequential(s)	EPS(s)	Seq./EPS	EPSrd(s)	Seq./EPSrd	EPS/EPSrd
a280	8.1	1.7	4.8	1.6	5.0	1.0
ali535	13,013.9	4,056.3	3.2	463.9	28.1	8.7
d198	7.0	14.5	0.5	13.3	0.5	1.1
d493	4,989.9	2,247.2	2.2	1,084.5	4.6	2.1
gil262	2,042.2	113.6	18.0	30.6	66.7	3.7
gr229	41.2	41.4	1.0	16.2	2.5	2.6
gr431	520.6	169.4	3.1	82.6	6.3	2.0
gr666	14,719.6	1,555.2	9.5	926.0	15.9	1.7
kroA200	63.7	6.1	10.5	7.3	8.7	0.8
kroB150	8.7	2.5	3.5	2.6	3.3	1.0
kroB200	15.1	4.4	3.4	6.6	2.3	0.7
lin318	6.4	3.6	1.8	3.4	1.9	1.1
pcb442	91,137.7	5,166.8	17.6	159.5	571.5	32.4
pr136	8.9	1.7	5.1	1.7	5.2	1.0
pr264	3.8	12.8	0.3	16.5	0.2	0.8
pr299	1,596.5	3,160.6	0.5	211.3	7.6	15.0
rat195	17.9	4.8	3.7	5.7	3.1	0.8
rd400	6,721.0	560.9	12.0	135.9	49.5	4.1
si175	100.7	15.0	6.7	5.0	20.2	3.0
tsp225	23.5	8.8	2.7	9.6	2.4	0.9
mean	6,752.3	857.4	5.5	159.2	40.3	4.2

In Table 5, we show the general results obtained by EPS and EPSrd with their best configurations. Compared to the mean sequential solving times, an improvement by a factor 5.5 is observed for EPS and an improvement by a factor 40.3 is observed for EPSrd. Therefore, EPSrd is much more efficient than EPS (the mean improvement ratio is of 4.2).

5.7 Robustness

In Table 6, we compare the robustness of EPS and EPSrd. We run several times each instance and we compare the minimum solving time, the maximum solving time, the mean solving time and the standard deviation of the solving times. For EPS, most instances have quite a large variation between min and max and therefore a huge standard deviation. For instance, with EPS the min and the max solving times of gr666 are respectively 1,003 s and 1,555 s. It leads to a standard deviation of 229.1. Moreover, the max solving time is 55% slower than the min solving time. With EPSrd, the min and the max solving times of gr666 are respectively 855 s and 930 s with a standard deviation of 30.1. Then, the max solving time is 9% slower than the min solving time. Finally, the ratio of max/min for EPS is between 1.0 and 7.2 whereas it is between 1.1 and 1.3 for EPSrd. Then, EPSrd brings more robustness in the solving times than EPS.

Table 6. Comparison of the robustness of EPS and EPSrd.

	EPS					EPSrd				
	min	max	mean	s.d.	max/min	min	max	mean	s.d.	max/min
a280	1.5	2.0	1.7	0.2	1.3	1.6	1.9	1.7	0.1	1.2
ali535	3,827.2	4,400.1	4,072.5	241.2	1.1	447.5	561.7	478.3	47.0	1.3
d198	13.4	40.0	19.2	11.6	3.0	13.1	14.7	13.8	0.6	1.1
d493	2,247.2	4,784.6	3,229.2	1,175.3	2.1	858.0	1,092.2	979.8	104.8	1.3
gil262	106.0	125.3	115.3	7.0	1.2	28.1	35.1	30.9	2.8	1.2
gr229	39.6	41.4	40.9	0.7	1.0	16.2	18.0	17.0	0.7	1.1
gr431	169.4	180.8	173.8	4.4	1.1	78.9	84.8	82.2	2.6	1.1
gr666	1,003.2	1,555.2	1,180.0	229.1	1.6	854.8	930.0	904.2	30.1	1.1
kroA200	6.1	8.1	6.6	0.9	1.3	5.9	7.3	6.6	0.7	1.2
kroB150	2.5	2.6	2.5	0.0	1.0	2.5	3.0	2.7	0.2	1.2
kroB200	4.2	4.6	4.4	0.1	1.1	5.4	6.6	6.0	0.4	1.2
lin318	3.4	3.6	3.5	0.1	1.1	3.3	4.2	3.5	0.4	1.3
pcb442	3,288.8	23,783.8	11,776.5	8,244.6	7.2	159.5	195.1	180.8	16.4	1.2
pr136	1.7	1.7	1.7	0.0	1.0	1.3	1.7	1.5	0.2	1.3
pr264	12.8	14.2	13.6	0.5	1.1	15.3	17.5	16.3	0.8	1.1
pr299	1,412.1	3,160.6	2,222.7	656.9	2.2	189.9	211.3	198.8	10.1	1.1
rat195	4.5	4.9	4.8	0.2	1.1	5.3	6.4	5.7	0.4	1.2
rd400	560.9	759.9	644.9	74.7	1.4	130.6	142.3	137.6	4.5	1.1

6 Conclusion

In this paper, we focused on the parallel solving of difficult TSP problems with EPS. We identified three major issues: unstable decomposition, non-monotonic decomposition and extremely heterogeneous sub-problems. It has been shown that the use of a stopping criterion while decomposing can solve the non-monotonic decomposition issue [12]. We have shown that EPSrd, which uses a re-decomposition, improves EPS by solving the two other issues. Indeed, a re-decomposition allows the use of a small first decomposition and it allows avoiding the case of too small a proportion of workers working because of some extremely heterogeneous problems. Experimentally, we have been able to define that a small number of sub-problems per worker leads the best results when using EPSrd because the load balancing is managed by the re-decomposition.

References

1. Burton, F.W., Sleep, M.R.: Executing functional programs on a virtual tree of processors. In: Proceedings of the 1981 Conference on Functional Programming Languages and Computer Architecture, FPCA 1981, pp. 187–194. ACM, New York (1981)
2. Fages, J., Lorca, X., Rousseau, L.: The salesman and the tree: the importance of search in CP. Constraints **21**(2), 145–162 (2016)
3. Galea, F., Le Cun, B.: Bob++: a framework for exact combinatorial optimization methods on parallel machines. In: International Conference High Performance Computing & Simulation 2007 (HPCS 2007) and in conjunction with The 21st European Conference on Modeling and Simulation (ECMS 2007), pp. 779–785, June 2007

4. Halstead, R.: Implementation of multilisp: lisp on a multiprocessor. In: Proceedings of the 1984 ACM Symposium on LISP and Functional Programming, LFP 1984, pp. 9–17. ACM, New York (1984)
5. Haralick, R., Elliot, G.: Increasing tree search efficiency for constraint satisfaction problems. Artif. Intell. **14**, 263–313 (1980)
6. Held, M., Karp, R.M.: The traveling-salesman problem and minimum spanning trees. Oper. Res. **18**(6), 1138–1162 (1970)
7. Held, M., Karp, R.M.: The traveling-salesman problem and minimum spanning trees: Part II. Math. Program. **1**(1), 6–25 (1971)
8. Helsgaun, K.: An effective implementation of the Lin-Kernighan traveling salesman heuristic. Eur. J. Oper. Res. **126**(1), 106–130 (2000)
9. Isoart, N.: The traveling salesman problem in constraint programming. Ph.D. thesis, Université Côte d'Azur (2021)
10. Isoart, N., Régin, J.-C.: Integration of structural constraints into TSP models. In: Schiex, T., de Givry, S. (eds.) CP 2019. LNCS, vol. 11802, pp. 284–299. Springer, Cham (2019). https://doi.org/10.1007/978-3-030-30048-7_17
11. Isoart, N., Régin, J.-C.: Adaptive CP-based Lagrangian relaxation for TSP solving. In: Hebrard, E., Musliu, N. (eds.) CPAIOR 2020. LNCS, vol. 12296, pp. 300–316. Springer, Cham (2020). https://doi.org/10.1007/978-3-030-58942-4_20
12. Isoart, N., Régin, J.-C.: Parallelization of TSP solving in CP. In: Simonis, H. (ed.) CP 2020. LNCS, vol. 12333, pp. 410–426. Springer, Cham (2020). https://doi.org/10.1007/978-3-030-58475-7_24
13. Isoart, N., Régin, J.: A k-opt based constraint for the TSP. In: Michel, L.D. (ed.) 27th International Conference on Principles and Practice of Constraint Programming, CP 2021, Montpellier, France (Virtual Conference), 25–29 October 2021. LIPIcs, vol. 210, pp. 30:1–30:16. Schloss Dagstuhl - Leibniz-Zentrum für Informatik (2021)
14. Isoart, N., Régin, J.: A linear time algorithm for the k-cutset constraint. In: Michel, L.D. (ed.) 27th International Conference on Principles and Practice of Constraint Programming, CP 2021, Montpellier, France (Virtual Conference), 25–29 October 2021. LIPIcs, vol. 210, pp. 29:1–29:16. Schloss Dagstuhl - Leibniz-Zentrum für Informatik (2021)
15. Korf, R.: Depth-first iterative-deepening: an optimal admissible tree search. Artif. Intell. **27**, 97–109 (1985)
16. Le Cun, B., Menouer, T., Vander-Swalmen, P.: Bobpp (2007). http://forge.prism.uvsq.fr/projects/bobpp
17. Lecoutre, C., Sais, L., Tabary, S., Vidal, V., et al.: Nogood recording from restarts. In: IJCAI, vol. 7, pp. 131–136 (2007)
18. Lecoutre, C., Saïs, L., Tabary, S., Vidal, V.: Reasoning from last conflict(s) in constraint programming. Artif. Intell. **173**(18), 1592–1614 (2009)
19. Lee, J., Schulte, C., Zhu, Z.: Increasing nogoods in restart-based search. In: Proceedings of the AAAI Conference on Artificial Intelligence, vol. 30 (2016)
20. Lin, S., Kernighan, B.: An effective heuristic algorithm for the traveling-salesman problem. Oper. Res. **21**, 498–516 (1973)
21. Malapert, A., Régin, J., Rezgui, M.: Embarrassingly parallel search in constraint programming. J. Artif. Intell. Res. (JAIR) **57**, 421–464 (2016)
22. Perron, L.: Search procedures and parallelism in constraint programming. In: Jaffar, J. (ed.) CP 1999. LNCS, vol. 1713, pp. 346–360. Springer, Heidelberg (1999). https://doi.org/10.1007/978-3-540-48085-3_25

23. Régin, J.-C., Malapert, A.: Parallel constraint programming. In: Hamadi, Y., Sais, L. (eds.) Handbook of Parallel Constraint Reasoning, pp. 337–379. Springer, Cham (2018). https://doi.org/10.1007/978-3-319-63516-3_9

24. Régin, J.-C., Rezgui, M., Malapert, A.: Improvement of the embarrassingly parallel search for data centers. In: O'Sullivan, B. (ed.) CP 2014. LNCS, vol. 8656, pp. 622–635. Springer, Cham (2014). https://doi.org/10.1007/978-3-319-10428-7_45

25. Régin, J.-C., Rezgui, M., Malapert, A.: Embarrassingly parallel search. In: Schulte, C. (ed.) CP 2013. LNCS, vol. 8124, pp. 596–610. Springer, Heidelberg (2013). https://doi.org/10.1007/978-3-642-40627-0_45

26. Reinelt, G.: TSPLIB-a traveling salesman problem library. ORSA J. Comput. **3**(4), 376–384 (1991)

27. Vidal, V., Bordeaux, L., Hamadi, Y.: Adaptive K-parallel best-first search: a simple but efficient algorithm for multi-core domain-independent planning. In: Proceedings of the Third International Symposium on Combinatorial Search. AAAI Press (2010)

Efficient Operations Between MDDs and Constraints

Victor Jung[(✉)] and Jean-Charles Régin[ⒾⒹ]

Université Côte d'Azur, CNRS, I3S, Sophia-Antipolis, France
{victor.jung,jean-charles.regin}@univ-cotedazur.fr

Abstract. Many problems can be solved by performing operations between Multi-valued Decision Diagrams (MDDs), for example in music or text generation. Often these operations involve an MDD that represents the result of past operations and a new constraint. This approach is efficient, but it is very difficult to implement with some constraints such as ALLDIFFERENT or cardinality constraints because it is often impossible to represent them by an MDD because of their size (e.g. a permutation constraint involving n variables requires 2^n nodes).

In this paper, we propose to build on-the-fly MDDs of structured constraints as the operator needs them. For example, we show how to realise the intersection between an MDD and an ALLDIFFERENT constraint by never constructing more than the parts of the ALLDIFFERENT constraint that will be used to perform the intersection. In addition we show that we can anticipate some reductions (i.e. merge of MDD nodes) that normally occur after the end of the operation.

We prove that our method can be exponentially better than building the whole MDD beforehand and we present a direct application of our method to construct constraint MDDs without having to construct some intermediate states that will be removed by the reduction process.

At last, we give some experimental results confirming the gains of our approach in practice.

1 Introduction

Multi-valued and binary decision diagrams (MDDs/BDDs) took an important place in modern optimisation techniques. From the theory to the applications, MDDs have shown a large interest in operational research and optimisation [1,2,7,10,15,18]. They offer a broad range of modeling and solving possibilities, from being a basic block of constraint solvers [6,13], to the development of MDD-based solvers [3,9,19].

MDDs are a very efficient graph-based data structure to represent a set of solutions in a compressed way. The fundamental reasons of their use is their exponential compression power. A polynomial size MDD have the capacity to represent an exponential size set of tuples. For example in a music scheduling problem [16], an MDD having $14,000$ nodes and $600,000$ arcs stored 10^{90} meaningful tuples of size one hundred. Unlike trees, where each leaf represent a solution, an MDD can have much fewer nodes and arcs than solutions.

© Springer Nature Switzerland AG 2022
P. Schaus (Ed.): CPAIOR 2022, LNCS 13292, pp. 173–189, 2022.
https://doi.org/10.1007/978-3-031-08011-1_13

MDDs are also often reduced, that is a reduction operation is applied to them. The reduction operation of an MDD merges equivalent nodes until a fix point is reached. It may reduce the size of an MDD by an exponential factor.

Several other operators are available to combine MDDs without decompressing them. The most important are the intersection, which corresponds to a conjunction of two constraints, and the union, which corresponds to a disjunction of two constraints and the negation.

Some problems can be solved by a succession of operations applied on MDDs [8,14,16]. In other words, there is no search procedure that is used. To do so, the different constraints are represented by MDDs, and they are combined by applying operators between these different MDDs. In this way, all solutions can be computed at once. However, even if the final solution can fit into memory, it is possible that the memory explodes during the intermediate computation steps - worse, it is even quite frequent that the MDD of the constraints are themselves too big to fit in memory because they are exponential, as for most cardinality constraints (e.g. ALLDIFFERENT and global cardinality constraints). One solution to be able to represent such constraints is to relax them [4,10]: we gain memory in exchange for the loss of information. Usually, a relaxed MDD is an MDD representing a super set of the solutions of an exact MDD. Preferably, these relaxed MDDs are smaller than their exact versions. In general in such techniques, the total size is a fixed given parameter, which has a strong impact on the quality of the relaxation. Even if this approach is efficient, it can still be unsatisfactory to a certain extent. The ideal would be to be able to perform the computations while remaining exact. This is what interests us in this article: to be exact. In particular, we are interested in being able to perform operations without having to represent the whole constraint's MDD in order to avoid the problem of the intermediate representation.

More precisely the question we consider in this paper is: how can we compute an operation between a given MDD and the MDD of a constraint without consuming too much memory? We need to answer this question even if the MDD of the constraint cannot fit into memory. A simple example is the intersection between an MDD involving fifty variables and the MDD representing an ALLDIFFERENT constraints on these variables. This latter MDD will have 2^{50} nodes (i.e. $1,000,000$ Giga nodes) and so is too big to fit in memory.

The main ideas of this paper are to avoid building the MDD of the constraint before applying the operation and to anticipate the reduction of the obtained MDD.

The first idea can be implemented by using operators that proceed by layer [14] because we can avoid building in advance the MDD of the constraint. During the operation, if a node can hold all the information necessary to build its children in a way that satisfies the constraint, then we do not need to retain all previous nodes. Thus, we can perform the construction having only at most two layers in memory (the current layer and the next one being built). Furthermore, if an arc does not exist in the first MDD, then it does not need to exist in the constraint's MDD: building the MDD during the operation allows us to have more gain by only representing what is necessary to be represented.

The second idea is based on the remark that a constraint is useless when we can make sure that it will always be satisfied. For instance if 3 variables remains and if we know that they have disjoint domains then an ALLDIFFERENT constraints between these variables is useless. Thus, we can avoid defining the constraint's MDD and we can immediately merge some nodes that would have been merged by the reduction process and so gaining some space in memory.

The advantage of this approach is that it can provably gain an exponential factor in space. It has also a direct application to construct constraint MDDs without having to construct all intermediate states. This allows either to build constraint MDDs that cannot be built otherwise, or to build them much faster.

We also show that processing several constraints simultaneously is not advantageous compared to doing the operations successively for each constraint. This is due to the lack of reduction which is normally performed after each operation and which can strongly reduce the resulting.

The paper is organised as follows. First, we enrich the classical internal data of an MDD with several notions: we introduce the notion of node states allowing to represent the information associated with the node with respect to a certain constraint, the notion of transition function $\delta_C(s, v)$ and the notion verification function $V_C(s)$ allowing respectively to make the state of a node evolve by performing a transition of value v and to verify if a state is satisfying the constraint (absence of violation). Then, we elaborate on the importance of performing merges to have some control over the growing behavior of some constraints, by giving for each constraint described in the article the conditions to perform a merge. In addition, we try to convey some intuition of the potential gain behind these merges, which might greatly depend on the constraint's parameters. Next, we give for each of these constraints a possible implementation of the state and functions $\delta_C(s, v)$ and $V_C(s)$, as well as a generic algorithm allowing to perform an on-the-fly intersection operation based on the notions described. We also give the size of the MDD representing each constraint. Afterwards, we prove that building the constraint's MDD on the fly can be exponentially better than building the whole MDD beforehand. Finally, we present a direct application to compute constraint MDDs and we give some experimental results, notably for the car sequencing problem and we study the generalisation of our method for a set of constraints. At last, we conclude.

2 Preliminaries

2.1 Constraint Programming

A finite constraint network $\mathcal{N}$ is defined as a set of n variables $X = \{x_1, \ldots, x_n\}$, a set of current domains $\mathcal{D} = \{D(x_1), \ldots, D(x_n)\}$ where $D(x_i)$ is the finite set of possible values for variable x_i, and a set $\mathcal{C}$ of constraints between variables. We introduce the particular notation $\mathcal{D}_0 = \{D_0(x_1), \ldots, D_0(x_n)\}$ to represent the set of initial domains of $\mathcal{N}$ on which constraint definitions were stated. A constraint C on the ordered set of variables $X(C) = (x_{i_1}, \ldots, x_{i_r})$ is a subset $T(C)$ of the Cartesian product $D_0(x_{i_1}) \times \cdots \times D_0(x_{i_r})$ that specifies the allowed combinations of values for the variables $x_{i_1}, \ldots, x_{i_r}$. An element of $D_0(x_{i_1}) \times \cdots \times D_0(x_{i_r})$ is called a tuple on $X(C)$. A value a for a variable x is often denoted by (x, a). Let C be a constraint. A tuple τ on $X(C)$ is valid if $\forall (x, a) \in \tau, a \in D(x)$. C is consistent iff there exists a tuple τ of $T(C)$ which is valid. A value $a \in D(x)$ is consistent with C iff $x \notin X(C)$ or there exists a valid tuple τ of $T(C)$ with $(x, a) \in \tau$. We denote by $\#(a, \tau)$ the number of occurrences of the value a in a tuple τ.

We present some constraints that we will use in the rest of this paper.

Definition 1. *Given X a set of variables and $[l, u]$ a range, the* **sum constraint** *ensures that the sum of all variables $x \in X$ is at least l and at most u.*
$\text{SUM}(l, u) = \{\tau \mid \tau \text{ is a tuple on } X(C) \text{ and } l \leq \sum_{i=0} \tau_i \leq u\}.$

A global cardinality constraint (GCC) constrains the number of times every value can be taken by a set of variables. This is certainly one of the most useful constraints in practice. Note that the ALLDIFFERENT constraint corresponds to a GCC in which every value can be taken at most once.

Definition 2. *A* **global cardinality constraint** *is a constraint C in which each value $a_i \in D(X(C))$ is associated with two positive integers l_i and u_i with $l_i \leq u_i$ defined by*
$\text{GCC}(X, l, u) = \{\tau \mid \tau \text{ is a tuple on } X(C) \text{ and } \forall a_i \in D(X(C)) : l_i \leq \#(a_i, \tau) \leq u_i\}.$

2.2   Multi-valued Decision Diagram

The decision diagrams considered in this paper are reduced, ordered multi-valued decision diagrams (MDD) [2,12,17], which are a generalisation of binary decision diagrams [5]. They use a fixed variable ordering for canonical representation and shared sub-graphs for compression obtained by means of a reduction operation. an MDD is a rooted directed acyclic graph (DAG) used to represent some multi-valued functions $f : \{0 \ldots d - 1\}^n \rightarrow true, false$. Given the n input variables, the DAG contains $n + 1$ layers of nodes, such that each variable is represented at a specific layer of the graph. Each node on a given layer has at most d outgoing arcs to nodes in the next layer of the graph. Each arc is labeled by its corresponding integer. The arc (u, a, v) is from node u to node v and labeled by a. Sometimes it is convenient to say that v is a child of u. All outgoing arcs of the layer n reach tt, the true terminal node (the false terminal node is typically omitted). There is an equivalence between $f(a_1, \ldots, a_n) = true$ and the existence of a path from the root node to the tt whose arcs are labeled $a_1, \ldots, a_n$.

The reduction of an MDD is an important operation that may reduce the MDD size by an exponential factor. It consists in removing nodes that have no successor and merging equivalent nodes, i.e. nodes having the same set of children associated with the same labels. This means that only nodes of the same layer can be merged.

3 Generalisation of the Construction Process

Perez and Régin [14] have explained how an MDD can be built directly from functions defining an automaton or a pattern. The general principle is to define states and to link them by a transition function, which can be defined globally or for each level. In order to be more general and less dependent on automata theory, we propose to generalize the previous concepts and to introduce a verification function. This is similar to what is done in [9,11].

By doing so, the notion of *state*, the function of transition $\delta_C(s, v)$ as well as the verification function $V_C(s)$, form a lightweight and general scheme allowing an efficient on the fly construction of a constraint's MDD.

3.1 State, Transition and Verification

The notion of *state* holds some information in an MDD node about the constraint representation, giving it an actual meaning: for the SUM constraint, a state will hold the value of the sum for the current node. Given that piece of information, we will be able to build all the valid successors of a node.

In order to build the MDD layer by layer, i.e. to build the children of a node, we define a transition function on the nodes: this transition function takes into account the current state s of a node and the constraint C. Given a certain state, we need to build all successors of a node such that the state of each successor satisfies the constraint C.

Let $\delta_C(s, v)$ be the *transition* function that builds a new state from a state s and a label v and $V_C(s)$ the *verification* function that checks whether a state s satisfies the constraint or not.

Thanks to these notions we can define the following property:

Property 1. *Nodes of the same layer with the exact same state can be merged.*

Proof. The transition function δ_C takes into account the constraint C and the current state s to build the successors of a node. The constraint C being invariant during the construction process, it means that for a given layer if $s_1 = s_2$, then $\delta_C(s_1) = \delta_C(s_2)$. In an MDD, two nodes can be merged if they have the same successors: therefore, we can merge two nodes having the same state.   □

Immediate Merges. If during the construction of the MDD, we can clearly see that some nodes will be merged during the reduce operation, then we can immediately merge some states. However, merging states will result in the loss of precision concerning the information represented. For instance, if two nodes representing a sum s_1 and s_2 are merged, the result of this merge is the set $\{s_1, s_2\}$. The node holding this state could be considered to be in both state s_1 and s_2 at the same time. Now, let's imagine that we must have a sum satisfying the range $[10, 15]$, and that we have a merged state of $[14, 15]$ for a given node: can we add 1? The answer is yes when considering 14, and no when considering 15. This shows that, in order not to lose information about the constraint (i.e.

solutions), we have to be careful about what we merge - we cannot do it blind-folded. In a general way, a node holding a merged state would mean that the state of this node represents less information than the initial constraint (which is the case for the GCC constraint), or in extreme cases that the state of the node does not matter anymore (which is the case for the SUM constraint).

For each constraint described in this article, we will detail: the total number of states for the constraint, the conditions to perform a merge and its consequences for the size of the MDD. Please note that the impact of the merges on the size highly depends on the constraint and its parameters.

We will now present a possible implementation for the SUM, GCC, and ALLD-IFFERENT constraints. Henceforth, we will refer to the transition function δ as CREATESTATE and the verification function V_C as ISVALID.

3.2   Sum

Representation. The SUM constraint is simply represented by the minimum min and maximum max value of the sum. For convenience purposes, we also add the minimum v_{min} and maximum v_{max} value of D, the union of the domains of the variables involved in the constraint.

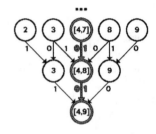

The state is represented by a single integer sum (Fig. 1).

Fig. 1. Merged states in the SUM, with $min = 4$ and $max = 9$. The merged path (red) necessarily lead to tt, whatever the value taken.

Transition and Validity. Creating a state means to add the label's value to the current sum, and a transition is valid if the obtained sum belongs in the constraint's bounds (See Algorithm 1).

Number of States. The number of states at layer i is at most $i \times (v_{max} - v_{min} + 1)$. The total number of states in the MDD is therefore at most:

$$2 + \sum_{i=1}^{n-1} i \times (v_{max} - v_{min} + 1) = 2 + \frac{(n-1)^2 + (n-1)}{2} \times (v_{max} - v_{min} + 1)$$

This upper bound is achieved when the set of values D is an integer range. This number can also be bounded by the upper and lower bounds of the sum.

Merging Condition. Let s be the state representing a sum. If $min \le s + (n - layer) \times v_{min}$ and $s + (n - layer) \times v_{max} \le max$, then we are certain to satisfy the constraint no matter what values we assign next. Therefore, we can drop the state of the node. Table 1 gives some experimental results of immediate merges.

Algorithm 1. Sum State

CREATESTATE(*constraint, state, label, layer, n*): state
| *nextState.sum* ← *state.sum* + *label*;
|_ **return** *nextState*;

ISVALID(*constraint, state, label, layer, n*): boolean
| *minReach* ← *state.sum* + *label* + $(n - layer) \times constraint.v_{min}$;
| *maxReach* ← *state.sum* + *label* + $(n - layer) \times constraint.v_{max}$;
| // We cannot reach the minimum sum or go below the maximum
| **if** (*maxReach* < *constraint.min* ∨ *minReach* > *constraint.max*) **then**
| |_ **return false**;
|_ **return true**;

Table 1. Impact of immediate merges on the number of nodes created and memory consumption for the SUM constraint

				With merges		Without merges	
min	max	n	|V|	#nodes	memory (MB)	#nodes	memory (MB)
20	200	50	10	6204	29	7914	35
124	480	200	14	82349	448	87047	472
500	1000	200	10	117864	456	131504	506
500	1000	200	20	161720	1192	168051	1242

3.3 GCC

Representation. To represent the GCC constraint, we need: the set V of constrained values, and the minimum lb_v and maximum ub_v occurrences of each specific value $v \in V$. To represent the state, we only need a array *count* that contains the number of occurrences of each value $v \in V$. We also define another variable named *minimum* that counts the minimum number of layers required to reach all the lower bounds of values.

Transition and Validity. Creating a new state means adding 1 to the counter of the value of the label we take if this value is constrained. A transition is valid if we can reach the lower bounds with the remaining layers, and if the added value of the label is not greater than its upper bound (always true if the value is not constrained, of course). Algorithm 2 is a possible implementation.

Notation 1

- n is number of variables and layer the index of the current layer.
- $\forall v \in V : c_v$ is the number of times v is assigned, lb_v the lower bound of v and ub_v the upper bound of v.

Number of States. The number of states in a GCC is at most:

$$\prod_{v\in V} ub_v$$

because this is the number of count tuples that can be represented by the numeral system defined by the GCC.

Merging Condition

Property 2. *We can remove the count c_v of value v from the state iff:*

$$(n - layer \leq ub_v - c_v) \wedge (lb_v \leq c_v + max(0, (n - layer) - \sum_{i\neq v}(u_i - c_i)))$$

If the number of variables left to assign is less than the number of times we can assign the value v, and if the lower bound lb_v is reached, then it means that, no matter how many times we assign the value v to the future variables, we are certain to be in the range $[lb_v, ub_v]$.

Example: Let the bound $[lb_v, ub_v] = [10, 20]$, $c_v = 10$ and $n - layer = 10$. Then, we can have a merged state $[c_v, c_v + i] = [10, 10 + i]$ up to $i = n - layer = 10$. In that case, the value v can be ignored (i.e. deleted) by the state because, no matter what choices we make, we are assured to satisfy the constraint: there is therefore no need to take into account v.

Table 2 gives some experimental results of immediate merges.

Table 2. Impact of immediate merges on the number of nodes created and memory consumption for the GCC constraint

With merges		Without merges	
#nodes	memory (MB)	#nodes	memory (MB)
405081	3308	543196	4300
58385	450	100341	715
5064	44	40558	266
430076	3584	470801	3849

3.4 AllDifferent

Representation. The ALLDIFFERENT constraint is simply a GCC constraint for which the set of values V contains all values and each value can only be assigned once. We represent the state by the set of previously assigned values.

Transition and Validity. A transition is valid if the *label* is not already assigned. To create a new state, we simply copy the current state (i.e. the set of assigned values) and add to it the new *label* (See Algorithm 3).

Algorithm 2. GCC State

CREATESTATE(*constraint, state, label, layer, n*): state

   count $\leftarrow \emptyset$;

   min $\leftarrow$ *state.minimum*;

   potential $\leftarrow n - layer - 1$;

   for each *value* $v \in$ *state.count* **do**

      if *state.count*[v] $<$ *constraint.min*[v] **then** *count*[v] $\leftarrow$ *state.count*[v]

      if *state.count*[v] $+$ *potential* $>$ *constraint.max*[v] **then**

         $\lfloor$ *count*[v] $\leftarrow$ *state.count*[v]

   if *label* $\in$ *count* **then**

      if *state.count*[*label*] $<$ *constraint.min*[*label*] **then** *min* $\leftarrow$ *min* $- 1$;

      // If we are sure to satisfy the constraint for the label

      if (*constraint.min*[*label*] $\leq$ *count*[*label*]$+1$)$\wedge$(*count*[*label*]$+n-layer \leq$ *constraint.max*[*label*]) **then** *count.remove*(*label*);

      else *count*[*label*] $\leftarrow$ *count*[*label*] $+ 1$

   nextState.count $\leftarrow$ *count*;

   nextState.minimum $\leftarrow$ *min*;

   return *nextState*;

ISVALID(*constraint, state, label, layer, n*): boolean

   potential $\leftarrow n - layer - 1$;

   min $\leftarrow$ *state.minimum*;

   if *label* $\notin$ *state.count.values* **then** **return** *min* $\leq$ *potential*;

   value $\leftarrow$ *state.count*[*label*];

   if *value* $<$ *constraint.min*[*label*] **then** *min* $\leftarrow$ *min* $- 1$;

   return (*min* $\leq$ *potential*) $\wedge$ (*value* $+ 1 \leq$ *constraint.max*[*label*]);

Number of States. The number of states in the ith layer for a given set of constrained values V is $\binom{|V|}{i}$. Therefore, the total number of nodes in the MDD is:

$$2^n \leq \sum_{i=0}^{n} \binom{|V|}{i} \leq 2^{|V|}$$

Merging Condition. The ALLDIFFERENT constraint can be seen as a GCC constraint where all values $v \in V$ are associated with the range $[0, 1]$. Thus, the merging conditions for the ALLDIFFERENT constraint is the same as the GCC under the described parameters: it means that we can only merge during the last layer, which is negligible.

3.5 Generic Constraint Intersection Function

This function is a possible implementation of the generic constraint intersection function. It is based on Functions defined in [14]. It shows that we do not need to store the full MDD$_c$ and that we build the next layer only knowing the current layer.

Algorithm 3. ALLDIFFERENT State

CREATESTATE($constraint, state, label, layer, n$): state
> $values \leftarrow \emptyset$;
> **for each** $v \in state.values$ **do** $values$.APPEND(v);
> **if** $label \in constraint.values$ **then** $values$.APPEND($label$);
> $nextState.values \leftarrow values$;
> **return** $nextState$;

ISVALID($constraint, state, label, layer, n$): boolean
> **return** $label \notin state.values$

Algorithm 4. Generic Constraint Intersection Function.

APPLYINTER($mdd_1, constraint, root_c$): MDD
> // When creating a node, we associate it with one node from each MDD
> $root \leftarrow$ CREATENODE($root(mdd_1), root_c$) ;
> $L[0] \leftarrow \{root\}$ // $L[i]$ is the set of nodes in layer i. ;
> $C[0] \leftarrow \{root_c\}$;
> **for each** $i \in 1..n$ **do**
>> $L[i] \leftarrow \emptyset; C[i] \leftarrow \emptyset$;
>> **for each** $node\ x \in L[i-1]$ **do**
>>> get x_1 and x_2 from $x = (x_1, x_2)$;
>>> **for each** $v \in$ CHILDOF(x_1) **do**
>>>> **if** $v \notin$ CHILDOF(x_2)$\wedge$ ISVALID($constraint, x2.state, v, i-1, r$)
>>>> **then**
>>>>> $y_2 \leftarrow$ CREATENODE();
>>>>> $y_2.state \leftarrow$ CREATESTATE($constraint, x_2.state, v, i-1, r$);
>>>>> ADDCHILD(x_2, v, y_2);
>>>>> $C[i]$.APPEND(y_2);
>>>>
>>>> // Add the arc between x and the node defined by $y = (y_1, y_2)$
>>>> // The node y will be added to the MDD if it is not yet in it.
>>>> ADDARCANDNODE(L, i, x, v, y_1, y_2) ;
>>
>> DESTROY($C[i-1]$) // Remove previous constraint layer from memory ;
> merge all nodes of $L[n]$ into t;
> pREDUCE(L) ;
> **return** $root$;

---

4   Exponential Gain

Theorem 1. *Building the MDD of a constraint on the fly can be exponentially better in terms of space and time than building the whole MDD beforehand.*

Proof. We show how to perform an intersection between MDD_x, an MDD, and MDD_{AD} the MDD of an ALLDIFFERENT constraint. Let x be a number of sets, MDD_x is built as follows (See Fig. 2 for $x = 3$ and $|X| = 3$):

– **Step 1** - Generate $MDD_U(X)$ a universal MDD with domain X. This means that the variables can take any value in X.

- **Step 2** - Copy the MDD from step 1 for x sets having a cardinality equal to $|X|$ and make the union of them. We denote by MDD$_V$ the obtained MDD.
- **Step 3** - Copy x times MDD_V, and concatenate them.

The size of MDD$_{AD}$ is exponential (i.e. at least 2^n). So, when n is large, it is not possible to build it. However, the size of MDD$_x$ ∩ MDD$_{AD}$ is exponentially smaller than the size of MDD$_{AD}$, and our method is able to compute this intersection as shown by the following process:

1. The number of nodes in the MDD$_{AD}$ involving all the variables is $2^{|X| \times x}$, being $(2^{|X|})^x$.
2. MDD$_V$ is the union of x universal MDDs. The intersection between a universal MDD and MDD$_C$ (i.e. the MDD of a constraint C) is MDD$_C$. So, for any set X, the intersection between MDD$_U(X)$ and MDD$_{AD}$ is equal to MDD$_{AD}$ which has $2^{|X|}$ nodes.
3. We can simplify by stating that each MDD$_U(X)$ is an arc. The shape of our MDD is therefore the one of a universal MDD. Thus, the same observation that in 2. applies: the simplified MDD intersecting with the ALLDIFFERENT constraint is the MDD of the ALLD-IFFERENT constraint. It is denoted by $metaMDD_{AD}$.
4. If we know the number of arcs in our $metaMDD_{AD}$, we can deduce the number of nodes created during the intersection.
5. The layer i of $metaMDD_{AD}$ contains $\binom{x}{i}$ nodes.
6. Each node in the layer i has $(x - i)$ out-going arcs (because we already chose i values of the x possible).
7. By combining (5) and (6), the total number of arcs in our $metaMDD_{AD}$ is $\sum_{i=0}^{x} \binom{x}{i} \times (x - i) = x \times 2^{x-1}$
8. By combining (2) and (7), the total number of nodes in our $metaMDD_{AD}$ is $x \times 2^{x-1} \times 2^{|X|} = x \times 2^{|X|+x-1}$, because each arc of $metaMDD_{AD}$ is an MDD$_{AD}(X)$ involving $|X|$ variables.
9. The difference between (8) $x \times 2^{|X|+x-1}$ and (1) $(2^{|X|})^x$ is exponential. □

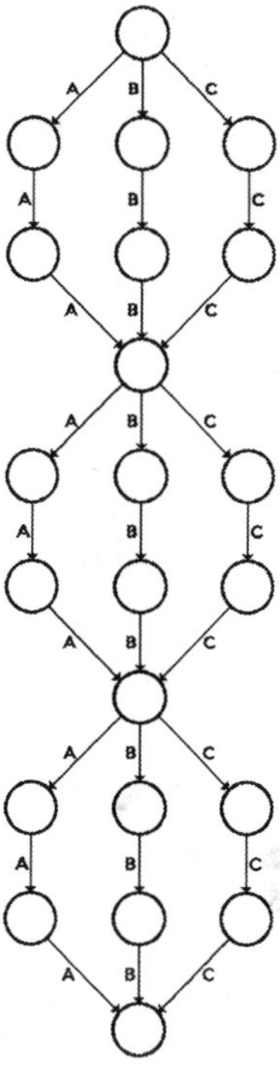

Fig. 2. Final step. For convenience purposes, we represent only one arc for a whole set of values.

We perform some benchmarks to experimentally confirm this gain.

4.1 Building the AllDifferent's MDD

Table 3 shows that the ALLDIFFERENT constraint quickly becomes impossible to construct: after barely 24 values it is impossible to represent the constraint.

Table 3. Construction of the ALLDIFFERENT MDD with size variation

|V|	Memory (MB)	Time (s)	Layer
20	8840	12.363	20
21	18789	27.476	21
22	39675	62.850	22
23	83619	134.696	23
24	Out of memory ($\geq 100\,$GB)	160.894	13

4.2 Performing the Construction on the Fly

The results of Table 4 show that by constructing the ALLDIFFERENT's MDD on the fly, it is possible to compute intersections with a lot of values (here between $|V| = 25$ and $|V| = 100$) very efficiently. We notice that, when we increase the number of sets, the intersection becomes more and more difficult to compute: this testifies to the exponential behaviour of the constraint.

Table 4. Evolution of Time (ms) and Memory (MB) consumption for MDD_{AD} intersection according to the variation of the number and size of sets (A, B, C in Fig. 2). The number of variables is equal to $Number \times Size$.

		Time (ms)						Memory (MB)				
Number \\ Size	5	6	7	8	9	10	5	6	7	8	9	10
5	8	16	36	84	192	424	50	22	53	128	284	665
6	19	36	132	212	492	1096	53	58	167	348	852	2042
7	44	96	224	524	1232	2748	101	151	374	930	2312	5613
8	101	257	548	1273	2993	8129	216	403	984	2456	6116	15433
9	241	564	1397	3384	7814	15957	543	1010	3042	6605	16012	50836
10	556	1348	3144	7116	16649	39073	1211	3079	6600	15308	37723	137917

The second test (Table 5) is a variant of the first one (Table 4). Arcs are added randomly between several sets, which as a consequence drastically increases the number of states in the MDD. The result is that intersection becomes impossible very quickly: for 6 sets and 6 values by set, we have a factor of 5 200 in time.

Table 5. Evolution of Time (ms) and Memory (MB) consumption for the ALLDIFFERENT MDD intersection according to the variation of the number and size of sets with random arcs added between sets

Number \\ Size	5			6		
	Time	Memory	#Arcs	Time	Memory	#Arcs
5	322	107	6	2753	505	6
6	7059	2513	6	187061	53981	7

5   Application: Construction of the MDD of Constraints

In this section, we show that our method can be useful to build the MDD of some constraints and not only to perform some intersections.

Consider C a constraint. Suppose that the construction of MDD_C, the MDD of C, is problematic because a very large number of intermediate states are generated but do not appear in the reduced MDD_C. As the reduction can gain an exponential factor this case is quite conceivable. It occurs, for example, with a bounded sum of variables that can take very different values. The number of states created is therefore huge, but it is quite possible that the reduction induced by the bounds on the sum removes a large part of them.

This kind of constraint can either prevent us from building the MDD due to lack of memory to store all the intermediate states, or require a lot of time to compute. To remedy these problems we propose to use successively our method on relaxations of MDD_C allowing to deal with smaller MDDs.

This approach assumes that it is possible to define different relaxations of MDD_C more or less strong. We recall that an MDD is a relaxation of an MDD if it represents a super set of the solutions of the exact MDD. In addition, we assume that the relaxation has fewer nodes. This is achievable by merging nodes for relaxing the MDD, which is quite usual. Thus, we suppose that we have MDDs noted $Relax(MDD_C,p)$ which are relaxations of MDD_C according to a parameter p such that $p < q$ implies $Relax(MDD_C,p)$ is smaller than or equal to $Relax(MDD_C,q)$. The value of p can be ad hoc. For example, for a sum constraint, a relaxation is simply to consider the numbers up to a given p precision. Thus we can merge many more states and the greater the precision the less the MDD is relaxed.

For convenience we will consider that $Relax(MDD_C,n)$ is MDD_C. Then, we can compute MDD_C by applying the following process named OTF Inc:

1. Let $M \leftarrow Relax(MDD_C,p)$
2. Compute M' by performing the intersection on the fly between M and $Relax(MDD_C,p+1)$
3. Set $p \leftarrow p+1$, and $M \leftarrow M'$
4. If $p < n$ then goto 2 else return M

6   Experiments

6.1   Constraint Building

We consider the following stochastic problem: there are n variables with domains having the same size. The values represent the chance for an event to appear. A solution is a combination of events such that their chance to happen simultaneously is above a certain threshold, for instance 75%. This problem is equivalent to a bounded product of variables. The goal is to build the MDD containing all the solutions. It is equivalent to building the MDD of C_Π the constraint defining a bounded product of variables. The difficulty is that values are quite different (because computed from other elements) leading to the lack of collision.

We propose to compute the MDD of this problem by using the process OTF Inc defined in Sect. 5. We define Relax(MDD$_{C_{II}}$,p) by the MDD of C_{II} for which the variables have been rounded to a precision p (i.e. the number of decimal places after the decimal point).

We test the method on different sets of data (available upon request). Each set involves values between 0.95 and 1 with at least 4 digits. The combined probability must be greater than 0.9. We compare the time and memory needed to compute the MDD using OTF Inc and the MDD directly computed (Base), both for a final precision $p = 8$, starting from $p = 0$. Table 6 shows that we obtain a factor of at least 9 both in time and in memory. We achieve up to a factor of 77.55 in time (181.5 s vs 2.6 s) and 19 in memory (405 MB vs 7 617 MB) for the hardest dataset (set 5).

Table 6. Time (ms) and memory (MB) needed to compute the exact MDD.

Set	Time (ms)			Memory (MB)		
	OTF Inc	Base	Ratio	OTF Inc	Base	Ratio
Set 1	1 131	10 235	9.05	136	1 423	10.46
Set 2	1 877	17 970	9.57	266	2 584	9.71
Set 3	1 405	25 528	18.17	189	1 869	9.89
Set 4	1 974	61 904	31.36	289	3 854	13.34
Set 5	2 642	181 468	77.55	405	7 617	18.81

6.2   The Car Sequencing Problem

A number of cars are to be produced. There are different options available to customise a car, and it is possible that a car has to be built with several options (paint job, sunroof, ABS, etc.). Each option is installed in a station that has a maximum handling capacity: if, for example, a station installing an option A can only handle one car in any two, then the assembly line must be designed in such a way that there are never two cars in a row requiring the option A. This constraint must be satisfied for each station. This problem is NP-complete.

All instances used in this article are available on csplib: https://www.csplib. org/Problems/prob001/data/data.txt.html

We will use the following methods:

- OTF: On The Fly, the method presented in this article.
- OTF$_\times$: OTF performing operations with multiple constraints at once.
- Classic: The method that builds the MDD then performs the operation.

For the car sequencing problem it means that the final MDD is built as follows: for the classic method, we build the MDD containing all SEQUENCE constraints, then we build the MDD of the GCC, and we perform the intersection between these two MDDs. For the OTF method, we do the same as the classic method, but the intersection with the GCC is done on the fly (we do not build the GCC MDD). Finally, for the OTF$_\times$, we directly compute the intersection between the SEQUENCE and GCC, without building them explicitly.

Table 7. Problem 4/72 (Regin & Puget #1), Problem 19/71 (Regin & Puget #4) and Problem 60-02 from CSPLib. Time measured in ms and Memory in MB.

		Problem 4/72			Problem 19/71		
#options	Method	Time	Memory	Layer	Time	Memory	Layer
	OTF	**1042**	**187**	100	**915**	**127**	100
2	OTF$_\times$	3435	712	100	5858	1274	100
	Classic	5177	1684	100	4845	1356	100
	OTF	**665 628**	**90 614**	100	**73 019**	**15 724**	100
3	OTF$_\times$	2 736 002	224 931	69^{1*}	847 006	114 448	100
	Classic	957 063	231 213	38^{2*}	819 905	214 445	38^{2*}
		Problem 60-02					
	OTF	**41 129**	**9 043**	100			
2	OTF$_\times$	315 825	33 986	100			
	Classic	51 975	11 059	100			
	OTF	1 960 854	212 669	41^{3*}			
3	OTF$_\times$	2 763 521	**172 543**	31^{1*}			
	Classic	**957 063**	231 213	38^{2*}			

^{1*}: MO during intersection of SEQUENCE + GCC
^{2*}: MO during the construction of the GCC, after the SEQUENCE intersection
^{3*}: MO during the GCC intersection

Table 7 shows that it is possible to build an MDD that is **not** possible to build otherwise (because the GCC explodes in memory). Thus, we can conclude that an instance has no solution (*Problem 19/71*), which we could not do before. In the case of larger problems, we still manage to observe a strong progression, even if it remains insufficient: where we could only build 38 layers of the GCC, we manage to carry out the intersection with it up to layer 41 (*Problem 60-02*). These results clearly show the advantage of this intersection method.

We notice that OTF$_\times$ is systematically worse than the method doing them one by one (*Problem 4/72, Problem 60-02*), and even worse in some cases than Classic (*Problem 60-02*). This can be explained by the fact that the complexity is proportional to the number of states constructed for each constraint. However, by making successive intersections, we observe a reduction in the number of solutions, which can imply (and does imply in a general case) a reduction of states. Performing several operations at the same time is therefore not interesting, especially if the MDDs are easy to construct, i.e. they are not exponential in memory like would be ALLDIFFERENT or GCC.

7   Conclusion

This article shows that building constraints' MDD during an operation is more advantageous in every way than building the complete constraint's MDD first, even if it does not prevent an explosion of memory. Moreover, this method shows a major impact in performance for solving some well known problems or building MDDs of constraints. However, doing multiple constraints at once is not necessarily better, and is shown to be worse most of times.

References

1. Andersen, H.R.: An Introduction to Binary Decision Diagrams (1999)
2. Bergman, D., Ciré, A.A., van Hoeve, W., Hooker, J.N.: Decision Diagrams for Optimization. Artificial Intelligence: Foundations, Theory, and Algorithms. Springer, Heidelberg (2016). https://doi.org/10.1007/978-3-319-42849-9
3. Bergman, D., Cire, A.A., Van Hoeve, W.J., Hooker, J.N.: Discrete optimization with decision diagrams. INFORMS J. Comput. **28**(1), 47–66 (2016)
4. Bergman, D., van Hoeve, W.-J., Hooker, J.N.: Manipulating MDD relaxations for combinatorial optimization. In: Achterberg, T., Beck, J.C. (eds.) CPAIOR 2011. LNCS, vol. 6697, pp. 20–35. Springer, Heidelberg (2011). https://doi.org/10.1007/978-3-642-21311-3_5
5. Bryant, R.E.: Graph-based algorithms for Boolean function manipulation. IEEE Trans. Comput. **35**(8), 677–691 (1986). https://doi.org/10.1109/TC.1986.1676819
6. Cheng, K.C.K., Yap, R.H.C.: An MDD-based generalized arc consistency algorithm for positive and negative table constraints and some global constraints. Constraints **15**(2), 265–304 (2010). https://doi.org/10.1007/s10601-009-9087-y
7. Davarnia, D., van Hoeve, W.: Outer approximation for integer nonlinear programs via decision diagrams. Math. Program. **187**(1), 111–150 (2021). https://doi.org/10.1007/s10107-020-01475-4
8. Demassey, S.: Compositions and hybridizations for applied combinatorial optimization. Habilitation à Diriger des Recherches (2017)
9. Gentzel, R., Michel, L., van Hoeve, W.-J.: HADDOCK: a language and architecture for decision diagram compilation. In: Simonis, H. (ed.) CP 2020. LNCS, vol. 12333, pp. 531–547. Springer, Cham (2020). https://doi.org/10.1007/978-3-030-58475-7_31
10. Hadzic, T., Hooker, J.N., O'Sullivan, B., Tiedemann, P.: Approximate compilation of constraints into multivalued decision diagrams. In: Stuckey, P.J. (ed.) CP 2008. LNCS, vol. 5202, pp. 448–462. Springer, Heidelberg (2008). https://doi.org/10.1007/978-3-540-85958-1_30
11. Hoda, S., van Hoeve, W.-J., Hooker, J.N.: A systematic approach to MDD-based constraint programming. In: Cohen, D. (ed.) CP 2010. LNCS, vol. 6308, pp. 266–280. Springer, Heidelberg (2010). https://doi.org/10.1007/978-3-642-15396-9_23
12. Kam, T., Brayton, R.K.: Multi-valued decision diagrams. Technical report. UCB/ERL M90/125, EECS Department, University of California, Berkeley. http://www2.eecs.berkeley.edu/Pubs/TechRpts/1990/1671.html
13. Perez, G., Régin, J.-C.: Improving GAC-4 for table and MDD constraints. In: O'Sullivan, B. (ed.) CP 2014. LNCS, vol. 8656, pp. 606–621. Springer, Cham (2014). https://doi.org/10.1007/978-3-319-10428-7_44

14. Perez, G., Régin, J.C.: Efficient operations on MDDs for building constraint programming models. In: International Joint Conference on Artificial Intelligence, IJCAI 2015, Argentina, pp. 374–380 (2015)
15. Perez, G., Régin, J.C.: Soft and cost MDD propagators. In: The Thirty-First AAAI Conference on Artificial Intelligence (AAAI 2017) (2017)
16. Roy, P., Perez, G., Régin, J.-C., Papadopoulos, A., Pachet, F., Marchini, M.: Enforcing structure on temporal sequences: the Allen constraint. In: Rueher, M. (ed.) CP 2016. LNCS, vol. 9892, pp. 786–801. Springer, Cham (2016). https://doi.org/10.1007/978-3-319-44953-1_49
17. Srinivasan, A., Ham, T., Malik, S., Brayton, R.K.: Algorithms for discrete function manipulation. In: 1990 IEEE International Conference on Computer-Aided Design. Digest of Technical Papers, pp. 92–95 (1990). https://doi.org/10.1109/ICCAD.1990.129849
18. Tjandraatmadja, C., van Hoeve, W.-J.: Incorporating bounds from decision diagrams into integer programming. Math. Program. Comput. **13**(2), 225–256 (2020). https://doi.org/10.1007/s12532-020-00191-6
19. Verhaeghe, H., Lecoutre, C., Schaus, P.: Compact-MDD: efficiently filtering (s) MDD constraints with reversible sparse bit-sets. In: IJCAI, pp. 1383–1389 (2018)

Deep Policy Dynamic Programming for Vehicle Routing Problems

Wouter Kool[1,2]([✉]) [iD], Herke van Hoof[1] [iD], Joaquim Gromicho[1,2] [iD],
and Max Welling[1] [iD]

[1] University of Amsterdam, Amsterdam, The Netherlands
w.w.m.kool@uva.nl
[2] ORTEC, Zoetermeer, The Netherlands

Abstract. Routing problems are a class of combinatorial problems with many practical applications. Recently, end-to-end deep learning methods have been proposed to learn approximate solution heuristics for such problems. In contrast, classical dynamic programming (DP) algorithms guarantee optimal solutions, but scale badly with the problem size. We propose *Deep Policy Dynamic Programming* (DPDP), which aims to combine the strengths of learned neural heuristics with those of DP algorithms. DPDP prioritizes and restricts the DP state space using a policy derived from a deep neural network, which is trained to predict edges from example solutions. We evaluate our framework on the travelling salesman problem (TSP), the vehicle routing problem (VRP) and TSP with time windows (TSPTW) and show that the neural policy improves the performance of (restricted) DP algorithms, making them competitive to strong alternatives such as LKH, while also outperforming most other 'neural approaches' for solving TSPs, VRPs and TSPTWs with 100 nodes.

Keywords: Dynamic Programming · Deep Learning · Vehicle Routing

1 Introduction

Dynamic programming (DP) [7] is a powerful framework for solving optimization problems by solving smaller subproblems through the principle of optimality [4]. Famous examples are Dijkstra's algorithm [16] for the shortest route between two locations, and the classic Held-Karp algorithm for the travelling salesman problem (TSP) [5,26]. Despite their long history, dynamic programming algorithms for vehicle routing problems (VRPs) have seen limited use in practice, primarily due to their bad scaling performance. More recently, a line of research has attempted the use of machine learning (especially deep learning) to automatically learn heuristics for solving routing problems [6,9,36,50,64]. While the results are promising, most learned heuristics are not (yet) competitive to 'traditional' algorithms such as LKH [27] and lack (asymptotic) guarantees on their performance.

© Springer Nature Switzerland AG 2022
P. Schaus (Ed.): CPAIOR 2022, LNCS 13292, pp. 190–213, 2022.
https://doi.org/10.1007/978-3-031-08011-1_14

In this paper, we propose *Deep Policy Dynamic Programming* (DPDP) as a framework for solving vehicle routing problems. The key of DPDP is to combine the strengths of deep learning and DP, by restricting the DP state space (the search space) using a policy derived from a neural network. In Fig. 1 it can be seen how the neural network indicates promising parts of the search space as a *heatmap* over the edges of the graph. This heatmap used by the DP algorithm to find a good solution. DPDP is more powerful than some related ideas [8, 25, 42, 69, 70] as it combines supervised training of a large neural network with just a *single* model evaluation at test time, to enable running a large scale guided search using DP. The DP framework is flexible as it can model a variety of realistic routing problems with difficult practical constraints [22]. We illustrate this by testing DPDP on the TSP, the capacitated VRP and the TSP with (hard) time window constraints (TSPTW).

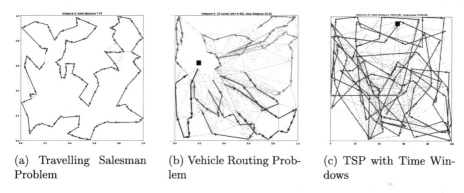

(a) Travelling Salesman Problem

(b) Vehicle Routing Problem

(c) TSP with Time Windows

Fig. 1. Heatmap predictions (red) and solutions (colored) by DPDP (VRP depot edges omitted for clarity). The heatmap indicates only a small fraction of all edges as promising, while including (almost) all edges from the solution. (Color figure online)

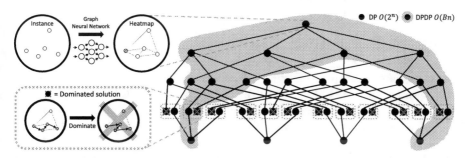

Fig. 2. DPDP for the TSP. A GNN creates a (sparse) heatmap indicating promising edges, after which a tour is constructed using forward dynamic programming. In each step, at most B solutions are expanded according to the heatmap policy, restricting the size of the search space. Partial solutions are dominated by shorter (lower cost) solutions with the same DP state: the same nodes visited (marked grey) and current node (indicated by dashed rectangles).

In more detail, the starting point of our proposed approach is a *restricted dynamic programming* algorithm [22,46], which heuristically reduces the search space by retaining at most B solutions per iteration. The selection process is important as it defines the part of the DP state space considered and, thus, the quality of the solution found (see Fig. 2). DPDP defines the selection using a (sparse) heatmap of promising route segments, obtained by pre-processing the problem instance using a (deep) graph neural network (GNN) [32]. This brings the power of neural networks to DP, inspired by the success of neural networks that improved tree search [57] or branch-and-bound algorithms [21,49].

In this work, we thus aim for a 'neural boost' of DP algorithms, by using a GNN for scoring partial solutions. Prior work on 'neural' vehicle routing has focused on auto-regressive models [6,15,36,64], but they have high computational cost when combined with (any form of) search, as the model needs to be evaluated for each partial solution considered. Instead, we use a model to predict a heatmap indicating promising edges [32], and define the *score* of a partial solution as the 'heat' of the edges it contains (plus an estimate of the 'heat-to-go' or *potential* of the solution). As the neural network only needs to be evaluated *once* for each instance, this enables a *much larger search* (defined by B), making a good trade-off between quality and computational cost. Additionally, we can apply a threshold to the heatmap to define a sparse graph on which to run the DP algorithm, reducing the runtime by eliminating many solutions.

Figure 2 illustrates DPDP. In Sect. 4, we show that DPDP significantly improves over 'classic' restricted DP algorithms. Additionally, we show that DPDP outperformes most other 'neural' approaches for TSP, VRP and TSPTW and is competitive with the highly-optimized LKH solver [27] for VRP, while achieving similar results much faster for TSP and TSPTW. For TSPTW, DPDP also outperforms the best open-source solver we could find [12], illustrating the power of DPDP to handle difficult hard constraints (time windows).

2   Related Work

DP [7] has a long history as an exact solution method for routing problems [38,59], e.g. the TSP with time windows [17] and precedence constraints [48], but is limited to small problems due to the curse of dimensionality. Restricted DP (with heuristic policies) has been used to address, e.g., the time dependent TSP [46], and has been generalized into a flexible framework for VRPs with different types of practical constraints [22]. DP approaches have also been shown to be useful in settings with difficult practical issues such as time-dependent travel times and driving regulations [35] or stochastic demands [51]. For more examples of DP for routing (and scheduling), see [28]. For sparse graphs, alternative, but less flexible, formulations can be used [10].

Despite the flexibility, DP methods have not gained much popularity compared to heuristic approaches such as R&R [56], ALNS [55], LKH [27], HGS [62,63] or FILO [1], which, while effective, have limited flexibility as special operators are needed for different types of problems. While restricted DP was shown to have superior performance on *realistic* VRPs with many constraints [22], the performance gap of around 10% for standard (benchmark) VRPs (with time windows) is too large to popularize this approach. We argue that the missing ingredient is a strong but computationally cheap policy for selecting which solutions to consider, which is the motivation behind DPDP.

In the machine learning community, deep neural networks (DNNs) have recently boosted performance on various tasks [39]. After the first DNN model was trained (using example solutions) to construct TSP tours [64], many improvements have been proposed, e.g. different training strategies such as reinforcement learning (RL) [6,14,33,37] and model architectures, which enabled the same idea to be used for other routing problems [15,18,36,45,50,54,67]. Most constructive neural methods are *auto-regressive*, evaluating the model many times to predict one node at the time, but other works have considered predicting a heatmap of promising edges *at once* [19,32,52], which allows a tour to be constructed (using sampling or beam search) without further evaluating the model. An alternative to constructive methods is 'learning to search', where a neural network is used to guide a search procedure such as local search [9,20,29,30,34,41,43,66,68]. Scaling to instances beyond 100 nodes remains challenging [19,44].

The combination of machine learning with DP has been proposed in limited settings [25,69,70]. Most related to our approach, a DP algorithm for TSPTW, guided by an RL agent, was implemented using an existing solver [8], which is less efficient than DPDP (see Sect. 4.3). Also similar to our approach, a neural network predicting edges has been combined with tree search and local search for maximum independent set (MIS) [42]. Whereas DPDP directly builds on the idea of predicting promising edges [32,42], it uses these more efficiently through a policy with *potential function* (see Sect. 3.2), and by using DP rather than tree search or beam search, we exploit known problem structure in a principled and general manner. As such, DPDP obtains strong performance without using extra heuristics such as local search. For a wider view on machine learning for routing problems and combinatorial optimization, see [3,47,61].

3   Deep Policy Dynamic Programming

DPDP uses an existing graph neural network [32], suitably adapted for VRP and TSPTW, to predict a heatmap of promising edges. This heatmap is used in the DP algorithm in two ways: 1) to exclude edges with a value below the *heatmap threshold* of 10^{-5} from the graph and 2) to define a *scoring policy* to select candidate solutions in each iteration. In more detail, as illustrated in Fig. 2, the DP algorithm starts with a *beam* of a single initial (empty) solution, and proceeds by iterating the following steps: (1) all solutions on the beam are expanded, (2) dominated solutions are removed for each *DP state*, (3) the B best solutions according to the scoring policy define the beam for the next iteration. The objective function is used to select the best solution from the final beam. The resulting algorithm is a *beam search* over the *DP state space*, with *beam size*

B. This is different from a 'standard' beam search, which considers the *solution space* by not removing dominated solutions. DPDP is asymptotically optimal as using $B = n \cdot 2^n$ for a TSP with n nodes guarantees optimal results, but by choosing a smaller B, DPDP can trade off performance for computational cost.

DPDP is a generic framework that can be applied to different problems, by defining the following ingredients: (1) the **variables** to track while constructing solutions, (2) the **initial solution**, (3) **feasible actions** to expand solutions, (4) rules to define **dominated solutions** and (5) the **scoring policy**, based on the neural network, for selecting the B solutions to keep. A solution is always defined by a sequence of actions, which allows the DP algorithm to construct the final solution by backtracking. In the next sections, we describe the neural network and define the DPDP ingredients for the TSP, VRP and TSPTW.

3.1   The Graph Neural Network

We use the original (pre-trained) model from [32] (which we describe in detail in Appendix 1 for self-containment) for the TSP, but we modify the neural network architecture and train new models to support the VRP and TSPTW, as we describe in Sects. 3.3 and 3.4. In general, the resulting model uses problem-specific node input features and edge input features, which get transformed into initial representations of the nodes and edges. These representations then get updated sequentially using a number of *graph convolutional layers*, which exchange information between the nodes and edges. The final edge representation is used to make the prediction whether the edge is promising, i.e. whether it has a high probability of being part of the optimal solution.

The model is trained using a large training dataset of problem instances with optimal (or high-quality) solutions, obtained using an existing solver. While it takes a significant amount of resources to create this dataset and train the model (each of which can take up to a number of days on a single machine), training of the model is, in principle, only required once given a specific distribution of problem instances. We consider only instances with $n = 100$ nodes, but the model can handle instances of different graph sizes, although good generalization may be limited to graphs with sizes close to the size trained for [33,36].

3.2   Travelling Salesman Problem

We implement DPDP for Euclidean TSPs with n nodes on a (sparse) graph, where the cost for edge (i, j) is given by c_{ij}, the Euclidean distance between the nodes i and j. The objective is to construct a tour that visits all nodes (and returns to the start node) and minimizes the total cost of its edges.

For each partial solution, defined by a sequence of actions $\boldsymbol{a}$, the **variables** we track are $\text{cost}(\boldsymbol{a})$, the total *cost* (distance), $\text{current}(\boldsymbol{a})$, the current node, and $\text{visited}(\boldsymbol{a})$, the set of visited nodes (including the start node). Without loss of generality, we let 0 be the start node, so we initialize the beam at step $t = 0$ with the empty **initial solution** with $\text{cost}(\boldsymbol{a}) = 0$, $\text{current}(\boldsymbol{a}) = 0$ and $\text{visited}(\boldsymbol{a}) = \{0\}$. At step t, the action $a_t \in \{0, ..., n-1\}$ indicates the next node to visit, and is a **feasible action** for a partial solution $\boldsymbol{a} = (a_0, ..., a_{t-1})$

if (a_{t-1}, a_t) is an edge in the graph and $a_t \notin$ visited($\boldsymbol{a}$), or, when all nodes are visited, if $a_t = 0$ to return to the start node. When expanding the solution to $\boldsymbol{a}' = (a_0, ..., a_t)$, we can compute the tracked variables incrementally as:

$$\text{cost}(\boldsymbol{a}') = \text{cost}(\boldsymbol{a}) + c_{\text{current}(\boldsymbol{a}), a_t}, \; \text{current}(\boldsymbol{a}') = a_t, \; \text{visited}(\boldsymbol{a}') = \text{visited}(\boldsymbol{a}) \cup \{a_t\}. \tag{1}$$

A (partial) solution $\boldsymbol{a}$ is a **dominated solution** if there exists a (dominating) solution $\boldsymbol{a}^*$ such that visited($\boldsymbol{a}^*$) = visited($\boldsymbol{a}$), current($\boldsymbol{a}^*$) = current($\boldsymbol{a}$) and cost($\boldsymbol{a}^*$) < cost($\boldsymbol{a}$). We refer to the tuple (visited($\boldsymbol{a}$), current($\boldsymbol{a}$)) as the *DP state*, so removing all dominated partial solutions, we keep exactly one minimum-cost solution for each unique DP state[1]. A solution can only dominate other solutions with the same set of visited nodes, so we only need to remove dominated solutions from sets of solutions with the same number of actions. This is why the DP algorithm can be executed in iterations (as explained): at step t all solutions in the beam have t actions and $t + 1$ visited nodes (including the start node). The resulting memory need is thus limited to $O(B)$ states, with B the beam size.

We define the **scoring policy** using the pretrained model from [32], which takes as input node coordinates and edge distances to predict a raw heatmap value $\hat{h}_{ij} \in (0, 1)$ for each edge (i, j). The model was trained to predict optimal solutions, so $\hat{h}_{ij}$ can be seen as the probability that edge (i, j) is in the optimal tour. We force the heatmap to be symmetric thus we define $h_{ij} = \max\{\hat{h}_{ij}, \hat{h}_{ji}\}$. The policy is defined using the heatmap values, in such a way to select the (partial) solutions with the largest total *heat*, while also taking into account the (heat) *potential* for the unvisited nodes. The policy thus selects the B solutions which have the highest *score*, defined as score($\boldsymbol{a}$) = heat($\boldsymbol{a}$) + potential($\boldsymbol{a}$), with heat($\boldsymbol{a}$) = $\sum_{i=1}^{t-1} h_{a_{i-1}, a_i}$, i.e. the sum of the heat of the edges, which can be computed incrementally when expanding a solution. The potential is added as an estimate of the 'heat-to-go' (similar to the heuristic in A^* search) for the remaining nodes, and avoids the 'greedy pitfall' of selecting the best edges while skipping over nearby nodes, which would prevent good edges from being used later. It is defined as potential($\boldsymbol{a}$) = potential$_0$($\boldsymbol{a}$) + $\sum_{i \notin \text{visited}(\boldsymbol{a})}$ potential$_i$($\boldsymbol{a}$) with potential$_i$($\boldsymbol{a}$) = $w_i \sum_{j \notin \text{visited}(\boldsymbol{a})} \frac{h_{ji}}{\sum_{k=0}^{n-1} h_{ki}}$, where w_i is the node *potential weight* given by $w_i = (\max_j h_{ji}) \cdot (1 - 0.1(\frac{c_{i0}}{\max_j c_{j0}} - 0.5))$. By normalizing the heatmap values for incoming edges, the (remaining) potential for node i is initially equal to w_i but decreases as good edges become infeasible due to neighbors being visited. The node potential weight w_i is equal to the maximum incoming edge heatmap value (an upper bound to the heat contributed by node i), which gets multiplied by a factor 0.95 to 1.05 to give a higher weight to nodes closer to the start node, which we found helps to encourage the algorithm to keep edges that enable to return to the start node. The overall heat + potential function identifies promising partial solutions and is computationally cheap. It is a heuris-

---

[1] If we have multiple partial solutions with the same state and cost, we can arbitrarily choose one to dominate the other(s), for example the one with the lowest index of the current node.

tic estimate of the total heat of the complete solution, but it is not an estimate of the cost objective (which has a different unit), neither it is a *bound* on the total heat or cost objective.

3.3 Vehicle Routing Problem

For the VRP, we add a special depot node DEP to the graph. Node i has a demand d_i, and the goal is to minimize the cost for a set of routes that visit all nodes. Each route must start and end at the depot, and the total demand of its nodes cannot exceed the vehicle capacity denoted by CAPACITY.

Additionally to the TSP **variables** cost(a), current(a) and visited(a), we keep track of capacity(a), which is the *remaining* capacity in the current route/vehicle. A solution starts at the depot, so we initialize the beam at step $t = 0$ with the empty **initial solution** with cost(a) = 0, current(a) = DEP, visited(a) = $\emptyset$ and capacity(a) = CAPACITY. For the VRP, we do not consider visiting the depot as a separate action. Instead, we define $2n$ actions, where $a_t \in \{0, ..., 2n - 1\}$. The actions $0, ..., n - 1$ indicate a *direct* move from the current node to node a_t, whereas the actions $n, ..., 2n - 1$ indicate a move to node $a_t - n$ *via the depot*. **Feasible actions** are those that move to unvisited nodes via edges in the graph and obey the following constraints. For the first action a_0 there is no choice and we constrain (for convenience of implementation) $a_0 \in \{n, ..., 2n - 1\}$. A direct move ($a_t < n$) is only feasible if $d_{a_t} \leq$ capacity(a) and updates the state similar to TSP but reduces remaining capacity by d_{a_t}. A move via the depot is always feasible (respecting the graph edges and assuming $d_i \leq$ CAPACITY $\forall i$) as it resets the vehicle CAPACITY before subtracting demand, but incurs the 'via-depot cost' $c_{ij}^{\text{DEP}} = c_{i,\text{DEP}} + c_{\text{DEP},j}$. When all nodes are visited, we allow a special action to return to the depot. This somewhat unusual way of representing a VRP solution has desirable properties similar to the TSP formulation: at step t we have exactly t nodes visited, and we can run the DP in iterations, removing dominated solutions at each step t.

For VRP, a partial solution a is a **dominated solution** dominated by a^* if visited(a^*) = visited(a) and current(a^*) = current(a) (i.e. a^* corresponds to the same DP state) and cost(a^*) $\leq$ cost(a) and capacity(a^*) $\geq$ capacity(a), with *at least one of the two inequalities being strict*. This means that for each DP state, given by the set of visited nodes and the current node, we do not only keep the (single) solution with lowest cost (as in the TSP algorithm), but keep the complete set of pareto-efficient solutions in terms of cost and remaining vehicle capacity. This is because a higher cost partial solution may still be preferred if it has more remaining vehicle capacity, and vice versa.

For the VRP **scoring policy**, we modify the model [32] (described in Appendix 1) to include the depot node and demands. We mark the depot as a special node type, which affects the initial node representation similarly to edge types, and we add additional edge types for connections to the depot. Additionally, each node gets an extra input (next to its coordinates) corresponding to d_i/CAPACITY (where we set $d_{\text{DEP}} = 0$). The model is trained on example solutions from LKH [27] (see Sect. 4.2), which are not optimal, but still provide a useful training signal. Compared to TSP, the definition of the heat is slightly changed to accommodate for the 'via-depot actions' and is best defined incrementally using the 'via-depot heat' $h_{ij}^{\text{DEP}} = h_{i,\text{DEP}} \cdot h_{\text{DEP},j} \cdot 0.1$, where multiplication is used to keep heat values interpretable as probabilities and in the range $(0,1)$. The additional penalty factor of 0.1 for visiting the depot encourages the algorithm to minimize the number of vehicles/routes. The heat of the initial state is 0 and when expanding a solution $\boldsymbol{a}$ to $\boldsymbol{a}'$ using action a_t, the heat is incremented with either $h_{\text{current}(\boldsymbol{a}),a_t}$ (if $a_t < n$) or $h_{\text{current}(\boldsymbol{a}),a_t-n}^{\text{DEP}}$ (if $a_t \geq n$). The potential is defined similarly to TSP, replacing the start node 0 by DEP.

3.4 Travelling Salesman Problem with Time Windows

For the TSPTW, we also have a special depot/start node 0. The goal is to create a single tour that visits each node i in a time window defined by (l_i, u_i), where the travel time from i to j is equal to the cost/distance c_{ij}, i.e. we assume a speed of 1 (w.l.o.g. as we can rescale time). It is allowed to wait if arrival at node i is before l_i, but arrival cannot be after u_i. We minimize the total *cost* (*excluding* waiting time), but to minimize *makespan* (including waiting time), we only need to train on different example solutions. Due to the hard constraints, TSPTW is typically considered more challenging than plain TSP, for which every solution is feasible.

The **variables** we track and **initial solution** are equal to TSP except that we add time$(\boldsymbol{a})$ which is initially 0 ($= l_0$). **Feasible actions** $a_t \in \{0, ..., n-1\}$ are those that move to unvisited nodes via edges in the graph such that the arrival time is no later than u_{a_t} and do not directly eliminate the possibility to visit other nodes in time[2]. Expanding a solution $\boldsymbol{a}$ to $\boldsymbol{a}'$ using action a_t updates the time as time$(\boldsymbol{a}') = \max\{\text{time}(\boldsymbol{a}) + c_{\text{current}(\boldsymbol{a}),a_t}, l_{a_t}\}$.

For each DP state, we keep all efficient solutions in terms of cost and time, so a partial solution $\boldsymbol{a}$ is a **dominated solution** dominated by $\boldsymbol{a}^*$ if $\boldsymbol{a}^*$ has the same DP state (visited$(\boldsymbol{a}^*)$ = visited$(\boldsymbol{a})$ and current$(\boldsymbol{a}^*)$ = current$(\boldsymbol{a})$) and is strictly better in terms of cost and time, i.e. cost$(\boldsymbol{a}^*) \leq$ cost$(\boldsymbol{a})$ and time$(\boldsymbol{a}^*) \leq$ time$(\boldsymbol{a})$, with *at least one of the two inequalities being strict*.

The model [32] for the **scoring policy** is adapted to include the time windows (l_i, u_i) as node features (scaled to correspond to a speed of 1 for the input distances and coordinates, which are scaled to the range $[0,1]$), and we use a special embedding for the depot similar to VRP. Due to the time dimension, a TSPTW solution is *directed*, and edge (i,j) may be good whereas (j,i) may be not, so we adapt the model to enable predictions $h_{ij} \neq h_{ji}$ (see Appendix 1). We generated example training solutions using (heuristic) DP with a large beam size, which was faster than LKH. Given the heat predictions, the score (heat + potential) is exactly as for TSP.

---

[2] E.g., arriving at node i at $t = 10$ is not feasible if node j has $u_j = 12$ and $c_{ij} = 3$.

4   Experiments

We implement DPDP using PyTorch [53] to leverage GPU computation. For details, see Appendix 2. Our code is publicly available.[3] DPDP has very few hyperparameters, but the heatmap threshold of 10^{-5} and details like the functional form of e.g. the scoring policy are 'educated guesses' or manually tuned on a few validation instances and can likely be improved. The runtime is influenced by implementation choices which were tuned on a few validation instances.

4.1   Travelling Salesman Problem

In Table 1 we report our main results for DPDP with beam sizes of 10K (10 thousand) and 100K, for the TSP with 100 nodes on a commonly used test set of 10000 instances [36]. We report cost and *gap* to the optimal solution found using Concorde [2] (following [36]) and compare against LKH [27] and Gurobi [24], as well as recent results of the strongest methods using neural networks ('neural approaches') from literature. Running times for solving 10000 instances *after training* should be taken as rough indications as some are on different machines, typically with 1 GPU or a many-core CPU (8 - 32). The costs indicated with * are not directly comparable due to slight dataset differences [19]. Times for generating heatmaps (if applicable) is reported separately (as the first term) from the running time for MCTS [19] or DP. DPDP achieves close to optimal results, strictly outperforming the neural baselines achieving better results in less time (except the Attention Model trained with POMO [37], see Sect. 4.2).

4.2   Vehicle Routing Problem

For the VRP, we train the model using 1 million instances of 100 nodes, generated according to the distribution described by [50] and solved using one run of LKH [27]. We train using a batch size of 48 and a learning rate of 10^{-3} (selected as the result of manual trials to best use our GPUs), for (at most) 1500 epochs of 500 training steps (following [32]) from which we select the saved checkpoint with the lowest validation loss. We use the validation and test sets by [36].

Table 1 shows the results, where the gap is relative to Hybrid Genetic Search (HGS)[4], a SOTA heuristic VRP solver [62,63]. HGS is faster and improves around 0.5% over LKH [27], which is typically considered the baseline in related work. We present the results for LKH, as well as the strongest neural approaches and DPDP with beam sizes up to 1 million. Some results used 2000 (different) instances [43] and cannot be directly compared[5]. DPDP outperforms all other neural baselines, except the Attention Model trained with POMO [37], which delivers good results very quickly by exploiting symmetries in the problem.

---

[3] https://github.com/wouterkool/dpdp.

[4] https://github.com/vidalt/HGS-CVRP.

[5] The running time of 4000 h (167 days) is estimated from 24 min/instance [43].

Table 1. Mean cost, gap and *total time* to solve 10000 TSP/VRP test instances.

Problem Method	TSP100 Cost	Gap	Time	VRP100 Cost	Gap	Time
Concorde [2]	7.765	0.000%	6M			
Hybrid Genetic Search [62,63]				15.563	0.000%	6H11M
Gurobi [24]	7.776	0.151%	31M			
LKH [27]	7.765	0.000%	42m	15.647	0.536%	12H57M
GNN Heatmap + Beam Search [32]	7.87	1.39%	40M			
Learning 2-opt heuristics [11]	7.83	0.87%	41M			
Merged GNN Heatmap + MCTS [19]	7.764*	0.04%	4M + 11M			
Attention Model + Sampling [36]	7.94	2.26%	1H	16.23	4.28%	2H
Step-wise Attention Model [67]	8.01	3.20%	29s	16.49	5.96%	39s
Attn. Model + Coll. Policies [34]	7.81	0.54%	12H	15.98	2.68%	5H
Learning improv. heuristics [66]	7.87	1.42%	2H	16.03	3.00%	5H
Dual-Aspect Coll. Transformer [45]	7.77	0.09%	5h	15.71	0.94%	9H
Attention Model + POMO [37]	7.77	0.14%	1M	15.76	1.26%	2M
NeuRewriter [9]				16.10	3.45%	1H
Dynamic Attn. Model + 2-opt [54]				16.27	4.54%	6H
Neur. Lrg. Neighb. Search [30]				15.99	2.74%	1H
Learn to improve [43]				15.57*	-	4000H
DPDP 10K	7.765	0.009%	10M + 16M	15.830	1.713%	10M + 50M
DPDP 100K	7.765	0.004%	10M + 2H35M	15.694	0.843%	10M + 5H48M
DPDP 1M				15.627	0.409%	10M + 48H27M

However, as it cannot (easily) improve further with additional runtime, we consider this contribution orthogonal to DPDP. DPDP is competitive to LKH (see also Sect. 4.4).

More Realistic Instances. We also train the model and run experiments with instances with 100 nodes from a more realistic and challenging data distribution [60]. This distribution, commonly used in the routing community, has greater variability, in terms of node clustering and demand distributions. LKH failed to solve two of the test instances, which is because LKH by default uses a fixed number of routes equal to a lower bound, given by $\left\lceil \frac{\sum_{i=0}^{n-1} d_i}{\text{CAPACITY}} \right\rceil$, which may be infeasible[6]. Therefore we solve these instances by rerunning LKH with an unlimited number of allowed routes (which gives worse results, see Sect. 4.4).

DPDP was run on a machine with 4 GPUs, but we also report (estimated) runtimes for 1 GPU (1080Ti), and we compare against 16 or 32 CPUs for HGS and LKH. In Table 2 it can be seen that the difference with LKH is, as expected,

---

[6] For example, three nodes with a demand of two cannot be assigned to two routes with a capacity of three.

slightly larger than for the simpler dataset, but still below 1% for beam sizes of 100K–1M. We also observed a higher validation loss, so it may be possible to improve results using more training data. Nevertheless, finding solutions within 1% of the specialized SOTA HGS algorithm, and even closer to LKH, is impressive for these challenging instances, and we consider the runtime (for solving 10K instances) acceptable, especially when using multiple GPUs.

Table 2. Mean cost, gap and *total time* to solve 10000 realistic VRP100 instances.

METHOD	COST	GAP	TIME (1 GPU OR 16 CPUS)	TIME (4 GPUS OR 32 CPUS)
HGS [62,63]	18050	0.000%	7H53M	3H56M
LKH [27]	18133	0.507%	25H32M	12H46M
DPDP 10K	18414	2.018%	10M + 50M	2M + 13M
DPDP 100K	18253	1.127%	10M + 5H48M	2M + 1H27M
DPDP 1M	18168	0.659%	10M + 48H27M	2M + 12H7M

4.3   TSP with Time Windows

For the TSP with hard time window constraints, we use the data distribution by [8] and use their set of 100 test instances with 100 nodes. These were generated with small time windows, resulting in a small feasible search space, such that even with very small beam sizes, our DP implementation solves these instances optimally, eliminating the need for a policy. Therefore, we also consider a more difficult distribution similar to [12], which has larger time windows which are more difficult as the feasible search space is larger[7] [17]. For details, see Appendix 1. For both distributions, we generate training data and train the model exactly as we did for the VRP.

Table 3 shows the results for both data distributions, which are reported in terms of the difference to General Variable Neighborhood Search (GVNS) [12], the best open-source solver for TSPTW we could find[8], using 30 runs. For the small time window setting, both GVNS and DPDP find optimal solutions for all 100 instances in just 7 s (in total, either on 16 CPUs or a single GPU). LKH fails to solve one instance, but finds close to optimal solutions, but around 50 times slower. BaB-DQN* and ILDS-DQN* [8], methods combining an existing solver with an RL trained neural policy, take around 15 min *per instance* (orders of magnitudes slower) to solve most instances to optimality. Due to complex setup, we were unable to run BaB-DQN* and ILDS-DQN* ourselves for the setting with larger time windows. In this setting, we find DPDP outperforms both LKH (where DPDP is orders of magnitude faster) and GVNS, in both speed and solution quality. This illustrates that DPDP, due to its nature, is especially well suited to handle constrained problems.

---

[7] Up to a limit, as making the time windows infinite size reduces the problem to plain TSP.

[8] https://github.com/sashakh/TSPTW.

Table 3. Mean cost, gap and *total time* to solve TSPTW100 instances.

| PROBLEM | SMALL TIME WINDOWS [8] (100 INST.) | | | | LARGE TIME WINDOWS [12] (10K INST.) | | | |
METHOD	COST	GAP	FAIL	TIME	COST	GAP	FAIL	TIME
GVNS 30x [12]	5129.58	0.000%		7S	2432.112	0.000%		37M15S
GVNS 1x [12]	5129.58	0.000%		<1S	2457.974	1.063%		1M4S
LKH 1x [27]	5130.32	0.014%	1.00%	5M48S	2431.404	−0.029%		34H58M
BaB-DQN* [8]	5130.51	0.018%		25H				
ILDS-DQN* [8]	5130.45	0.017%		25H				
DPDP 10K	5129.58	0.000%		6S + 1S	2431.143	−0.040%		10M + 8M7S
DPDP 100K	5129.58	0.000%		6S + 1S	2430.880	−0.051%		10M + 1H16M

4.4 Ablations

Scoring Policy. To evaluate the value of different components of DPDP's **GNN Heat + Potential** scoring policy, we compare against other variants. **GNN Heat** is the version without the potential, whereas **Cost Heat + Potential** and **Cost Heat** are variants that use a 'heuristic' $\hat{h}_{ij} = \frac{c_{ij}}{\max_k c_{ik}}$ instead of the GNN. **Cost** directly uses the current cost of the solution, and can be seen as 'classic' restricted DP. Finally, **BS GNN Heat + Potential** uses beam search without dynamic programming, i.e. without removing dominated solutions. To evaluate only the scoring policy, each variant uses the fully connected graph (no heatmap threshold). Figure 3a shows the value of DPDP's potential function, although even without it results are still significantly better than 'classic' heuristic DP variants using cost-based scoring policies. Also, it is clear that using DP significantly improves over a standard beam search (by removing dominated solutions). Lastly, the figure illustrates how the time for generating the heatmap using the neural network, despite its significant value, only makes up a small portion of the total runtime.

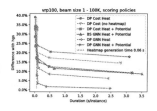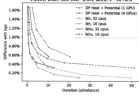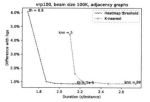

(a) Different scoring policies, as well as 'pure' beam search, for beam sizes 1, 10, 100, 1000, 10K, 100K.

(b) Beam sizes 10K, 25K, 50K, 100K, 250K, 500K, 1M, 2.5M compared against LKH(U) with 1, 2, 5 and 10 runs.

(c) Sparsities with heatmap thresholds 0.9, 0.5, 0.2, 0.1, 10^{-2}, 10^{-3}, 10^{-4}, 10^{-5} and knn = 5, 10, 20, 50, 99. Beam size 100K.

Fig. 3. DPDP ablations on 100 validation instances of VRP with 100 nodes.

Beam Size. With DPDP, we can trade off the performance vs. the runtime using the beam size B (and the graph sparsity, see below). Figure 3b illustrates this trade-off, where we evaluate DPDP on 100 validation instances for VRP, with different beam sizes from 10K to 2.5M. We also report the trade-off curve for LKH(U), which is the strongest baseline that can also solve different problems. We vary the runtime using 1, 2, 5 and 10 runs (returning the best solution). LKHU(nlimited) is the version which allows an unlimited number of routes (see Sect. 4.2). It is hard to compare GPU vs CPU, so we report (estimated) runtimes for different hardware, i.e. 1 or 4 GPUs (with 3 CPUs per GPU) and 16 or 32 CPUs. We report the difference (i.e. the gap) with HGS, analogous to how results are reported in Table 1. We emphasize that in most related work (e.g. [36]), the strongest baseline considered is one run of LKH, so we compare against a much stronger baseline. Also, our goal is not to outperform HGS (which is SOTA and specific to VRP) or LKH, but to show DPDP has reasonable performance, while being a flexible framework for other (routing) problems.

Graph Sparsity. Using the heatmap threshold, the DP algorithm uses a sparse graph to define feasible expansions, which reduces the runtime but may also sacrifice solution quality. For most edges, the model confidently predicts close to 0, such that they are ruled out, even using the default (low) heatmap threshold of 10^{-5}. We may rule out even more edges by increasing the threshold, which can be seen as a secondary way (besides varying the beam size) to trade off the performance and computational cost of DPDP. While this can be seen as a form of learned *problem reduction* [58], we also consider a heuristic alternative of using the K-nearest neighbor (KNN) graph.[9] In Fig. 3c, we experiment with different heatmap thresholds from 10^{-5} to 0.9 and different values for KNN from 5 to 99 (fully connected). The heatmap threshold strategy clearly outperforms the KNN strategy as it yields the same results using sparser graphs (and lower runtimes). This illustrates that the heatmap threshold strategy is more informed than the KNN strategy, confirming the value of the neural network predictions.

5   Discussion

In this paper we introduced Deep Policy Dynamic Programming, which combines machine learning and dynamic programming for solving vehicle routing problems. The method yields close to optimal results for TSPs with 100 nodes and is competitive to the highly optimized LKH [27] solver for VRPs with 100 nodes. On the TSPTW, DPDP also outperforms LKH, being significantly faster, as well as GVNS [12], the best open source solver we could find. Given that DPDP was not specifically designed for TSPTW, and thus can likely be improved, we consider this an impressive and promising achievement.

---

[9] For the symmetric TSP and VRP, we add KNN edges in both directions. For the VRP, we also connect each node to the depot (and vice versa) to ensure feasibility.

The constructive nature of DPDP (combined with search) naturally supports hard constraints such as time windows, which are typically considered challenging in neural combinatorial optimization [6,36] and are also difficult for local search heuristics (as they need to maintain feasibility while adapting a solution). Given our results on TSP, VRP and TSPTW, and the flexibility of DP as a framework, we think DPDP has great potential for solving many more variants of routing problems, and possibly even other problems that can be formulated using DP (e.g. job shop scheduling [23]). We hope that our work brings machine learning research for combinatorial optimization closer to the operations research (especially vehicle routing) community, by combining machine learning with DP and evaluating the resulting new framework on different data distributions used by different communities [8,12,50,60].

Scope, Limitations and Future Work. Deep learning for combinatorial optimization is a recent research direction, which could significantly impact the way practical optimization problems get solved in the future. Currently, however, it is still hard to beat most SOTA problem specific solvers from the OR community. Despite our success for TSPTW, DPDP is not yet a practical alternative in general, but we do consider our results as highly encouraging for further research. We believe such research could yield significant further improvement by addressing key current limitations: (1) the scalability to larger instances, (2) the dependency on example solutions and (3) the heuristic nature of the scoring function. First, while 100 nodes is not far from the size of common benchmarks (100–1000 for VRP [60] and 20–200 for TSPTW [12]), scaling is a challenge, mainly due to the 'fully-connected' $O(n^2)$ graph neural network. Future work could reduce this complexity following e.g. [40]. The dependency on example solutions from an existing solver also becomes more prominent for larger instances, but could potentially be removed by 'bootstrapping' using DP itself as we, in some sense, have done for TSPTW (see Sect. 3.4). Future work could iterate this process to train the model 'tabula rasa' (without example solutions), where DP could be seen analogous to MCTS in *AlphaZero* [57]. Lastly, the heat + potential score function is a well-motivated but heuristic function that was manually designed as a function of the predicted heatmap. While it worked well for the three problems we considered, it may need suitable adaption for other problems. Training this function end-to-end [13,65], while keeping a low computational footprint, would be an interesting topic for future work.

Acknowledgement. We would like to thank Jelke van Hoorn and Johan van Rooij for helpful discussions. Also we would like to thank anonymous reviewers for helpful suggestions. This work was carried out on the Dutch national e-infrastructure with the support of SURF Cooperative.

Appendix 1 The Graph Neural Network Model

For the TSP, we use the exact model from [32], which we describe here for self-containment. The model uses node input features and edge input features, which get transformed into initial representations of the nodes and edges. These representations then get updated sequentially using a number of graph convolutional layers, which exchange information between nodes and edges, after which the final edge representation is used to predict whether the edge is part of the optimal solution.

Input Features and Initial Representation. The model uses input features for the nodes, consisting of the (x, y)-coordinates, which are then projected into H-dimensional initial embeddings $\mathbf{x}_i^0$ ($H = 300$). The initial edge features $\mathbf{e}_{ij}^0$ are a concatenation of a $\frac{H}{2}$-dimensional projection of the cost (Euclidean distance) c_{ij} from i to j, and a $\frac{H}{2}$-dimensional embedding of the edge type: 0 for normal edges, 1 for edges connecting K-nearest neighbors ($K = 20$) and 2 for *self-loop* edges connecting a node to itself (which are added for ease of implementation).

Graph Convolutional Layers. In each of the $L = 30$ layers of the model, the node and edge representations $\mathbf{x}_i^\ell$ and $\mathbf{e}_{ij}^\ell$ get updated into $\mathbf{x}_i^{\ell+1}$ and $\mathbf{e}_{ij}^{\ell+1}$ [32]:

$$\mathbf{x}_i^{\ell+1} = \mathbf{x}_i^\ell + \mathrm{ReLU}\left(\mathrm{BN}\left(W_1^\ell \mathbf{x}_i^\ell + \sum_{j \in \mathcal{N}(i)} \frac{\sigma(\mathbf{e}_{ij}^\ell)}{\sum_{j' \in \mathcal{N}(i)} \sigma(\mathbf{e}_{ij'}^\ell)} \odot W_2^\ell \mathbf{x}_j^\ell\right)\right) \quad (2)$$

$$\mathbf{e}_{ij}^{\ell+1} = \mathbf{e}_{ij}^\ell + \mathrm{ReLU}\left(\mathrm{BN}\left(W_3^\ell \mathbf{e}_{ij}^\ell + W_4^\ell \mathbf{x}_i^\ell + W_5^\ell \mathbf{x}_j^\ell\right)\right). \quad (3)$$

Here $\mathcal{N}(i)$ is the set of neighbors of node i (in our case all nodes, including i, as we use a fully connected input graph), $\odot$ is the element-wise product and σ is the sigmoid function, applied element-wise to the vector $\mathbf{e}_{ij}^\ell$. $\mathrm{ReLU}(\cdot) = \max(\cdot, 0)$ is the rectified linear unit and BN represents batch normalization [31]. W_1, W_2, W_3, W_4 and W_5 are trainable parameter matrices, where we fix $W_4 = W_5$ for the symmetric TSP.

Output Prediction. After L layers, the final prediction $h_{ij} \in (0, 1)$ is made independently for each edge (i, j) using a multi-layer perceptron (MLP), which takes $\mathbf{e}_{ij}^L$ as input and has two H-dimensional hidden layers with ReLU activation and a 1-dimensional output layer, with sigmoid activation. We interpret h_{ij} as the predicted probability that the edge (i, j) is part of the optimal solution, which indicates how promising this edge is when searching for the optimal solution.

Training. For TSP, the model is trained on a dataset of 1 million optimal solutions, found using Concorde [2], for randomly generated TSP instances. The training loss is a weighted binary cross-entropy loss, that maximizes the prediction quality when h_{ij} is compared to the ground-truth optimal solution. Generating the dataset takes between half a day and a few days (depending on number of CPU cores), and training the model takes a few days on one or multiple GPUs, but both are only required once given a desired data distribution.

1.1 Predicting Directed Edges for the TSPTW

The TSP is an undirected problem, so the neural network implementation[10] by [32] shares the parameters W_4^l and W_5^l in Eq. (3), i.e. $W_4^l = W_5^l$, resulting in $e_{ij}^l = e_{ji}^l$ for all layers l, as for $l = 0$ both directions are initialized the same. While the VRP also is an undirected problem, the TSPTW is directed as the direction of the route determines the times of arrival at different nodes. To allow the model to make different predictions for different directions, we implement W_5^l as a separate parameter, such that the model can have different representations for edges (i,j) and (j,i). We define the training labels accordingly for directed edges, so if edge (i,j) is in the directed solution, it will have a label 1 whereas the edge (j,i) will not (for the undirected TSP and VRP, both labels are 1).

1.2 Dataset Generation for the TSPTW

We found that using our DP formulation for TSPTW, the instances by [8] were all solved optimally, even with a very small beam size (around 10). This is because there is very little overlap in the time windows as a result of the way they are generated, and therefore very few actions are feasible as most of the actions would 'skip over other time windows' (advance the time so much that other nodes can no longer be served)[11]. We conducted some quick experiments with a weaker DP formulation, that only checks if actions *directly* violate time windows, but does not check if an action causes other nodes to be no longer reachable in their time windows. Using this formulation, the DP algorithm can run into many dead ends if just a single node gets skipped, and using the GNN policy (compared to a cost based policy as in Sect. 4.4) made the difference between good solutions and no solution at all being found.

We made two changes to the data generation procedure by [8] to increase the difficulty and make it similar to [12], defining the 'large time window' dataset. First, we sample the time windows around arrival times when visiting nodes in a random order without any waiting time, which is different from [8] who 'propagate' the waiting time (as a result of time windows sampled). Our modification causes a tighter schedule with more overlap in time windows, and is similar to [12]. Secondly, we increase the maximum time window size from 100 to 1000, which makes that the time windows are in the order of 10% of the horizon[12]. This doubles the maximum time window size of 500 used by [12] for instances with 200 nodes, to compensate for half the number of nodes that can possibly overlap the time window.

To generate the training data, for practical reasons we used DP with the heuristic 'cost heat + potential' strategy and a large beam size (1M), which in many cases results in optimal solutions being found.

---

[10] https://github.com/chaitjo/graph-convnet-tsp/blob/master/models/gcn_layers.py.

[11] If all time windows are disjoint, there is only one feasible solution. Therefore, the amount of overlap in time windows determines to some extent the 'branching factor' of the problem and the difficulty.

[12] Serving 100 customers in a 100×100 grid, empirically we find the total schedule duration including waiting (the makespan) is around 5000.

Appendix 2 Implementation

We implement the dynamic programming algorithm on the GPU using PyTorch [53]. While mostly used as a Deep Learning framework, it can be used to speed up generic (vectorized) computations.

2.1 Beam Variables

For each solution in the beam, we keep track of the following variables (storing them for all solutions in the beam as a vector): the cost, current node, visited nodes and (for VRP) the remaining capacity or (for TSPTW) the current time. As explained, these variables can be computed incrementally when generating expansions. Additionally, we keep a variable vector *parent*, which, for each solution in the current beam, tracks the index of the solution in the previous beam that generated the expanded solution. To compute the score of the policy for expansions efficiently, we also keep track of the score for each solution and the potential for each node for each solution incrementally.

We do not keep past beams in memory, but at the end of each iteration, we store the vectors containing the parents as well as last actions for each solution on the *trace*. As the solution is completely defined by the sequence of actions, this allows to backtrack the solution after the algorithm has finished. To save GPU memory (especially for larger beam sizes), we store the $O(Bn)$ sized trace on the CPU memory.

For efficiency, we keep the set of visited nodes as a bitmask, packed into 64-bit long integers (2 for 100 nodes). Using bitwise operations with the packed adjacency matrix, this allows to quickly check feasible expansions (but we need to *unpack* the mask into boolean vectors to find all feasible expansions explicitly). Figure 4a shows an example of the beam (with variables related to the policy and backtracking omitted) for the VRP.

2.2 Generating Non-dominated Expansions

A solution a can only dominate a solution a' if visited(a) = visited(a') and current(a) = current(a'), i.e. if they correspond to the same *DP state*. If this is the case, then, if we denote by parent(a) the parent solution from which a was expanded, it holds that

$$\text{visited}(\text{parent}(a)) = \text{visited}(a) \setminus \{\text{current}(a)\}$$
$$= \text{visited}(a') \setminus \{\text{current}(a')\}$$
$$= \text{visited}(\text{parent}(a')).$$

This means that only expansions from solutions with the same set of visited nodes can dominate each other, so we only need to check for dominated solutions among groups of expansions originating from parent solutions with the same set of visited nodes. Therefore, before generating the expansions, we group the

Cost	Capacity	Visited	Current	Direct 0 1 2 3 4	Via-depot 0 1 2 3 4
10	5	01101	1	1 0 0 0 0	1 0 0 1 0
12	8	01101	1	1 0 0 1 0	1 0 0 1 0
13	7	01101	2	1 0 0 1 0	0 0 0 0 0
8	3	01101	4	0 0 0 0 0	1 0 0 1 0
11	7	10101	0	0 1 0 1 0	0 0 0 1 0
12	6	10101	2	0 0 0 1 0	0 0 0 1 0
13	7	10101	2	0 0 0 1 0	0 0 0 1 0

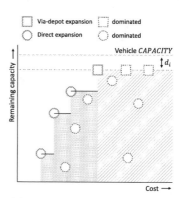

Via-depot expansion · dominated
Direct expansion · dominated

(a) Example beam for VRP with variables, grouped by set of visited nodes (left) and feasible, non-dominated expansions (right), with $2n$ columns corresponding to n direct expansions and n via-depot expansions. Some expansions to unvisited nodes are infeasible, e.g. due to the capacity constraint or a sparse adjacency graph. The shaded areas indicate groups of candidate expansions among which dominances should be checked: for each set of visited nodes there is only one non-dominated via-depot expansion (indicated by solid green square), which must necessarily be an expansion of the solution that has the lowest cost to return to the depot (indicated by the dashed green rectangle ; note that the cost displayed excludes the cost to return to the depot). Direct expansions can be dominated (indicated by red dotted circles) by the single non-dominated via-depot expansion or other direct expansions with the same DP state (set of visited nodes and expanded node, as indicated by the shaded areas). See also Figure 4b for (non-)dominated expansions corresponding to the same DP state.

(b) Example of a set of dominated and non-dominated expansions (direct and via-depot) corresponding to the same DP state (set of visited nodes and expanded node i) for VRP. Non-dominated expansions have lower cost or higher remaining capacity compared to all other expansions. The right striped area indicates expansions dominated by the (single) non-dominated via-depot expansion. The left (darker) areas are dominated by individual direct expansions. Dominated expansions in this area have remaining capacity lower than the cumulative maximum remaining capacity when going from left to right (i.e. in sorted order of increasing cost), indicated by the black horizontal lines.

Fig. 4. Implementation of DPDP for VRP (Color figure online)

current beam (the parents of the expansions) by the set of visited nodes (see Fig. 4). This can be done efficiently, e.g. using a lexicographic sort of the packed bitmask representing the sets of visited nodes[13].

Travelling Salesman Problem. For TSP, we can generate (using boolean operations) the $B \times n$ matrix with boolean entries indicating feasible expansions

---

[13] For efficiency, we use a custom function similar to TORCH.UNIQUE, and argsort the returned inverse after which the resulting permutation is applied to all variables in the beam.

(with n action columns corresponding to n nodes, similar to the $B \times 2n$ matrix for VRP in Fig. 4), i.e. nodes that are unvisited and adjacent to the current node. If we find positive entries sequentially for each column (e.g. by calling TORCH.NONZERO on the transposed matrix), we get all expansions grouped by the combination of action (new current node) and parent set of visited nodes, i.e. grouped by the DP state. We can then trivially find the segments of consecutive expansions corresponding to the same DP state, and we can efficiently find the minimum cost solution for each segment, e.g. using TORCH_SCATTER[14].

Vehicle Routing Problem. For VRP, the dominance check has two dimensions (cost *and* remaining capacity) and additionally we need to consider $2n$ actions: n direct and n via the depot (see Fig. 4). Therefore, as we will explain, we check dominances in two stages: first we find (for each DP state) the *single* non-dominated 'via-depot' expansion, after which we find all non-dominated 'direct' expansions (see Fig. 4b).

The DP state of each expansion is defined by the expanded node (the new current node) and the set of visited nodes. For each DP state, there can be only *one*[15] non-dominated expansion where the last action was via the depot, since all expansions resulting from 'via-depot actions' have the same remaining capacity as visiting the depot resets the capacity (see Fig. 4b). To find this expansion, we first find, for each unique set of visited nodes in the current beam, the solution that can return to the depot with lowest total cost (thus including the cost to return to the depot, indicated by a dashed green rectangle in Fig. 4). The single non-dominated 'via-depot expansion' for each DP state must necessarily be an expansion of this solution. Also observe that this via-depot solution cannot be dominated by a solution expanded using a direct action, which will always have a lower remaining vehicle capacity (assuming positive demands) as can bee seen in Fig. 4b. We can thus generate the non-dominated via-depot expansion for each DP state efficiently and independently from the direct expansions.

For each DP state, all *direct* expansions with cost higher (or equal) than the via-depot expansion can directly be removed since they are dominated by the via-depot expansion (having higher cost and lower remaining capacity, see Fig. 4b). After that, we sort the remaining (if any) direct expansions for each DP state based on the cost (using a segmented sort as the expansions are already grouped if we generate them similarly to TSP, i.e. per column in Fig. 4). For each DP state, the lowest cost solution is never dominated. The other solutions should be kept only if their remaining capacity is strictly larger than the largest remaining capacity of all lower-cost solutions corresponding to the same DP state, which can be computed using a (segmented) cumulative maximum computation (see Fig. 4b).

---

[14] https://github.com/rusty1s/pytorch_scatter.
[15] Unless we have multiple expansions with the same costs, in which case can pick one arbitrarily.

TSP with Time Windows. For the TSPTW, the dominance check has two dimensions: cost and time. Therefore, it is similar to the check for non-dominated direct expansions for the VRP (see Fig. 4b), but replacing remaining capacity (which should be maximized) by current time (to be minimized). In fact, we could reuse the implementation, if we replace remaining capacity by time multiplied by -1 (as this should be minimized). This means that we sort all expansions for each DP state based on the cost, keep the first solution and keep other solutions only if the time is strictly lower than the lowest current time for all lower-cost solutions, which can be computed using a cumulative minimum computation.

2.3 Finding the Top B Solutions

We may generate all 'candidate' non-dominated expansions and then select the top B using the score function. Alternatively, we can generate expansions in batches, and keep a streaming top B using a priority queue. We use the latter implementation, where we can also derive a bound for the score as soon as we have B candidate expansions. Using this bound, we can already remove solutions before checking dominances, to achieve some speedup in the algorithm.[16]

2.4 Performance Improvements

There are many possibilities for improving the speed of the algorithm. For example, PyTorch lacks a segmented sort so we use a much slower lexicographic sort instead. Also an efficient GPU priority queue would allow much speedup, as we currently use sorting as PyTorch' top-k function is rather slow for large k. In some cases, a binary search for the k-th largest value can be faster, but this introduces undesired CUDA synchronisation points.

References

1. Accorsi, L., Vigo, D.: A fast and scalable heuristic for the solution of large-scale capacitated vehicle routing problems. Transp. Sci. **55**(4), 832–856 (2021)
2. Applegate, D., Bixby, R., Chvatal, V., Cook, W.: Concorde TSP Solver (2006). http://www.math.uwaterloo.ca/tsp/concorde
3. Bai, R., et al.: Analytics and machine learning in vehicle routing research. arXiv preprint arXiv:2102.10012 (2021)
4. Bellman, R.: On the theory of dynamic programming. Proc. Natl. Acad. Sci. U.S.A. **38**(8), 716 (1952)
5. Bellman, R.: Dynamic programming treatment of the travelling salesman problem. J. ACM (JACM) **9**(1), 61–63 (1962)
6. Bello, I., Pham, H., Le, Q.V., Norouzi, M., Bengio, S.: Neural combinatorial optimization with reinforcement learning. arXiv preprint arXiv:1611.09940 (2016)

---

[16] This may give slightly different results if the scoring function is inconsistent with the domination rules, i.e. if a better scoring solution would be dominated by a worse scoring solution but is not since that solution is removed using the score bound before checking the dominances.

7. Bertsekas, D.: Dynamic Programming and Optimal Control, vol. 1. Athena Scientific (2017)
8. Cappart, Q., Moisan, T., Rousseau, L.M., Prémont-Schwarz, I., Cire, A.: Combining reinforcement learning and constraint programming for combinatorial optimization. In: AAAI Conference on Artificial Intelligence (AAAI) (2021)
9. Chen, X., Tian, Y.: Learning to perform local rewriting for combinatorial optimization. In: Advances in Neural Information Processing Systems (NeurIPS), pp. 6281–6292 (2019)
10. Cook, W., Seymour, P.: Tour merging via branch-decomposition. INFORMS J. Comput. 15(3), 233–248 (2003)
11. da Costa, P.R.d.O., Rhuggenaath, J., Zhang, Y., Akcay, A.: Learning 2-opt heuristics for the traveling salesman problem via deep reinforcement learning. In: Asian Conference on Machine Learning (ACML) (2020)
12. Da Silva, R.F., Urrutia, S.: A general VNS heuristic for the traveling salesman problem with time windows. Discret. Optim. 7(4), 203–211 (2010)
13. Daumé, H., III., Marcu, D.: Learning as search optimization: approximate large margin methods for structured prediction. In: International Conference on Machine Learning (ICML), pp. 169–176 (2005)
14. Delarue, A., Anderson, R., Tjandraatmadja, C.: Reinforcement learning with combinatorial actions: an application to vehicle routing. In: Advances in Neural Information Processing Systems (NeurIPS), vol. 33 (2020)
15. Deudon, M., Cournut, P., Lacoste, A., Adulyasak, Y., Rousseau, L.-M.: Learning heuristics for the TSP by policy gradient. In: van Hoeve, W.-J. (ed.) CPAIOR 2018. LNCS, vol. 10848, pp. 170–181. Springer, Cham (2018). https://doi.org/10.1007/978-3-319-93031-2_12
16. Dijkstra, E.W.: A note on two problems in connexion with graphs. Numer. Math. 1(1), 269–271 (1959)
17. Dumas, Y., Desrosiers, J., Gelinas, E., Solomon, M.M.: An optimal algorithm for the traveling salesman problem with time windows. Oper. Res. 43(2), 367–371 (1995)
18. Falkner, J.K., Schmidt-Thieme, L.: Learning to solve vehicle routing problems with time windows through joint attention. arXiv preprint arXiv:2006.09100 (2020)
19. Fu, Z.H., Qiu, K.B., Zha, H.: Generalize a small pre-trained model to arbitrarily large tsp instances. In: AAAI Conference on Artificial Intelligence (AAAI) (2021)
20. Gao, L., Chen, M., Chen, Q., Luo, G., Zhu, N., Liu, Z.: Learn to design the heuristics for vehicle routing problem. In: International Workshop on Heuristic Search in Industry (HSI) at the International Joint Conference on Artificial Intelligence (IJCAI) (2020)
21. Gasse, M., Chetelat, D., Ferroni, N., Charlin, L., Lodi, A.: Exact combinatorial optimization with graph convolutional neural networks. In: Advances in Neural Information Processing Systems (NeurIPS) (2019)
22. Gromicho, J., van Hoorn, J.J., Kok, A.L., Schutten, J.M.: Restricted dynamic programming: a flexible framework for solving realistic VRPs. Comput. Oper. Res. 39(5), 902–909 (2012)
23. Gromicho, J.A., Van Hoorn, J.J., Saldanha-da Gama, F., Timmer, G.T.: Solving the job-shop scheduling problem optimally by dynamic programming. Comput. Oper. Res. 39(12), 2968–2977 (2012)
24. Gurobi Optimization, LLC: Gurobi Optimizer Reference Manual (2021). https://www.gurobi.com
25. van Heeswijk, W., La Poutré, H.: Approximate dynamic programming with neural networks in linear discrete action spaces. arXiv preprint arXiv:1902.09855 (2019)

26. Held, M., Karp, R.M.: A dynamic programming approach to sequencing problems. J. Soc. Ind. Appl. Math. **10**(1), 196–210 (1962)
27. Helsgaun, K.: An extension of the Lin-Kernighan-Helsgaun TSP solver for constrained traveling salesman and vehicle routing problems: Technical report (2017)
28. van Hoorn, J.J.: Dynamic programming for routing and scheduling. Ph.D. thesis (2016)
29. Hottung, A., Bhandari, B., Tierney, K.: Learning a latent search space for routing problems using variational autoencoders. In: International Conference on Learning Representations (ICML) (2021)
30. Hottung, A., Tierney, K.: Neural large neighborhood search for the capacitated vehicle routing problem. In: European Conference on Artificial Intelligence (ECAI) (2020)
31. Ioffe, S., Szegedy, C.: Batch normalization: accelerating deep network training by reducing internal covariate shift. In: International Conference on Machine Learning (ICML), pp. 448–456 (2015)
32. Joshi, C.K., Laurent, T., Bresson, X.: An efficient graph convolutional network technique for the travelling salesman problem. In: INFORMS Annual Meeting (2019)
33. Joshi, C.K., Laurent, T., Bresson, X.: On learning paradigms for the travelling salesman problem. In: Graph Representation Learning Workshop at Neural Information Processing Systems (NeurIPS) (2019)
34. Kim, M., Park, J., Kim, J.: Learning collaborative policies to solve NP-hard routing problems. In: Advances in Neural Information Processing Systems (NeurIPS) (2021)
35. Kok, A., Hans, E.W., Schutten, J.M., Zijm, W.H.: A dynamic programming heuristic for vehicle routing with time-dependent travel times and required breaks. Flex. Serv. Manuf. J. **22**(1–2), 83–108 (2010)
36. Kool, W., van Hoof, H., Welling, M.: Attention, learn to solve routing problems! In: International Conference on Learning Representations (ICLR) (2019)
37. Kwon, Y.D., Choo, J., Kim, B., Yoon, I., Gwon, Y., Min, S.: Pomo: policy optimization with multiple optima for reinforcement learning. In: Advances in Neural Information Processing Systems (NeurIPS) (2020)
38. Laporte, G.: The vehicle routing problem: an overview of exact and approximate algorithms. Eur. J. Oper. Res. (EJOR) **59**(3), 345–358 (1992)
39. LeCun, Y., Bengio, Y., Hinton, G.: Deep learning. Nature **521**(7553), 436–444 (2015)
40. Lee, J., Lee, Y., Kim, J., Kosiorek, A., Choi, S., Teh, Y.W.: Set transformer: a framework for attention-based permutation-invariant neural networks. In: International Conference on Machine Learning (ICML), pp. 3744–3753. PMLR (2019)
41. Li, S., Yan, Z., Wu, C.: Learning to delegate for large-scale vehicle routing. In: Advances in Neural Information Processing Systems (NeurIPS) (2021)
42. Li, Z., Chen, Q., Koltun, V.: Combinatorial optimization with graph convolutional networks and guided tree search. In: Advances in Neural Information Processing Systems (NeurIPS), p. 539 (2018)
43. Lu, H., Zhang, X., Yang, S.: A learning-based iterative method for solving vehicle routing problems. In: International Conference on Learning Representations (2020)
44. Ma, Q., Ge, S., He, D., Thaker, D., Drori, I.: Combinatorial optimization by graph pointer networks and hierarchical reinforcement learning. In: AAAI International Workshop on Deep Learning on Graphs: Methodologies and Applications (DLGMA) (2020)

45. Ma, Y., et al.: Learning to iteratively solve routing problems with dual-aspect collaborative transformer. In: Advances in Neural Information Processing Systems (NeurIPS) (2021)

46. Malandraki, C., Dial, R.B.: A restricted dynamic programming heuristic algorithm for the time dependent traveling salesman problem. Eur. J. Oper. Res. (EJOR) **90**(1), 45–55 (1996)

47. Mazyavkina, N., Sviridov, S., Ivanov, S., Burnaev, E.: Reinforcement learning for combinatorial optimization: a survey. arXiv preprint arXiv:2003.03600 (2020)

48. Mingozzi, A., Bianco, L., Ricciardelli, S.: Dynamic programming strategies for the traveling salesman problem with time window and precedence constraints. Oper. Res. **45**(3), 365–377 (1997)

49. Nair, V., et al.: Solving mixed integer programs using neural networks. arXiv preprint arXiv:2012.13349 (2020)

50. Nazari, M., Oroojlooy, A., Snyder, L., Takac, M.: Reinforcement learning for solving the vehicle routing problem. In: Advances in Neural Information Processing Systems (NeurIPS), pp. 9860–9870 (2018)

51. Novoa, C., Storer, R.: An approximate dynamic programming approach for the vehicle routing problem with stochastic demands. Eur. J. Oper. Res. (EJOR) **196**(2), 509–515 (2009)

52. Nowak, A., Villar, S., Bandeira, A.S., Bruna, J.: A note on learning algorithms for quadratic assignment with graph neural networks. In: Principled Approaches to Deep Learning Workshop at the International Conference on Machine Learning (ICML) (2017)

53. Paszke, A., et al.: Pytorch: an imperative style, high-performance deep learning library. In: Advances in Neural Information Processing Systems (NeurIPS), vol. 32, pp. 8026–8037 (2019)

54. Peng, B., Wang, J., Zhang, Z.: A deep reinforcement learning algorithm using dynamic attention model for vehicle routing problems. In: Li, K., Li, W., Wang, H., Liu, Y. (eds.) ISICA 2019. CCIS, vol. 1205, pp. 636–650. Springer, Singapore (2020). https://doi.org/10.1007/978-981-15-5577-0_51

55. Ropke, S., Pisinger, D.: An adaptive large neighborhood search heuristic for the pickup and delivery problem with time windows. Transp. Sci. **40**(4), 455–472 (2006)

56. Schrimpf, G., Schneider, J., Stamm-Wilbrandt, H., Dueck, G.: Record breaking optimization results using the ruin and recreate principle. J. Comput. Phys. **159**(2), 139–171 (2000)

57. Silver, D., et al.: A general reinforcement learning algorithm that masters chess, shogi, and go through self-play. Science **362**(6419), 1140–1144 (2018)

58. Sun, Y., Ernst, A., Li, X., Weiner, J.: Generalization of machine learning for problem reduction: a case study on travelling salesman problems. OR Spectr. **43**(3), 607–633 (2020). https://doi.org/10.1007/s00291-020-00604-x

59. Toth, P., Vigo, D.: Vehicle Routing: Problems, Methods, and Applications. SIAM (2014)

60. Uchoa, E., Pecin, D., Pessoa, A., Poggi, M., Vidal, T., Subramanian, A.: New benchmark instances for the capacitated vehicle routing problem. Eur. J. Oper. Res. (EJOR) **257**(3), 845–858 (2017)

61. Vesselinova, N., Steinert, R., Perez-Ramirez, D.F., Boman, M.: Learning combinatorial optimization on graphs: a survey with applications to networking. IEEE Access **8**, 120388–120416 (2020)

62. Vidal, T.: Hybrid genetic search for the CVRP: open-source implementation and swap* neighborhood. arXiv preprint arXiv:2012.10384 (2020)

63. Vidal, T., Crainic, T.G., Gendreau, M., Lahrichi, N., Rei, W.: A hybrid genetic algorithm for multidepot and periodic vehicle routing problems. Oper. Res. **60**(3), 611–624 (2012)
64. Vinyals, O., Fortunato, M., Jaitly, N.: Pointer networks. In: Advances in Neural Information Processing Systems (NeurIPS), pp. 2692–2700 (2015)
65. Wiseman, S., Rush, A.M.: Sequence-to-sequence learning as beam-search optimization. In: Conference on Empirical Methods in Natural Language Processing (EMNLP), pp. 1296–1306 (2016)
66. Wu, Y., Song, W., Cao, Z., Zhang, J., Lim, A.: Learning improvement heuristics for solving routing problems. IEEE Trans. Neural Netw. Learn. Syst. (2021)
67. Xin, L., Song, W., Cao, Z., Zhang, J.: Step-wise deep learning models for solving routing problems. IEEE Trans. Ind. Inform. (2020)
68. Xin, L., Song, W., Cao, Z., Zhang, J.: NeuroLKH: combining deep learning model with Lin-Kernighan-Helsgaun heuristic for solving the traveling salesman problem. In: Advances in Neural Information Processing Systems (NeurIPS) (2021)
69. Xu, S., Panwar, S.S., Kodialam, M., Lakshman, T.: Deep neural network approximated dynamic programming for combinatorial optimization. In: AAAI Conference on Artificial Intelligence (AAAI), vol. 34, pp. 1684–1691 (2020)
70. Yang, F., Jin, T., Liu, T.Y., Sun, X., Zhang, J.: Boosting dynamic programming with neural networks for solving np-hard problems. In: Asian Conference on Machine Learning (ACML), pp. 726–739. PMLR (2018)

Learning a Propagation Complete Formula

Petr Kučera[✉][iD]

Department of Theoretical Computer Science and Mathematical Logic,
Faculty of Mathematics and Physics, Charles University, Prague, Czech Republic
kucerap@ktiml.mff.cuni.cz

Abstract. Propagation complete formulas were introduced by Bordeaux and Marques-Silva (2012) as a possible target language for knowledge compilation. A CNF formula is propagation complete (PC) if for every partial assignment, the implied literals can be derived by unit propagation. Bordeaux and Marques-Silva (2012) proposed an algorithm for compiling a CNF formula into an equivalent PC formula which is based on incremental addition of so-called empowering implicates. In this paper, we propose a compilation algorithm based on the implicational structure of propagation complete formulas described by Kučera and Savický (2020) and the algorithm for learning a definite Horn formula with closure and equivalence queries introduced by Atserias et al. (2021). We have implemented both approaches and compared them experimentally. Babka et al. (2013) showed that checking if a CNF formula admits an empowering implicate is an NP-complete problem. We propose a particular CNF encoding which allows us to use a SAT solver to check propagation completeness, or to find an empowering implicate.

Keywords: propagation complete formula · satisfiability · consistency checking · empowering clause · learning algorithm

1 Introduction

The class of propagation complete formulas was introduced in [11] as a possible target language for knowledge compilation. A CNF formula φ is propagation complete (PC), if for any partial assignment α, we can check the consistency of $\varphi \wedge \alpha$ by unit propagation and if $\varphi \wedge \alpha$ is consistent, then unit propagation derives the literals implied by $\varphi \wedge \alpha$.

PC encodings with existentially quantified auxiliary variables were used as a target compilation language for a compilation from decision diagrams in [1]. Decomposable negation normal forms (DNNFs) introduced in [17] can be compiled into a PC encoding in polynomial time [28] and [30] showed that PC encodings have the same properties as DNNFs with respect to query answering and transformations according to the knowledge compilation map [19].

Propagation completeness can be characterized using the notion of empowerment [11,35]. In particular, a formula is not propagation complete if and only

© Springer Nature Switzerland AG 2022
P. Schaus (Ed.): CPAIOR 2022, LNCS 13292, pp. 214–231, 2022.
https://doi.org/10.1007/978-3-031-08011-1_15

if it admits an empowering implicate [11]. To compile a CNF into a PC formula, we can thus use an incremental approach suggested in [11] in which we keep adding empowering clauses to a formula, until it becomes propagation complete. In this approach, we can use the fact that the clauses learned by a CDCL SAT solver when called on the formula are empowering [35]. We can also look for an empowering implicate using a SAT solver [7]. To keep a formula compact, we can remove the clauses that are absorbed [4] by the others. Following [11], we refer to this step as minimization.

Two other approaches to the automatic construction of a PC formula were described in [14,21]. Both of these approaches are based on considering all partial assignments and are usable only for formulas on a very small number of variables.

The authors of [29] introduced the notion of an implicational dual rail encoding associated with a given CNF formula φ. It uses the well-known dual rail encoding of partial assignments [9,15,25,32,34] to simulate unit propagation in φ in the same way as [8,10]. If ψ is a PC formula equivalent to φ, then the definite Horn part of the implicational dual rail encoding of ψ defines a particular definite Horn function and the compilation of φ into a PC formula can be described as learning a representation of this definite Horn function. An algorithm for learning a definite Horn formula using membership and equivalence queries was described in [2]. Later, [3] described an algorithm that learns a definite Horn formula using closure and equivalence queries. In our case, a closure query corresponds to finding all literals implied by a given partial assignment, this query can be thus answered using a SAT solver on the input CNF. In an equivalence query, we are looking for a negative example in form of a partial assignment α and a literal l which is implied by $\varphi \wedge \alpha$, but it is not derived by unit propagation. This is same as asking if φ an empowering implicate. We describe an encoding that allows us to use a SAT solver to answer an equivalence query. Our encoding exploits the structure of a directed graph associated with the implicational dual rail encoding in a way similar to the encodings of ASP to SAT (see for example [22]). The encoding is usually substantially harder to solve than the input CNF. The equivalence query is thus more "expensive" than the closure query. We verify experimentally that the learning approach based on the algorithm from [3] uses substantially smaller number of the SAT based checks of propagation completeness than the incremental approach and is thus a more efficient approach to the compilation of a CNF into a PC formula.

2   Propagation Complete Formulas

We assume the reader is familiar with the basics of propositional logic, especially with the notion of entailment $\models$ and the notation related to formulas in conjunctive normal form (*CNF formulas*). We treat a clause as a set of literals and a CNF formula as a set of clauses. We use lit($\mathbf{x}$) to denote the set of literals (x, $\neg x$) over the set of variables $\mathbf{x}$. A set of literals α that contains a complementary pair of literals is called *contradictory*, otherwise, it is a *partial assignment*. If an assignment is clear from the context, we use simply $x = 0$ or $x = 1$ to denote the value of a variable x in the assignment. We use $\perp$ to represent the empty clause or a contradiction.

We use $\varphi \vdash_1 l$ to denote the fact that a literal l can be derived from formula φ by unit propagation or, in other words, by an iterative use of unit resolution rule which derives clause C given clauses $C \vee e$ and $\neg e$ for a literal e. Let φ be a CNF formula on variables $\mathbf{x}$ and let $\alpha \subseteq \mathrm{lit}(\mathbf{x})$ be a set of literals. We define the *unit propagation closure* of α with respect to φ as the set of literals we can derive from $\varphi \wedge \alpha$ with unit propagation, i.e.

$$\mathcal{U}_\varphi(\alpha) = \{l \in \mathrm{lit}(\mathbf{x}) \mid \varphi \wedge \alpha \vdash_1 l\}.$$

Note that if φ does not contain the empty clause and $\varphi \wedge \alpha \vdash_1 \bot$, then $\mathcal{U}_\varphi(\alpha)$ is a contradictory set of literals. It is not hard to check that $\mathcal{U}$ is a closure operator (extensive, monotone, and idempotent).

We say that a clause C is an *implicate* of a formula φ if $\varphi \models C$. If f is the boolean function represented by φ, then we also say that C is an implicate of f. Clause C is a *prime implicate* of φ (resp. f) if no proper subclause of C is an implicate of φ (resp. f). A CNF is *prime* if it consists only of prime implicates. We say that implicate C is 1-provable by φ if $\varphi \wedge \bigwedge_{l \in C} \neg l \vdash_1 \bot$.

The notion of a propagation complete CNF formula was introduced in [11].

Definition 1 (Propagation complete formula). *Let φ be a CNF formula on a set of variables $\mathbf{x}$. We say that φ is* propagation complete (PC)*, if for every partial assignment $\alpha \subseteq \mathrm{lit}(\mathbf{x})$ and for each $l \in \mathrm{lit}(\mathbf{x})$, such that $\varphi \wedge \alpha \models l$ we have $\varphi \wedge \alpha \vdash_1 l$ or $\varphi \wedge \alpha \vdash_1 \bot$.*

It was shown in [11] that propagation complete formulas can be characterized using the notions of empowerment [35] and absorption [4]. We extend these notions to non-implicates, since this is useful for algorithmic purposes. A non-empty clause C is *empowering* with respect to a CNF formula φ if for some literal (called *empowered literal*) $l \in C$ we have that $\varphi \wedge \bigwedge_{e \in C \setminus \{l\}} \neg e \nvdash_1 l$ and $\varphi \wedge \bigwedge_{e \in C \setminus \{l\}} \neg e \nvdash_1 \bot$, and C is *absorbed* by φ otherwise. It was shown in [11] that a CNF formula φ is propagation complete if and only if it does not admit any empowering implicate or, equivalently, every non-empty implicate of φ is absorbed by φ. Moreover, if φ is a PC formula and $C \in \varphi$ is such that C is absorbed by $\varphi' = \varphi \setminus \{C\}$, then φ' is a PC formula equivalent to φ. Since testing absorption can be done in polynomial time, this can be used to obtain in polynomial time an inclusion-wise minimal subformula φ' of φ that is equivalent to φ and which defines the same closure operator $\mathcal{U}_{\varphi'}$ as $\mathcal{U}_\varphi$. Following [11], we refer to this step as *minimization* of φ. The clauses learned by a CDCL SAT solver when run on φ are empowering [35].

We say that a boolean function $f(\mathbf{x})$ is *nontrivial* if it is not a constant 0 or 1 function. Assume $f(\mathbf{x})$ is a nontrivial boolean function on variables $\mathbf{x}$ and assume φ is a CNF representation of f. We say that a partial assignment $\alpha \subseteq \mathrm{lit}(\mathbf{x})$ is a *partial model* of f, if it can be extended to a model of f, in other words, $\varphi \wedge \alpha$ is satisfiable. The *semantic closure* of a set of literals $\alpha \subseteq \mathrm{lit}(\mathbf{x})$ is defined as the set of literals implied by $\varphi \wedge \alpha$ (the *backbone literals* of $\varphi \wedge \alpha$)

$$\mathcal{S}_f(\alpha) = \{l \mid \varphi \wedge \alpha \models l\}.$$

Note that if α is not a partial model (i.e. $\varphi \wedge \alpha$ is unsatisfiable), then $\mathcal{S}_f(\alpha) = \text{lit}(\mathbf{x})$. It follows that $\mathcal{U}_\varphi(\alpha) \subseteq \mathcal{S}_f(\alpha)$. It is not hard to check that φ is propagation complete if and only if for every set of literals α we have that $\mathcal{U}_\varphi(\alpha)$ is contradictory, or $\mathcal{U}_\varphi(\alpha) = \mathcal{S}_f(\alpha)$. The following stronger property holds.

Lemma 1. *Let $f(\mathbf{x})$ be a nontrivial boolean function on variables $\mathbf{x}$. Let φ be a CNF representation of $f(\mathbf{x})$. Then φ is PC if and only if $\mathcal{U}_\varphi(\alpha) = \mathcal{S}_f(\alpha)$ for every partial model $\alpha \subseteq \text{lit}(\mathbf{x})$ of f.*

Proof. (*only if*) Assume φ is PC and α is a partial model of f. It follows that $\varphi \wedge \alpha \not\vdash_1 \bot$ and thus $\mathcal{U}_\varphi(\alpha)$ is not contradictory and thus $\mathcal{U}_\varphi(\alpha) = \mathcal{S}_f(\alpha)$.

(*if*) Assume that φ is not PC. We will show that there is a partial model $\alpha \subseteq \text{lit}(\mathbf{x})$ of f such that $\mathcal{U}_\varphi(\alpha) \subsetneq \mathcal{S}_f(\alpha)$. Since φ is not PC, there is an implicate C of f which is empowering with respect to φ. Consider a prime subimplicate $C' \subseteq C$, it was shown in [7] that C' is empowering with respect to φ. It follows that there is a literal $l \in C'$ such that $\varphi \wedge \bigwedge_{e \in C' \setminus \{l\}} \neg e \not\vdash_1 l$. On the other hand, since C' is an implicate of φ, we have $\varphi \wedge \bigwedge_{e \in C' \setminus \{l\}} \neg e \models l$. Set $\alpha = \{\neg e \mid e \in C' \setminus \{l\}\}$. Because C' is a prime implicate of f, we have that α is a partial model of f. It follows that $l \in \mathcal{S}_f(\alpha) \setminus \mathcal{U}_\varphi(\alpha)$ and thus $\mathcal{U}_\varphi(\alpha) \subsetneq \mathcal{S}_f(\alpha)$.

The authors of [30] considered PC encodings with existentially quantified auxiliary variables as a target compilation language in the sense of the knowledge compilatiom map [19]. PC encodings allow consistency checking (CO), clausal entailment (CE), and model enumeration (ME) queries. If we consider PC formulas without existentially quantified variables, we also gain validity (VA) and implicant check (IM) (because we consider CNF formulas), sentential entailment (SE) and equality (EQ) (because the formulas are propagation complete). PC formulas support conditioning (CD) and singleton forgetting (SFO), but for forgetting (FO) and (bounded) disjunction, auxiliary variables are needed. In addition, PC formulas can be used to compute the backbone literals relative to a partial assignment in linear time. Model counting is hard for PC formulas (as it is already hard for monotone CNF formulas), however, we believe that in some applications, ME and VA might be sufficient.

3   Implicational Systems

Let us briefly recall the notion of an implicational dual rail encoding introduced in [29]. Assume a CNF formula φ that does not contain an empty clause. We associate a variable $[\![l]\!]$ with every literal l on a variable from φ. The *implicational dual rail* encoding $\text{DR}(\varphi)$ associated with φ then contains definite Horn clause representing the implications $\bigwedge_{e \in C \setminus \{l\}} [\![\neg e]\!] \implies [\![l]\!]$ for every clause $C \in \varphi$ and every literal $l \in C$. This implication represents the fact that C is used to derive l by unit propagation. In addition, $\text{DR}(\varphi)$ contains clauses $\neg[\![x]\!] \vee \neg[\![\neg x]\!]$ to capture the fact that any model of $\text{DR}(\varphi)$ represents a consistent set of literals of φ. It was shown in [29] that if ψ is a PC formula equivalent to φ, then φ is PC if and only if $\text{DR}(\varphi)$ is equivalent to $\text{DR}(\psi)$.

Assume a CNF φ, the definite Horn part of $\mathrm{DR}(\varphi)$ can be also interpreted as an *implicational system* on literals Φ which consists of *implications* $\sigma \to \tau$ where both the *body* σ and *tail* τ are sets of literals. For every clause $C \in \varphi$ and every literal $l \in C$, we introduce body $\sigma = \{\neg e \mid e \in C \setminus \{l\}\}$. For each such body, Φ contains implication $\sigma \to \tau$ with tail $\tau = \{l \mid (\bigvee_{e \in \sigma} \neg e \vee l) \in \varphi\}$. A single implication $\sigma \to \tau$ in Φ thus corresponds to a conjunction of clauses $\bigwedge_{l \in \tau} (\bigvee_{e \in \sigma} \neg e \vee l)$. For example, clauses $\neg a \vee \neg b \vee c$ and $\neg a \vee \neg b \vee d$ introduce an implication $\{a, b\} \to \{c, d\}$ with body $\{a, b\}$ and tail $\{c, d\}$. Also, if $l_1, \ldots, l_k$ are all literals in the unit clauses of φ, then Φ contains implication $\emptyset \to \{l_1, \ldots, l_k\}$.

We define the *forward chaining closure* $\mathcal{F}_\Phi(\alpha)$ of a set of literals $\alpha \subseteq \mathrm{lit}(\mathbf{x})$ in Φ as a minimal set of literals which satisfies that $\alpha \subseteq \mathcal{F}_\Phi(\alpha)$ and if $\sigma \subseteq \mathcal{F}_\Phi(\alpha)$ for some implication $\sigma \to \tau \in \Phi$, then also $\tau \subseteq \mathcal{F}_\Phi(\alpha)$. If implication $\sigma \to \tau$ is used to add a literal $e \in \tau$ during an iterative construction of the closure, we say that it *fires* for e. It follows that $\mathcal{F}_\Phi(\alpha) = \mathcal{U}_\varphi(\alpha)$. Implicational systems on literals Φ and Ψ are *equivalent* if and only if $\mathcal{F}_\Phi(\alpha) = \mathcal{F}_\Psi(\alpha)$ for every set of literals $\alpha \subseteq \mathrm{lit}(\mathbf{x})$.

Note that an implicational system can be represented using different bases. The one that is so-called left-saturated and has a minimum number of implications is called Guigues-Duquenne (GD) basis and was introduced in [24]. A body minimal representation can be constructed in polynomial time [31] and it can be made irredundant as well [13].

We will use the following characterization of propagation completeness.

Lemma 2. *Assume φ is a prime CNF representation of a boolean function $f(\mathbf{x})$. Let ψ be a prime PC representation of f and let Φ and Ψ be the implicational systems associated with φ and ψ respectively. Then the following propositions are equivalent:*

(i) φ is PC.
(ii) Φ is equivalent to Ψ.
(iii) $\mathcal{F}_\Phi(\alpha) = \mathcal{F}_\Psi(\alpha)$ for every partial model $\alpha \subseteq \mathrm{lit}(\mathbf{x})$ of f.

Proof. The equivalence of (i) and (ii) follows from the results of [12, 29] as noted above. The equivalence of (i) and (iii) follows by Lemma 1.

4   Checking Propagation Completeness by SAT

Let us fix a boolean function $f(\mathbf{x})$ on n variables $\mathbf{x}$ and a CNF representation φ of $f(\mathbf{x})$. We describe a CNF encoding $\mathcal{E}$ which is satisfiable if and only if φ admits a 1-provable empowering implicate C. By [7], this is equivalent to the fact that φ is not PC. The encoding is built using the implicational system Φ defined in Sect. 3. We will assume that $\Phi = \{\sigma_j \to \tau_j \mid j = 1, \ldots, m\}$.

Given the implicational system Φ, we can rephrase the task as looking for a partial assignment α and a literal $l \in \alpha$ such that $\mathcal{F}_\Phi(\alpha)$ is contradictory (i.e. clause $C = \bigvee_{e \in \alpha} \neg e$ is 1-provable) and $\alpha' = \alpha \setminus \{l\}$ is closed under forward chaining in Φ (i.e. $\alpha' = \mathcal{F}_\Phi(\alpha')$). Since $\neg l \notin \alpha' = \mathcal{F}_\Phi(\alpha')$, we have that C is empowering with empowered literal $\neg l$. We can assume that α' is closed under forward chaining without loss of generality, because adding literals derivable by forward chaining to α' does not change its forward chaining closure.

To encode the partial assignment α, we introduce the set of variables $\mathbf{v} = \{v[e] \mid e \in \mathrm{lit}(\mathbf{x})\}$. Literal l is selected by picking a variable in this literal using $\mathbf{s} = \{s[x] \mid x \in \mathbf{x}\}$. We use the following clauses to encode this semantics.

(C1) $\neg v[x] \vee \neg v[\neg x]$ for every $x \in \mathbf{x}$.
(C2) Clauses encoding the exactly one constraint on variables $\mathbf{s}$.

A model of $\mathcal{E}$ then specifies a partial assignment $\alpha_{\mathcal{E}} = \bigwedge_{e:v[e]=1} e$, a clause $C_{\mathcal{E}} = \bigvee_{e:v[e]=1} \neg e$ and a unique variable $x_{\mathcal{E}}$ such that $s[x_{\mathcal{E}}] = 1$. The encoding implies that there is a literal $l_{\mathcal{E}} \in \alpha$ on variable $x_{\mathcal{E}}$.

Section 4.1 describes clauses representing the fact that $C_{\mathcal{E}}$ is empowering with respect to $\neg l_{\mathcal{E}}$. Section 4.2 describes the encoding of 1-provability of $C_{\mathcal{E}}$. In Sect. 4.3, we will discuss how to put these parts together.

4.1   Encoding Empowerment

In this section, we describe the part of the encoding that ensures that $C_{\mathcal{E}}$ is empowering with empowered literal $\neg l_{\mathcal{E}}$. Denote $\alpha'_{\mathcal{E}} = \alpha_{\mathcal{E}} \setminus \{x_{\mathcal{E}}, \neg x_{\mathcal{E}}\}$, we encode the fact that $\alpha'_{\mathcal{E}}$ is not contradictory and it is closed under unit propagation in φ (in particular, $\mathcal{F}_{\Phi}(\alpha'_{\mathcal{E}}) = \alpha'_{\mathcal{E}}$). To this end, we introduce variable $p[l]$ for every literal $l \in \mathrm{lit}(\mathbf{x})$ and clauses representing the following subformulas.

(E1) $p[l] \iff \neg s[x] \wedge v[l]$ for every variable $x \in \mathbf{x}$ and every literal $l \in \{x, \neg x\}$.
(E2) $\bigwedge_{e \in \sigma_j} p[e] \implies p[l]$ for every implication $\sigma_j \to \tau_j \in \Phi$ and every $l \in \tau_j$.

Assume a model of $\mathcal{E}$. Clauses of group (E1) ensure that the set of literals l for which $p[l] = 1$ is exactly $\alpha'_{\mathcal{E}}$. Since $\alpha'_{\mathcal{E}} \subseteq \alpha_{\mathcal{E}}$, we have that $\alpha'_{\mathcal{E}}$ is not contradictory by using clauses (C1). Clauses (E2) ensure that $\mathcal{F}_{\Phi}(\alpha'_{\mathcal{E}}) = \alpha'_{\mathcal{E}}$.

4.2   Encoding 1-Provability

In this section, we describe the part of the encoding which ensures that $\mathcal{F}_{\Phi}(\alpha_{\mathcal{E}})$ is contradictory, in other words $C_{\mathcal{E}}$ is 1-provable. The idea of our encoding is based on the encodings of ASP to SAT (see for example [22]).

We associate a directed graph $G = (V, E)$ with the system of implications Φ. The set of vertices of G is $V = \mathrm{lit}(\mathbf{x})$ and a pair of literals $l_1, l_2 \in \mathrm{lit}(\mathbf{x})$ forms an edge $(l_1, l_2) \in E$ iff $l_1 \in \sigma_j$ and $l_2 \in \tau_j$ for some implication $\sigma_j \to \tau_j$ in Φ.

A strongly connected component (SCC) of G is an inclusion-wise maximal set of nodes S such that G contains a directed path between any pair of nodes of S. Let $S_1, \ldots, S_q$ be the list of all strongly connected components (SCCs) of G. Due to maximality requirement, the SCCs in the list are pairwise disjoint. Every literal $e \in \mathrm{lit}(\mathbf{x})$ thus belongs to exactly one SCC whose index is denoted scc(e). We will assume that the SCCs are topologically ordered, in particular, for every edge $(l_1, l_2) \in E$ we have scc(l_1) < scc(l_2).

For an SCC S_i, $i = 1, \ldots, q$, we use $n_i = |\text{var}(S_i)|$ to denote the number of different variables in the literals within S_i. For a literal $e \in \text{lit}(\mathbf{x})$, we define

$$\ell(e) = \begin{cases} n_{\text{scc}(e)} & \text{scc}(e) = \text{scc}(\neg e) \\ n_{\text{scc}(e)} - 1 & \text{otherwise} \end{cases}$$

For an SCC S_i we also define $n_i' = \max_{e \in S_i} \ell(e)$. In particular, $n_i' = n_i$ if S_i is contradictory, otherwise $n_i' = n_i - 1$.

A *level mapping* λ assigns every literal $e \in \text{lit}(\mathbf{x})$ an integer satisfying $0 \leq \lambda(e) \leq \ell(e)$. Consider a set of literals $L \subseteq \text{lit}(\mathbf{x})$. We say that a literal $e \in L$ is *proven* by implication $\sigma_j \to \tau_j \in \Phi$ with λ, if

(i) $e \in \tau_j$,
(ii) $\sigma_j \subseteq L$, and
(iii) $\lambda(a) < \lambda(e)$ for every literal $a \in \sigma_j \cap S_{\text{scc}(e)}$.

We say that λ *proves* L with respect to a partial assignment $\alpha \subseteq \text{lit}(\mathbf{x})$, if every literal $e \in L \backslash \alpha$ is proven by some implication from Φ. The encoding of 1-provability is based on the following characterization.

Theorem 1. *Let* $\alpha \subseteq \text{lit}(\mathbf{x})$ *be a partial assignment. Then* $\mathcal{F}_\Phi(\alpha)$ *is contradictory, if and only if there is a contradictory set of literals* $L \subseteq \mathcal{F}_\Phi(\alpha)$ *and a level mapping* λ *such that* λ *proves* L *with respect to* α.

A similar characterization is known in the area of ASP [22], however, there are two differences. Firstly, the bound $\ell(e)$ can be strictly smaller than $|S_{\text{scc}(e)}|$. Secondly, our goal is to detect if a contradiction is derived, not a particular literal. Theorem 1 directly follows from Lemma 3 and Lemma 4 below.

Lemma 3. *Assume* $\alpha \subseteq \text{lit}(\mathbf{x})$ *is a partial assignment and let* L *be a set of literals which is proven by some level mapping* λ *with respect to* α. *Then* $L \subseteq \mathcal{F}_\Phi(\alpha)$.

Proof. Consider a literal $e \in L$. If $e \in L \cap \alpha$, then clearly $e \in \mathcal{F}_\Phi(\alpha)$. Otherwise, there is an implication $\sigma_j \to \tau_j$ that proves e. We shall show by induction on scc(e) and the value of $\lambda(e)$ that $e \in \mathcal{F}_\Phi(\alpha)$.

Assume first that $e \in S_1$ and $\lambda(e) = 0$. Since S_1 is the first in the topological order, we have $\sigma_j \subseteq S_1$ and by (iii) we have $\sigma_j = \emptyset$. It follows that $e \in \mathcal{F}_\Phi(\alpha)$.

Suppose now that $e \in S_i$ for $i > 1$ or $\lambda(e) > 0$. If $\sigma_j = \emptyset$, then trivially $e \in \mathcal{F}_\Phi(\alpha)$. Otherwise, let $a \in \sigma_j$ be arbitrary and let us show that $a \in \mathcal{F}_\Phi(\alpha)$. By (ii) we have that $a \in L$. Since $(a, e) \in E$, we get by (iii) that scc(a) < scc(e) or $\lambda(a) < \lambda(e)$. By induction hypothesis, we can conclude that $\sigma_j \subseteq \mathcal{F}_\Phi(\alpha)$ and using $\sigma_j \to \tau_j$ we obtain $e \in \mathcal{F}_\Phi(\alpha)$.

Lemma 4. *Assume* $\alpha \subseteq \text{lit}(\mathbf{x})$ *is a partial assignment and assume that* $\mathcal{F}_\Phi(\alpha)$ *is contradictory. Then there is a contradictory set of literals* $L \subseteq \mathcal{F}_\Phi(\alpha)$ *and a level mapping* λ *such that* λ *proves* L *with respect to* α.

Proof. Set $\lambda(a) = 0$ for every literal $a \in \alpha$ and initialize $L = \alpha$. Then start forward chaining derivation. While L is not contradictory, pick an implication $\sigma_j \rightarrow \tau_j$ such that $\sigma_j \subseteq L$ and $\tau_j \not\subseteq L$. Every literal $e \in \tau_j \backslash L$ is added to L and we set $\lambda(e) = 0$ if $\sigma_j \cap S_{\text{scc}(e)} = \emptyset$ and $1 + \max_{a \in \sigma_j \cap S_{\text{scc}(e)}} \lambda(a)$ otherwise. After L becomes contradictory, we set $\lambda(b) = 0$ for every literal $b \notin L$.

Let S_i be an SCC such that $S_i \cap L \neq \emptyset$ and let l be the first literal of S_i added to L. Then either $l \in \alpha$, or l is added due to implication $\sigma_j \rightarrow \tau_j$ such that $\sigma_j \cap S_i = \emptyset$ and $l \in \tau_j$. In both cases we have $\lambda(l) = 0$. If $S_i \cap L$ is not contradictory, then the literals in $S_i \cap L$ are on pairwise disjoint variables. These literals thus have at most n_i different levels. Since the smallest level is 0, we have that $\lambda(e) \leq n_i - 1 \leq \ell(e)$ for every literal $e \in S_i \cap L$. Assume now that $S_i \cap L$ is contradictory and assume that e is the last literal from S_i added to L. Denote $S_i' = (S_i \cap L) \backslash \{e\}$. Then S_i' is not contradictory and all literals $a \in S_i'$ have levels $\lambda(a) \leq n_i - 1 \leq \ell(a)$. Since S_i' becomes contradictory after adding e, we have that $\neg e \in S_i$ and that $\lambda(e) \leq n_i = \ell(e)$.

The encoding of 1-provability encodes the fact that there is a level mapping which proves a contradictory set of literals L with respect to α. We consider two ways of representing a level of a literal. In the unary representation, we use an indicator variable of property $\lambda(e) \leq k$ for every literal e and every possible level $\lambda(e) \in \{0, \ldots, \ell(e)\}$. In the binary representation, we represent the binary representation of level $\lambda(e)$ with a logarithmic number of bits. In both cases, we associate an output variable out$[e]$ with every literal $e \in \text{lit}(\mathbf{x})$ which indicates if $e \in L$. Set L is required to be contradictory by adding clausal encoding of the following subformula.

(L1) $\bigvee_{x \in \mathbf{x}} \text{out}[x] \wedge \text{out}[\neg x]$

We use the following observation in the description of the encodings.

Lemma 5. *For every implication $\sigma_j \rightarrow \tau_j \in \Phi$, there is at most one SCC S_i such that $S_i \cap \sigma_j \neq \emptyset$ and $S_i \cap \tau_j \neq \emptyset$.*

Proof. Assume there are two different SCCs S_{i_1} and S_{i_2} satisfying the assumption. Let $a_1 \in S_{i_1} \cap \sigma_j$, $a_2 \in S_{i_2} \cap \sigma_j$, $b_1 \in S_{i_2} \cap \tau_j$, and $b_2 \in S_{i_2} \cap \tau_j$. By having $\sigma_j \rightarrow \tau_j \in \Phi$ there is a path from a_1 to b_2, using S_{i_2}, there is a path from b_2 to a_2. By having $\sigma_j \rightarrow \tau_j \in \Phi$, there is a path from a_2 to b_1 and using S_{i_1}, there is a path from b_1 to a_1. Nodes a_1, b_2, a_2, b_1 thus belong to the same SCC which is a contradiction with the assumption that S_{i_1} is not equal to S_{i_2}.

We shall call the SCC S_i satisfying the assumption of Lemma 5 the *main SCC of implication $\sigma_j \rightarrow \tau_j$*. We shall use ms$(j)$ to denote the main SCC of implication $\sigma_j \rightarrow \tau_j$, or $\emptyset$ if the implication does not have a main SCC.

In both versions of the encoding, we introduce a variable fire$[j]$ for every implication $\sigma_j \rightarrow \tau_j$, such that $\tau_j \backslash \text{ms}(j) \neq \emptyset$. It is an indicator variable of the property that $\sigma_j \rightarrow \tau_j$ can fire for literals in $\tau_j \backslash \text{ms}(j)$ regardless of the level. This semantic is captured by the following clauses.

(F1) Unit clause fire$[j]$
    ... for every implication $\sigma_j \to \tau_j$ with $\sigma_j = \emptyset$.
(F2) fire$[j] \implies \bigwedge_{a \in \sigma_j}$ out$[a]$
    ... for every implication $\sigma_j \to \tau_j$ such that $\tau_j \setminus \mathrm{ms}(j) \neq \emptyset$.

Note that if $\sigma_j = \{a\}$ for a single literal a, then fire$[j]$ can be identified with out$[a]$ and the corresponding subformula (F2) is not needed.

Unary Representation of Levels. We introduce variable drv$[e, k]$ for every literal $e \in \mathrm{lit}(\mathbf{x})$ and every level $k = 0, \ldots, \ell(e)$. It is an indicator variable of the property that e was derived at level $\lambda(e) \leq k$. Variable drv$[e, \ell(e)]$ is identified with the output variable out$[e]$.

We also introduce variable fire$_\mathrm{U}[j, k]$ for every implication $\sigma_j \to \tau_j \in \Phi$ which has a main SCC S_i and every $k = 0, \ldots, n'_i - 1$. It is an indicator variable of the property that $\sigma_j \to \tau_j$ can fire for any literal $e \in \tau_j \cap \mathrm{ms}(j)$ at level $\lambda(e) \geq k+1$.

We use the clauses representing the following subformulas in the encoding.

(U1) fire$_\mathrm{U}[j, k] \implies \bigwedge_{a \in \sigma_j \setminus \mathrm{ms}(j)}$ out$[a] \wedge \bigwedge_{b \in \sigma_j \cap \mathrm{ms}(j)}$ drv$[b, k]$
    ... for every $k = 0, \ldots, n'_i - 1$ assuming $\mathrm{ms}(j) = S_i$.
(U2) drv$[e, 0] \implies v[e] \vee \bigvee_{j \colon e \in \tau_j \setminus \mathrm{ms}(j)}$ fire$[j]$
    ... for every literal $e \in \mathrm{lit}(\mathbf{x})$.
(U3) drv$[e, k] \implies$ drv$[e, k-1] \vee \bigvee_{j \colon e \in \tau_j \cap \mathrm{ms}(j)}$ fire$_\mathrm{U}[j, k-1]$
    ... for every literal $e \in \mathrm{lit}(\mathbf{x})$ and every level value $k = 1, \ldots, \ell(e)$.

Recall that out$[e]$ is identified with drv$[e, \ell(e)]$ for every literal $e \in \mathrm{lit}(\mathbf{x})$ and thus the respective implication (U3) captures the necessary requirements on the derivability of e when out$[e] = 1$. Note that if $\sigma_j = \{a\}$ for a single literal a, then fire$_\mathrm{U}[j, k]$ can be identified with drv$[a, k]$ and subformula (U1) is not needed.

Binary Encoding with Direct Inequalities. With every literal $e \in \mathrm{lit}(\mathbf{x})$, we associate a bitvector $B[e]$ which represents the binary representation of level $\lambda(e)$. The width of bitvectors of literals in an SCC S_i is set to $w_i = \lceil \log_2(n'_i + 1) \rceil$.

In addition to the previously defined variables, we introduce variable fire$_\mathrm{B}[j, e]$ for every implication $\sigma_j \to \tau_j$ and every literal $e \in \tau_j \cap \mathrm{ms}(j)$. It is an indicator variable of the fact that the implication $\sigma_j \to \tau_j$ can fire for literal e.

We use the clauses representing the following subformulas in the encoding.

(D1) fire$_\mathrm{B}[j, e] \implies \bigwedge_{a \in \sigma_j}$ out$[a] \wedge \bigwedge_{a \in \sigma_j \cap \mathrm{ms}(j)} B[e] > B[a]$
    ... for every implication $\sigma_j \to \tau_j$ and every literal $e \in \tau_j \cap \mathrm{ms}(j)$.
(D2) out$[e] \implies v[e] \vee \bigvee_{j \colon e \in \tau_j \setminus \mathrm{ms}(j)}$ fire$[j] \vee \bigvee_{j \colon e \in \tau_j \cap \mathrm{ms}(j)}$ fire$_\mathrm{B}[j, e]$
    ... for every literal $e \in \mathrm{lit}(\mathbf{x})$.

It is also possible to require $B[e] \leq n'_{\mathrm{scc}(e)}$ for every literal $e \in \mathrm{lit}(\mathbf{x})$, but this inequality is not needed for the correctness.

4.3 Putting the Parts Together

We define two CNF encodings. They share the *base clauses* for subformulas (C1), (C2), (E1)–(E2), (L1), and (F1)–(F2). The *unary* encoding $\mathcal{E}_u$ contains the base clauses and the clauses for subformulas (U1)–(U3). The *binary direct* encoding $\mathcal{E}_d$ contains the base clauses and the clauses for subformulas (D1)–(D2).

Recall that $n = |\mathbf{x}|$ and m denotes the number of implications in Φ (which is bounded by $\|\varphi\| = \sum_{C \in \varphi} |C|$). Denote $\ell_m = \max\{n_i' \mid i = 1, \dots, q\}$. Then $\mathcal{E}_u$ has $\Theta(n + (m+n)\ell_m)$ variables and $\mathcal{E}_d$ has $O(n + t + (n + |E|) \log_2 \ell_m)$ variables where $t = \sum_{j=1}^m \tau_j$ and E is the set of edges of the graph G associated with the implicational system Φ (the estimate includes the auxiliary variables needed to encode the inequalities on bitvectors in subformulas (D1)).

Value of ℓ_m can be as much as n, but can also be much smaller if the graph G has only small SCCs. The exact sizes of the encodings thus depend on the structure of the formula φ and its associated implicational system Φ. To increase the efficiency of a SAT-based PC check, we use the following approach. The compiler maintains a bound δ initialized with value 1. When a PC check should be executed, we use δ as a level bound in the following way. In the unary encoding $\mathcal{E}_u$, we use additional restriction $\lambda(e) \leq \delta$. In the binary direct encoding $\mathcal{E}_d$, δ limits the bit width of the bitvectors $B[e]$. If no empowering literal with the restricted encoding is found, another attempt is performed without using the additional restriction. If no empowering literal is found in this second check, then the formula is propagation complete, otherwise, we use the empowering implicate found in this way and we increase the bound by 1. The process ends when δ reaches the maximum level bound—ℓ_m in the case of the unary encoding $\mathcal{E}_u$, and $\max\{w_i \mid i = 1, \dots, q\} = \lceil \log_2(\ell_m + 1) \rceil$ in the case of the binary direct encoding $\mathcal{E}_d$.

Using this approach, we limit the size of the encoding in the most SAT based PC checks which makes the process much more efficient.

5 The Learning Approach to Compilation

In this section, we shall describe how the algorithm for learning a GD basis of an implicational system with equivalence and closure queries [3] can be used to construct a PC representation of a given boolean function $f(\mathbf{x})$ given by its CNF representation φ. Consider a PC formula ψ representing $f(\mathbf{x})$ and let Ψ be the implicational system associated with ψ. Following Lemma 2, our goal is to modify φ in such a way that its associated implicational system Φ is equivalent to Ψ. To answer a closure query, we need to compute $\mathcal{F}_\Psi(\alpha)$ for a partial assignment α. This step can be carried out by a SAT solver using φ, because $\mathcal{F}_\Psi(\alpha) = \mathcal{S}_f(\alpha)$ due to propagation completeness of ψ. To answer the equivalence query, we can use SAT on the encoding described in Sect. 4. The overall structure of our approach is described in Algorithm 1 which closely follows the algorithm from [3].

Preprocessing in Step 1 starts with finding backbone literals, propagating and removing them from the CNF. The next step involves turning φ into a prime CNF and minimizing φ by removing absorbed clauses.

Input: A CNF φ representing a boolean function $f(\mathbf{x})$
Output: A propagation complete CNF representing $f(\mathbf{x})$.
1 $\varphi \leftarrow$ preprocess(φ)
2 $\Gamma \leftarrow$ empty list of implications
3 **while** next_negative(α) **do**
4   **forall** $\sigma \rightarrow \tau \in \Gamma$ *in order* **do**
5          $\beta \leftarrow \alpha \cap \sigma$
6   **if** $\beta \subsetneq \alpha$ **and** $\beta \subsetneq \mathcal{S}_f(\beta)$ **then**
7   Replace $\sigma \rightarrow \tau$ with $\beta \rightarrow \mathcal{S}_f(\beta)$ in Γ
8   add_implication$(\varphi,\ \beta \rightarrow \mathcal{S}_f(\beta))$
9   **break**

10   **if** *no implication was replaced* **then**
11   Add $\alpha \rightarrow \mathcal{S}_f(\alpha)$ as the last implication of Γ
12   add_implication$(\varphi,\ \alpha \rightarrow \mathcal{S}_f(\alpha))$

13 **return** φ

Algorithm 1: Learning a PC representation

The algorithm maintains the hypothesis in two forms. The CNF φ and the implicational system Γ represented as a list of implications. The implications in Γ are kept in the order in which they are added to the hypothesis. The order is important for the proof of correctness and complexity of the algorithm, see [3] for more details. Implicational system Γ is initially empty. During the work of the algorithm, both the CNF φ and the implicational system Γ are being updated. Implicational system Γ is not necessarily equivalent to the implicational system Φ associated to φ. However, at any moment we have for every partial assignment α that $\mathcal{F}_\Gamma(\alpha) \subseteq \mathcal{F}_\Phi(\alpha) \subseteq \mathcal{S}_f(\alpha)$. In particular, when Γ is equivalent to Ψ, then also Φ is equivalent to Ψ and thus φ is propagation complete. To avoid SAT based PC checks which would improve Γ but not Φ, we try to keep Γ as close to Φ as possible. We use φ to answer closure queries, i.e. to compute $\mathcal{S}_f(\alpha)$.

Function next_negative(α) represents the equivalence query. It looks for a negative example in form of a partial model α for which $\alpha = \mathcal{F}_\Gamma(\alpha) \subsetneq \mathcal{S}_f(\alpha)$. If no negative example is found, the compilation finishes and φ is PC.

An important feature of the learning algorithm is that if α is a negative example, then the corresponding implication $\alpha \rightarrow \mathcal{S}_f(\alpha)$ is not directly added to Γ, but the algorithm first tries to use α to refine an existing implication of Γ at steps 4–9. Only if no existing implication can be refined, the implication is added to Γ using function add_implication().

Function add_implication$(\varphi,\ \sigma \rightarrow \tau)$ is used to add the clauses corresponding to the implication $\sigma \rightarrow \tau$ to φ. This step is important to have that Φ always lies between Γ and Ψ. For every literal $l \in \tau$ we construct clause $C = (\bigvee_{e \in \sigma} \neg e \vee l)$. We check if C is empowering for φ and if so, then a prime subimplicate of C is added to φ. After adding a certain amount of clauses to φ, we run minimization on φ to remove clauses that are absorbed by the others.

We use the following sources of negative examples in our implementation.

(i) The initial list of candidates on negative examples is extracted from the input CNF φ. The default is to use all bodies in a body minimal representation of the implicational system Φ.

(ii) Whenever a clause C is being added to φ using add_implication(), we check for every $l \in C$, if a partial assignment $\alpha_l = \bigwedge_{e \in C \setminus \{l\}} \neg e$ is a negative example. Note that α_l is a partial model, because we only add prime implicates C to φ. This way, we keep Γ to Φ as close as possible.

(iii) When an implication of Γ is being replaced during the refinement (steps 4–9), both σ and α are checked if they are negative examples.

(iv) When a SAT solver is used on φ, we collect the learned clauses and use them as a source of possible negative examples as in (ii).

(v) We use random checks of propagation completeness in which we try a random partial assignment α' and check if we can obtain a negative example of it.
   - If α' is a partial model, then we check if $\alpha = \mathcal{F}_\Gamma(\alpha') \subsetneq \mathcal{S}_f(\alpha')$.
   - If $C = \bigvee_{e \in \alpha'} \neg e$ is an implicate of f, then we find a prime subimplicate C' of C and we try to find a negative example based on C' as in (ii).

(vi) We use SAT based checks of propagation completeness using the encoding described in Sect. 4. In the construction of the encoding, we use Γ to encode empowerment as described in Sect. 4.1. However, the encoding of 1-provability (described in Sect. 4.2) is based on the implicational system Φ associated with φ as it is potentially stronger than Γ.

It follows by the results of [3] that the number of equivalence queries needed by the algorithm is bounded by $O(mn)$ where m is the number of implications in a GD basis of Ψ and $n = |\mathbf{x}|$. Each negative example found by one of the ways listed above counts as an equivalence query. The fact that we have a bound on the number of equivalence queries needed to compile the formula is the main motivation for using Algorithm 1, because our goal is to minimize the number of SAT based checks of propagation completeness.

6   Experiments

We have implemented the learning algorithm as part of the tool pccompile which can be accessed through the project web pages [27]. The program uses SAT solver Glucose 4.1 [5,6] which is based on MiniSat 2.2 [20]. All experiments were executed on a computational server with CPU Intel® Xeon® CPU E5–2420 v2 @ 2.20 GHz and with 32 of memory.

We have also implemented the incremental approach which follows the idea proposed in [11]. After the preprocessing (which is the same as in Algorithm 1), the incremental algorithm keeps adding empowering implicates to the input CNF formula φ, until it becomes PC. In addition to the SAT based PC checks, we use the clauses learned during the SAT calls on φ (e.g. when finding a prime subimplicate of a clause) as a source of empowering implicates. Formula is regularly

Table 1. Results on random CNF formulas. See the description of the instances and columns in the text.

set	A	t	t_c	p^+	t_p^+	p^-	t_p^-
modgen60.1	I	66.11 s	42.07 s	431	37.26 s	4	26.38 s
	L	54.66 s	30.98 s	134	25.76 s	4	25.79 s
	LR	49.40 s	26.20 s	54	18.68 s	4	25.11 s
modgen60.2	I	541.10 s	373.08 s	1084	322.20 s	5	193.75 s
	L	484.75 s	280.57 s	309	241.67 s	5	223.77 s
	LR	525.30 s	237.80 s	171	197.58 s	5	245.46 s
modgen80	I	1000.09 s	671.66 s	1478	582.33 s	5	356.45 s
	L	842.53 s	537.15 s	438	466.29 s	5	308.42 s
	LR	751.48 s	448.24 s	237	373.98 s	5	322.84 s

minimized by removing absorbed clauses. We have not performed any experiments with the algorithms introduced in [14,21], because these approaches are only suitable for formulas with a small number of variables.

In the first set of experiments, we used three sets of random 3-CNF formulas generated using the modularity based generator [23]. Each set contains 50 instances generated with the following settings: modgen60.1 (60 variables, 120 clauses, 4 communities, PC formulas with 497 clauses on average), modgen60.2 (60 variables, 150 clauses, 3 communities, PC formulas with 1008 clauses on average), modgen80 (80 variables, 200 clauses, 4 communities, PC formulas with 1377 clauses on average).

Table 1 contains the results of experiments on these instances. Column "A" contains the version of the algorithm where "I" stands for incremental, "L" for the learning algorithm. We have run the learning algorithm twice—with ("LR") and without ("L") using the randomly generated negative examples (obtained by (v)). Column t contains the total running time, t_c contains the time of compilation without including the last SAT based PC check (as the last SAT based PC check is present in both algorithms, it makes sense to compare the time of compilation without it). Columns p^+ and t_p^+ contain the number and total time of the SAT based PC checks with result SAT (i.e. those which found an empowering implicate or a negative example). Columns p^- and t_p^- contains the number of total time of the SAT based PC checks with result UNSAT. In all cases, we used the unary encoding $\mathcal{E}_u$, the binary direct encoding $\mathcal{E}_d$ was about 3 to 4 times slower.

We can see that the learning algorithm has substantially smaller number of SAT based PC checks, which is even smaller with using randomly generated negative examples. The difference in t_c is also significant, although not as much as the difference in the number of SAT based PC checks would suggest. This is because the learning algorithm is more complex than the incremental algorithm.

In the second set of experiments, we have run `pccompile` on some of the benchmark formulas obtained from [16]. The results are contained in Table 2.

Table 2. Results on benchmark CNF formulas. See the description of the instances and columns in the text.

instance	A	n	m	n_b	m_{out}	t	t_c	p^+	t_p^+
C169_FV	I	1411	1982	50	1404	0.03 s	0.01 s	1	0.01 s
	L	1411	1982	50	1404	0.06 s	0.04 s	0	0 s
C169_FW	I	1411	1982	50	1404	0.04 s	0.02 s	2	0.01 s
	L	1411	1982	50	1404	0.06 s	0.04 s	0	0 s
C171_FR	I	1758	4005	451	2921	184.61 s	159.11 s	177	145.59 s
	L	1758	4005	451	2921	117.25 s	83.75 s	11	13.69 s
C211_Fs	I	1635	3662	247	3068	1049.31 s	1031.47 s	1972	758.04 s
	L	1635	3662	247	3070	93.51 s	71.49 s	3	1.53 s
C211_FW	I	1665	5929	341	5347	6980.68 s	6789.70 s	3433	3760.19 s
	L	1665	5929	341	5354	2852.89 s	2633.81 s	73	443.54 s
C250_FV	I	1465	2356	129	1470	4.31 s	4.07 s	52	3.48 s
	L	1465	2356	129	1471	1.02 s	0.73 s	0	0 s
C250_FW	I	1465	2356	129	1470	4.19 s	3.93 s	52	3.33 s
	L	1465	2356	129	1471	1.03 s	0.75 s	0	0 s
sat-grid-pbl-0010	I	110	191	102	297	4.42 s	4.00 s	127	3.01 s
	L	110	191	102	297	2.46 s	1.52 s	1	0.04 s
sat-grid-pbl-0015	L	240	436	232	15984	7261.54 s	4249.99 s	17	2112.25 s
log-1	I	939	3785	308	1294	1412.17 s	655.78 s	386	564.69 s
	L	939	3785	308	1280	587.50 s	189.45 s	33	113.51 s
ais6	I	61	581	59	1955	2971.50 s	2927.32 s	4025	973.85 s
	L	61	581	59	1948	966.39 s	928.25 s	922	634.62 s

Formulas C169_FV to C250_FW belong to the Configuration category, formulas sat-grid-pbl-0010 and sat-grid-pbl-0015 belong to BayesianNetwork, log-1 belongs to Planning, and ais6 belongs to the Handmade category. We have used the unary encoding $\mathcal{E}_u$ on ais6 (we were not able to compile the formula with using $\mathcal{E}_d$). For the rest of formulas, we list the results for the binary-direct encoding $\mathcal{E}_d$ as it performed better. In addition to the columns already used in Table 1, we use the following columns: Columns n and m contain the number of variables and clauses in the instance. Column n_b contains the number of variables after propagating and removing the backbone literals (the instances we consider usually have lots of backbones which are removed in the preprocessing and ignored for the rest of the compilation). Column m_{out} contains the number of clauses in the output PC formula (including the unit clauses for the backbone literals).

In all cases, the learning algorithm outperforms the incremental one. The difference in the number of SAT based PC checks is quite significant in all cases as well, this is especially true for the C211_FS and C211_FW instances. We have been able to compile sat-grid-pbl-0015 only using the learning algorithm. The incremental algorithm did not finish computation within 15 h (it made 12 518 SAT based PC checks within this time).

7   Conclusion

Although the learning approach generally outperforms the incremental one, it has the following main limitations.

(A) The output formula can be much bigger than the input formula.
(B) Checking propagation completeness with SAT is hard.

Related to (A) is the fact that there are even Horn CNF formulas, for which any equivalent PC formula is necessarily exponentially bigger [29]. By using the learning algorithm, we make sure that the size of the output PC formula is within a polynomial factor from the size of a smallest equivalent PC formula. To tackle (B), we have designed an encoding that limits the necessary number of levels of unit propagation. It is possible that a better upper bound on the number of levels of unit propagation could be derived from the structure of the formula. This would lead to a smaller encoding of 1-provability, possibly making SAT based PC checks easier.

The algorithm is based on an assumption that the SAT on the input formula is easy and thus answering the closure query is easy. This is satisfied by most of the instances in [16]. Let us also note that if a CNF formula is hard for SAT, it is also presumably far from being PC which is related to (A). When a SAT solver is run on a hard formula, it takes a long time, but the solver also learns a lot of clauses that can be used as a source of negative examples in the algorithm.

We have to admit that most of the benchmark formulas in [16] are currently out of reach for the method presented in the paper. For this reason, we have included the results of experiments only on a handful of formulas which we were able to successfully compile. For the same reason, we used random formulas to compare the incremental approach with the learning approach. Modularity based generator [23] was used to introduce some structure into the formulas.

The compilers from a CNF into a d-DNNF such as D4 [26] or c2d [18] can successfully compile much more formulas than pccompile. Let us note, however, that if we run D4 on C171_FR, we obtain a Decision DNNF with 483486 nodes while the formula output by pccompile has length (i.e. the sum of the lengths of the clauses) only 4731. Similarly, for C211_FW we get a Decision DNNF with 635993 nodes and PC CNF has length 15563. Thus for some instances, our approach can produce substantially smaller formulas than D4.

As a direction of future research, we would like to consider adding auxiliary variables to the formula, e.g. using a variant of bounded variable addition [33].

Acknowledgements. The author acknowledges the support by Grant Agency of the Czech Republic (grant No. GA19-19463S). The author would like to thank Pierre Marquis, Jean-Marie Lagniez, Gilles Audemard, and Stefan Mengel (from CRIL, U. Artois, Lens, France) who proposed the problem of compilation of a CNF into a PC formula. Last, but not least, the author would like to thank Petr Savický for lots of discussions that helped to finish this paper and, in particular, for the suggestion to use the learning algorithm for the compilation.

References

1. Abío, I., Gange, G., Mayer-Eichberger, V., Stuckey, P.J.: On CNF encodings of decision diagrams. In: Quimper, C.-G. (ed.) CPAIOR 2016. LNCS, vol. 9676, pp. 1–17. Springer, Cham (2016). https://doi.org/10.1007/978-3-319-33954-2_1
2. Angluin, D., Frazier, M., Pitt, L.: Learning conjunctions of Horn clauses. Mach. Learn. 9(2), 147–164 (1992). https://doi.org/10.1007/BF00992675
3. Arias, M., Balcázar, J.L., Tîrnăucă, C.: Learning definite Horn formulas from closure queries. Theor. Comput. Sci. 658, 346–356 (2017). https://doi.org/10.1016/j.tcs.2015.12.019, https://www.sciencedirect.com/science/article/pii/S0304397515011809. Horn formulas, directed hypergraphs, lattices and closure systems: related formalism and application
4. Atserias, A., Fichte, J.K., Thurley, M.: Clause-learning algorithms with many restarts and bounded-width resolution. J. Artif. Int. Res. 40(1), 353–373 (2011)
5. Audemard, G., Simon, L.: Predicting learnt clauses quality in modern SAT solvers. In: Twenty-first International Joint Conference on Artificial Intelligence (2009)
6. Audemard, G., Simon, L.: Glucose 4.1. https://www.labri.fr/perso/lsimon/glucose. Accessed 14 Apr 2022
7. Babka, M., Balyo, T., Čepek, O., Gurský, Š., Kučera, P., Vlček, V.: Complexity issues related to propagation completeness. Artif. Intell. 203, 19–34 (2013). https://doi.org/10.1016/j.artint.2013.07.006, http://www.sciencedirect.com/science/article/pii/S0004370213000726
8. Bessiere, C., Katsirelos, G., Narodytska, N., Walsh, T.: Circuit complexity and decompositions of global constraints. In: Proceedings of the Twenty-First International Joint Conference on Artificial Intelligence (IJCAI-09), pp. 412–418 (2009)
9. Bonet, M.L., Buss, S., Ignatiev, A., Marques-Silva, J., Morgado, A.: MaxSAT resolution with the dual rail encoding. In: Proceedings of the AAAI Conference on Artificial Intelligence, vol. 32, no. 1, April 2018. https://ojs.aaai.org/index.php/AAAI/article/view/12204
10. Bordeaux, L., Janota, M., Marques-Silva, J., Marquis, P.: On unit-refutation complete formulae with existentially quantified variables. In: Proceedings of the Thirteenth International Conference on Principles of Knowledge Representation and Reasoning, KR 2012, pp. 75–84. AAAI Press (2012). http://dl.acm.org/citation.cfm?id=3031843.3031854
11. Bordeaux, L., Marques-Silva, J.: Knowledge compilation with empowerment. In: Bieliková, M., Friedrich, G., Gottlob, G., Katzenbeisser, S., Turán, G. (eds.) SOFSEM 2012. LNCS, vol. 7147, pp. 612–624. Springer, Heidelberg (2012). https://doi.org/10.1007/978-3-642-27660-6_50
12. Boros, E., Čepek, O., Kogan, A., Kučera, P.: Exclusive and essential sets of implicates of boolean functions. Discrete Appl. Math. 158(2), 81–96 (2010). https://doi.org/10.1016/j.dam.2009.08.012, http://www.sciencedirect.com/science/article/B6TYW-4XBP93T-1/2/dfeacfe4911d5f1e5aed44f1ea8dc6bb
13. Boros, E., Čepek, O., Makino, K.: Strong duality in horn minimization. In: Klasing, R., Zeitoun, M. (eds.) FCT 2017. LNCS, vol. 10472, pp. 123–135. Springer, Heidelberg (2017). https://doi.org/10.1007/978-3-662-55751-8_11
14. Brain, M., Hadarean, L., Kroening, D., Martins, R.: Automatic generation of propagation complete SAT encodings. In: Jobstmann, B., Leino, K.R.M. (eds.) VMCAI 2016. LNCS, vol. 9583, pp. 536–556. Springer, Heidelberg (2016). https://doi.org/10.1007/978-3-662-49122-5_26

15. Bryant, R.E., Beatty, D., Brace, K., Cho, K., Sheffler, T.: COSMOS: a compiled simulator for MOS circuits. In: Proceedings of the 24th ACM/IEEE Design Automation Conference, DAC 1987, pp. 9–16. Association for Computing Machinery, New York (1987). https://doi.org/10.1145/37888.37890

16. Compile! Project: Benchmarks. https://www.cril.univ-artois.fr/KC/benchmarks.html. Accessed 14 Apr 2022

17. Darwiche, A.: Compiling knowledge into decomposable negation normal form. In: Proceedings of the 16th International Joint Conference on Artifical Intelligence, IJCAI 1999, vol. 1, pp. 284–289. Morgan Kaufmann Publishers Inc., San Francisco (1999)

18. Darwiche, A.: New advances in compiling CNF to decomposable negation normal form. In: Proceedings of the 16th European Conference on Artificial Intelligence, ECAI 2004, pp. 318–322. IOS Press, Amsterdam (2004). http://dl.acm.org/citation.cfm?id=3000001.3000069

19. Darwiche, A., Marquis, P.: A knowledge compilation map. J. Artif. Intell. Res. **17**, 229–264 (2002)

20. Eén, N., Sörensson, N.: An extensible SAT-solver. In: Giunchiglia, E., Tacchella, A. (eds.) SAT 2003. LNCS, vol. 2919, pp. 502–518. Springer, Heidelberg (2004). https://doi.org/10.1007/978-3-540-24605-3_37

21. Ehlers, R., Palau Romero, F.: Approximately propagation complete and conflict propagating constraint encodings. In: Beyersdorff, O., Wintersteiger, C.M. (eds.) SAT 2018. LNCS, vol. 10929, pp. 19–36. Springer, Cham (2018). https://doi.org/10.1007/978-3-319-94144-8_2

22. Fandinno, J., Hecher, M.: Treewidth-aware complexity in ASP: not all positive cycles are equally hard. In: Thirty-Fifth AAAI Conference on Artificial Intelligence, AAAI 2021, Thirty-Third Conference on Innovative Applications of Artificial Intelligence, IAAI 2021, The Eleventh Symposium on Educational Advances in Artificial Intelligence, EAAI 2021, Virtual Event, 2–9 February 2021, pp. 6312–6320. AAAI Press (2021). https://ojs.aaai.org/index.php/AAAI/article/view/16784

23. Giráldez-Cru, J., Levy, J.: A modularity-based random SAT instances generator. In: Twenty-Fourth International Joint Conference on Artificial Intelligence (2015)

24. Guigues, J.L., Duquenne, V.: Familles minimales d'implications informatives résultant d'un tableau de données binaires. Math. Sci. Hum. **95**, 5–18 (1986)

25. Ignatiev, A., Morgado, A., Marques-Silva, J.: On tackling the limits of resolution in SAT solving. In: Gaspers, S., Walsh, T. (eds.) SAT 2017. LNCS, vol. 10491, pp. 164–183. Springer, Cham (2017). https://doi.org/10.1007/978-3-319-66263-3_11

26. Jean-Marie Lagniez, P.M.: An improved decision-DNNF compiler. In: Proceedings of the Twenty-Sixth International Joint Conference on Artificial Intelligence, IJCAI-17, pp. 667–673 (2017). https://doi.org/10.24963/ijcai.2017/93

27. Kučera, P.: Program pccompile. http://ktiml.mff.cuni.cz/~kucerap/pccompile (2021), accessed: 2022–04-14

28. Kučera, P., Savický, P.: Propagation complete encodings of smooth DNNF theories. arXiv preprint arXiv:1909.06673 (2019)

29. Kučera, P., Savický, P.: Bounds on the size of PC and URC formulas. J. Artif. Intell. Res. **69**, 1395–1420 (2020)

30. Kučera, P., Savický, P.: Backdoor decomposable monotone circuits and propagation complete encodings. In: Proceedings of the AAAI Conference on Artificial Intelligence, vol. 35, no. 5, pp. 3832–3840 (2021). https://ojs.aaai.org/index.php/AAAI/article/view/16501

31. Maier, D.: Minimal covers in the relational database model. J. ACM **27**, 664–674 (1980)

32. Manquinho, V.M., Flores, P.F., Silva, J.P.M., Oliveira, A.L.: Prime implicant computation using satisfiability algorithms. In: Proceedings Ninth IEEE International Conference on Tools with Artificial Intelligence, pp. 232–239, November 1997. https://doi.org/10.1109/TAI.1997.632261
33. Manthey, N., Heule, M.J.H., Biere, A.: Automated reencoding of boolean formulas. In: Biere, A., Nahir, A., Vos, T. (eds.) HVC 2012. LNCS, vol. 7857, pp. 102–117. Springer, Heidelberg (2013). https://doi.org/10.1007/978-3-642-39611-3_14
34. Morgado, A., Ignatiev, A., Bonet, M.L., Marques-Silva, J., Buss, S.: DRMaxSAT with MaxHS: first contact. In: Janota, M., Lynce, I. (eds.) SAT 2019. LNCS, vol. 11628, pp. 239–249. Springer, Cham (2019). https://doi.org/10.1007/978-3-030-24258-9_17
35. Pipatsrisawat, K., Darwiche, A.: On the power of clause-learning SAT solvers as resolution engines. Artif. Intell. **175**(2), 512–525 (2011). https://doi.org/10.1016/j.artint.2010.10.002, http://dx.doi.org/10.1016/j.artint.2010.10.002

A FastMap-Based Algorithm for Block Modeling

Ang Li[1]([✉]), Peter Stuckey[2], Sven Koenig[1], and T. K. Satish Kumar[1]

[1] University of Southern California, Los Angeles, CA 90007, USA
{ali355,skoenig}@usc.edu, tkskwork@gmail.com
[2] Monash University, Wellington Road, Clayton, VIC 3800, Australia
peter.stuckey@monash.edu

Abstract. Block modeling algorithms are used to discover important latent structures in graphs. They are the graph equivalent of clustering algorithms. However, existing block modeling algorithms work directly on the given graphs, making them computationally expensive and less effective on large complex graphs. In this paper, we propose a FastMap-based algorithm for block modeling on single-view undirected graphs. FastMap embeds a given undirected graph into a Euclidean space in near-linear time such that the pairwise Euclidean distances between vertices approximate a desired graph-based distance function between them. In the first phase, our FastMap-based block modeling (FMBM) algorithm uses FastMap with a probabilistically-amplified shortest-path distance function between vertices. In the second phase, it uses Gaussian Mixture Models (GMMs) for identifying clusters (blocks) in the resulting Euclidean space. FMBM outperforms other state-of-the-art methods on many benchmark and synthetic instances, both in efficiency and solution quality. It also enables a perspicuous visualization of clusters (blocks) in the graphs, not provided by other methods.

Keywords: Community Detection and Block Modeling · Graph Embeddings · FastMap

1 Introduction

Finding inherent groups in graphs, i.e., the "graph" clustering problem, has important applications in many real-world domains, such as identifying communities in social networks [8], analyzing the diffusion of ideas in them [13], identifying functional modules in protein-protein interactions [11], and understanding the modular design of brain networks [2]. In general, identifying the

This work at the University of Southern California is supported by DARPA under grant number HR001120C0157 and by NSF under grant numbers 1409987, 1724392, 1817189, 1837779, 1935712, and 2112533. The views, opinions, and/or findings expressed are those of the author(s) and should not be interpreted as representing the official views or policies of the sponsoring organizations, agencies, or the U.S. Government. This research was partially supported by the OPTIMA ARC training centre IC200100009.

© Springer Nature Switzerland AG 2022
P. Schaus (Ed.): CPAIOR 2022, LNCS 13292, pp. 232–248, 2022.
https://doi.org/10.1007/978-3-031-08011-1_16

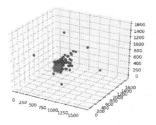

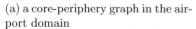

(a) a core-periphery graph in the air-port domain

(b) a FastMap embedding of the graph on the left

Fig. 1. The left side shows a core-periphery graph in the airport domain with edges representing flight connections, red vertices representing "hub" airports at the core, and blue vertices representing "local" airports at the periphery. The right side shows a FastMap embedding of the graph in Euclidean space, in which the red and blue vertices correctly appear in the core and periphery, respectively. (Color figure online)

groups involves mapping each vertex in the graph to a group (cluster), where vertices in the same group share important properties in the underlying graph.

The conditions under which two vertices are deemed to be similar and there-fore belonging to the same group are popularly studied in community detection and block modeling [1]. In community detection, a group (community) implicitly requires its vertices to be more connected to each other than to vertices of other groups. Although this is justified in many real-world domains, such as social networks, it is not always justified in general.

Block modeling uses more general criteria for identifying groups (blocks) where community detection fails. For example, block modeling can be used to correctly identify groups in *core-periphery* graphs characterized by a core of ver-tices tightly connected to each other and a peripheral set of vertices loosely con-nected to each other but well connected to the core.[1] Core-periphery graphs are common in many real-world domains, such as financial networks and flight net-works [1,19]. Figure 1a shows a core-periphery graph in an air flight domain [5].

Existing block modeling algorithms work directly on the given graphs and are inefficient. They typically use matrix operations that incur cubic time complex-ities even within their inner loops. For example, FactorBlock [3], a state-of-the-art block modeling algorithm, uses matrix multiplications in its inner loop and an expectation-maximization-style outer loop. Due to their inefficiency, existing block modeling algorithms are not scalable and result in poor solution qualities on large complex graphs.

In this paper, we propose a FastMap-based algorithm for block modeling on single-view[2] undirected graphs. FastMap embeds a given undirected graph into a Euclidean space in near-linear time such that the pairwise Euclidean distances

---

[1] The conditions used in community detection prevent the proper identification of peripheral groups.

[2] In a single-view graph, there is at most one edge between any two vertices.

between vertices approximate a desired graph-based distance function between them. In general, graph embeddings have been used in many different contexts, such as for shortest-path computations [4], multi-agent meeting problems [12], and social network analysis [18]. They are useful because they facilitate geometric interpretations and algebraic manipulations in vector spaces.

FastMap [4,12] is a recently developed graph embedding algorithm that runs in near-linear time[3]. While it has thus far been used to create Euclidean embeddings that approximate pairwise shortest-path distances between vertices, it can also be extended to creating Euclidean embeddings that approximate more general pairwise graph-based distances between vertices. In particular, for the purpose of block modeling in this paper, we propose a novel distance function that probabilistically amplifies the shortest-path distances between vertices.

Our FastMap-based block modeling algorithm (FMBM) works in two phases. In the first phase, FMBM uses FastMap to efficiently embed the vertices of a given graph in a Euclidean space, preserving the probabilistically-amplified shortest-path distances between them. In the second phase, FMBM identifies clusters (blocks) in the resulting Euclidean space using standard methods from unsupervised learning. Therefore, the first phase of FMBM efficiently reformulates the block modeling problem from a graphical space to a Euclidean space, as illustrated in Fig. 1b; and the second phase of FMBM leverages any technique that is already known or can be developed for clustering in Euclidean space. In our current implementation of FMBM, we use Gaussian Mixture Models (GMMs) for identifying clusters in the Euclidean space.

We empirically show that, in addition to the theoretical advantages of FMBM, it outperforms other state-of-the-art methods on many benchmark and synthetic test cases. We report on the superior performance of FMBM both in terms of efficiency and solution quality. We also show that it enables a perspicuous visualization of clusters in the graphs, beyond the capabilities of other methods.

2   Preliminaries and Background

In this section, we review some preliminaries of block modeling and provide a background description of FastMap.

2.1   Block Modeling

Let $G = (V, E)$ be an undirected graph with vertices $V = \{v_1, v_2 \ldots v_n\}$ and edges $E = \{e_1, e_2 \ldots e_m\} \subseteq V \times V$. Let $A \in \{0, 1\}^{n \times n}$ be the adjacency matrix representation of G, where $A_{ij} = 1$ iff $(v_i, v_j) \in E$.

A *block model* decomposes G into a set of k vertex partitions representing the blocks (groups), for a given value of k. The partitions are represented by the *membership matrix* $C \in \{0, 1\}^{n \times k}$, where $C_{ij} = 0$ and $C_{ij} = 1$ represent vertex v_i being absent from and being present in partition j, respectively. Therefore,

---

[3] i.e., linear time after ignoring logarithmic factors

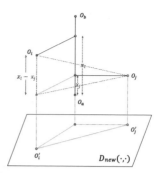

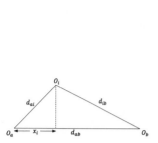

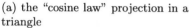

(a) the "cosine law" projection in a triangle

(b) projection onto a hyperplane that is perpendicular to $\overline{O_a O_b}$

Fig. 2. The figure, borrowed from [4], illustrates how coordinates are computed and recursion is carried out in FastMap. (Color figure online)

$\sum_{j=1}^{k} C_{ij} = 1$ for all $1 \leq i \leq n$. An *image matrix* is a matrix $M \in [0,1]^{k \times k}$, where M_{ij} represents the likelihood of an edge between a vertex in partition i and a vertex in partition j. The block model decomposition of G, as discussed in [3], tries to approximate A by $CMC^{\top}$ with the best choice for C and M. In other words, the objective is

$$\min_{C,M} \|A - CMC^{\top}\|_F^2, \tag{1}$$

where $\| \cdot \|_F$ is the Frobenius norm. An improved objective function is also considered in [3] to account for the imbalance of edges to non-edges[4] in A, since real-world graphs are typically sparse with significantly more non-edges than edges. The revised objective is

$$\min_{C,M} \|(A - CMC^{\top}) \circ (A - R)\|_F^2, \tag{2}$$

where $R \in [0,1]^{n \times n}$, $R_{ij} = \frac{m}{n^2}$, and $\circ$ represents element-wise multiplication.

The above formalization can be generalized to directed graphs and multi-view graphs [19]. It can also be generalized to soft partitioning, where each vertex partially belongs to each partition, i.e., $C \in [0,1]^{n \times k}$ with $\sum_{j=1}^{k} C_{ij} = 1$ for all $1 \leq i \leq n$.

2.2 FastMap

FastMap [6] was introduced in the Data Mining community for automatically generating Euclidean embeddings of abstract objects. For many complex objects (such as long DNA strings), multi-media datasets (like voice excerpts or images), or medical datasets (like ECGs or MRIs), there is no geometric space in which

---

[4] i.e., a pair of vertices not connected by an edge

they can be naturally visualized. However, there is often a well-defined distance function between every pair of objects in the problem domain. For example, the *edit distance* between two DNA strings is well-defined although an individual DNA string cannot be conceptualized in geometric space.

FastMap embeds a collection of abstract objects in an artificially created Euclidean space to enable geometric interpretations, algebraic manipulations, and downstream machine learning algorithms. It gets as input a collection of abstract objects $\mathcal{O}$, where $D(O_i, O_j)$ represents the domain-specific distance between objects $O_i, O_j \in \mathcal{O}$. A Euclidean embedding assigns a K-dimensional point $\vec{p_i} \in \mathbb{R}^K$ to each object O_i. For $\vec{p_i} = ([\vec{p_i}]_1, [\vec{p_i}]_2 \ldots [\vec{p_i}]_K)$ and $\vec{p_j} = ([\vec{p_j}]_1, [\vec{p_j}]_2 \ldots [\vec{p_j}]_K)$, we define the Euclidean distance $\chi_{ij} = \sqrt{\sum_{r=1}^{K}([\vec{p_j}]_r - [\vec{p_i}]_r)^2}$. A good Euclidean embedding is one in which χ_{ij} between any two points $\vec{p_i}$ and $\vec{p_j}$ closely approximates $D(O_i, O_j)$.

FastMap creates a K-dimensional Euclidean embedding of the abstract objects in $\mathcal{O}$ for a user-specified value of K. In the very first iteration, FastMap heuristically identifies the farthest pair of objects O_a and O_b in linear time. Once O_a and O_b are determined, every other object O_i defines a triangle with sides of lengths $d_{ai} = D(O_a, O_i)$, $d_{ab} = D(O_a, O_b)$, and $d_{ib} = D(O_i, O_b)$, as shown in Fig. 2a. The sides of the triangle define its entire geometry, and the projection of O_i onto the line $\overline{O_a O_b}$ is given by

$$x_i = (d_{ai}^2 + d_{ab}^2 - d_{ib}^2)/(2d_{ab}). \tag{3}$$

FastMap sets the first coordinate of $\vec{p_i}$, the embedding of O_i, to x_i. In the subsequent $K - 1$ iterations, the same procedure is followed for computing the remaining $K - 1$ coordinates of each object. However, the distance function is adapted for different iterations. For example, for the first iteration, the coordinates of O_a and O_b are 0 and d_{ab}, respectively. Because these coordinates fully explain the true distance between these two objects, from the second iteration onward, the rest of $\vec{p_a}$ and $\vec{p_b}$'s coordinates should be identical. Intuitively, this means that the second iteration should mimic the first one on a hyperplane that is perpendicular to the line $\overline{O_a O_b}$, as shown in Fig. 2b. Although the hyperplane is never constructed explicitly, its conceptualization implies that the distances for the second iteration should be changed for all i and j so that:

$$D_{new}(O_i', O_j')^2 = D(O_i, O_j)^2 - (x_i - x_j)^2. \tag{4}$$

Here, O_i' and O_j' are the projections of O_i and O_j, respectively, onto this hyperplane, and $D_{new}(\cdot, \cdot)$ is the new distance function.

FastMap can also be used to embed the vertices of a graph in a Euclidean space to preserve the pairwise shortest-path distances between them. The idea is to view the vertices of a given graph $G = (V, E)$ as the objects to be embedded. As such, the Data Mining FastMap algorithm cannot be directly used for generating an embedding in near-linear time. This is so because it assumes that the distance d_{ij} between any two objects O_i and O_j can be computed in constant time, independent of the number of objects in the problem domain. However, computing the shortest-path distance between two vertices depends on the size of the graph.

The near-linear time complexity of FastMap can be retained as follows: In each iteration, after we heuristically identify the farthest pair of vertices O_a and O_b, the distances d_{ai} and d_{ib} need to be computed for *all* other vertices O_i. Computing d_{ai} and d_{ib} for any single vertex O_i can no longer be done in constant time but requires $O(|E| + |V| \log |V|)$ time instead [7]. However, since we need to compute these distances for all vertices, computing two shortest-path trees rooted at each of the vertices O_a and O_b yields all necessary distances in one shot. The complexity of doing so is also $O(|E| + |V| \log |V|)$, which is only linear in the size of the graph[5]. The amortized complexity for computing d_{ai} and d_{ib} for vertex O_i is therefore near-constant time.

The foregoing observations are used in [12] to build a graph-based version of FastMap that embeds the vertices of a given undirected graph in a Euclidean space in near-linear time. The Euclidean distances approximate the pairwise shortest-path distances between vertices. A slight modification of this FastMap algorithm, presented in [4], can also be used to preserve *consistency* and *admissibility* of the Euclidean distance approximation, which is important when using it as a heuristic in A* search for shortest-path computations. In both [4] and [12], K is user-specified, but a threshold parameter ϵ is introduced to terminate with a smaller value of K once diminishing returns on the accuracy of approximating pairwise shortest-path distances are detected.

3   FastMap-Based Block Modeling Algorithm (FMBM)

In this section, we describe FMBM, our novel algorithm for block modeling based on FastMap [12]. As mentioned before, FMBM works in two phases. In the first phase, FMBM uses FastMap to efficiently embed vertices in a K-dimensional Euclidean space, preserving the probabilistically-amplified shortest-path distances between them. In the second phase, FMBM identifies the required blocks in the resulting Euclidean space using GMM clustering.

To facilitate the description of FMBM, we first examine what happens when FastMap is used naively in the first phase, i.e., when it is used to embed the vertices of a given undirected graph in a K-dimensional Euclidean space for preserving the pairwise shortest-path distances. This naive attempt fails even in relatively simple cases. For example, Figs. 3a–3d show that it fails on a bipartite graph and a core-periphery graph. This is so because preserving the pairwise shortest-path distances in Euclidean space does not necessarily help GMM clustering to identify the two blocks (partitions). In fact, in a bipartite graph, the closest neighbors of a vertex are in the other partition.

---

[5] unless $|E| = O(|V|)$, in which case the complexity is near-linear in the size of the input because of the $\log |V|$ factor

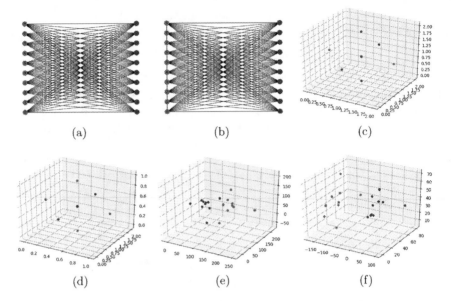

Fig. 3. The figure shows two simple graphs that guide the design of a proper FastMap distance function for block modeling. (a) shows a fully-connected bipartite graph with the red and blue vertices indicating the two partitions. (b) shows a core-periphery graph with the red vertices indicating the core and the blue vertices indicating the periphery. All pairs of red vertices are connected by edges (not all shown to avoid clutter). (c) and (d) show the FastMap Euclidean embeddings of the graphs in (a) and (b), respectively, using the shortest-path distance function. This naive FastMap distance function fails for block modeling. Red and blue points correspond to red and blue vertices of the graphs, respectively. Many vertices are mapped to the same point. (e) and (f) show the FastMap Euclidean embeddings of the graphs in (a) and (b), respectively, when using the probabilistically-amplified shortest-path distance function. This FastMap distance function is appropriate for block modeling. (Color figure online)

3.1   Probabilistically-Amplified Shortest-Path Distances

From the foregoing discussion, it is clear that the shortest-path distance between two vertices v_i and v_j is not a viable distance function for block modeling. Therefore, in this subsection, we create a new distance function $D(v_i, v_j)$ for pairs of vertices based on the following intuitive guidelines: (a) the smaller the *shortest-path distance* between v_i and v_j, the smaller the distance $D(v_i, v_j)$ should be; (b) the more *paths* exist between v_i and v_j, the smaller the distance $D(v_i, v_j)$ should be; and (c) the *complement graph*[6] $\bar{G}$ of the given graph G should yield the same distance function as G: The distance function should be independent of the arbitrary choice of representing a relationship between two vertices as either an edge or a non-edge. Intuitively, these guidelines capture an effective "resistance"

---

[6] The complement graph $\bar{G}$ has the same vertices as the original graph G but represents every edge in G as a non-edge and every non-edge in G as an edge.

between vertices and facilitate the subsequent embedding to represent relative "potentials" of vertices in Euclidean space. The effectiveness of these guidelines is validated through test cases in this section and comprehensive experiments in the next section.

We define a new distance function $D_P(v_i, v_j)$, referred to as the *probabilistically-amplified shortest-path distance* (PASPD) between v_i and v_j, as:

$$\sum_{\mathbb{G} \in G_{set}} d_{\mathbb{G}}(v_i, v_j). \tag{5}$$

Here, $d_{\mathbb{G}}(v_i, v_j)$ represents the shortest-path distance between v_i and v_j in an undirected graph $\mathbb{G}$. G_{set} represents a collection of undirected graphs derived from the given graph G or its complement $\bar{G}$. In particular, each graph in G_{set} is an edge-induced subgraph of either G or $\bar{G}$.[7] The edge-induced subgraphs are created by probabilistically dropping edges from G or $\bar{G}$.

Intuitively, the use of shortest-path distances on multiple graphs that are probabilistically derived from the same input graph G accounts for $D_P(\cdot, \cdot)$. Indeed, the smaller $d_G(v_i, v_j)$, the smaller $D_P(v_i, v_j)$ also is. Similarly, the more paths between v_i and v_j in G, the more likely it is for such paths to survive in its edge-induced subgraphs, and the smaller $D_P(v_i, v_j)$ consequently is. Moreover, since the subgraphs in G_{set} are derived from both G and $\bar{G}$, $D_P(\cdot, \cdot)$ satisfies all these intuitive guidelines mentioned above. From an efficiency perspective, the use of multiple graphs does not create much overhead if the number of graphs does not depend on the size of G. However, $\bar{G}$ can have significantly more edges than G if G is sparse. In such cases, if G has n vertices and $m < \binom{n}{2}/2$ edges, $\bar{G}$ itself is probabilistically derived from G by randomly retaining only m out of the $\binom{n}{2} - m$ edges that it would otherwise have. This keeps the size of $\bar{G}$ upper-bounded by the size of the input.

Although more details on FMBM are presented in the next subsection, the benefits of using a probabilistically-amplified distance function are visually apparent in Figs. 3e and 3f. In both cases, the red and blue vertices are mapped to linearly-separable red and blue points, respectively, in Euclidean space. Its benefits can also be seen in Fig. 1b, where the core red vertices are mapped to a core set of red points and the peripheral blue vertices are mapped to a peripheral set of blue points, respectively, in Euclidean space. In this case, although the red and blue points are not linearly separable, GMM clustering [15] in the second phase of FMBM is capable of separating them using two overlapping but different Gaussian distributions.

3.2   Main Algorithm

Algorithm 1 shows the pseudocode for computing the PASPD function $D_P(\cdot, \cdot)$ parameterized by L and F. Like the shortest-path distance function, it, too, can be computed efficiently (in one shot) for all pairs (v_s, v_i), for a specified source v_s and all $v_i \in V$. On Lines 3–15, the algorithm populates G_{set} with L lineages

---

[7] An edge-induced subgraph of G has the same vertices as G but a subset of its edges.

Algorithm 1. SS-PASPD: Single-Source Probabilistically-Amplified Shortest-Path Distance Function

---

Input: $G = (V, E)$ and $v_s \in V$
Parameters: L and F
Output: d_{si} for each $v_i \in V$

1: Let $\bar{G} = (V, \bar{E})$ be the complement graph of G.
2: $G_{set} \leftarrow \{\}$ and $T_{set} \leftarrow \{\}$.
3: **for** $l = 1, 2 \ldots L$ **do**
4:     $\mathbb{G} \leftarrow G$ and $\bar{\mathbb{G}} \leftarrow \bar{G}$.
5:     **if** $|\bar{E}| > |E|$ **then**
6:         Drop $|\bar{E}| - |E|$ randomly chosen edges from $\bar{\mathbb{G}}$.
7:     **end if**
8:     $G_{set} \leftarrow G_{set} \cup \{\mathbb{G}\}$ and $f \leftarrow |E|/F$.
9:     **while** $\mathbb{G}$ has edges **do**
10:         Drop f randomly chosen edges from $\mathbb{G}$ to obtain $\hat{G}$.
11:         $G_{set} \leftarrow G_{set} \cup \{\hat{G}\}$.
12:         $\mathbb{G} \leftarrow \hat{G}$.
13:     **end while**
14:     Repeat lines 8-13 for $\bar{\mathbb{G}}$.
15: **end for**
16: **for** $G_i \in G_{set}$ **do**
17:     $T_i \leftarrow$ SS-SHORTESTPATHDISTANCE(G_i, v_s).
18:     $T_{set} \leftarrow T_{set} \cup \{T_i\}$.
19: **end for**
20: **for** each $v_j \in V$ **do**
21:     $d_{sj} \leftarrow \sum_{T_i \in T_{set}} T_i(v_j)$.
22: **end for**
23: **return** d_{si} for each $v_i \in V$.

---

of F nested edge-induced subgraphs of G and $\bar{G}$. On Lines 5–7, the algorithm constructs the complement graph $\bar{G}$ but probabilistically retains at most $|E|$ of its edges. On Lines 16–23, it uses the single-source shortest-path distance function to compute and return the sum of the shortest-path distances from v_s to v_i in all $\mathbb{G} \in G_{set}$, for all $v_i \in V$. If v_s and v_i are disconnected in any graph $\mathbb{G} \in G_{set}$, $d_{\mathbb{G}}(v_s, v_i)$ is technically equal to $+\infty$. However, for practical reasons in such cases, $d_{\mathbb{G}}(v_s, v_i)$ is set to twice the maximum shortest-path distance from v_s to any other vertex connected to it in $\mathbb{G}$. T_i on Line 17 refers to the array of shortest-path distances from v_s in G_i. $T_i(v_j)$ on Line 21 is the array element that corresponds to vertex v_j.

Algorithm 2 shows the pseudocode for FMBM. On Lines 3–25, it essentially implements FastMap as described in [12] but calls the SS-PASPD distance function in Algorithm 1 instead of the regular single-source shortest-path distance function. As in FastMap [12], K represents the user-specified upper bound on the dimensionality of the Euclidean embedding, ϵ represents the user-specified threshold to recognize an accurate embedding, and Q represents a small constant number of pivot changes used to heuristically identify the farthest pair of ver-

Algorithm 2. FMBM: FastMap-Based Block Modeling

Input: $G = (V, E)$ and k
Parameters: L, F, T, K, Q, and ϵ
Output: c_i for each $v_i \in V$

1: $MinObj \leftarrow +\infty$ and $BestC \leftarrow \emptyset$.
2: **for** $t = 1, 2 \ldots T$ **do**
3:     **for** $r = 1, 2 \ldots K$ **do**
4:         Choose $v_a \in V$ randomly and let $v_b \leftarrow v_a$.
5:         **for** $q = 1, 2 \ldots Q$ **do**
6:             $\{d_{ai} : v_i \in V\} \leftarrow$ SS-PASPD(G, v_a).
7:             $v_c \leftarrow \text{argmax}_{v_i}\{d_{ai}^2 - \sum_{j=1}^{r-1}([\vec{p}_a]_j - [\vec{p}_i]_j)^2\}$.
8:             **if** $v_c == v_b$ **then**
9:                 Break.
10:            **else**
11:                 $v_b \leftarrow v_a$ and $v_a \leftarrow v_c$.
12:            **end if**
13:         **end for**
14:         $\{d_{ai} : v_i \in V\} \leftarrow$ SS-PASPD(G, v_a).
15:         $\{d_{ib} : v_i \in V\} \leftarrow$ SS-PASPD(G, v_b).
16:         $d_{ab}' \leftarrow d_{ab}^2 - \sum_{j=1}^{r-1}([\vec{p}_a]_j - [\vec{p}_b]_j)^2$.
17:         **if** $d_{ab}' < \epsilon$ **then**
18:            Break.
19:         **end if**
20:         **for** each $v_i \in V$ **do**
21:             $d_{ai}' \leftarrow d_{ai}^2 - \sum_{j=1}^{r-1}([\vec{p}_a]_j - [\vec{p}_i]_j)^2$.
22:             $d_{ib}' \leftarrow d_{ib}^2 - \sum_{j=1}^{r-1}([\vec{p}_i]_j - [\vec{p}_b]_j)^2$.
23:             $[\vec{p}_i]_r \leftarrow (d_{ai}' + d_{ab}' - d_{ib}')/(2\sqrt{d_{ab}'})$.
24:         **end for**
25:     **end for**
26:     $P \leftarrow [\vec{p}_1, \vec{p}_2 \ldots \vec{p}_{|V|}]$.
27:     $C \leftarrow$ GMM(P, k).
28:     $Obj \leftarrow \text{GETOBJECTIVEVALUE}(G, C)$.
29:     **if** $Obj \leq MinObj$ **then**
30:         $MinObj \leftarrow Obj$.
31:         $BestC \leftarrow C$.
32:     **end if**
33: **end for**
34: **return** c_i for each $v_i \in V$ according to $BestC$.

---

tices. L and F are simply passed to Algorithm 1 in the function call SS-PASPD. Because Algorithm 2 employs randomization, it qualifies as a Monte-Carlo algorithm. It implements an outer loop to boost the performance of FMBM using T independent trials. On Lines 26–32, each trial invokes the GMM clustering algorithm and evaluates the results on the objective function in Eq. 2,[8] keeping

---

[8] M can be computed from A and C in $O(|E| + k^2)$ time while evaluating the objective function in Eq. 2.

record of the best value. The results of the best trial, i.e., the block assignment c_i for each $v_i \in V$, are returned on Line 34.[9]

A formal time complexity analysis of FMBM is evasive since Line 27 of Algorithm 2 calls the GMM clustering procedure, which has no defined time complexity. Therefore, we only claim to be able to reformulate the block modeling problem on graphs to its Euclidean version in $O(LFK(|E| + |V| \log |V|))$ time in each of the T iterations. Here, the factor LF comes from the cardinality of G_{set} in Algorithm 1, and the factor $K(|E| + |V| \log |V|)$ comes from the complexity of FastMap, that uses SS-PASPD on Lines 3–25 of Algorithm 2. The time complexity of GETOBJECTIVEVALUE on Line 28 is technically $O(|V|^2 k + |V| k^2)$, where k is the user-specified number of blocks, also passed to the GMM clustering algorithm. This time complexity comes from the matrix multiplication $CMC^\top$ in Eq. 2. The factor $|V|^2$ in this matrix multiplication, and more generally in Eq. 2, can be reduced to $O(|E|)$ by evaluating $|E|$ entries corresponding to edges and $\min(|E|, \binom{|V|}{2} - |E|)$ randomly chosen entries corresponding to non-edges in the matrix expression $(A - CMC^\top) \circ (A - R)$. The matrix multiplication CM takes $O(|V| k^2)$ time and results in a $|V| \times k$ matrix. $|E| + \min(|E|, \binom{|V|}{2} - |E|)$ entries in the multiplication of this matrix with $C^\top$ can be computed in $O(|E| k)$ time. Overall, therefore, the reformulation to Euclidean space can be done in near-linear time, i.e., linear in $|V|$ and $|E|$, after ignoring logarithmic factors.

4   Experiments

In this section, we present empirical results on the comparative performances of FMBM and three other state-of-the-art solvers for block modeling: Graph-Tool, DANMF, and CPLNS. We also compared against two other solvers for block modeling: FactorBlock [3] and ASBlock [19]. However, they are not competitive with the other solvers; and we exclude them from Tables 1, 2, 3, 4, and 5 to save column space. Graph-Tool [17] uses an agglomerative multi-level Markov Chain Monte Carlo algorithm and has been largely ignored in the computer science literature on block modeling; DANMF [20] uses deep autoencoders; and CPLNS [14] uses constraint programming with large neighborhood search.

We used the following hyperparameter values for FMBM:[10] $L = 4$, $F = 10$, $T = 10$, $K = 4$, $Q = 10$, and $\epsilon = 10^{-4}$. The value of k, i.e., the number of blocks, was given as input for all the solvers in the experiments.[11] We used three metrics for comparison: the value of the objective function stated in Eq. 2, the

---

[9] The domain of each c_i is $\{1, 2 \dots k\}$. Block $\mathcal{B}_h$ refers to the collection of all vertices $v_i \in V$ such that $c_i = h$.

[10] These values are only important as ballpark estimates. We observed that the performance of FMBM often stays stable within broad ranges of hyperparameter values, imparting robustness to FMBM. Moreover, only a few different hyperparameter settings had to be examined to determine the best one.

[11] Although Graph-Tool does not require a user-specified value of k, it has a tendency to produce trivial solutions with $k = 1$, resulting in 0 NMI values when the value of k is not explicitly specified.

Table 1. Real-World Single-View Undirected Graphs.

Test Case	Size (|V|, |E|)	FMBM			Graph-Tool			DANMF			CPLNS		
		Objective	NMI	Time	Objective	NMI	Time	Objective	NMI	Time	Objective	NMI	Time
adjnoun	(112, 425)	616.86	0.0025	6.11	612.98	**0.2978**	**0.04**	636.75	0.0083	1.62	**591.76**	0.0154	1.51
baboons	(14, 23)	11.97	0.0158	0.54	**11.49**	**0.2244**	**0.00**	15.49	0.1341	0.97	12.81	0.0172	0.87
football	(115, 613)	665.97	0.5608	9.22	343.32	**0.9150**	**0.03**	863.91	0.2574	1.55	558.94	0.6991	83.33
karate	(34, 78)	74.66	**0.6127**	1.47	**64.67**	0.2512	**0.00**	81.94	0.1672	0.77	75.43	0.2228	1.06
polblogs	(1,490, 16,715)	98,788.53	0.0098	239.33	99,014.21	**0.4668**	**2.14**	101,195.89	0.0465	404.02	**95,859.73**	0.0543	506.29
polbooks	(105, 441)	522.33	0.5329	6.20	**496.02**	**0.5462**	**0.02**	590.20	0.3177	1.98	531.48	0.2073	2.09

Table 2. Complement Graphs of the Graphs in Table 1.

Test Case	Size (|V|, |E|)	FMBM			Graph-Tool			DANMF			CPLNS		
		Objective	NMI	Time	Objective	NMI	Time	Objective	NMI	Time	Objective	NMI	Time
adjnoun	(112, 5,791)	611.04	0.0048	1.41	636.34	0.0168	**0.41**	641.40	0.0000	8.34	**591.54**	0.0169	1.85
baboons	(14, 68)	**12.64**	**0.0547**	0.62	12.86	0.0416	**0.01**	15.46	0.0500	1.23	13.35	0.0316	0.86
football	(115, 5,944)	595.52	0.5899	27.53	344.38	**0.9111**	**0.17**	815.71	0.2229	9.56	525.54	0.7040	82.11
karate	(34, 483)	**72.73**	**0.7625**	2.46	77.31	0.2065	**0.02**	84.23	0.0914	1.78	75.00	0.2439	1.04
polblogs	(1,490, 1,094,951)	26,896.52	0.0153	3155.33	26,048.04	0.0454	**49.90**	-	-	> 1 hour	**25,871.42**	**0.0541**	470.48
polbooks	(105, 5,019)	**509.88**	**0.5409**	21.55	606.96	0.0867	**0.13**	631.77	0.0141	8.09	531.65	0.2056	2.15

Table 3. Sparse Single-View Undirected Graphs Using Generative Model 1.

Test Case	Size (|E|)	FMBM			Graph-Tool			DANMF			CPLNS		
		Objective	NMI	Time	Objective	NMI	Time	Objective	NMI	Time	Objective	NMI	Time
V0400b04	6,722	9,355.91	0.1622	83.22	**8,385.61**	**0.6565**	**1.49**	9,465.48	0.0111	10.88	9,400.46	0.0534	39.74
V0800b04	14,723	24,201.35	0.1775	199.14	**22,848.79**	**0.6599**	**3.99**	24,387.24	0.0019	62.50	24,303.43	0.0394	195.08
V1600b04	25,103	45,849.52	0.2357	391.92	**44,598.02**	**0.9667**	**7.04**	46,379.74	0.0043	753.62	46,292.59	0.0206	1018.54
V3200b04	70,973	134,217.82	0.0348	1,376.41	**131,751.34**	**0.6654**	**108.08**	-	-	> 1 hour	-	-	> 1 hour
V0400b10	3,246	5,461.99	0.1217	40.9	**4,728.69**	**0.8542**	**0.63**	5,489.8	0.0509	7.25	5,364.07	0.1289	101.01
V0800b10	7,499	13,623.69	0.0425	108.69	**12,612.44**	**0.9596**	**2.01**	13,636.29	0.0156	47.49	13,485.93	0.0734	423.03
V1600b10	15,118	28,782.35	0.0691	272.65	**27,828.70**	**0.8556**	**4.82**	28,829.58	0.0117	537.27	28,682.31	0.0384	2019.76
V3200b10	36,170	70,292.36	0.0369	782.44	**68,653.51**	**0.9173**	**17.62**	70,315.35	0.0074	3,272.65	-	-	> 1 hour
V0400b20	2,297	4,048.03	0.1639	29.66	**3,632.92**	**0.6256**	**0.54**	4,064.77	0.1859	8.03	-	-	> 1 hour
V0800b20	5,049	9,451.30	0.0848	72.92	**8,960.16**	**0.5857**	**1.90**	9,460.80	0.0828	62.32	-	-	> 1 hour
V1600b20	11,575	22,305.01	0.0457	251.06	**21,591.83**	**0.6718**	**3.91**	22,315.63	0.0445	444.49	-	-	> 1 hour
V3200b20	24,639	48,321.14	0.0212	579.90	**47,650.73**	**0.6067**	**12.89**	-	-	> 1 hour	-	-	> 1 hour

Table 4. Dense Single-View Undirected Graphs Using Generative Model 1.

Test Case	Size (|E|)	FMBM			Graph-Tool			DANMF			CPLNS		
		Objective	NMI	Time	Objective	NMI	Time	Objective	NMI	Time	Objective	NMI	Time
V0100b06	4.125	815.58	0.1308	20.06	818.83	0.1291	**0.46**	829.20	0.0829	5.85	**750.28**	**0.2727**	3.50
V0300b06	41.771	**4,702.6**	**0.2061**	176.65	4,819.43	0.0311	1.93	4,828.49	0.0149	143.10	4,746.32	0.0629	18.14
V0500b06	118.536	10,473.0	**0.0537**	513.31	10,489.79	0.0193	4.86	10,497.63	0.0053	658.45	**10,419.98**	0.0411	41.13
V0100b08	4.212	782.91	0.2182	19.23	761.68	0.3024	**0.33**	807.05	0.1532	6.02	**712.92**	**0.3331**	4.59
V0300b08	42.078	4,468.67	**0.0944**	175.55	4,487.12	0.0443	1.48	4,498.23	0.0151	144.96	**4,397.28**	0.0895	19.95
V0500b08	120.166	8,188.55	**0.1503**	505.41	8,278.07	0.0267	4.94	8,290.82	0.0056	680.16	**8,175.52**	0.0582	59.04
V0100b10	4.268	731.61	0.3446	19.22	720.38	0.3290	**0.32**	779.34	0.1570	6.41	**662.42**	**0.4100**	6.93
V0300b10	42.385	4,069.42	**0.1652**	171.93	4,126.01	0.0569	1.47	4,141.94	0.0292	146.23	**4,009.88**	0.1160	29.28
V0500b10	120.366	7,931.97	**0.1174**	516.16	7,985.47	0.0347	4.03	8,000.60	0.0000	616.74	**7,872.77**	0.0761	60.22

Normalized Mutual Information (NMI) value with respect to the ground truth, and the running time in seconds. Unlike other methods, FMBM is an anytime algorithm since it uses multiple trials. Each trial takes roughly $(1/T)$'th, i.e., one-tenth, of the time reported for FMBM in the experimental results. For each method and test case, we averaged the results over 10 runs. All experiments were conducted on a laptop with a 3.1GHz Quad-Core Intel Core i7 processor and 16GB LPDDR3 memory. Our implementation of FMBM was done in Python3 with NetworkX [10].

Table 5. Single-View Undirected Graphs Using Generative Model 2.

Test Case	Size ($|E|$)	FMBM			Graph-Tool			DANMF			CPLNS		
		Objective	NMI	Time	Objective	NMI	Time	Objective	NMI	Time	Objective	NMI	Time
V0400b04	7,176	9,843.84	0.0327	75.09	**9,444.48**	**0.8214**	1.77	9,855.97	0.0116	12.23	9,786.67	0.0246	34.55
V0800b04	16,780	27,040.21	0.0087	192.97	26,982.65	**0.0698**	5.67	27,053.24	0.0008	69.68	**26,945.04**	0.0087	243.28
V1600b04	28,183	51,531.41	**0.0146**	386.32	51,556.38	0.0043	8.04	51,559.96	0.0025	684.58	**51,467.00**	0.0073	1,098.12
V3200b04	72,182	136,376.53	0.0087	1,196.30	**135,967.24**	**0.5287**	32.92	-		$->1$ hour	-		$->1$ hour
V0400b10	7,069	9,729.49	0.0639	73.76	**9,349.76**	**0.4827**	1.52	9,753.04	0.0625	10.69	9,585.60	0.0837	127.68
V0800b10	15,613	25,528.33	0.0385	181.56	**25,022.25**	**0.4893**	4.05	25,553.52	0.0274	86.58	25,353.08	0.0418	494.10
V1600b10	32,294	58,279.85	0.0195	428.40	58,244.14	**0.0305**	13.92	58,305.57	0.0157	687.84	**58,103.92**	0.0224	2459.38
V3200b10	79,173	**148,741.84**	**0.0082**	1,307.21	148,745.65	0.0065	**32.88**	-		$->1$ hour	-		$->1$ hour
V0400b20	6,829	9,453.81	0.1425	72.78	**9,139.14**	**0.2039**	1.38	9,506.35	0.1790	9.98	-		$->1$ hour
V0800b20	15,106	24,810.19	0.0784	176.60	**24,521.35**	**0.1251**	3.49	24,866.54	0.0918	62.93	-		$->1$ hour
V1600b20	30,462	55,265.42	0.0436	420.49	**55,207.39**	0.0366	9.38	55,310.7	**0.0440**	606.39	-		$->1$ hour
V3200b20	67,675	128,267.45	**0.0232**	1,199.32	**128,234.10**	0.0168	40.91	-		$->1$ hour	-		$->1$ hour

Although the underlying theory of FMBM can be generalized to directed graphs with weighted edges [9] and to multi-view graphs, the current version of FMBM is operational only on singe-view undirected graphs, sufficient to illustrate the power of FastMap embeddings. Therefore, only such test cases are borrowed from other commonly used datasets [16,19]. However, we also created new synthetic test cases to be able to do a more comprehensive analysis.[12]

The synthetic test cases were generated according to two similar stochastic block models [1] as follows. In Generative Model 1, given a user-specified number of vertices $|V|$ and a user-specified number of blocks k, we first assign each vertex to a block chosen uniformly at random to obtain the membership matrix C, representing the ground truth. The image matrix M is drafted using certain "block structural characteristics" designed for that instance with a parameter p. Each entry M_{ij} is set to either p or $10p$ according to a rule explained below. If M_{ij} is set to p ($10p$), the two blocks $\mathcal{B}_i$ and $\mathcal{B}_j$ are weakly (strongly) connected to each other with respect to p. The adjacency matrix A, representing the entire graph, is constructed from C and M by connecting any two vertices $v_s \in \mathcal{B}_i$ and $v_t \in \mathcal{B}_j$ with probability M_{ij}. In Generative Model 2, each entry M_{ij} is set to cp, where c is an integer chosen uniformly at random from the interval $[1, 10]$.

Tables 1, 2, 3, 4, and 5 show the comparative performances of FMBM, Graph-Tool, DANMF, and CPLNS.[13] Table 1 contains commonly used real-world test cases from [16] and [19]. Here, FMBM outperforms DANMF and CPLNS with respect to the value of the objective function on 3 out of 6 instances, despite the fact that it uses the expression in Eq. 2 only for evaluation on Line 28 of Algorithm 2. Graph-Tool performs well on all the instances. Table 2 shows the comparative performances on the complement graphs of the graphs in Table 1. This is done to test the robustness of the solvers against encoding the same relationships between vertices as either edges or non-edges. While the value of the objective function and the running time are expected to change, the NMI value is expected to be stable. We observe that FMBM and CPLNS are the only solvers that convincingly pass this test. Moreover, FMBM outperforms the other

---

[12] https://github.com/leon-angli/Synthetic-Block-Modeling-Dataset

[13] DANMF did not assign any block membership to a few vertices in some synthetic test cases. We assign Block $\mathcal{B}_1$ by default to such vertices.

solvers on more instances than in Table 1. Tables 1 and 2 do not test scalability since $|V|$ is small in these test cases.

Table 3 contains synthetic sparse test cases from Generative Model 1, named "Vnbk", where n indicates the number of vertices and k indicates the number of blocks. These test cases have the following block structural characteristics. Each block is strongly connected to two other randomly chosen blocks and weakly connected to the remaining ones (including itself). We set $p = (\ln |V|)/|V|$, making $|E| = O(|V| \log |V|)$ in expectation. After generating A, we also add some noise to it by flipping each of its entries independently with probability $0.05/|V|$. FMBM outperforms DANMF and CPLNS with respect to both the value of the objective function and the NMI value on 8 out of 12 instances. We also begin to see FMBM's advantages in scalability. However, Graph-Tool outperforms all other methods by a significant margin on all the instances. Table 4 contains synthetic dense test cases from Generative Model 1 constructed by setting $p = (\ln |V|)/|V|$, modifying each entry M_{ij} to $1 - M_{ij}$, and adding noise, as before. We observe that the performance of Graph-Tool is poor on such dense graphs. FMBM outperforms DANMF and CPLNS with respect to the NMI value on 6 out of 9 instances. Although CPLNS produces marginally better values of the objective function, its performance on large sparse graphs in Table 3 is bad.

Table 5 contains synthetic test cases from Generative Model 2 constructed by setting $p = (\ln |V|)/|V|$. FMBM outperforms DANMF and CPLNS with respect to the value of the objective function on 6 out of 12 instances. It also outperforms DANMF and CPLNS with respect to the NMI value on a different set of 6 instances. Graph-Tool performs comparatively well on all the instances but occasionally produces low NMI values.

4.1   Visualization

In addition to identifying blocks, their visualization is important for uncovering trends, patterns, and outliers in large graphs. A good visualization aids human intuition for gauging the spread[14] of blocks, both individually and relative to each other. In market analysis, for example, a representative element can be chosen from each block with proper visualization. Figure 4 shows that FMBM provides a much more perspicuous visualization compared to a standard graph visualization procedure in NetworkX[15] used with Graph-Tool, even though Graph-Tool shows good overall performance in Tables 1, 3, and 5. This is so because FMBM solves the block modeling problem in Euclidean space, while other approaches use abstract methods that are harder to visualize.

---

[14] The spread here refers to how a block extends from its center to its periphery.
[15] https://networkx.org/documentation/stable/reference/generated/networkx.
drawing.nx_pylab.draw.html

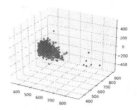

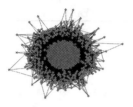

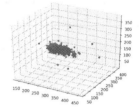

(a) standard graph visualization of blocks in an instance with 1,600 vertices and 4,353 edges

(b) FMBM visualization of blocks in Euclidean space for the instance from 4a

(c) standard graph visualization of blocks in an instance with 1,600 vertices and 25,103 edges

(d) FMBM visualization of blocks in Euclidean space for the instance from 4c

Fig. 4. The left column shows a visualization of two different instances with four blocks obtained by a standard graph visualization procedure in NetworkX used with Graph-Tool. The right column shows a visualization of the same two instances obtained in the Euclidean embedding by FMBM. Four different colors are used to indicate the four different blocks. The FMBM visualization is more helpful for gauging the spread of blocks, both individually and relative to each other. (Color figure online)

5   Conclusions and Future Work

In this paper, we proposed FMBM, a FastMap-based algorithm for block modeling. In the first phase, FMBM adapts FastMap to embed a given undirected graph into a Euclidean space in near-linear time such that the pairwise Euclidean distances between vertices approximate a probabilistically-amplified graph-based distance function between them. In doing so, it avoids having to directly work on the given graphs and instead reformulates the graph block modeling problem to a Euclidean version. In the second phase, FMBM uses GMM clustering for identifying clusters (blocks) in the resulting Euclidean space. Empirically, FMBM outperforms other state-of-the-art methods like FactorBlock, Graph-Tool, DANMF, and CPLNS on many benchmark and synthetic test cases. FMBM also enables a perspicuous visualization of blocks in the graphs, not provided by other methods.

In future work, we will generalize FMBM to work on directed graphs and multi-view graphs. We will also apply FMBM and its generalizations to real-world graphs from various domains, including social and biological networks.

References

1. Abbe, E.: Community detection and stochastic block models: recent developments. J. Mach. Learn. Res. **18**, 6446–6531 (2017)
2. Antonopoulos, C.G.: Dynamic range in the C. elegans brain network. Chaos: Interdisc. J. Nonlinear Sci. **26**(1), 013102 (2016)
3. Chan, J., Liu, W., Kan, A., Leckie, C., Bailey, J., Ramamohanarao, K.: Discovering latent blockmodels in sparse and noisy graphs using non-negative matrix factorisation. In: Proceedings of the ACM International Conference on Information & Knowledge Management (2013)
4. Cohen, L., Uras, T., Jahangiri, S., Arunasalam, A., Koenig, S., Kumar, T.K.S.: The FastMap algorithm for shortest path computations. In: Proceedings of the International Joint Conference on Artificial Intelligence (2018)
5. Davis, T.: USAir97 (2014). https://www.cise.ufl.edu/research/sparse/matrices/Pajek/USAir97
6. Faloutsos, C., Lin, K.I.: FastMap: a fast algorithm for indexing, data-mining and visualization of traditional and multimedia datasets. In: Proceedings of the ACM SIGMOD International Conference on Management of Data (1995)
7. Fredman, M.L., Tarjan, R.E.: Fibonacci heaps and their uses in improved network optimization algorithms. J. ACM (JACM) **34**, 596–615 (1987)
8. Girvan, M., Newman, M.E.: Community structure in social and biological networks. Natl. Acad. Sci. **99**, 7821–7826 (2002)
9. Gopalakrishnan, S., Cohen, L., Koenig, S., Kumar, T.K.S.: Embedding directed graphs in potential fields using FastMap-D. In: Proceedings of the International Symposium on Combinatorial Search (2020)
10. Hagberg, A., Swart, P., Chult, D.S.: Exploring network structure, dynamics, and function using NetworkX. Technical report, Los Alamos National Lab, Los Alamos, NM (United States) (2008)
11. Lee, J., Gross, S.P., Lee, J.: Improved network community structure improves function prediction. Sci. Rep. **3**, 1–9 (2013)
12. Li, J., Felner, A., Koenig, S., Kumar, T.K.S.: Using FastMap to solve graph problems in a Euclidean space. In: Proceedings of the International Conference on Automated Planning and Scheduling (2019)
13. Lin, S., Hu, Q., Wang, G., Yu, P.S.: Understanding community effects on information diffusion. In: Cao, T., Lim, E.-P., Zhou, Z.-H., Ho, T.-B., Cheung, D., Motoda, H. (eds.) PAKDD 2015. LNCS (LNAI), vol. 9077, pp. 82–95. Springer, Cham (2015). https://doi.org/10.1007/978-3-319-18038-0_7
14. Mattenet, A., Davidson, I., Nijssen, S., Schaus, P.: Generic constraint-based block modeling using constraint programming. J. Artif. Intell. Res. **70**, 597–630 (2021)
15. Murphy, K.P.: Machine Learning: A probabilistic perspective. The MIT Press, Cambridge (2012)
16. Newman, M.E.: Finding community structure in networks using the eigenvectors of matrices. Phys. Rev. E **74**(3), 036104 (2006)

17. Peixoto, T.P.: Efficient Monte Carlo and greedy heuristic for the inference of stochastic block models. Phys. Rev. E **89**(1), 012804 (2014)
18. Perozzi, B., Al-Rfou, R., Skiena, S.: DeepWalk: online learning of social representations. In: Proceedings of the ACM SIGKDD International Conference on Knowledge Discovery and Data Mining (2014)
19. Ramteke, R., et al.: Improving single and multi-view blockmodelling by algebraic simplification. In: Proceedings of the International Joint Conference on Neural Networks (IJCNN) (2020)
20. Ye, F., Chen, C., Zheng, Z.: Deep autoencoder-like nonnegative matrix factorization for community detection. In: Proceedings of the ACM International Conference on Information and Knowledge Management (2018)

Packing by Scheduling: Using Constraint Programming to Solve a Complex 2D Cutting Stock Problem

Yiqing L. Luo and J. Christopher Beck[✉]

Department of Mechanical and Industrial Engineering, University of Toronto,
Toronto, ON M5S 3G8, Canada
{louisluo,jcb}@mie.utoronto.ca

Abstract. We investigate the novel Two-stage Cutting Stock Problem with Flexible Length and Flexible Demand (2SCSP-FF): orders for rectangular items must be cut from rectangular stocks using guillotine cuts with the objective to minimize waste. Motivated by our industrial partner and different from problems in the literature, the 2SCSP-FF allows both the length of individual items and the total area of orders to vary within customer-specified intervals. We develop constraint programming (CP) and mixed-integer programming models, with the most successful coming from the adaptation of CP scheduling techniques. Numerical results show that this CP model has orders of magnitude smaller memory requirements and is the only model-based approach investigated that can solve industrial instances.

Keywords: Cutting Stock Problem · Guillotine Cuts · Constraint Programming · Mixed-Integer Linear Programming · Optimization

1 Introduction

As scheduling is one of the most successful application areas of Constraint Programming (CP) [17,19], we are interested in investigating whether CP scheduling approaches can be adapted to other combinatorial problems that share similar substructure. In this paper, we explore this idea for a complex, novel packing problem from the rolled-metal industry: the Two-stage Cutting Stock Problem with Flexible Length and Flexible Demand (2SCSP-FF), a generalization of the classic Two-stage Two-Dimensional Cutting Stock Problem with Guillotine Constraints (2SCSP). While the literature on 2SCSP is rich, our problem considers flexibility in item dimensions and total fulfillment, two characteristics that are commonplace in this industry [34], but have received limited attention in the literature. We approach these complications from a scheduling perspective, drawing inspiration from batch scheduling and the single resource transformation, a recently proposed CP modelling technique to handle the choice of alternative resources [3]. We conduct experiments over both generated and real-life

© Springer Nature Switzerland AG 2022
P. Schaus (Ed.): CPAIOR 2022, LNCS 13292, pp. 249–265, 2022.
https://doi.org/10.1007/978-3-031-08011-1_17

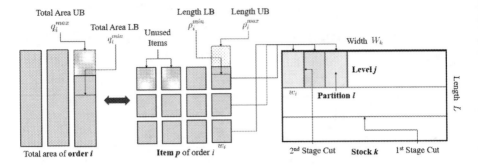

Fig. 1. A visualization of 2SCSP-FF. Each order is represented by its total quantity (left) and the partitions assigned to it (middle), as illustrated by the double-arrow. Dashed lines and the dotted fills indicate flexibility in the associated parameter. Orange and blue lines represent the first and second stage cuts, respectively. (Color figure online)

instances and demonstrate our approach's computational advantages over alternative modelling approaches in CP, mixed integer programming (MIP), and a custom greedy heuristic.

Our contributions are as follows:

1. We introduce the novel 2SCSP-FF problem.
2. We adapt the single resource CP model to the 2SCSP-FF. This compact formulation increases the size of the instances that can be solved within memory and time limits by an order of magnitude.
3. We propose and experiment with alternative MIP and CP models and a two-stage heuristic.

2   Problem Definition

The 2SCSP-FF (Fig. 1) is a novel generalization of the Two-stage Two-Dimensional Cutting Stock Problem with Guillotine Constraints (2SCSP). Given a set of *orders* for rectangular *items* and a set of larger *stock rectangles*, the classic Two-Dimensional Cutting Stock Problem (2DCSP) fulfills orders by cutting items from stocks. A more constrained variant, the 2SCSP only allows stocks to be processed using *guillotine cuts*, a cut that runs from one edge of the object to another. All cuts must also be executed in two stages, each consisting of a set of parallel guillotine cuts performed on a rectangle obtained from the previous stage. Without loss of generality, we let the direction of the first stage cuts be widthwise and that of the second stage ones lengthwise. The rectangles produced in the first stage are referred to as *levels*, following the literature [7], and those produced in the second stage as *partitions*. On top of 2SCSP, the 2SCSP-FF has the following characteristics arising from our application:

- **Flexible Length:** The length of an item is flexible within some integer interval. If a level contains items from different orders, its length must lie in the intersection of the item-length intervals. In our application, items are subsequently rolled into cylindrical coils used as feedstock for downstream processing. The maximum length requirement ensures a maximum coil diameter to enable mounting it on a downstream machine. The minimum length requirement comes from the desire to limit the number of coils.
- **Flexible Demand:** Consistent with real-world manufacturing practices, each order can tolerate a percentage deviation from the total area demanded. For example, an order may request items totalling $10000 \pm 15\%$ units of area.
- **Maximum Partition Count per Level:** To reflect the limitations of an industrial cutter, the maximum number of partitions on each level is fixed.
- **Limited Stocks with Variable Widths:** Stock rectangles of various widths are available in limited quantities.
- **Cost Minimization:** The goal is to minimize cost: a weighted difference between the area of the stocks used and the area of the orders fulfilled.

Formally, we are given a set of stock rectangles K, whose types are characterized by set H. Each rectangle $k \in K$ has width W_k and length L. Stock rectangles with identical dimensions belong to the same type, K_h, $K = \bigcup_{h \in H} K_h$. We are also given a set of orders, N, where each order $i \in N$ has a required area interval of $[q_i^{min}, q_i^{max}]$. Each item belonging to order i should have a fixed width w_i and a length within the interval $[\rho_i^{min}, \rho_i^{max}]$. Due to the flexible length, the total number of items belonging to order i must be within an integer interval $[n_i^{min}, n_i^{max}] = [\lceil \frac{q_i^{min}}{\rho_i^{max}} \rceil, \lfloor \frac{q_i^{max}}{\rho_i^{min}} \rfloor]$. For order i, we denote its set of necessary items as $A_i = \{1, \ldots, n_i^{min}\}$, its set of possible, but not necessary items as $B_i = \{n_i^{min} + 1, \ldots, n_i^{max}\}$, and all possible items as $C_i = A_i \bigcup B_i$. Lastly, we let α and β be the weights associated with the area of stocks used and the area of orders fulfilled, respectively, and seek to minimize this weighted difference.

Since all stocks share the same length, a stock k can take on at most $\bar{j} = \lfloor \frac{L}{\min_{i \in N} \rho_i^{min}} \rfloor$ levels; we denote the set of possible numbers of levels of any stock as $J = \{0, \ldots, \bar{j}\}$. There must also be no more than η partitions on each level. We let $P = \{1, \ldots, \eta\}$ be the set of partitions on a given level, and $P_i = \{l \in P \mid l \leq n_i^{max}\}$ be the set of partitions on a given level assuming they are all assigned to order i. A partition of a stock that is assigned to an order becomes an item.

The 2SCSP-FF can easily be reduced to the 2SCSP if the quantity demanded of each order and the length of each item are fixed. As the 2SCSP is NP-hard [7], the 2SCSP-FF problem is at least NP-hard.

3 Literature Review

The 2SCSP was formalized by Gilmore and Gomory [9], who proposed a dynamic program and the well-known exponential-sized model solved via column generation. Since then, MIP has been the dominant approach in model-based studies. Limited to a single stock rectangle, Lodi and Monaci [22] proposed a compact formulation that restricts levelwise assignments for each item to reduce symmetry. Silva et al. [29] proposed a pseudo-polynomial formulation based on the idea of cuts and residual plates for a 2SCSP with identical stocks. Furini et al. [7] extended both of these models to include different stock sizes while also proposing a branch-and-price algorithm based on Gilmore and Gomory's model. Macedo et al. [24] developed an arc flow formulation based on item positioning.

Dincbas and Simonis proposed the first CP-based approach [6] to the 2SCSP, generating stock patterns using a combination of backtrack search and a finite domain model. Later, Beldiceanu and Contejean introduced DIFFN [1,2], a global constraint with an option to enforce guillotine cuts; however, no experimental results related to guillotine cuts were provided. Since then, CP has largely been investigated in other packing contexts. For the two-dimensional optimal rectangle packing problem, Korf [15,16] considered solving a constraint satisfaction problem using the absolute positions of items. Moffitt and Pollack [26] studied the same satisfaction problem from a relative placement perspective, focusing on the pairwise relationships between items. For the same problem, Clautiaux et al. [5] considered a scheduling approach, representing the width and length of items as two interval variables. This was improved by Mesyagutov et al. [25], who integrated linear-programming-based pruning rules to propagate the constraints. Simonis and O'Sullivan investigated CP search strategies to pack squares into rectangles using the CUMULATIVE global constraint [30,31]. For 1D packing, Shaw [28] proposed a global constraint PACK. For a comprehensive review of 2D packing problems, we refer the reader to surveys by Lodi et al. [21], Wäscher et al. [33] and Iori et al. [12].

While the 2SCSP has been widely studied, we could find only one work addressing item flexibility in the 2D setting. Lee et al. [20] considered a variant of the 2SCSP with flexible width and length and proposed a multi-stage heuristic to iteratively pack items and adjust level dimensions. They also proposed a nonlinear model but did not investigate its performance.

CP techniques have been widely adopted in scheduling [17,19]. In relation to our main approach, we discuss the literature on two types of problem: batch scheduling and vehicle routing. Batch scheduling arises when a set of jobs with common characteristics need to be processed together. Tang and Beck [32] proposed a CP formulation using interval variables for a multi-stage tool layup line problem. Ham and Cakici [10] used interval variables and state functions to represent a flexible job shop scheduling problem with parallel batch processing machines. Vehicle Routing Problems (VRP) optimize the routes of vehicles while some criteria associated with each route are satisfied. A number of CP models for VRP [4,8,13] represent the problem from a scheduling perspective with the trip-to-vehicle assignments modelled with some form of the ALTERNATIVE constraint. Recently, Booth and Beck [3] introduced the single resource model, where multiple resources are unified into a single resource on an expanded time horizon. They show that their formulation yields computation advantage over traditional modelling constructs in a capacity- and time-constrained routing problem.

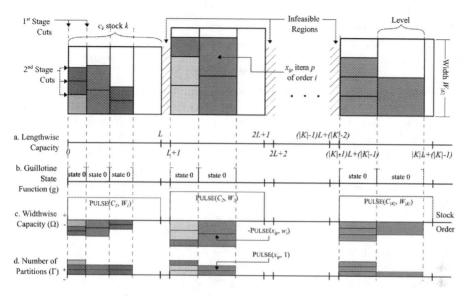

Fig. 2. Illustration of the CP_{SR} model. The length of the stock rectangles are concatenated along the horizontal axis. Here, a level is a vertical strip. The smaller rectangles and the dashed lines represent items and guillotine cuts, respectively.

4   The Single Resource CP Formulation

In this section, we present our main contribution: the Single Resource CP model, CP_{SR}. Our model poses the 2SCSP-FF as a scheduling problem composed of three main components: a unified domain of stock length, a state function for guillotine cuts, and cumulative functions tracking widthwise resources.

Unified Lengthwise Domain. Our model adapts the single resource transformation [3], a CP modelling technique that unifies alternative resources into a single horizon, to the 2SCSP-FF. CP_{SR} concatenates the stock rectangles so that the total length of the stocks is analogous to a temporal horizon on which items belonging to all orders need to be allocated (Fig. 2a). For each possible item $p \in C_i$ belonging to order i, we introduce an *optional interval variable* x_{ip}. In CP scheduling, an optional interval variable is a variable whose domain is a subset of $\{\bot\} \bigcup \{[s, \epsilon) | s, \epsilon \in \mathbb{Z}, s \leq \epsilon\}$, where s and ϵ are the start and end times of the interval, and $\bot$ is a special value indicating absence. In our case, the start time of x_{ip} represents an item's leftmost lengthwise coordinate, and the duration of x_{ip} its length, which we further restrict to be within $[\rho_i^{min}, \rho_i^{max}]$. For necessary items (i.e., in set A_i), we remove $\{\bot\}$ from the domain of x_{ip} for all $p \in A_i$ and simply declare them as *interval variables*.

To avoid an item spanning multiple stocks in the unified horizon, we insert a dummy unit of forbidden space between adjacent stocks to create an infeasible region (Fig. 2a, hatched). The horizon is thus augmented from $L|K|$ to $(L +$

1)$|K|$, and no items can be placed in the infeasible region $\bar{F} = \bigcup_{k \in K}[Lk, (L + 1)(k)]$. We denote the augmented horizon as $\mathcal{H} = [0, L|K| + (|K| - 1)]$ and use FORBIDEXTENT to exclude $\bar{F}$ from the domain of the item variables x_{ip}.

Guillotine State Function. We draw inspiration from batch scheduling to model guillotine cuts: we treat each level in a stock rectangle as a batch so that the level's lengthwise endpoints coincide with the corresponding endpoints of the items. We first introduce a *state function*, g, a variable whose domain is a set of non-overlapping intervals (Fig. 2b). Then, we associate the items with a level using an ALWAYSCONSTANT constraint, which coerces their interval variables to align with an interval in g. Thus, items can only be on the same level if they belong to the same interval in the state function.

Cumulative Resource Function Expressions. The width of stocks and a level's partition count limit are interpreted as widthwise resources. Typical of CP scheduling, we use *cumulative functions* and *pulses*: the former are expressions that represent the sum of individual contributions of intervals over time, while the latter are expressions that indicate each interval's contribution. We let Ω, a cumulative function, be the net widthwise capacity over the horizon. In Ω, we generate a pulse with magnitude W_k for each stock rectangle k and a pulse with magnitude $-w_i$ for every item belonging to order i (Fig. 2c). As long as Ω is non-negative, the widthwise capacity is satisfied. A similar construct is used to express the limit on the number of partitions on each level (Fig. 2d), where a positive pulse with unit magnitude is generated for each item. We constrain the total cumulative function of these unit pulses, Γ, to be within η.

Overall, our decision variables are as follows:

- $x_{ip} :=$ (interval) lengthwise interval of item p belonging to order i.
- $c_k :=$ (interval) lengthwise interval representing stock k.
- $g :=$ (state function) guillotine state function.

CP_{SR} is defined in Model 1, using the syntax of IBM's CP Optimizer [11]. Objective (1a) defines our cost, the weighted difference between the areas of stocks used and orders fulfilled. Expressions PRESENCEOF and SIZEOF are used to access the presence and the duration of an interval variable. If the variable is not present, both expressions evaluate to 0. Constraints (1b) and (1c) define the widthwise usage of each stock. The last two parameters in the ALWAYSIN constraint respectively dictate the minimum and maximum values that the cumulative function Ω can take on over the horizon $\mathcal{H}$. Constraints (1d) and (1e) define the restriction on the number of partitions on each level. Constraint (1f) defines the guillotine cut restrictions. The last two parameters in ALWAYSCONSTANT ensure that the start and end times of the variables x_{ip} are aligned with those of the intervals within the state function g. Constraints (1g) and (1h) ensure that the total quantity of the order fulfilled is within the demand tolerance. Constraint (1i) ensures that no partition is assigned across two stock rectangles. The remaining constraints declare the decision variables.

$$\min \alpha \sum_{k \in K} LW_k \text{PRESENCEOF}(c_k) \qquad (CP_{SR}) \qquad (1a)$$

$$-\beta \sum_{i \in N} \sum_{p \in C_i} w_i \text{PRESENCEOF}(x_{ip}) \text{SIZEOF}(x_{ip})$$

$$\text{s.t. } \Omega = \sum_{k \in K} \text{PULSE}(c_k, W_k) - \sum_{i \in N} \sum_{p \in C_i} \text{PULSE}(x_{ip}, w_i) \qquad (1b)$$

$$\text{ALWAYSIN}(\Omega, \mathcal{H}, 0, \max_{k \in K} W_k) \qquad (1c)$$

$$\Gamma = \sum_{i \in N} \sum_{p \in C_i} \text{PULSE}(x_{ip}, 1) \qquad (1d)$$

$$\text{ALWAYSIN}(\Gamma, \mathcal{H}, 0, \eta) \qquad (1e)$$

$$\text{ALWAYSCONSTANT}(g, x_{ip}, True, True) \qquad \forall i \in N, p \in C_i \quad (1f)$$

$$\sum_{p \in C_i} \text{SIZEOF}(x_{ip}) \geq q_i^{min}/w_i \qquad \forall i \in N \qquad (1g)$$

$$\sum_{p \in C_i} \text{SIZEOF}(x_{ip}) \leq q_i^{max}/w_i \qquad \forall i \in N \qquad (1h)$$

$$\text{FORBIDEXTENT}(x_{ip}, \bar{F}) \qquad \forall i \in N, p \in C_i \quad (1i)$$

$$x_{ip} : \text{INTERVALVAR}(\mathcal{H}, [\rho_i^{min}, \rho_i^{max}]) \qquad \forall i \in N, p \in A_i \quad (1j)$$

$$x_{ip} : \text{OPTINTERVALVAR}(\mathcal{H}, [\rho_i^{min}, \rho_i^{max}]) \qquad \forall i \in N, p \in B_i \quad (1k)$$

$$c_k : \text{INTERVALVAR}([(L+1)(k-1), (L+1)k-1], L) \, \forall k \in K \qquad (1l)$$

$$g : \text{STATEFUNCTION}() \qquad (1m)$$

5   Alternative Approaches

We also propose integer-based CP and MIP models and a two-stage heuristic.

5.1   Integer-Based CP Formulations

Counting-Based CP Model. Due to the two-stage cuts, partitions assigned to the same order on a level must be identical; hence, we can count them. For a given level j from stock k, we use an integer variable x_{ijk} to denote the number of partitions assigned to order i. We use an integer variable y_{jk} to represent the length of level j on stock k. As the position of the lengthwise cut is not restricted to be integral, we magnify the domain using a precision parameter $\mathcal{P}$ equal to some power of ten, so that y_{jk} represents the first $\log_{10} \mathcal{P}$ decimal places of the actual length. More formally, our decision variables are as follows:

- $x_{ijk} \coloneqq$ (integer) # of partitions on level j of stock k assigned to order i
- $y_{jk} \coloneqq$ (integer) length of level j of stock k magnified by $\mathcal{P}$

$$\min \ \alpha \sum_{k \in K} LW_k c_k - \beta \sum_{i \in N} \sum_{j \in J} \sum_{k \in K} w_i x_{ijk} y_{jk}/\mathcal{P} \qquad (CP_{CO}) \qquad (2a)$$

$$\text{s.t.} \ \sum_{i \in N} x_{ijk} w_i \leq W_k s_{jk} \qquad\qquad\qquad \forall j \in J, k \in K \qquad (2b)$$

$$y_{jk}/\mathcal{P} \leq \rho_i^{max} + (x_{ijk} == 0) \max_{n \in N}(\rho_n^{max}) \qquad \forall i \in N, j \in J, k \in K \qquad (2c)$$

$$y_{jk}/\mathcal{P} \geq \rho_i^{min}(x_{ijk} \geq 1) \qquad\qquad \forall i \in N, j \in J, k \in K \qquad (2d)$$

$$\sum_{j \in J} \sum_{k \in K} y_{jk} x_{ijk}/\mathcal{P} \geq q_i^{min}/w_i \qquad\qquad \forall i \in N \qquad (2e)$$

$$\sum_{j \in J} \sum_{k \in K} y_{jk} x_{ijk}/\mathcal{P} \leq q_i^{max}/w_i \qquad\qquad \forall i \in N \qquad (2f)$$

$$\sum_{j \in J} y_{jk}/\mathcal{P} \leq Lc_k \qquad\qquad\qquad \forall k \in K \qquad (2g)$$

$$s_{jk} = \text{ANY}([x_{ijk} > 0, \forall i \in N]) \qquad\qquad \forall j \in J, k \in K \qquad (2h)$$

$$c_k = \text{ANY}([s_{jk} = 1, \forall j \in J]) \qquad\qquad \forall k \in M \qquad (2i)$$

$$x_{ijk} \in \{0, ..., \eta\} \qquad\qquad\qquad \forall i \in N, j \in J, k \in K \qquad (2j)$$

$$y_{jk} \in \{0, \min_{i \in N} \rho_i^{min}\mathcal{P}, \dots, \max_{i \in N} \rho_i^{max}\mathcal{P}\} \qquad \forall j \in J, k \in K \qquad (2k)$$

Model 2 formalizes CP_{CO}. Objective (2a) describes the cost. Since y_{jk} is magnified, we divide it by $\mathcal{P}$ to recover its actual length. Constraint (2b) restricts the width of the stocks. Constraints (2c) and (2d) restrict the length of a level by the tightest interval determined by the allotted orders. Constraints (2e) and (2f) ensure that partitions of each order fulfilled satisfy the total quantity range demanded. Constraint (2g) restricts the length of the stocks in use. Constraints (2h) and (2i) describe if a level and a stock is used, respectively.

Stock-Based CP Model. Extending the standard integer-based CP model for one-dimensional bin packing [14], the stock-based CP model, CP_{ST}, takes advantage of the limited number of possible partitions on a level, matching each partition to some order. Specifically, we define integer variables x_{jkl} representing the index of the order to which the lth partition of the jth level on the kth stock is assigned. As not all partitions are always needed, we define a dummy order that serves as a placeholder. Formally, the dummy order, indexed by $D = |N| + 1$, has width $w_D = 0$ and length interval $[\rho_D^{min}, \rho_D^{max}] = [0, \max_{i \in N}(\rho_i^{max})]$. We use $\overline{N} = N \bigcup \{D\}$ to denote the set of original orders plus the dummy order; $\overline{w}$ to denote the set of widths of original orders union the dummy width w_D; $\overline{\rho^{min}}$ and $\overline{\rho^{max}}$ to denote the lengthwise bounds of orders union the dummy bounds ρ_D^{min} and ρ_D^{max}. Similar to CP_{CO}, we let y_{jk} be the length of level j on stock k and magnify its domain using $\mathcal{P}$.

$$\text{minimize} \quad \sum_{k \in K} LW_k c_k - \sum_{j \in J} \sum_{k \in K} \sum_{l \in P} \overline{w}_{x_{jkl}} y_{jk}/\mathcal{P} \qquad (CP_{ST}) \qquad (3a)$$

s.t.

$$\sum_{l \in P} \overline{w}_{x_{jkl}} \leq W_k s_{jk} \qquad\qquad \forall j \in J, k \in K \qquad (3b)$$

$$y_{jk}/\mathcal{P} \leq \overline{\rho^{max}}_{x_{jkl}} \qquad\qquad \forall j \in J, k \in K, l \in P \quad (3c)$$

$$y_{jk}/\mathcal{P} \geq \overline{\rho^{min}}_{x_{jkl}} \qquad\qquad \forall j \in J, k \in K, l \in P \quad (3d)$$

$$\sum_{j \in J} \sum_{k \in K} \sum_{l \in P} (x_{jkl} == i) y_{jk}/\mathcal{P} \geq q_i^{min}/w_i \quad \forall i \in N \qquad (3e)$$

$$\sum_{j \in J} \sum_{k \in K} \sum_{l \in P} (x_{jkl} == i) y_{jk}/\mathcal{P} \leq q_i^{max}/w_i \quad \forall i \in N \qquad (3f)$$

$$s_{jk} = \text{ANY}([x_{jkl} \neq D, \forall l \in P]) \qquad\qquad \forall j \in J, k \in K \qquad (3g)$$

$$x_{jkl} \in \overline{N} \qquad\qquad \forall j \in J, k \in K, l \in P \,(3h)$$

$$(2g), (2i), (2k)$$

Model 3 formalizes CP_{ST}. Objective (3a) minimizes the cost. Constraint (3b) ensures that the widthwise capacity is satisfied on each stock. In particular, $\overline{w}$ is indexed by x_{jkl} using the ELEMENT constraint. Constraints (3c) and (3d) constrain the length of a level by the items assigned on it. Constraints (3e) and (3f) satisfy the total area of each order. Constraint (3g) instantiates intermediate parameters indicating level usage.

Modelling Considerations

Symmetry-Breaking: The problem has a number of inherent symmetries due to the homogenous items, levels, and stock rectangles. Hence, we augment CP_{CO} and CP_{ST} with the following symmetry-breaking constraints:

$$y_{jk} \geq y_{(j+1)k} \quad \forall j \in J', k \in K \qquad\qquad (4a)$$

$$c_k \geq c_{k+1} \qquad \forall k \in K_h', h \in H \qquad\qquad (4b)$$

These constraints break the symmetry between the lengths of consecutive levels on the same stock and the presence of homogeneous stocks, respectively. We use a prime to indicate an ordered set without its last element: $J' = J \setminus \{|J|\}$. For CP_{ST}, we also specify a lexicographic ordering of the order indices on consecutive levels of the same stock via LEXICOGRAPHIC$([x_{jkl}, \forall l \in P], [x_{(j+1)kl}, \forall l \in P])$.

Item-Based CP Model: Using PACK [28], we can also construct an integer-based CP model that decides on the level that an item is assigned to. Two such structures exist in 2SCSP-FF: the packing of items into levels and that of levels into stocks. While the former can be represented by PACK, the latter cannot due to the lengthwise flexibility and is represented by constraints that decompose PACK. The model is omitted due to poor computational performance.

5.2 Mixed-Integer Formulation

We also introduce a mixed-integer program, MIP, that uses binary variables to assign partitions on each level to orders. Formulating a strong MIP model is challenging, as determining the area of each order requires information related to two independent decisions: the order-to-level assignment and the level length given order assignments. In MIP, we linearize this relationship at the expense of introducing new variables, each one packing an item of an order into a level of a stock. While it is tempting to decompose them into independent orders-to-levels and levels-to-stocks decisions similar to the compact formulation in Furini et al. [7], representing both the area of each order and the variable length of each level using linear constraints is nontrivial. Here, we do not investigate this further.

More formally, our decision variables are as follows:

- $x_{ijkl} :=$ (binary) 1 if the lth partition from the jth level of the kth stock is assigned to the ith order, else 0.
- $y_{jk} :=$ (continuous) the length of the jth level on the kth stock.
- $a_{ijkl} :=$ (continuous) the area occupied by the lth partition of the jth level on the kth stock belonging to order i.
- $c_k :=$ (binary) 1 if the kth stock is used, else 0.

MIP is defined in Model 5. Objective (5a) describes the cost. Constraint (5b) restricts the stocks' width. Constraint (5c) limits the number of lengthwise cuts. Constraint (5d) ensures that the lengthwise capacity of each stock is satisfied. Constraints (5e) and (5f) assert that the level's length must respect the minimum and maximum length of items assigned to it. Constraints (5g) and (5h) ensure that the quantity of each order assigned across all stocks is satisfactory. Constraints (5i), (5j), and (5k) define the area of each partition on a level.

$$\min \ \alpha \sum_{k \in K} LW_k c_k - \beta \sum_{i \in N, j \in J, k \in K, l \in P_i} a_{ijkl} \qquad (MIP) \qquad (5a)$$

$$\text{s.t.} \sum_{i \in N, l \in P_i} w_i x_{ijkl} \leq W_k c_k \qquad \forall j \in J, k \in K \qquad (5b)$$

$$\sum_{i \in N, l \in P_i} x_{ijkl} \leq \eta c_k \qquad \forall j \in J, k \in K \qquad (5c)$$

$$\sum_{j \in J} y_{jk} \leq L c_k \qquad \forall k \in K \qquad (5d)$$

$$y_{jk} \geq \rho_i^{min} x_{ijkl} \qquad \forall i \in N, j \in J, k \in K, l \in P_i \qquad (5e)$$

$$y_{jk} \leq \rho_i^{max} x_{ijkl} + \max_{i' \in N}(\rho_{i'}^{max})(1 - x_{ijkl}) \qquad \forall i \in N, j \in J, k \in K, l \in P_i \qquad (5f)$$

$$\sum_{l \in P_i, j \in J, k \in K} a_{ijkl} \geq q_i^{min} \qquad \forall i \in N \qquad (5g)$$

$$\sum_{l \in P_i, j \in J, k \in K} a_{ijkl} \leq q_i^{max} \qquad \forall i \in N \qquad (5h)$$

$$a_{ijkl} \leq w_i y_{jk} \qquad \forall i \in N \qquad (5i)$$

$$a_{ijkl} \geq w_i y_{jk} - \rho_i^{max} w_i (1 - x_{ijkl}) \qquad \forall i \in N \qquad (5j)$$

$$a_{ijkl} \leq \rho_i^{max} w_i x_{ijkl} \qquad \forall i \in N, j \in J, k \in K, l \in P_i \qquad (5k)$$

$$x_{ijkl} \in \{0,1\} \qquad\qquad \forall i \in N, j \in J, k \in K, l \in P_i \qquad (5l)$$

$$y_{jk} \in \mathbb{R}^+ \qquad\qquad \forall j \in J, k \in K \qquad\qquad (5m)$$

$$a_{ijkl} \in \mathbb{R}^+ \qquad\qquad \forall i \in N, j \in J, k \in K, l \in P_i \qquad (5n)$$

$$c_k \in \{0,1\} \qquad\qquad \forall k \in K \qquad\qquad (5o)$$

Modelling Considerations

Symmetry-Breaking: We can again add symmetry-breaking constraints (4a) and (4b) to *MIP* similar to CP_{CO} and CP_{ST}. Furthermore, we add constraints (6a) and (6b) to break the symmetry between partitions on the same level belonging to the same order and the length of the first level of identical stocks, respectively.

$$x_{ijkl} \geq x_{ijk(l+1)} \quad \forall i \in N, j \in J, k \in K, l \in P_i' \qquad (6a)$$

$$y_{0k} \geq y_{0(k+1)} \qquad \forall k \in K_h', h \in H \qquad\qquad (6b)$$

One-Hot Encoded Formulation: In order to retain linearity, *MIP* treats the assignment of different partitions on the same level to an order as individual decisions. Alternatively, we can one-hot encode the number of partitions assigned to the level so that each binary variable x_{ijkl} takes the value 1 if and only if there are l partitions (with identical dimensions) on level j of stock k assigned to order i. This formulation underperforms *MIP* in our experiments and is omitted.

5.3   First-Fit Based Heuristic

In addition to the mathematical models, we develop a two-stage first-fit-based heuristic, *FFMH*. The first stage sorts the orders' items in a lexicographically decreasing order based on their width and length interval size and packs each one into a level. The intuition is that orders with less lengthwise flexibility and larger width should be packed into a level first, as they can be more difficult to pack into a partial solution. A new level or stock is opened if an item cannot fit into the previous level or stock. Packing an item into a stock's level only narrows its length interval: another decision is required to obtain its exact length and thereafter each order's total area. For simplicity, we pack items of an order until the sum of the average possible area of each item is not less than the middle of the required area interval for that order. In the second stage, given the complete item-to-level assignment, we solve a linear program (Model 7) to determine each level's length, while minimizing cost.[1]

---

[1] The form of this two-stage heuristic suggests that a classical Benders decomposition approach may be worth investigation in future work.

$$\max \quad \sum_{i \in N, j \in J, k \in K} w_i \Phi_{ijk} y_{jk} \tag{7a}$$

$$\text{s.t.} \quad \sum_{j \in J} y_{jk} \leq L \qquad\qquad\qquad\qquad\qquad \forall k \in K \tag{7b}$$

$$y_{jk} \geq \Phi_{ijk}^{ind} \rho_i^{min} \qquad\qquad\qquad \forall i \in N, j \in J, k \in K \tag{7c}$$

$$y_{jk} \leq \Phi_{ijk}^{ind} \rho_i^{max} + (1 - \Phi_{ijk}^{ind}) \max_{i' \in N}(\rho_{i'}^{max}) \quad \forall i \in N, j \in J, k \in K \tag{7d}$$

$$\sum_{j \in J, k \in K} w_i \Phi_{ijk} y_{jk} \geq q_i^{min} \qquad\qquad \forall i \in N \tag{7e}$$

$$\sum_{j \in J, k \in K} w_i \Phi_{ijk} y_{jk} \leq q_i^{max} \qquad\qquad \forall i \in N \tag{7f}$$

$$y_{jk} \in \mathbb{R}^+ \qquad\qquad\qquad\qquad \forall j \in J, k \in K \tag{7g}$$

The only variables in Model 7 are the continuous variables y_{jk} describing the length of the jth level on the kth stock. The parameter Φ_{ijk} is the number of partitions on the jth level of the kth stock that belongs to order i, and the parameter Φ_{ijk}^{ind} is a 0-1 indicator for $\Phi_{ijk} > 0$. We simplify the cost minimization objective to maximize total fulfillment (7a) because the number of stocks used is fixed given the item-to-level assignment. Constraint (7b) constrains the stock length. Constraints (7c) and (7d) satisfy the length specifications of the partitions. Constraints (7e) and (7f) ensure the total fulfillment of each order to be within tolerance limits. Constraint (7g) declares the variable domain.

6   Numerical Results

We conduct our analysis on a combination of 50 generated problem instances and 4 real-life instances provided by our industry partner (Table 1).

For the generated instances, we draw from distributions provided by our industrial collaborator (Table 2), generating 10 instances for each parameter combination in the set $\{(|N|, |K|)\} \in \{(4, 8), (8, 16), (16, 32), (32, 64), (64, 128)\}$. For 5 out of these 10 instances, we halve the total area tolerances to provide variability. In the rare case that, for some order i, $\rho_i^{max} \leq \rho_i^{min}$ is generated, we swap the two values.

Table 1. Mean of the parameter combinations from the four industrial instances.

$|N|$	$|K|$	W_k	w_i	ρ_i^{min}	ρ_i^{max}	q_i^{min}	q_i^{max}
19	42	48	22.2	112.9	180.2	1.3e5	1.7e5
21	172	45.2	16.3	56.6	94.6	9.7e4	1.3e5
47	149	43.3	10.4	68.2	134.1	1.6e5	2.1e5
149	636	44.6	13.3	74.4	134.4	2.5e5	3.4e5

Table 2. Data distributions for each parameter.

Parameter	Distribution
∀ order $i \in N$	
q_i	Exponential(λ=5.608e-0.5)
q_i^{max}	Constant, 0.85 q_i
q_i^{min}	Constant, 1.15 q_i
w_i	Integer Uniform(a=1, b=20)
ρ_i^{max}	Integer Uniform(a=70, b=115)
ρ_i^{min}	Integer Uniform(a=85, b=130)
∀ stock $k \in K$	
W_k	Integer Uniform(a=36, b=50) with 50% chance of duplicating previous stock
L	Constant, 400

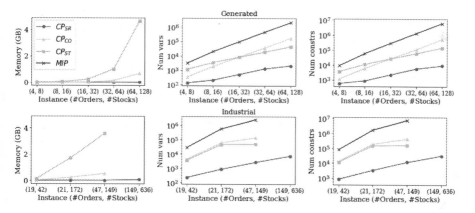

Fig. 3. Comparison of the number of variables, the number of constraints, and the model memory before search over instance sizes. Note the scales of the y-axes.

All experiments are implemented in Python 3.8, and computations were performed on individual nodes of the SciNet Niagara cluster [23, 27]. We use CPLEX and CP Optimizer from the CPLEX Optimization Studio version 20.1.0 via the DOcplex library. Each model is given 16 GB of RAM and runs that exceed this size are aborted. All experiments are single-threaded with default search and inference settings. A one-hour time limit is used.

For the CP integer-based models, we set $\mathcal{P}$, the magnifying parameter, to 1, as increasing it led to poor performance. We set α and β, the objective weights, to 0.3 and 0.7, respectively, to reflect the industrial use case.

Model Size Comparison. Figure 3 compares the mean model sizes based on the number of variables, the number of constraints, and the model memory (before search). The memory usage of *MIP* is not accessible from the solver. A data point is omitted if the corresponding model fails to initialize in memory within the one hour time limit. The CP_{SR} formulation is significantly smaller than the

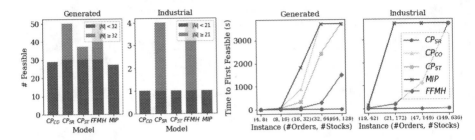

Fig. 4. (Left) Number of generated and industrial instances with a feasible solution by each model. (Right) The average run time required to find a feasible solution for a given model at an instance size.

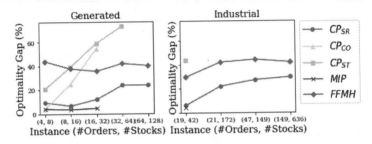

Fig. 5. % optimality gap for generated and industrial instances. Instances that are not solved by an approach are not included in that approach's measure.

other models across all three measures, especially as the instances scale up. For the largest industrial instances, no other models could be loaded before timing out. For the largest generated instances, $(64, 128)$, the CP_{CO} model requires about 800MB of memory, while CP_{SR} only needs 20MB.

Feasibility Analysis. Figure 4 reports the number of instances for which a feasible solution was found and the average time to feasibility or termination. Only CP_{SR} and *FFMH* found a feasible solution to all instances. In particular, CP_{SR} reached feasibility the fastest amongst all methods, requiring less than 100 s for the largest industrial instance, while *FFMH*, the second fastest, needed almost the entire one-hour run time. Notably, *MIP* failed to find feasible solutions for generated and industrial instances with $|N| \geq 32$ and $|N| \geq 21$, respectively.

Solution Quality. Figure 5 displays the optimality gap of each approach calculated from Eq. (8). Here, $z(n; i)$ is the objective value of approach n for instance i, and $lb(i)$ is the best lower bound of instance i across all approaches. For a given solution approach, we omit any unsolved instances from the visualization; hence, CP_{SR} is penalized for finding solutions to harder instances compared to approaches that did not do so.

$$\%\mathrm{OptGap}(n; i) = 100\frac{z(n; i) - lb(i)}{lb(i)} \qquad (8)$$

MIP demonstrated the strongest performance for generated instances with $|N| \leq 16$. It proved optimality for three generated instances, the only ones proven optimal across all models. For larger instances ($|N| \geq 32$), MIP scaled poorly, failing to find a feasible solution within the time limit. Both CP_{CO} and CP_{ST} struggle to find competitive solutions to the generated instances past $N = 4$, eventually encountering loading time issues for larger instances. Notably, CP_{CO} found similar solutions to MIP for the smallest generated instances, but could not prove optimality due to a weaker lower bound. CP_{SR} consistently outperformed MIP and the other CP models for all but the smaller instances. A similar trend is observed on the industrial instances, where CP_{SR} was the only model-based approach that found a feasible solution to more than one instance. For both sets of instances, CP_{SR} consistently found better solutions than the heuristic in less time. We also observe that the optimality gaps of the larger instances that only CP_{SR} can solve are of the same order of magnitude as the gaps of the smaller ones. The lower bounds of these large instances are generated by CP_{SR}, but their values are non-trivial, a rare feat for typical CP approaches. We note that CP Optimizer computes the lower bound using an automatic LP-based relaxation of the scheduling constraints [18], a feature not available in some other CP solvers.

7 Discussion and Conclusion

In this paper, we create a CP scheduling approach for a novel packing problem: the Two-stage Cutting Stock Problem with Flexible Length and Flexible Demand (2SCSP-FF). Using optional intervals, state functions, and cumulative functions, our model, CP_{SR}, has significant computational and performance advantages over two alternative CP models, a MIP model, and a two-stage heuristic on large generated and industrial instances.

The memory efficiency of CP_{SR} can be attributed to the compact representation of the complicated substructures. To represent the guillotine cuts, CP_{SR} is the only model that does not enumerate over the set of levels J, instead using just a state function and a ALWAYSCONSTANT constraint. Similarly, CP_{SR} restricts the widthwise capacities and the partition counts without levelwise constraints. By using the fewest variables and constraints, the CP_{SR} model has at least an order-of-magnitude savings in its memory usage. As the instances scale up, this advantage increases.

Accordingly, only CP_{SR} found a feasible solution to more than one industrial-scale instance. The short time-to-feasibility, however, differs from the results for routing problems [3], where the model struggled to find feasible solutions quickly. We suspect that this disparity is due to the looser constraints on interval variables for our problem compared to the routing formulation.

Overall, our success here suggests that the flexibility of CP scheduling tools provides a promising approach to attacking complex real-world problems beyond traditional scheduling ones.

Acknowledgements. We thank anonymous reviewers for their valuable feedback. This research was partially supported by Visual Thinking International Ltd (Visual8) and the Natural Sciences and Engineering Research Council of Canada.

References

1. Beldiceanu, N., Carlsson, M., Flener, P., Pearson, J.: On the reification of global constraints. Constraints **18**, 1–6 (2012)
2. Beldiceanu, N., Contejean, E.: Introducing global constraints in chip. Math. Comput. Model. **20**(12), 97–123 (1994)
3. Booth, K.E.C., Beck, J.C.: A constraint programming approach to electric vehicle routing with time windows. In: Rousseau, L.-M., Stergiou, K. (eds.) CPAIOR 2019. LNCS, vol. 11494, pp. 129–145. Springer, Cham (2019). https://doi.org/10.1007/978-3-030-19212-9_9
4. Cappart, Q., Schaus, P.: Rescheduling railway traffic on real time situations using time-interval variables. In: Salvagnin, D., Lombardi, M. (eds.) CPAIOR 2017. LNCS, vol. 10335, pp. 312–327. Springer, Cham (2017). https://doi.org/10.1007/978-3-319-59776-8_26
5. Clautiaux, F., Jouglet, A., Carlier, J., Moukrim, A.: A new constraint programming approach for the orthogonal packing problem. Comput. Oper. Res. **35**(3), 944–959 (2008). Part Special Issue: New Trends in Locational Analysis
6. Dincbas, M., Simonis, H., Hentenryck, P.V.: Solving a cutting-stock problem with the constraint logic programming language CHIP. Math. Comput. Model. **16**, 95–105 (1992)
7. Furini, F., Malaguti, E.: Models for the two-dimensional two-stage cutting stock problem with multiple stock size. Comput. Oper. Res. **40**(8), 1953–1962 (2013)
8. Gedik, R., Kirac, E., Milburn, A.B., Rainwater, C.: A constraint programming approach for the team orienteering problem with time windows. Comput. Ind. Eng. **107**, 178–195 (2017)
9. Gilmore, P.C., Gomory, R.E.: Multistage cutting stock problems of two and more dimensions. Oper. Res. **13**(1), 94–120 (1965)
10. Ham, A.M., Cakici, E.: Flexible job shop scheduling problem with parallel batch processing machines: MIP and CP approaches. Comput. Ind. Eng. **102**, 160–165 (2016)
11. IBM: CP optimizer user manual. https://www.ibm.com/docs/en/icos/20.1.0?topic=optimizer-cp-users-manual
12. Iori, M., de Lima, V.L., Martello, S., Miyazawa, F.K., Monaci, M.: Exact solution techniques for two-dimensional cutting and packing. Eur. J. Oper. Res. **289**(2), 399–415 (2021)
13. Kinable, J., van Hoeve, W.-J., Smith, S.F.: Optimization models for a real-world snow plow routing problem. In: Quimper, C.-G. (ed.) CPAIOR 2016. LNCS, vol. 9676, pp. 229–245. Springer, Cham (2016). https://doi.org/10.1007/978-3-319-33954-2_17
14. Kong, V.L.: IBMDecisionOptimization: Docplex-Examples/Trimloss.py (2020). https://github.com/IBMDecisionOptimization/docplex-examples/blob/master/examples/cp/basic/trimloss.py
15. Korf, R.E.: Optimal rectangle packing: initial results. In: Proceedings of the Thirteenth International Conference on Automated Planning and Scheduling (ICAPS 2003), pp. 287–295. AAAI (2003)

16. Korf, R.E.: Optimal rectangle packing: new results. In: Proceedings of the Fourteenth International Conference on Automated Planning and Scheduling (ICAPS 2004), pp. 142–149. AAAI (2004)
17. Ku, W., Beck, J.C.: Mixed integer programming models for job shop scheduling: a computational analysis. Comput. Oper. Res. **73**, 165–173 (2016)
18. Laborie, P., Rogerie, J.: Temporal linear relaxation in IBM ILOG CP optimizer. J. Sched. **19**(4), 391–400 (2016)
19. Laborie, P., Rogerie, J., Shaw, P., Vilím, P.: IBM ILOG CP optimizer for scheduling - 20+ years of scheduling with constraints at IBM/ILOG. Constraints **23**(2), 210–250 (2018)
20. Lee, J., Kim, B.I., Johnson, A.L.: A two-dimensional bin packing problem with size changeable items for the production of wind turbine flanges in the open die forging industry. IIE Trans. **45**, 1332–1344 (2013)
21. Lodi, A., Martello, S., Monaci, M.: Two-dimensional packing problems: a survey. Eur. J. Oper. Res. **141**(2), 241–252 (2002)
22. Lodi, A., Monaci, M.: Integer linear programming models for 2-staged two-dimensional knapsack problems. Math. Program. **94**(2–3), 257–278 (2003)
23. Loken, C., et al.: SciNet: lessons learned from building a power-efficient top-20 system and data centre. In: Journal of Physics: Conference Series, vol. 256, p. 012026 (2010)
24. Macedo, R., Alves, C., de Carvalho, J.M.V.: Arc-flow model for the two-dimensional guillotine cutting stock problem. Comput. Oper. Res. **37**(6), 991–1001 (2010)
25. Mesyagutov, M., Scheithauer, G., Belov, G.: LP bounds in various constraint programming approaches for orthogonal packing. Comput. Oper. Res. **39**(10), 2425–2438 (2012)
26. Moffitt, M.D., Pollack, M.E.: Optimal rectangle packing: a meta-CSP approach. In: Proceedings of the Sixteenth International Conference on Automated Planning and Scheduling, (ICAPS 2006), pp. 93–102. AAAI (2006)
27. Ponce, M., et al.: Deploying a top-100 supercomputer for large parallel workloads: the Niagara supercomputer. In: Proceedings of the Practice and Experience in Advanced Research Computing on Rise of the Machines (Learning), PEARC 2019, Chicago, IL, USA, 28 July–01 August 2019, pp. 34:1–34:8. ACM (2019)
28. Shaw, P.: A constraint for bin packing. In: Wallace, M. (ed.) CP 2004. LNCS, vol. 3258, pp. 648–662. Springer, Heidelberg (2004). https://doi.org/10.1007/978-3-540-30201-8_47
29. Silva, E., Alvelos, F., Valério de Carvalho, J.: An integer programming model for two- and three-stage two-dimensional cutting stock problems. Eur. J. Oper. Res. **205**(3), 699–708 (2010)
30. Simonis, H., O'Sullivan, B.: Search strategies for rectangle packing. In: Stuckey, P.J. (ed.) CP 2008. LNCS, vol. 5202, pp. 52–66. Springer, Heidelberg (2008). https://doi.org/10.1007/978-3-540-85958-1_4
31. Simonis, H., O'Sullivan, B.: Almost square packing. In: Achterberg, T., Beck, J.C. (eds.) CPAIOR 2011. LNCS, vol. 6697, pp. 196–209. Springer, Heidelberg (2011). https://doi.org/10.1007/978-3-642-21311-3_19
32. Tang, T.Y., Beck, J.C.: CP and hybrid models for two-stage batching and scheduling. In: Hebrard, E., Musliu, N. (eds.) CPAIOR 2020. LNCS, vol. 12296, pp. 431–446. Springer, Cham (2020). https://doi.org/10.1007/978-3-030-58942-4_28
33. Wäscher, G., Haußner, H., Schumann, H.: An improved typology of cutting and packing problems. Eur. J. Oper. Res. **183**(3), 1109–1130 (2007)
34. Yang, D., et al.: Flexibility in metal forming. CIRP Ann. **67**(2), 743–765 (2018)

Dealing with the Product Constraint

Steve Malalel[(✉)], Victor Jung, Jean-Charles Régin, and Marie Pelleau

Université Côte d'Azur, CNRS, I3S, Nice, France
{steve.malalel,victor.jung,jean-charles.regin,
marie.pelleau}@univ-cotedazur.fr

Abstract. The product constraint ensures that the product of some variables will be greater than a given value, that is $\Pi_{i=1}^{n} x_i \geq w$. With the emergence of stochastic problems, this constraint appears more and more frequently in practice. The variables are most often probability variables that represent the probability that an event will occur and the minimum bound is the minimum probability that must be satisfied. This is often done to guarantee a certain level of security or a certain quality of service. To deal with this constraint, it is tempting as proposed by many authors to take the logarithm of the sum and the bound in order to transform the product into a sum. In this article we show that this idea creates many problems and forbids an exact calculation. We propose and compare different representations allowing to compute the set of solutions of this problem exactly or up to a certain precision. We also give an efficient method to represent that constraint by a Multi-valued Decision Diagram (MDD) in order to combine this constraint with some others MDDs.

1 Introduction

More and more problems involve uncertain data associated with probabilities. For quality of service or security reasons, it is frequently imposed that any solution must be associated with a minimum probability. This kind of problem is naturally modeled by defining for each variable x representing uncertain values, a variable p_x which represents the probabilities of these values. Then, the variables x and p_x are linked together (i.e. $x = a \Leftrightarrow p_x = p(a)$) and the constraint $\Pi_{i=1}^{n} p_{x_i} \geq w$ is added to the model in order to guarantee that each solution will be associated with a probability higher than a given value.

We are mainly interested in defining the multi-valued decision diagram (MDD) of this constraint as it is classically made for a sum constraint. We assume that the values of variables are decimal with a given precision.

Usually, the product constraint is modeled by taking the logarithm of both terms, and so by transforming it into the sum $\sum_{i=1}^{n} \log(p_{x_i}) \geq \log(w)$. It seems more convenient because the product of variables is not easy to manage in constraint programming solvers due to overflows. However, using a logarithm has a major drawback: we lose the possibility to make exact calculations because the logarithm function cannot be represented exactly in a computer as it can

© Springer Nature Switzerland AG 2022
P. Schaus (Ed.): CPAIOR 2022, LNCS 13292, pp. 266–281, 2022.
https://doi.org/10.1007/978-3-031-08011-1_18

return a transcendental number. Thus, floating-point numbers have to be used and errors in the representation have to be managed.

In this paper we study several methods for defining the MDD of this constraint. The first one is based on the sum of the logarithm. The second one computes the exact MDD of the product of variables. Unfortunately, this method may need a lot of memory. Therefore we propose to relax the previous MDD up to a certain precision. On the other hand, we present a method that builds the MDDs by successive iterations and further compresses it, taking into account the bound imposed on the constraint. Each iteration corresponds to a precision that is higher than the previous one. The main idea is to stop considering the parts of the MDD that will always be satisfied when the precision of the computation is increased. For instance, no matter the precision, we will always have $(0.95... \times 0.95...) > 0.9$.

The paper is organised as follows. First, we recall some definitions. Then, we present different methods to compute the MDD of the product constraint. Next, we experiment with these methods. At last, we conclude.

2   Preliminaries

2.1   Constraint Programming

A finite constraint network $\mathcal{N}$ is defined as a set of n variables $X = \{x_1, \ldots, x_n\}$, a set of current domains $\mathcal{D} = \{D(x_1), \ldots, D(x_n)\}$ where $D(x_i)$ is the finite set of possible values for variable x_i, and a set $\mathcal{C}$ of constraints between variables. We introduce the particular notation $\mathcal{D}_0 = \{D_0(x_1), \ldots, D_0(x_n)\}$ to represent the set of initial domains of $\mathcal{N}$ on which constraint definitions were stated. A constraint C on the ordered set of variables $X(C) = (x_{i_1}, \ldots, x_{i_r})$ is a subset $T(C)$ of the Cartesian product $D_0(x_{i_1}) \times \cdots \times D_0(x_{i_r})$ that specifies the allowed combinations of values for the variables $x_{i_1}, \ldots, x_{i_r}$. An element of $D_0(x_{i_1}) \times \cdots \times D_0(x_{i_r})$ is called a tuple on $X(C)$ denoted by τ. In a tuple τ, the assignment of the ith variable is denoted by τ_i.

We present some constraints that we will use in the rest of this paper.

Definition 1. *Given X a set of variables and w a value, the* sum constraint *ensures that the sum of all variables $x \in X$ is greater than or equal to w.*
SUM$(l, u) = \{\tau \mid \tau$ *is a tuple on $X(C)$ and* $\sum_{i=0} \tau_i \geq w\}$

Definition 2. *Given X a set of variables and w a value, the* product constraint *ensures that the product of all variables $x \in X$ is greater than or equal to w.*
PRODUCT$(l, u) = \{\tau \mid \tau$ *is a tuple on $X(C)$ and* $\Pi_{i=0} \tau_i \geq w\}$

Use of Logarithm Function. Using mathematical functions in CP imposes to have some guarantees on the values that are computed. In this case, two important concepts are considered: the correctness and the completeness.

Consider M a model of a problem P, that can use different types of numbers like floating-point and real numbers. We say that a model M satisfies the correctness condition iff all solutions found by the solver are solutions of P. In other words, there is no solution of M that is not a solution of P. The completeness condition is satisfied iff all solutions of P are solutions of M. If both conditions are met then we say that P is exactly solved.

Unfortunately, the log function may be very complex to calculate exactly, mainly because the discrete logarithm problem (given real numbers a and b, the logarithm $\log_b(a)$ is a number x such that $b^x = a$) is considered to be computationally intractable. Note that computing the log function with a fixed precision is equivalent to the discrete logarithm problem. Therefore, we need to compute an approximation of log. Consider $\underline{\log}$ (resp. $\overline{\log}$) a lower bound (resp. upper bound) of the log function. If the product constraint is modeled by $\sum_{i=1}^{n} \underline{\log}(p_{x_i}) \geq \overline{\log}(w)$ then the model is correct, because $\sum_{i=1}^{n} \log(p_{x_i}) \geq \sum_{i=1}^{n} \underline{\log}(p_{x_i}) \geq \overline{\log}(w) \geq \log(w)$. If the product constraint is modeled by $\sum_{i=1}^{n} \overline{\log}(p_{x_i}) \geq \underline{\log}(w)$ then the model is complete because $\sum_{i=1}^{n} \overline{\log}(p_{x_i}) \geq \sum_{i=1}^{n} \log(p_{x_i})$ and $\log(w) \geq \underline{\log}(w)$.

As the filtering algorithms in CP are based on the deletion of values that do not belong to a solution, it is fundamental to guarantee that at least all solutions of the problem are considered and thus that the model is complete. This means that we need to be able to compute both a lower and an upper bound of the log. Unfortunately, the properties of the log functions available in a language, like Java, or in a library is not always provided. However, modern implementation of elementary functions are at least faithful [7], i.e., they return one of the two floating-point numbers surrounding the exact value $\log(x)$. Thus, with floating-point representation, a lower bound can be obtained by subtracting one machine epsilon to the value returned by the log function and an upper bound can be obtained by adding one machine epsilon to the value returned by the log function.

Use of Decimal Variables. Decimal variables impose a certain precision in their representation and in the calculations involving them. If they are represented by floating-point variables IEEE754 rounding modes can be managed for ensuring the safeness of some operations (in order to guarantee the completeness of the model). Note that some programming languages, like Java, do not offer this possibility.

We will use the following notations:

Notation 1

- δ *is the precision of a decimal variable, i.e. the number of decimal digits that are taking in account (after the decimal separator).*
- ϵ *is the computational precision between decimal numbers.*

2.2   Multi-valued Decision Diagram

The decision diagrams considered in this paper are reduced ordered multi-valued decision diagrams (MDD) [1,6,8], which are a generalisation of binary decision diagrams [2]. They use a fixed variable ordering for canonical representation and shared sub-graphs for compression obtained by means of a reduction operation. An MDD is a rooted directed acyclic graph (DAG) used to represent some multi-valued functions $f : \{0...d-1\}^n \rightarrow true, false$. Given the n input variables, the DAG contains $n+1$ layers of nodes, such that each variable is represented at a specific layer of the graph. Each node on a given layer has at most d outgoing arcs to nodes in the next layer of the graph. Each arc is labeled by its corresponding integer. The arc (u, a, v) is from node u to node v and labeled by a. Sometimes it is convenient to say that v is a child of u. The set of outgoing arcs from node u is denoted by $\omega^+(u)$. All outgoing arcs of the layer n reach tt, the true terminal node (the false terminal node is typically omitted). There is an equivalence between $f(a_1, ..., a_n) = true$ and the existence of a path from the root node to the tt whose arcs are labeled $a_1, ..., a_n$.

The reduction of an MDD is an important operation that may reduce the MDD size by an exponential factor. It consists in removing nodes that have no successor and merging equivalent nodes, i.e., nodes having the same set of neighbors associated with the same labels. This means that only nodes of the same layer can be merged.

Construction of MDDs. The classical approach to build MDD(C), the MDD of a constraint C, is to use states. When building MDD(C), we assign an information representing the current state $s(x)$ of the constraint C to each node x. Given (u, a, v) an arc, $s(u)$ the state of the node u and a transition function, we are able to produce $s(v)$ the state of the node v and to know if this state satisfies the constraint C or not. If two different nodes a and b have the same state (i.e. $s(a) = s(b)$), they can be merged into one node ab with $s(ab) = s(a) = s(b)$ during the building process. As we try to build the MDD of a constraint we add a validity function that anticipates the fact that a state cannot lead to at least one solution. This function, noted ISVALID returns false if we can guarantee that a state will never satisfy the constraint. This function is called before creating a state. For a given layer i it computes the maximum value from the next layer to tt denoted by $vMax[i+1]$.

In the case of the SUM (resp. PRODUCT) constraint, the state represents the current sum (resp. product) of the current node. The creation of a new state is given by function CREATESTATE. The transition function is simply the addition (resp. multiplication) between the current sum (resp. product) and the label, therefore we will not explicit them.

3   Product Constraint as a Sum of Logarithms

There are two possible ways to build the MDD of the constraint depending on the representation of the decimal/floating-point variable: either by floating-point or by integers. In any case, a faithful log function has to be used and the rounding errors of the sum have to be managed. In the first case, the MDD of the constraint is just the MDD of the SUM constraint on floating-point variables. The latter case deserves more attention.

Algorithm 1: Integer Logarithm Representation

CREATESTATE($state$, $label$) : State
   │ $newState \leftarrow$ CREATESTATE()
   │ $newState.sum \leftarrow state.sum + label$
   └ **return** $newState$

ISVALID(C, $state$, $label$, $layer$) : Boolean
   │ $newSum \leftarrow state.sum + label$
   │ $maxPotential \leftarrow newSum + vMax[layer + 1]$
   └ **return** $maxPotential \geq C.min$

SIGNATURE(C, $state$, $label$, $layer$, $size$) : Integer
   └ **return** $\lceil (state.sum + label)/10^{\delta - \epsilon} \rceil$

MERGE(C, $state_1$, $state_2$)
   └ **if** $state_2.sum < state_1.sum$ **then** $state_1.sum \leftarrow state_2.sum$

The logarithm is computed up to a certain precision λ, that is to say the number of digits taken into account after the decimal point [3]. This value can be different than δ (the precision of decimal variables). It is therefore important to take a precision value such that the integer representation of the sum does not cause an overflow: $n \times 10^{\lambda + d} \leq 2^b$, with d the maximum number of digits required to represent the integer value of the logarithm, n the number of variables and b the number of bits used to represent the integer (32 for an int, 64 for a long). We will denote this representation Integer Logarithm Representation. Algorithm 1 gives a possible implementation of the required functions to build the MDD of the product constraint, with $vMax[i] = \sum_{j=i}^{n} max(D_j)$. Algorithm 1 also gives a possible implementation of the merging conditions. To perform merges during the construction process, we introduce a SIGNATURE function that will behave as a hash: if two nodes have the same signature, then we merge them according to the MERGE function. This allows us to keep a slightly more precise sum variable, while still being able to merge nodes and save some space.

4   Exact Representation of the Product Constraint

This method aims to be as accurate as possible by representing the result of consecutive multiplications without loss. First, the decimal variables are transformed into integer variables. We choose a value for δ (i.e., the number of decimal we want to take into account for representing the values of variables) and then we turn decimals into their corresponding integers (for example with the decimal 0.98 and $\delta = 4$, we obtain 9800). Afterward, knowing w, the minimum value of the product of variables, and n the number of variables in the scope of the constraint, we calculate the minimal threshold $min = \lfloor w \times 10^{(n \times \delta)} \rfloor$.

The state of a node contains the value $prod$ which is the product of the labels of the arcs from the $root$ node to the current node. The $prod$ value of the $root$ is 1. Algorithm 2 gives a possible implementation for the CREATESTATE and ISVALID functions. ISVALID is defined using $vMax[i] = \prod_{j=i}^{n} max(D_j)$.

Algorithm 2: Exact Product Implementation

CREATESTATE($state$, $label$)
  | $newState \leftarrow$ CREATESTATE()
  | $newState.prod \leftarrow state.prod \times label$
  └ **return** $newState$
ISVALID(C, $state$, $label$, $layer$) : Boolean
  | $newProd \leftarrow state.prod \times label$
  | $maxPotential \leftarrow newProd \times vMax[layer + 1]$
  └ **return** $maxPotential \geq C.min$

Unfortunately, the use of integers is limited when representing big numbers: multiplying all the integers together can quickly cause an overflow. In order to address this issue, a specific data structure that can represent really big numbers must be used. Most programming languages offer such data structures (e.g. Big-Integer in Java). However, these data structures have two important drawbacks that can prevent their use in practice: they quickly run out of memory during the computation, and the computation of the multiplication takes more time as the product value increases. It is therefore necessary to study a more relaxed representation.

5 Relaxed Product Constraint

In order to decrease the memory consumption and the computation time, we can deliberately lose accuracy during the computation by truncating and rounding the result correctly, depending on a given precision ϵ. Therefore, instead of computing the exact MDD by performing basic multiplications, *relaxed* multiplications are performed, which will as a result compute a relaxed MDD. This means that the MDD can either be complete, correct, or both (but both cannot be guaranteed). We note this multiplication $\times^\epsilon$ with ϵ the precision such that $\lceil x \times^\epsilon y \rceil = \lceil (x \times y)/10^\epsilon \rceil$ and $\lfloor x \times^\epsilon y \rfloor = \lfloor (x \times y)/10^\epsilon \rfloor$. For example $\lceil 9800 \times^4 9780 \rceil = 9585$. Similarly we note $\prod^\epsilon$ the product using $\times^\epsilon$. Note that for the following part the rounding is done to guarantee only the completeness, but it can also be done to guarantee only the correctness by doing a switch between the floor rounding and the ceiling rounding. The MDD is built following the same first steps defined for the Exact Product: we choose the precision ϵ and turn floating-point numbers into their corresponding integers.

However, the minimal threshold is different: it is now defined as $min = \lfloor w \times 10^\epsilon \rfloor$. The way the value $vMax[i]$ is calculated for each layer is modified to take this change into account i, $vMax[i] = \lceil \prod^\epsilon{}_{j=i}^{n} max(D_j) \rceil$.

The state of a node x still holds the value *prod*, but it now represents the consecutive relaxed multiplications between the labels of the arc from the node *root* to the node x. We initialise the root value as 10^ϵ, which is the equivalent of 1 in ϵ precision.

Algorithm 3 gives a possible implementation of the CREATESTATE and ISVALID functions.

Algorithm 3: Relaxed Product Implementation

CREATESTATE(*state*, *label*) : State
   | *newState* ← CREATESTATE()
   | *newState.prod* ← $\lceil state.prod \times^{\epsilon} label \rceil$
   | **return** *newState*
ISVALID(*C*, *state*, *label*, *layer*) : Boolean
   | *newProd* ← $\lceil state.prod \times^{\epsilon} label \rceil$
   | *maxPotential* ← $\lceil newProd \times^{\epsilon} vMax[layer + 1] \rceil$
   | **if** *maxPotential* < *C.min* **then** **return** *false*
   | **return** *true*

Even if this method cannot ensure both the completeness and the correctness at the same time, it allows us to balance between performance and accuracy of the solutions. Indeed the smaller ϵ is, the more equivalent states we have. Therefore, we obtain more merges resulting in less computational resources to build the MDD.

6   Incremental Precision Refinement

In this section we present an Incremental Precision Refinement of the set of solutions (IPR). This method aims at computing the MDD of the constraint for a precision ϵ by computing successive MDDs of the constraint having a lower precision. Let MDD_{ϵ} be the MDD for the precision ϵ. The idea is to start by building MDD_1 and then build MDD_k from MDD_{k-1}. The advantage of this approach is that it avoids creating intermediate states that are not in the final reduced MDD.

We can classify the solutions computed in a relaxed MDD in two categories, the "sure solutions" and the "relaxed solutions". The "sure solutions" are the solutions that no matter the precision are valid. For instance $(0.95... \times 0.95...)$ is greater than 0.9 no matter the precision $k > 2$. The "relaxed solutions" are the solutions for which a higher precision is required in order to determine if they are solutions. For instance the fact that $(0.94... \times 0.95...)$ is greater than 0.9 is uncertain and requires a higher precision ($0.948 \times 0.955 > 0.9$ and $0.940 \times 0.950 < 0.9$).

This method is based on the fact that "sure solutions" in MDD_k are solutions in all the following iterations. Thus the method focuses only on the part of the MDD containing "uncertain solutions" to improve the precision at a lower cost.

6.1   Extracting Suspicious Arcs

The first step of the algorithm is to identify and extract arcs and nodes that are part of at least one "relaxed solution". In order to do so, we use the scheme introduced in [4] to propagate the bounds of the product constraint to each node of the MDD. In our case, the property associated with a node is the current possible sum interval for the SUM constraint and the current possible product interval for the PRODUCT constraint. A bottom-up propagation is performed

---

Algorithm 4: Extraction

EXTRACTION(mdd, w) : MDD

    $mdd_M \leftarrow$ CREATEMDD()

    $mdd.root.value \leftarrow 0$

    $mdd.root.x_1 \leftarrow mdd_M.root$

    $L[0] \leftarrow \{mdd_M.root\}$

    foreach $i \in 0..r - 1$ **do**

        foreach $Node\ x \in L[i]$ **do**

            foreach $label \in \omega^+(x)$ **do**

                $y \leftarrow x.$GETCHILD($label$)

                $v \leftarrow x.value + label$

                if $y.property[0] + v < w$ **then**

                    $x_1 \leftarrow x.x_1$

                    $y_1 \leftarrow y.x_1$

                    if y_1 *is nil* **then**

                        $y_1 \leftarrow$ CREATENODE()

                        $y.x_1 \leftarrow y_1$

                        ADDNODE($L, y_1, i + 1$)

                    ADDARC($x_1, label, y_1$)

                    $y.value \leftarrow$ MIN($y.value, v$)

    merge all nodes of $L[r]$ into t

    PREDUCE(L)

    return mdd_M

(instead of the top-down propagation presented in the cited paper). Starting from the *tt* node the propagation computes for each node the interval of the minimal values, called property, needed to be part of a "sure solution". After the bottom-up propagation, a top-down propagation of properties is performed. Starting from the root node, each outgoing arcs (*source, v, destination*) is checked. If the value v of the arc combined with the property p of the *destination* is below the threshold w, then we cannot be certain that this arc only leads to "sure solutions". This arc is thus added to the marked MDD$_M$ containing all arcs and nodes marked "suspicious". When such an arc is marked, the value of the node *destination* is updated to the lowest value between its current value and the value of *source* combined with the value of the arc v. At the end of the algorithm, we obtain the MDD$_M$ containing all arcs and nodes appearing in at least one "relaxed solution".

The implementation of the algorithm is given Algorithm 4. The algorithm takes as input an MDD *mdd* and the lower bound w.

Note that the algorithm only deals with a lower bound w, but can easily be adapted to deal with an upper bound, or both a lower and upper bound.

Proposition 1. *After executing Algorithm 4 MDD$_M$ contains all the "suspicious" arcs and nodes.*

Proof. Let the *root* node with property $p < w$. This means that at least one "relaxed solution" pass by the root node, which means that at least one "relaxed solution" pass by one of its children. Let p_c be the value of the property of the child c, and v be the value of the arc from c to *root*. If $p_c + v < w$, it means that at least one "relaxed solution" pass by the arc $(root, v, c)$, because taking it makes the property go below the threshold. Now, suppose that each node holds the value of the lowest path from the root. Consider the arc $(x, label, y)$ where x has a value v_x below the threshold w, and y has a property p_y. $p_y + label + v_x$ correspond to the lowest possible path taking the arc $(x, label, y)$. If this value is below the threshold w, then it means that at least one "relaxed solution" pass by this arc.

6.2   On the Fly Intersection

After extracting MDD_M containing all "relaxed solutions", we need to improve the precision in order to only have solutions. To do so, we perform the intersection between the MDD with higher precision and MDD_M, but without computing the entire constraint: we only need the parts in common with MDD_M. In order to do so, we perform the on the fly intersection [5]. This allows us to compute higher precision only for the parts of the MDD that need it, without wasting time and memory to recompute a large amount of solutions.

6.3   IPR Algorithm

We give a possible implementation of the global algorithm (Algorithm 5) that computes the MDD of the product constraint using the different schemes presented in this section. The algorithm will stop once it reaches an equilibrium, or when the maximum precision allowed is reached.

Algorithm 5: IPR Algorithm

$\text{IPR}(w, \epsilon, D) : \text{MDD}$
> $mdd_c \leftarrow \text{CREATEMDD}(0, w, D)$
> $mdd_M \leftarrow \text{EXTRACTION}(mdd_c, w)$
> $mdd_S \leftarrow mdd_c - mdd_M$
> **foreach** e *in 1..ϵ* **do**
>> $mdd_c \leftarrow \text{PERFORMINTERSECTION}(mdd_M, C, e)$
>> $mdd_M \leftarrow \text{EXTRACTION}(mdd_c, w)$
>> **if** mdd_M *is empty* **then return** mdd_S
>> $mdd_S \leftarrow mdd_S \cup (mdd_c - mdd_M)$
> **return** $mdd_S \cup mdd_c$

The first step of the algorithm is to compute the initial MDD_c (with lowest precision). Once this MDD is created, we execute the extract function (Algorithm 4) and retrieve the associated MDD_M. Then, we compute the difference between the MDD_c and MDD_M, basically filtering the initial MDD from all uncertain solutions. We store all good solutions in an accumulator MDD_S. We then repeat the same steps over MDD_M: we intersect it with the MDD of the constraint C with higher precision e (PERFORMINTERSECTION), then extract, filter, and add solutions to MDD_S. The algorithm stops when MDD_M is empty or when the maximum precision allowed is reached.

Note that, even if very unlikely, it is possible that $MDD_c = MDD_M$. This cannot be a stop criterion for the algorithm as it would still be possible that all solutions of MDD_c require a higher precision to decide if they belong to MDD_S or not.

7   Experiments

The algorithms presented in this paper have been implemented in Java 11. The experiments were performed on a machine having four E7-4870 Intel processors, each having 10 cores with 256 GB of memory and running under Scientific Linux. All the experiments were run in sequential.

First we use fixed data sets, then we study the impact of varying some parameters (number of variables, domain size, w value). The fixed data sets, denoted by data1... data10, involve 10 variables with a domain of size 10. Each value represents a probability between 95% and 100% to ensure that there exist solutions for the instances. Each resolution was made with our minimum threshold w representing 90%. All the data used in this paper are available upon request.

7.1   Exact Product Method

Table 1. Time (ms) and Memory (MB) needed to compute the exact MDD using the Exact Product method.

Data set	#Solutions	Time (ms)	Memory (MB)
data1	341 051	7 403	2 268
data2	902 485	14 134	3 383
data3	1 819 820	24 629	6 508
data4	489 297	5 469	1 807
data5	104 506	2 373	879
data6	882 970	13 567	3 394
data7	4 049 230	98 740	14 072
data8	510 291	23 077	4 121
data9	5 473 625	389 801	25 273
data10	797 484	37 497	4 689

The Exact Product method behaves exactly as expected: we obtain the exact MDD at a high cost both in term of memory consumption and time (Table 1). This method is nonetheless interesting because it serves as a proof of the total number of solutions, which will be helpful to compare the relative accuracy of other methods (Table 2).

7.2 Logarithm and Relaxed Methods

Comparison of the Methods. Concerning the number of solutions produced by the Logarithm method (both Integer and Library) and Relaxed Product method, we notice an interesting phenomenon: the more we increase the precision ϵ, the less we improve the lower bound and thus the accuracy of the solutions at each iteration (Table 2). Even worse, the time needed to compute the MDD does not scale at all with the number of solutions (Table 3). For instance, the time needed to compute 5 473 669 solutions is about 13 s (Table 2 and Table 3) for $\epsilon = 7$, while the time needed to compute 5 473 625 solutions is about 103 s for $\epsilon = 8$. On this data, the computation time is 8 times slower when $\epsilon = 8$ than when $\epsilon = 7$ for a difference of 44 solutions. It means that the trade-off between precision and computational resources is not worth it. It is nevertheless complicated to determine the correct ϵ such that we do not spend too much time on computation for a relatively good approximation of the exact MDD. Furthermore, the "relaxed" solutions introduced by the relaxation are very close to the defined threshold w (as shown by the lower bounds in Table 2). However, we notice that it is faster to build the MDD using the logarithm than using the Exact Method; we obtain the exact MDD at $\epsilon = 8$ for a time of 103 s (compared to almost 400 s). This difference is explained by the heaviness of the exact representation.

Table 2. Comparison of the number of solutions generated depending on the precision ϵ for data9. Number of exact solutions: 5 473 625.

ϵ	Logarithm	Lower Bound	Relaxed Product	Lower Bound
1	**9 800 000 000**	≈ 0.628	10 000 000 000	≈ 0.628
2	261 007 356	≈ 0.762	**213 455 660**	≈ 0.821
3	**9 193 737**	≈ 0.886	9 222 380	≈ 0.891
4	**5 731 323**	≈ 0.899	5 770 914	≈ 0.8992
5	**5 493 720**	≈ 0.8999	5 498 953	≈ 0.8999
6	**5 474 681**	≈ 0.89999	5 476 117	≈ 0.89999
7	**5 473 669**	≈ 0.899999	5 473 844	≈ 0.899999
8	**5 473 625**	≈ 0.90	5 473 649	≈ 0.8999999
9	**5 473 625**	≈ 0.90	5 473 627	≈ 0.89999999

Table 3 also shows that the method using a sum of logarithm based on a log function call from a library and using floats (Library Log), the method using a log sum represented as an integer and whose precision is controlled (Integer Log) and the method of relaxing the MDD of the exact product of variables constraint (RelaxProd) give very close results as soon as one chooses a computational precision higher than 4 decimals. However, the IntegerLog method seems to be faster and to consume a little less memory than the other two methods.

Table 3. Time (ms) and memory (MB) needed to compute the MDD of data9 for a given ϵ depending on the representation used.

ϵ	Time (ms)			Memory (MB)		
	Library Log	Integer Log	RelaxProd	Library Log	Integer Log	RelaxProd
1	49	49	**46**	4	**3**	**3**
2	60	**56**	68	4	**4**	**4**
3	92	**78**	109	6	**4**	7
4	197	**174**	240	19	**17**	31
5	566	**509**	801	110	**106**	166
6	2 752	**2 546**	4 558	609	**598**	830
7	13 604	**13 165**	25 848	2 556	**2 377**	2 915
8	103 460	**102 079**	181 015	6 852	**6 756**	7 251
9	324 239	**313 094**	337 803	9 929	9 838	**7 854**

7.3   Incremental Precision Refinement (IPR)

Tables 4 and 5 clearly show the advantages of the IPR routine. IPR L is the application of the IPR routine to the Library Log model, IPR IL is the application of the IPR routine to the Integer Log model and IPR RP is the application of the IPR routine to the RelaxProd model.

Table 4. Time (ms) and Memory (MB) comparison between the Exact Product method, the Library Log method with $\epsilon = 9$ and the methods with the IPR routine in order to compute the exact MDD.

Data set	Time (ms)					Memory (MB)				
	Lib Log	IPR L	IPR IL	IPR RP	Exact Prod	Lib Log	IPR L	IPR IL	IPR RP	Exact Prod
data1	5 708	**872**	904	1 145	7 403	808	**113**	**113**	173	2 268
data2	12 067	**1 075**	1 095	1 628	14 134	1 496	154	**153**	229	3 383
data3	23 069	1 883	**1 869**	2 420	24 629	2 062	258	**257**	397	6 508
data4	4 629	859	**858**	1 127	5 469	811	109	**105**	173	1 807
data5	1 930	616	**609**	825	2 373	378	**65**	**65**	105	879
data6	10 359	**980**	1 018	1 424	13 567	1 494	142	**137**	221	3 394
data7	81 257	**2 357**	2 526	2 945	98 740	5 775	**345**	**345**	495	14 072
data8	21 990	**954**	974	1 343	23 077	1 642	133	**129**	205	4 121
data9	324 239	**2 962**	3 085	3 715	389 801	9 929	438	**437**	619	25 273
data10	31 717	**1 182**	1 200	1 614	37 497	1 552	154	**153**	250	4 689

The IPR routine improves the computation up to a factor 131 in time and 60 in memory (data9 in Table 4). Furthermore, contrary to the direct computation of the MDD at a given precision ϵ, it is possible to guarantee the exactitude of the MDD if the IPR routine stopped before reaching the maximum allowed precision. Moreover, we can see that the differences between all IPR methods and the exact method or the Library Logarithm method are in the same order of magnitude. This shows that the IPR routine is generalisable to any form of relaxed representation.

Table 5. Time (ms) and Memory (MB) comparison between the different methods depending on the variations of the parameters, with n the number of variables, $|D|$ the size of each domain and w the threshold. The first line corresponds to data10 in other benchmarks. MO = 30 GB. All solving methods have the same number of solutions.

Parameters			Time (ms)				Memory (MB)			
n	$|D|$	w	IPR L	IPR IL	IPR RP	Exact	IPR L	IPR IL	IPR RP	Exact
10	10	0.77	80 932	**74 697**	91 505	-	**7 996**	8 640	11 357	MO
10	15	0.65	**8 779**	9 091	9 186	-	**1 262**	1 267	1 351	MO
10	15	0.90	9 090	**8 857**	13 773	-	1 421	**1 420**	1 893	MO
15	10	0.85	**248 596**	256 005	276 118	-	24 672	**24 557**	24 572	MO
15	10	0.90	**1 615**	1 684	2 222	49 986	**217**	218	277	7 176
15	15	0.92	12 139	**11 329**	16 273	-	1 647	**1 643**	2 136	MO
15	15	0.90	**178 285**	185 549	205 604	-	20 050	**20 046**	22 914	MO
20	5	0.9	**32 363**	32 598	44 197	-	4 270	**4 263**	5 108	MO

Variations of Data Set Parameters. Table 5 shows the behaviour of the building process depending on the different parameters such as the number of variables n, the size of the domains $|D|$ or the threshold w. The results seem to show that, the more we increase n and $|D|$, the more difficult the problem is to resolve, which is an expected result. The Exact Product method is only able to close one instance, which is the easiest one ($n = 15$, $|D| = 10$, $w = 0.90$). Nonetheless, the results are confirming yet again that the IPR method dominates the classical building approach. When comparing the efficiency of the methods, we find the same results as in Table 3: the logarithm approach is better than the relaxed product in terms of time and memory. Even though very close, the Library Logarithm (L) seems to be better at solving these instances than the Integer Logarithm (IL). However, the variation of the effect of w seems interesting: for some instances, lowering it makes the problem more difficult (1 615ms for $n = 10$ $|D| = 15$ $w = 0.90$ compared to 248 596 for $w = 0.85$) while it makes it easier for others ($n = 10$, $|D| = 15$).

When focusing particularly on the variation of w, we in fact observe a bell-shaped curve evolution for the time and memory (Figs. 1 and 2). The top of the curve seems to be achieved for the value w such that it cuts the set of all possible combinations in half (Table 3). For instance, the first line ($n = 10, |D| = 10, w = 0.77$) of Table 5 is very close to the top of the curve (Fig. 1), resulting in a factor 60 in time when comparing with $w = 0.90$ for the same data set (data10 in Table 4).

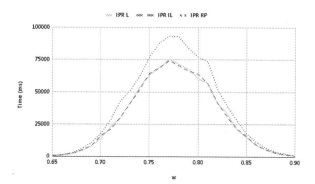

Fig. 1. Evolution of the time (ms) needed to compute the MDD for data1 depending on the parameter w.

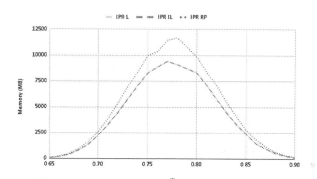

Fig. 2. Evolution of the memory (MB) needed to compute the MDD for data1 depending on the parameter w

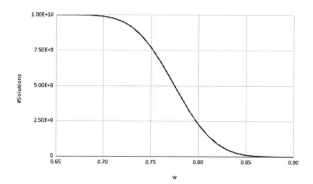

Fig. 3. Evolution of the number of solutions for data1 depending on the parameter w.

8 Conclusion

In this paper we studied several methods for defining the MDD of the PRODUCT constraint of decimal variables. The first and most popular one is based on the sum of the logarithm (using either floating point or integer numbers) with a given precision. The second one computes the exact MDD of the product of variables. The last one relaxes the previous MDD up to a certain precision.

We showed that an exact representation is not that expensive in terms of computational resources. More importantly we showed that models based on a precision can be accurate when using at least 5 decimals.

We also presented an incremental precision refinement method that efficiently computes an MDD for a given precision. It relies on the fact that if a solution is correct (a solution of the constraint) at a given precision it is also a correct solution at a higher precision. Thus this method only refines the precision on the uncertain parts of the MDD. In addition, when this method stops before reaching the fixed precision, it guarantees that the resulting MDD is exact. We showed that this method is very efficient both in terms of computational time and memory consumption no matter the method used to compute the logarithm.

In a future work it would be interesting to see if the obtained results on the logarithm are still the same when values are represented as interval of probabilities. Another very interesting development would be to study what are the necessary conditions for the use of the incremental precision refinement method, and what kind of constraints meet them.

Acknowledgments. This work has been supported by the French government, through the 3IA Côte d'Azur Investments in the Future project managed by the National Research Agency (ANR) with the reference number ANR-19-P3IA-0002.

References

1. Bergman, D., Ciré, A.A., van Hoeve, W., Hooker, J.N.: Decision Diagrams for Optimization. Artificial Intelligence: Foundations, Theory, and Algorithms, Springer, Heidelberg (2016). https://doi.org/10.1007/978-3-319-42849-9
2. Bryant, R.E.: Graph-based algorithms for boolean function manipulation. IEEE Trans. Comput. **35**(8), 677–691 (1986). https://doi.org/10.1109/TC.1986.1676819
3. Goldberg, M.: Computing logarithms digit-by-digit. Int. J. Math. Educ. Sci. Technol. **37**(1), 109–114 (2006)
4. Jung, V., Régin, J.-C.: Checking constraint satisfaction. In: Stuckey, P.J. (ed.) CPAIOR 2021. LNCS, vol. 12735, pp. 332–347. Springer, Cham (2021). https://doi.org/10.1007/978-3-030-78230-6_21
5. Jung, V., Régin, J.C.: Efficient operations between mdds and constraints. Technical report, Submitted to the International Conference on Integration of Constraint Programming, Artificial Intelligence, and Operations Research (2022)
6. Kam, T., Brayton, R.K.: Multi-valued decision diagrams. Technical report UCB/ERL M90/125, EECS Department, University of California, Berkeley (1990). http://www2.eecs.berkeley.edu/Pubs/TechRpts/1990/1671.html

7. Muller, J.M.: Elementary Functions, Algorithms and Implementation, 2nd edn. Birkhäuser (2006)
8. Srinivasan, A., Ham, T., Malik, S., Brayton, R.K.: Algorithms for discrete function manipulation. In: 1990 IEEE International Conference on Computer-Aided Design. Digest of Technical Papers, pp. 92–95 (1990). https://doi.org/10.1109/ICCAD.1990. 129849

Multiple-choice Knapsack Constraint in Graphical Models

Pierre Montalbano[1] , Simon de Givry[1(✉)] , and George Katsirelos[2]

[1] Université Fédérale de Toulouse, ANITI, INRAE, UR 875, 31326 Toulouse, France
{pierre.montalbano,simon.de-givry}@inrae.fr
[2] Université Fédérale de Toulouse, ANITI, INRAE, MIA Paris, AgroParisTech,
75231 Paris, France
gkatsi@gmail.com

Abstract. Graphical models, such as cost function networks (CFNs), can compactly express large decomposable functions, which leads to efficient inference algorithms. Most methods for computing lower bounds in Branch-and-Bound minimization compute feasible dual solutions of a specific linear relaxation. These methods are more effective than solving the linear relaxation exactly, with better worst-case time complexity and better performance in practice. However, these algorithms are specialized to the structure of the linear relaxation of a CFN and cannot, for example, deal with constraints that cannot be expressed in extension, such as linear constraints of large arity.

In this work, we show how to extend soft local consistencies, a set of approximate inference techniques for CFNs, so that they handle linear constraints, as well as combinations of linear constraints with at-most-one constraints. We embedded the resulting algorithm in TOULBAR2, an exact Branch-and-Bound solver for CFNs which has demonstrated superior results in several graphical model competitions and is state-of-the-art for solving large computational protein design (CPD) problems. We significantly improved performance of the solver in CPD with diversity guarantees. It also compared favorably with integer linear programming solvers on knapsack problems with conflict graphs.

Keywords: graphical model · cost function network · knapsack problem

1 Introduction

A Graphical Model (GM) may express an arbitrary complex function on several variables as a combination of smaller local functions on subsets of the variables. GMs have been used to reason about logic and probabilities. A deterministic GM can represent a Constraint Satisfaction Problem (CSP) where each local function

This research was funded by the French "Agence Nationale de la Recherche" through grants ANR-18-EURE-0021 and ANR-19-P3IA-0004.

P. Schaus (Ed.): CPAIOR 2022, LNCS 13292, pp. 282–299, 2022.
https://doi.org/10.1007/978-3-031-08011-1_19

is a constraint evaluating to true (satisfied) or false (unsatisfied) and the combination operator is Boolean conjunction. It can also represent a Cost Function Network (CFN), where each function evaluates to a cost and the combination operator is addition [14]. A probabilistic GM represents a probability distribution on random variables. Local functions may correspond to conditional probability distributions as in Bayesian Networks (BN) or potentials as in Markov Random Fields (MRF) [28]. They are combined by multiplication. In the following, we focus on CFNs. It can be shown that finding the Maximum A Posteriori assignment on MRFs (MAP/MRF) or the Most Probable Explanation on BNs (MPE/BN) can be cast as finding a solution of minimum cost in an appropriate CFN [13]. By allowing infinite costs to represent infeasibility, CFNs can be seen as a strict generalization of CSPs.

Exact methods to solve GMs/CFNs mostly rely on Branch-and-Bound (B&B) algorithms [20,36]. These methods have proved useful in many GM applications, such as resource allocation [10], image analysis [23], or computational biology [1,4]. For those, it has been shown to outperform other approaches, including Integer Linear Programming (ILP), MaxSAT and Constraint Programming (CP) [25].

CFNs have no native way to express linear constraints. This is in large part due to the algorithms used to compute lower bounds in B&B, which require that all constraints be expressed in extension. In many cases, having the ability to add such constraints would significantly improve the usefulness of CFNs. For example, when searching for diverse solutions, the Hamming distance constraint is naturally expressed as a linear constraint. There are ways to work around this [39,40] but, as we show later, they come with a non-trivial performance penalty. The lack of linear constraints is even more severe when the constraints have large coefficients, as in the knapsack problem with a conflict graph (KPCG). In this case, there is no workaround for the lack of linear constraints and CFN technology cannot be applied.

Contributions. Here, we show how to extend soft local consistency algorithms, a set of approximate inference techniques for CFNs, to deal with Pseudo-Boolean linear constraints (PB constraints for short), i.e., linear constraints over 0/1 variables. In the presence of *unary* cost functions (a cost function coupling a cost to each value), these correspond to knapsack constraints. We additionally consider the combination of PB constraints with Exactly-One (EO) or At-Most-One (AMO) constraints, which correspond to multiple-choice knapsack constraints [37], allowing finite-domain variables.

This new ability enables more modeling options for GM/CFN users, which we demonstrate by applying it to generating diverse solutions for Computational Protein Design (CPD). Here, the objective function has quadratic and linear terms that can be decomposed in a sum of *binary* cost functions on pairs of variables (the quadratic terms) and unary cost functions on single variables (linear terms). Searching for diverse solutions introduces linear constraints in the model (see Sect. 5.1). Our approach also compared favorably with a state-of-the-art ILP solver on knapsack problems with conflict graphs.

2   Related Work

Problems defined by PB constraints are a generalization of the SAT problem. Solvers for PB SAT typically use SAT-inspired constraint learning techniques, either by direct translation to Conjunctive Normal Form (CNF) [41] or by generalizing the clause learning mechanism to PB constraints [19]. These solvers typically do not compute lower bounds during search and have to rely on conflict reasoning only to prove bounds. A notable exception is RoundingSAT [26], which uses a Linear Programming (LP) solver to compute bounds during search and learn constraints from bound violations, but limits the number of iterations given to the LP solver in order to keep the runtime overhead of the solver reasonable. This is in contrast to our approach, which uses a suboptimal LP solver, but places no resource bounds on it. Also, PB solvers are usually restricted to a linear objective, whereas our approach can combine PB constraints with non-linear quadratic (or more) cost functions. PB solvers can also exploit the presence of AMO or EO constraints to strengthen propagation of PB constraints [5, 7].

ILP solvers are well suited to solve CFNs, given the local polytope. Their LP solving is not limited to a specific form of LP, like soft local consistency algorithms such as Existential Directional Arc Consistency (EDAC) [21] and Virtual AC (VAC) [12] are, therefore they have no issue reasoning with other linear constraints, as well as combinations with AMO/EO constraints. However, previous evaluations [25] showed that the size of the linear program that specifies that local polytope is often too large even for such highly optimized implementations and therefore they perform worse than a dedicated CFN solver in such problems.

On the CFN side, there has been work on clique constraints [22], a special case of PB constraints. Dlask and Werner [16, 17] have shown how to handle arbitrary LPs using BCA algorithms, based on a generalization of VAC. However, despite recent advances [49], BCA algorithms remain too costly for use at every node of a B&B. Many (soft) global constraints can be described by a set of (soft) linear constraints, but require an LP solver [34]. In addition, maintaining (weak) EDAC and the coupling with the other local cost functions can be costly in practice. This was also the case for other soft global constraints exploiting flow-based or dynamic programming algorithms [33, 35]. In our approach, we propose a simple and effective soft local consistency called Full ∅-Inverse Consistency for PB constraints. Finally, we can decompose linear constraints using cost functions of arity 3 and intermediate variables [3], similar to CNF encodings used by PB solvers. However, the size of the domains of the intermediate variables increases linearly with the value of the coefficients of the PB constraints.

3   Preliminaries

Definition 1. *A Cost Function Network (CFN) P is a tuple* $(\boldsymbol{X}, \boldsymbol{D}, \boldsymbol{C}, \top)$ *where* $\boldsymbol{X}$ *is a set of variables, with finite domain* $\boldsymbol{D}(x)$ *for* $x \in \boldsymbol{X}$. $\boldsymbol{C}$ *is a set of constraints. Each constraint* $c \in \boldsymbol{C}$ *is defined over a subset of variables called its scope* $(scope(c) \subseteq \boldsymbol{X})$. $\top$ *is a maximum cost indicating a forbidden assignment.*

The size of the scope of a constraint is its arity. Unary (resp. binary) cost functions have arity 1 (resp. 2). A partial assignment τ is an assignment of all the variables x_i in its scope ($scope(\tau)$) to a value of its domain $D(x_i)$. The set of all the partial assignments on a scope S is denoted $\tau(S)$. A constraint over a scope S is denoted c_S. The cost of a partial assignment τ for a constraint c_S is denoted $c_S(\tau)$ with $S \subseteq scope(\tau)$. Without loss of generality, we assume all costs are positive integers, bounded by $\top$, a special constant signifying infeasibility. Hence if $c_S(\tau) = \top$ then the assignment τ is not a feasible solution. A constraint c_S is hard if for all $\tau \in \tau(S)$, $c_S(\tau) \in \{0, \top\}$, otherwise it is soft. A CFN P that contains only hard constraints is a constraint network (CN). In the following, we use the term *cost function* interchangeably with the term constraint. An assignment τ with $scope(\tau) = X$ is a complete assignment. The cost of a complete assignment τ is given by $c_P(\tau) = \sum_{c_S \in C} c_S(\tau)$. The Weighted Constraint Satisfaction Problem (WCSP) asks, given a CFN P, to find a complete assignment minimizing $c_P(\tau)$. This task is NP-hard [14]. When the underlying CFN is a CN, the problem is the CSP, which we call crisp CSP here. In the following, we use WCSP to refer both to the optimization task and the underlying CFN.

In this paper, we assume there exists exactly one unary constraint for each variable and we say that the unary cost of $x_i = v$ for some $v \in D(x_i)$ is $c_i(v)$. We also assume the existence of a constraint $c_\varnothing$ with empty scope, which represents a constant in the objective function and, since there exist no negative costs, it is a lower bound on the cost of all possible assignments.

Exact methods to solve GMs/CFNs mostly rely on Branch-and-Bound (B&B) algorithms [20,36]. At every node of the B&B tree, the solver computes a bound and closes the node if that bound is higher than the cost of the incumbent solution or if it represents infeasibility. Typical bounding algorithms compute either static memory-intensive bounds [15] or memory-light ones [12] better suited to dynamic variable orderings. The latter, on which we focus here, are called Soft Arc Consistencies (SAC) because they reason on each non-unary cost function one by one, in a generalization of propagation in CSP.

Soft arc consistencies use $c_\varnothing$ as the lower bound and compute a reparameterization of the instance with a higher $c_\varnothing$. A reparameterization P' of a WCSP P is a WCSP with an identical structure, i.e., one where there exist constraints over the same scopes, the costs assigned by each individual cost function may differ, but $c_P(\tau) = c_{P'}(\tau)$ for all complete assignments τ.

Procedure MoveCost($c_{S_1}, c_{S_2}, \tau_1, \alpha$): Move α units of cost between the tuple τ_1 of scope S_1 and tuples τ_2 that extend τ_1 in scope S_2

Data: Scopes $S_1 \subset S_2$
Data: $\tau_1 \in \tau(S_1)$
Data: cost α to move
1   $c_{S_1}(\tau_1) \leftarrow c_{S_1}(\tau_1) + \alpha$
2   **foreach** $\tau_2 \in \tau(S_2) \mid \tau_2[S_1] = \tau_1$ **do**
3   $\quad \lfloor \ c_{S_2}(\tau_2) \leftarrow c_{S_2}(\tau_2) - \alpha$

All reparameterizations that we study here are computed as a sequence of local *Equivalence Preserving Transformations* (EPTs). Let $S_1 \subset S_2$ be two scopes with corresponding cost functions c_{S_1} and c_{S_2}. Procedure MoveCost describes how a cost α moves between the corresponding cost functions. To see its correctness, observe if τ_1 is used in a complete assignment, then exactly one extension of τ_1 to S_2 will be used. Therefore, the sum of c_{S_1} and c_{S_2} remains unaffected whether the cost α is attributed to τ_1 in c_{S_1} or to all of its extensions τ_2 in c_{S_2}. As an example, it is clear that adding a cost α on $c_x(a)$ and subtracting a cost α on $c_{\{x,y\}}(\{x = a, y = b\})$ for all $b \in D(y)$ preserves problem equivalence. Indeed, paying α when we assign $x = a$ (cost function $c_x(a) = \alpha$) or when we assign $x = a$ and $y = b$ ($\forall b \in D(y)$) (cost function $c_{\{x,y\}}(\{x = a, y = b\}) = \alpha$, $\forall b \in D(y)$) is equivalent. As a matter of terminology, when $\alpha > 0$, cost moves from the larger arity cost function c_{S_2} to the smaller arity c_{S_1} and the move is called a projection, denoted $project(c_{S_1}, c_{S_2}, \tau_1, \alpha)$. When $\alpha < 0$, cost moves to the larger arity cost function c_{S_2} and the move is called an extension, denoted $extend(c_{S_1}, \tau_1, c_{S_2}, -\alpha)$, equivalent to MoveCost$(c_{S_1}, c_{S_2}, \tau_1, \alpha)$. When $S_1 = \emptyset$ and $|S_2| = 1$, with $S_2 = \{x_i\}$, the move is called a unary projection, denoted $unaryProject(c_i, \alpha)$, equivalent to MoveCost$(c_\emptyset, c_i, \emptyset, \alpha)$. We never perform extensions from $c_\emptyset$, so it monotonically increases during the run of an algorithm and as we descend a branch of the search tree.

Finding which cost moves lead to an optimal reparameterization, which means one that derives the optimal increase in the lower bound, is not obvious. It has been shown that any reparameterization can be derived by a set of local cost moves [29] and that the optimal reparameterization (with α rational) – and, equivalently, the optimal set of cost moves – can be found from the optimal dual solution of the following linear relaxation of the WCSP [12], whose feasible region is called the *local polytope*:

$$\min \sum_{c_S \in C, \tau \in \tau(S)} c_S(\tau) \times y_\tau$$

$$s.t.$$

$$y_{\tau_1} = \sum_{\tau_2 \in \tau(S_2), \tau_2[S_1] = \tau_1} y_{\tau_2} \qquad \forall c_{S_1}, c_{S_2} \in C, S_1 \subset S_2,$$

$$\tau_1 \in \tau(S_1), |S_1| \geq 1$$

$$\sum_{\tau \in \tau(S)} y_\tau = 1 \qquad \forall c_S \in C, |S| \geq 1$$

However, solving this LP to optimality is often prohibitively expensive because the worst-case complexity of an exact LP algorithm is $O(N^{2.5})$ [50], with $N \in O(ed + nd)$ for binary WCSPs, where e is the number of distinct binary cost functions, n is the number of WCSP variables and d is the maximum domain size. The poor asymptotic complexity matches empirical observation [25]. Moreover, the particular structure of this LP does not allow for a more efficient solving algorithm, as it has been shown that solving LPs of this form is as hard as solving any LPs [38]. Instead, work has focused on producing good but potentially suboptimal feasible dual solutions. Various algorithms have been proposed for this, going all the way back to Schlesinger [44], who first expressed the problem as linear optimization and gave a specific algorithm for optimizing the dual. Since Schlesinger, a long line of algorithms has been pursued both in areas like image analysis [29,30,45–47,51], where Block-Coordinate Ascent (BCA) algorithms were developed, and constraint programming [12,21,31,43,53], where they are called *soft local consistencies*. Notably, the strongest algorithms from both lines of research, such as TRWS [29] and VAC [12] converge on fixpoints with the same properties.

We do not describe all the existing local consistency algorithms but we need the following consistency properties:

Definition 2. *A WCSP P is Node Consistent (NC) [31] if for every variable $x_i \in X$ there exists a value $v \in D(x_i)$ such that $c_i(v) = 0$ and for every value $v' \in D(x_i)$, $c_\varnothing + c_i(v') < \top$.*

In the following, we assume that a WCSP is NC before our propagator runs.

Definition 3. *A WCSP P is ∅-Inverse Consistent (∅IC) [53] if for every cost function $c_S \in C$ there exists a tuple $\tau \in \tau(S)$ such that $c_S(\tau) = 0$.*

Definition 4. *A WCSP P is Existential Arc Consistent (EAC) [21] if it is NC and for every $x_i \in X$ there exists a value $v \in D(x_i)$ such that $c_i(v) = 0$ and for every cost function $c_S \in C$, $x_i \in S$, $|S| > 1$, there exists a tuple $\tau \in \tau(S)$ verifying $\tau[x_i] = \{v\}$ (i.e., $x_i = v$ in τ) and $c_S(\tau) + \sum_{x_j \in S} c_j(\tau[x_j]) = 0$. Value v is called an EAC support.*

This last definition applies only to binary cost function networks.[1] A weaker notion of EAC has been defined on global cost functions in order to avoid cost oscillation [33]. Given a variable x_i, it relies on a partition of the unary cost functions $c_j(\tau[x_j]), x_j \in X$ such that each part is associated to some non-unary cost function c_S related to x_i ($x_i \in S$).

We follow another weakening approach related to ∅IC. We strengthen the previous definition to take into account unary costs as in EAC.

Definition 5. *A WCSP is Full ∅-Inverse Consistent (F∅IC) if for every cost function $c_S \in C$ there exists $\tau \in \tau(S)$ such that $c_S(\tau) + \sum_{x_j \in S} c_j(\tau[x_j]) = 0$.*

Compared to existing notions of consistency, F∅IC is weaker than T-DAC [2]. It is also weaker than EAC on binary cost function networks, but it is incomparable with weak EAC [33] on non-binary networks.

Example 1. Consider two variables x, y with $D(x) = D(y) = \{a, b, c\}$ and three cost functions $c_x, c_y, c_{\{x,y\}}$ such that the only non-zero costs are $c_x(a) = c_y(a) = 1$, $c_{\{x,y\}}(\{x = b, y = b\}) = c_{\{x,y\}}(\{x = b, y = c\}) = c_{\{x,y\}}(\{x = c, y = b\}) = 1$, and $c_{\{x,y\}}(\{x = c, y = c\}) = 2$.

---

[1] An extension to ternary cost functions has been proposed [42] but it requires managing all scope intersections and not only unary cost functions.

288   P. Montalbano et al.

Let $\forall u \in \boldsymbol{D}(x), \alpha_u = \min_{v \in \boldsymbol{D}(y)}(c_{\{x,y\}}(\{x = u, y = v\}) + c_y(v))$ and $\forall v \in \boldsymbol{D}(y), \beta_v = \max_{u \in \boldsymbol{D}(x)}(\alpha_u - c_{\{x,y\}}(\{x = u, y = v\}))$. We apply $project(c_x, c_{\{x,y\}}, \{x = u\}, \alpha_u)$ for each value $u \in \boldsymbol{D}(x)$ and $extend(c_y, \{y = v\}, c_{\{x,y\}}, \beta_v)$ for each value $v \in \boldsymbol{D}(y)$. These cost moves will result in adding a cost $\beta_v - \alpha_u$ to every tuple in $c_{\{x,y\}}$. We have $\alpha_a = 0, \alpha_b = \alpha_c = 1$ and $\beta_a = 1, \beta_b = \beta_c = 0$. All the costs remain positive (proof in [32]). The reparameterized cost functions are $c_x(a) = c_x(b) = c_x(c) = 1$, $c_{\{x,y\}}(\{x = a, y = a\}) = c_{\{x,y\}}(\{x = c, y = c\}) = 1$, the rest being equal to 0. We can now increase $c_{\varnothing}$ by 1 using $unaryProject(c_x, 1)$. The resulting WCSP is EAC.

For each of the consistencies we defined above, there exist corresponding algorithms that compute parametrization that satisfy them in polynomial-time[2]. Given the connection to linear programming, these reparameterizations map to feasible dual solutions of the local polytope. However, these algorithms rely on all constraints being expressed in extension, meaning that for all constraints the cost of every partial assignment must be explicitly written. This is not the case for many constraints that are typically used in modeling in CP, namely global constraints, i.e., those whose definition does not imply a fixed arity. In order to enforce these soft local consistencies in instances that contain global constraints, we need to define bespoke algorithms. In contrast with crisp CSPs, these algorithms must do more than prune values that appear in no feasible solution. They must compute a reparameterization such that the constraint satisfies the appropriate consistency, FØIC here.

Here, we deal with pseudo-Boolean (PB) linear constraints and their generalizations. These are constraints of the form $\sum_{x_i \in \boldsymbol{S}} w_i x_i \triangle C$, where $\boldsymbol{S}$ is a scope, all $x_i \in \boldsymbol{S}$ are Boolean variables, w_i and C are constants and $\triangle \in \{<, \leq, \neq, \geq, >\}$. A PB constraint is normalized if w_i, $C \geq 0$ and $\triangle$ is $\geq$. Any PB constraint can be written as a combination of normalized PB constraints. It is possible to detect in linear time whether this constraint is satisfiable in a crisp CSP, by testing if $\sum_{x_i \in \boldsymbol{S}} \max(0, w_i) \geq C$. It is also possible to detect values that appear in no solutions by computing all partial sums of $|\boldsymbol{S}| - 1$ variables, in linear time.

A PB constraint is an at-most-one (AMO) constraint if it has the form $\sum_{x_i \in \boldsymbol{S}} x_i \leq 1$, normalized as $\sum_{x_i \in \boldsymbol{S}} -x_i \geq -1$. It is an exactly-one (EO) constraint if it has the form $\sum_{x_i \in \boldsymbol{S}} x_i = 1$.

4   Pseudo-Boolean Constraints in CFNs

The specific constraint we consider here is a pseudo-Boolean constraint $\sum_{x_i \in \boldsymbol{S}} w_i x_i \geq C$ along with a partition of its variables into sets $A_1, \dots, A_k$ such that there exists an EO constraint among the variables of each partition A_i.

Reformulations. This formulation allows us to express PB constraints over multi-valued variables. Let $\boldsymbol{S}$ be a scope over a set of WCSP variables with arbitrary domains, and w_{iv} weights for each value. The constraint

---

[2] E.g., EDAC [21], an extension of EAC property, is maintained in $O(ed^2 \max(nd, \top))$ for a WCSP with n variables, maximum domain size d, and e binary cost functions.

$\sum_{x_i \in S, v \in D(x_i)} w_{iv} x_{iv} \geq C$, where x_{iv} is the 0/1 variable which takes the value 1 if $x_i = v$, matches the pattern described above, with partitions $A_i = \{x_{iv} \mid v \in D(X_i)\}$.

Finally, this formulation admits the case where there exists an AMO constraint over some partitions: we add another 0/1 variable in each such partition and give it weight 0, so that this partition now has an EO constraint.

Constraint Representation. We will focus here on $F\emptyset IC$ as the soft consistency we aim to enforce. But first, we need an appropriate encoding that can represent the state of the constraint after a series of cost moves to and from unary cost functions, without storing a cost for each of the exponentially (in the arity of the constraint) many tuples. Observe first that the cost of any given tuple starts out at 0 for allowed tuples and $\top$ for tuples that violate the constraint. After some cost moves, the cost of each tuple is the sum of costs that have been moved to or from the values it contains. Therefore, it can be expressed as a linear function. Let δ_{iv} be the total cost that has been moved between the constraint and the corresponding unary cost and $\delta_\emptyset$ the cost we have moved from this constraint to $c_\emptyset$. Therefore, initially $\delta_\emptyset = 0$ and $\delta_{iv} = 0$ for all i, v. We use the following integer program as the representation of the constraint.

$$\min \sum_{x_i \in S, v \in D(x_i)} \delta_{iv} x_{iv} - \delta_\emptyset \tag{1}$$
$$\text{s.t.}$$
$$\sum_{x_i \in S, v \in D(x_i)} w_{iv} x_{iv} \geq C \tag{2}$$
$$\sum_{v \in D(x_i)} x_{iv} = 1, \qquad \forall x_i \in S \tag{3}$$
$$x_{iv} \in \{0,1\}, \qquad \forall x_i \in S, v \in D(x_i) \tag{4}$$

We call this $ILP_\emptyset$. The main property of $ILP_\emptyset$ is that the cost of any feasible complete assignment is equal to the cost of the corresponding tuple in c_S after any sequence of cost moves. Hence, $opt(ILP_\emptyset) > 0$, if and only if c_S is not $\emptyset IC$, and we can move some cost to $c_\emptyset$: $project(c_\emptyset, c_S, \emptyset, opt(ILP_\emptyset))$.

However, for the purposes of detecting violations of $F\emptyset IC$, it is not enough to look at the cost of tuples of the constraint, as we must also take unary costs into account. Therefore, while $ILP_\emptyset$ remains the representation of the constraint, the propagator considers the problem with the modified objective

$$\min \sum_{x_i \in S, v \in D(x_i)} (\delta_{iv} + c_i(v)) x_{iv} - \delta_\emptyset \tag{5}$$

Let this problem be $ILP_{F\emptyset}$. c_S is $F\emptyset IC$ if and only if $opt(ILP_{F\emptyset}) = 0$. In the following, we write $p_{iv} = \delta_{iv} + c_i(v)$ for compactness, when it does not matter how much of the coefficient came from δ_{iv} and how much came from $c_i(v)$. In contrast with $ILP_\emptyset$, if $opt(ILP_{F\emptyset}) > opt(ILP_\emptyset)$, we cannot move $opt(ILP_{F\emptyset})$ units of cost to $c_\emptyset$. Instead, we first have to move some cost from unary cost functions into the constraint before we can project it to $c_\varnothing$. In this case, the composition of p_{iv} from δ_{iv} and $c_i(v)$ is significant.

Unfortunately, $ILP_\emptyset$ and $ILP_{F\emptyset}$ have the knapsack problem as a special case, hence it is NP-hard to determine whether a PB constraint is $\emptyset IC$ or $F\emptyset IC$. Therefore, we detect only a subset of cases where the constraint is not $F\emptyset IC$ by relaxing the integrality constraint (4) into $0 \le x_{iv} \le 1$ and solving the resulting linear programs, called $LP_\emptyset$ and $LP_{F\emptyset}$, respectively. This forgoes the guarantee that $opt(LP_{F\emptyset}) = 0$ if and only if the constraint is $F\emptyset IC$, and satisfies only the 'only if' part. More simply, if $opt(LP_{F\emptyset}) > 0$ then the constraint is not $F\emptyset IC$, and similarly for $LP_\emptyset$ and $\emptyset IC$.

$LP_{F\emptyset}$ has a special structure. It is a Multiple-Choice Knapsack Problem (MCKP) [37], or a *knapsack problem with special ordered sets* [27]. These can be solved more efficiently than arbitrary LPs, a fact that we use in our propagator.

4.1   Solving the Knapsack LP

We obtain an optimal solution $\boldsymbol{x}^*$ of the primal $LP_{F\emptyset}$ by applying Pisinger's greedy algorithm [37]. This gives a $\boldsymbol{x}^*$ in time $O(N \log N)^3$, with $N = |\boldsymbol{x}^*|$, such that either $\boldsymbol{x}^*$ has no fractional value or it has exactly two fractional values. In the latter case, the WCSP variable $x_k \in \boldsymbol{S}$, verifying $\exists s, s' \in \boldsymbol{D}(x_k)$ such that $0 < x_{ks}^*, x_{ks'}^* < 1$, is called a *split class* and $x_{ks}, x_{ks'}$ are the *split variables*. We denote by $o = \sum_{x_i \in \boldsymbol{S}, v \in \boldsymbol{D}(x_i)} p_{iv} x_{iv}^* - \delta_\emptyset$, the optimal solution cost of $LP_{F\emptyset}$. Consider now the dual of $LP_{F\emptyset}$:

$$\max C \times y_{cc} + \sum_{x_i \in \boldsymbol{S}} y_i \tag{6}$$

s.t.

$$y_{cc} \times w_{iv} + y_i \le p_{iv} \ \forall x_i \in \boldsymbol{S}, v \in \boldsymbol{D}(x_i)$$

$$y_{cc} \ge 0$$

where y_{cc} is the dual variable corresponding to the capacity constraint and y_i corresponds to the EO constraint of x_i. From the optimal primal solution, it is easy to compute the optimal dual solution. Let x_k be the split class, $x_{ks}, x_{ks'}$ the split variables and for $i \ne k$, define the variable x_{is} as the variable used in the optimal solution, *i.e.*, $x_{is}^* = 1$.

$$y_{cc} = \frac{p_{ks} - p_{ks'}}{w_{ks} - w_{ks'}}$$

$$y_k = p_{ks} - y_{cc} \times w_{ks} = p_{ks'} - y_{cc} \times w_{ks'}$$

$$y_i = p_{is} - y_{cc} \times w_{is} \ \forall x_i \in \boldsymbol{S} \backslash \{x_k\}$$

From the dual solution $\mathbf{y}$, we compute the reduced cost $rc^{\mathbf{y}}(x_{iv})$ of every variable x_{iv}, i.e., the slack of the dual constraint that corresponds to x. When context makes it clear, we omit $\mathbf{y}$ and write $rc(x_{iv})$.

---

[3] The Dyer-Zemel algorithm [18,52] can compute a solution in O(N) time, but we have not yet implemented it.

The reduced cost of a variable x can be interpreted as the amount by which we must decrease the coefficient of x in the objective function in order to have $x > 0$ in the optimal solution. We explain later that this implies that we can project some cost to unary cost functions.

In the specific case of $LP_{F\emptyset}$, we have:

$$rc(x_{ks}) = rc(x_{ks'}) = 0$$
$$rc(x_{is}) = 0 \ \forall x_i \in S \setminus \{x_k\}$$
$$rc(x_{iv}) = p_{iv} - y_{cc} \times w_{iv} - y_i \ \forall x_i \in S, v \neq s$$

Observation 1. Consider the linear program $LP'_{F\emptyset}$ which is identical to $LP_{F\emptyset}$ but has $p'_{iv} = p_{iv} - rc(x_{iv})$. Then $opt(LP'_{F\emptyset}) = opt(LP_{F\emptyset})$.

Proof. The optimal solution x^* of $LP_{F\emptyset}$ has the same cost o in $LP_{F\emptyset}$ and $LP'_{F\emptyset}$, as the coefficients of the variables that are greater than 0 are unchanged. The optimal dual solution x^* remains feasible in $LP'_{F\emptyset}$, as the slack in the dual of $LP_{F\emptyset}$ matches exactly the reduction in the right-hand side. Moreover, as the dual objective did not change, it has the same cost and matches the primal cost, so $opt(LP'_{F\emptyset}) = o = opt(LP_{F\emptyset})$. □

Example 2. Consider the following problem:

$$\min 40x_{11} + 55x_{12} + 85x_{13} + 47x_{21} + 95x_{22}$$
$$s.t.$$
$$4x_{11} + 14x_{12} + 24x_{13} + 16x_{21} + 40x_{22} \geq 40$$
$$\sum_{v \in D(x_i)} x_{iv} = 1 \quad \forall x_i \in \{x_1, x_2\}$$
$$0 \leq x_{iv} \leq 1 \quad \forall x_i \in \{x_1, x_2\}, v \in D(x_i)$$

Pisinger's algorithm gives the optimal primal solution $x^* = \{0, 1, 0, \frac{7}{12}, \frac{5}{12}\}$ with cost $o = 55 + \frac{7}{12} \times 47 + \frac{5}{12} \times 95 = 122$.

We deduce the following dual optimal solution : $y_{cc} = 2, y_1 = 55 - 2 \times 14 = 27, y_2 = 47 - 2 \times 16 = 15$.

The following reduced costs are obtained : $rc(x_{12}) = rc(x_{21}) = rc(x_{22}) = 0$ and $rc(x_{11}) = 5$, $rc(x_{13}) = 10$, we deduce that replacing the previous objective function by the following one does not change the cost of the optimal solution:

$$\min 35x_{11} + 55x_{12} + 75x_{13} + 47x_{21} + 95x_{22}$$

We observe that the solution $x^* = \{0, 1, 0, \frac{7}{12}, \frac{5}{12}\}$ is still optimal.

4.2   Propagation

Given a PB constraint and the associated unary costs, it is possible to increase the lower bound by at least $opt(LP_{F\emptyset})$. Our goal is to extend as little cost as possible from the unary cost functions in order to make $opt(LP_\emptyset) = o = opt(LP_{F\emptyset})$ and then project o to $c_\emptyset$.

Procedure TransformPB(c_S, y_{cc}, y_i, o)

Data: c_S: PB constraint
Data: y_{cc}, y_i, o: optimal dual solution of $LP_{F\emptyset}$

1 **for** *all the variables* x_{iv} **do**
2 $\quad$ $c_i(v) \leftarrow c_i(v) - y_{cc} \times w_{iv} - y_i + \delta_{iv}$
3 $\quad$ $\delta_{iv} \leftarrow y_{cc} \times w_{iv} + y_i$

4 $c_\emptyset \leftarrow c_\emptyset + o$
5 $\delta_\emptyset \leftarrow \delta_\emptyset + o$

If we move $|c_i(v) - rc(x_{iv})|$ between the constraint and each unary cost function and value, then $opt(LP_\emptyset) = o$ and we can project o to $c_\emptyset$. Indeed we have $|c_i(v) - rc(x_{iv})| = |(y_{cc} \times w_{iv} + y_i) - \delta_{iv}|$, we thus obtain the EPTs performed by Procedure TransformPB.

Theorem 1. *Algorithm TransformPB preserves equivalence.*

Proof. Recall that $p_{iv} = c_i(v) + \delta_{iv}$ and that $rc(x_{iv}) \geq 0$. If $c_i(v) - rc(x_{iv}) \geq 0$ then the cost move is an extension of less than $c_i(v)$, it is valid. If $c_i(v) - rc(x_{iv}) < 0$ then the cost move is a projection, while the cost of any solution x' with $x'_{iv} = 1$ is at least $o - c_i(v) + rc(x_{iv})$. This operation is also valid.

Finally, to check that our sequence of EPTs justifies the increase of $c_\emptyset$ by o, we compute the optimum of $LP_\emptyset$. From Observation 1, $opt(LP_\emptyset) = o$, which means we can project o to $c_\emptyset$ and increase $\delta_\emptyset$ to bring $opt(LP_\emptyset) = opt(LP_{F\emptyset}) = 0$. $\qquad\square$

We can improve on this by observing that the integer optimum must be integral. Therefore, we can increase $c_\emptyset$ by $\lceil o \rceil$. In this case, it is also necessary to round up all cost moves. By rounding up, we can no longer rely on Observation 1, but it still holds that $opt(LP_\emptyset) = 0$. We also approach $\emptyset IC$ by verifying that for any value x_{ab} we have $\delta_{ab} + \min \sum_{x_i \in S \setminus x_a, v \in D(x_i)} (\delta_{iv} + c_i(v)) x_{iv} - \delta_\emptyset = 0$. If this is not the case, we can project a positive cost to $c_a(b)$.

Procedure Propagate is the entry point to the propagator. It enforces domain consistency on the PB constraint, then solves $LP_{F\emptyset}$. If there is more than one optimal solution we prefer the one minimizing the reduced cost of the EAC support of each variable. Finally, it uses Procedure TransformPB, to perform cost moves.

Theorem 2. *Procedure Propagate runs in $O(nd \log nd)$ time where n is the number of WCSP variables involved and d the maximum domain size.*

Proof. Pisinger's algorithm dominates the complexity, as it runs in $O(N \log N)$, where N is the number of LP variables. In our case, $N = nd$, so it takes $O(nd \log nd)$ time. Domain consistency on the linear inequality can be performed in linear time. Finally, Procedure TransformPB iterates once over all variables and values and performs constant time operations on each. Hence, the total complexity is $O(nd \log nd)$. $\qquad\square$

Procedure Propagate(c_S)

Data: c_S: PB constraint with EO partitions
1 DomainConsistency(c_S)
2 $(y_{cc}, y_i, o) = \text{DualSolve}(LP_{F\emptyset})$
3 TransformPB(c_S, y_{cc}, y_i, o)

Example 3. Returning to Example 2, where c_S is the PB constraint with EO partitions over two WCSP variables x_1 and x_2, we had the following reduced costs: $rc(x_{12}) = rc(x_{21}) = rc(x_{22}) = 0$, $rc(x_{11}) = 5$, $rc(x_{13}) = 10$, the optimal cost was 122. We deduce the following cost moves:

- $extend(c_1, \{x_1 = 1\}, c_S, 35)$
- $extend(c_1, \{x_1 = 2\}, c_S, 55)$
- $extend(c_1, \{x_1 = 3\}, c_S, 75)$

- $extend(c_2, \{x_2 = 1\}, c_S, 47)$
- $extend(c_2, \{x_2 = 2\}, c_S, 95)$
- $project(c_\emptyset, c_S, \emptyset, 122)$

It implies the resulting costs: $\delta_{11} = 35$, $\delta_{12} = 55$, $\delta_{13} = 75$, $\delta_{21} = 47$, $\delta_{22} = 95$, $\delta_\emptyset = 122$. The unary costs after these operations are $c_1(2) = c_2(1) = c_2(2) = 0$, $c_1(1) = 5$, $c_1(3) = 10$. If we construct the table of possible assignments of $LP_\emptyset$ obtained after the extensions, we can see that $c_S(\{x_1 = 1, x_2 = 2\}) = 8$, $c_S(\{x_1 = 2, x_2 = 2\}) = 28$, $c_S(\{x_1 = 3, x_2 = 1\}) = 0$, $c_S(\{x_1 = 3, x_2 = 2\}) = 48$, and all the other assignments don't satisfy the constraint. We observe that the optimal solution is 0, hence our extensions justify the increase of $c_\emptyset$.

Now assume that other EPTs outside the PB constraint have modified the unary costs: $c_1(1) \rightarrow c_1(1) + 16 = 21$, $c_1(2) \rightarrow c_1(2) + 30 = 30$, $c_1(3) \rightarrow c_1(3) - 9 = 1$. We want to compute a new lower bound for the PB constraint by solving $LP_{F\emptyset}$:

$$\min 56x_{11} + 85x_{12} + 76x_{13} + 47x_{21} + 95x_{22} - 122$$

$$s.t.$$

$$4x_{11} + 14x_{12} + 24x_{13} + 16x_{21} + 40x_{22} \geq 40$$

$$\sum_{v \in D(x_i)} x_{iv} = 1 \quad \forall x_i \in \{x_1, x_2\}$$

$$0 \leq x_{ij} \leq 1 \quad \forall x_i \in \{x_1, x_2\}, v \in D(x_i)$$

The optimal solution is $x^* = \{0, 0, 1, 1, 0\}$ and its cost is $o = 76 + 47 - 122 = 1$. We deduce the dual optimal solution $y_{cc} = 1$, $y_1 = 52$, $y_2 = 31$ with reduced costs $rc(x_{11}) = rc(x_{13}) = rc(x_{21}) = 0$ and $rc(x_{12}) = 19$, $rc(x_{22}) = 24$. We carry out the following cost moves: $extend(c_1, \{x_1 = 1\}, c_S, 21)$, $extend(c_1, \{x_1 = 2\}, c_S, 11)$, $extend(c_1, \{x_1 = 3\}, c_S, 1)$, $project(c_2, c_S, \{x_2 = 2\}, 24)$, $project(c_\emptyset, c_S, \emptyset, 1)$, with $\delta_{11} = 56$, $\delta_{12} = 66$, $\delta_{13} = 76$, $\delta_{21} = 47$, $\delta_{22} = 71$, $\delta_\emptyset = 123$.

5   Experimental Results

We implemented our approach in TOULBAR2, an exact WCSP solver in C++,[4] winner of past UAI-2008, 2014 competitions. TOULBAR2 default variable ordering heuristic is the weighted degree heuristic [8], in order to gain information from PB constraints, we adapted an explanation-based weighted degree for linear inequality presented by Hebrard and Siala [24]. For all the tests we imposed a time limit of 30 min (except for CPD with 1 h) on a single core of an Intel Xeon E5-2680 v3 at 2.50 GHz and 256 GB of RAM. We compared our PB propagator with other modeling approaches in protein design. We compared TOULBAR2 to state-of-the-art ILP solver CPLEX 20.1 on knapsack problems with conflict graphs. We also compared TOULBAR2 on pseudo-Boolean Competition 2016 (previously out of reach by TOULBAR2) but the results were not competitive with recent PB solvers (not reported here for the lack of space).

5.1   Sequence of Diverse Solutions for CPD

A protein is a chain of simple molecules called amino acids. This sequence determines how the protein will *fold* into a specific 3D shape. The Computational Protein Design (CPD) [4] problem consists of identifying the sequence of amino acids that should fold into a given 3D shape. This problem can be modeled as a CFN[5] with unary and binary cost functions representing the energy of the protein but the criteria only approximate the reality, thus producing a sequence of diverse solutions increases the chance of finding the correct real sequence of amino acids. Each time a solution is found, a Hamming distance constraint is added to the model to enforce the next solution to be different from the previous ones. This Hamming distance can be directly encoded as a PB linear constraint, in the form of Eq. (2), with EO partitions associated to domains (Eq. (3)). For each variable, a negative weight of -1 is associated to the value found in the last solution (other values having a zero-weight) and the weighted sum in Eq. (2) must be greater than or equal to $-(|X| - \zeta)$, where ζ corresponds to the required minimum Hamming distance.

This has been implemented in TOULBAR2 and compared to previous automata-based encoding approaches (ternary, hidden, and dual encodings from [40]) on 30 instances [48].[6] Selected instances have from 23 to 97 *residues*/variables with maximum domain size going from 48 to 194

---

[4] https://github.com/toulbar2/toulbar2 version 1.2.

[5] Other paradigms such as ILP or Max-SAT have been tested but the experimental results using their corresponding state-of-the-art solvers were inferior to the CFN approach using TOULBAR2 [1,4]. E.g., for CPD instance *1BK2.matrix.24p.17aa.usingEref_self_digit2* ($n = 24$, $d = 182$, $e = 300$), CPLEX 20.1 solves it in 42.84 s, TOULBAR2 in 0.37 s. ROUNDINGSAT [26] timed out after 10 h.

[6] http://genoweb.toulouse.inra.fr/~tschiex/CPD-AIJ/Last35-instances. We removed 5 instances (*1ENH.matrix.36p.17aa*, *1STN.matrix.120p.18aa*, *HHR.matrix.115p.19aa*, *1PGB.matrix.31p.17aa*, *2CI2.matrix.51p.18aa*) on which TOULBAR2 timed out after 9,000 s even without diversity constraints [1].

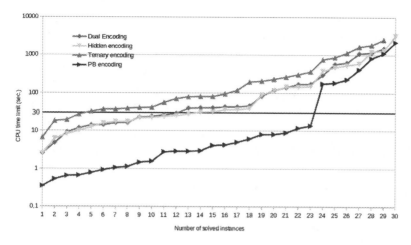

Fig. 1. Cactus plot of CPU solving time (log scale) for different encodings of Hamming distance constraints on CPD.

rotamers/values. The number of unary and binary cost functions goes from 276 to 4, 753. For each instance, the time limit was 1 h and the solver halts after finding a greedy sequence of 10 diverse solutions, the Hamming distance is $\zeta = 10$, and we enforce VAC on unary and binary cost functions in preprocessing (options -A -d: -a = 10 -div = 10 -divm = (0 for dual, 1 for hidden, 2 for ternary, and 3 for the PB encoding)). Figure 1 reports the solving time of each encoding. The dual and ternary encodings failed to give 10 diverse solutions for one instance, while the PB and hidden encodings didn't. Moreover the PB encoding is faster for 29 instances and it solves 23 of them in less than 30 s while dual, hidden, and ternary encodings solve respectively 12, 13, and 4 instances in less than 30 s. Note that we are computing a greedy sequence of solutions, the different encodings do not return the exact same sequence (except for 7/30 instances). We also compared for each instance the number of backtracks and time (not reported here) of the previous TOULBAR2 default encoding (dual) with the PB encoding. In all the instances the PB encoding needs fewer backtracks than the dual encoding and except for one instance, the PB encoding is also faster. Automata-based encodings have the flaw of introducing extra variables that can disturb the variable ordering heuristic (by default, *min domain size over weighted degree* [9])[7] and local consistency algorithm (by default, EDAC during search, except partial FØIC for PB constraints). While the PB encoding directly encodes the Hamming distance, it is heavier to propagate as we can see by comparing the number of backtracks per second (170 for PB encoding and 1060 for dual encoding).

---

[7] Additionally, the PB constraint provides finer-grain weights using explanations [24] when linear coefficients are not all equal as it is the case in the KPCG benchmark.

Table 1. Number of solved instances (left) and number of times a solver found the best solution within the time limit (right) for six different classes of KPCG.

	TB2	CPLEX$_t$	CPLEX$_d$		TB2	CPLEX$_t$	CPLEX$_d$
C1	718	689	**720**		**720**	701	**720**
C3	597	487	**614**		**718**	513	639
C10	**490**	318	457		**633**	346	547
R1	**720**	705	**720**		**720**	705	**720**
R3	**720**	573	682		**720**	589	691
R10	**571**	365	519		**665**	384	583

5.2 Knapsack Problem with a Conflict Graph

We compare here TOULBAR2 and CPLEX on Knapsack with Conflict Graph (KPCG) [6,11], a knapsack problem combined with binary constraints representing conflicts between pairs of variables. We use 6 different classes $C1,C3,C10,R1,$ $R3,R10$. In three of them the weight and the profit of each variable are correlated (class C) otherwise the profit is random between $[1, 100]$ (class R). The numbers $1, 3, 10$ correspond to a multiplying coefficient of the capacity, which has the effect of making the instances harder as the multiplier increases. In each class half of the instances have capacity 150, weights are uniformly distributed in $[20, 100]$, and the number of Boolean variables varies between $120, 250, 500$, and 1000. For the other half, the capacity is 1000, weights are uniformly distributed in $[250, 500]$, and the number of Boolean variables varies between $60, 120, 349,$ and 501. Additionally, the density of the conflict graph varies from 0.1 to 0.9. In total, each class has 720 instances. We used a direct encoding for TOULBAR2. For CPLEX, we tried with both tuple and direct encodings (tuple encoding corresponds to the local polytope with integer variables) [25]. Table 1 reports the number of instances solved by each solver. TOULBAR2 was more efficient than CPLEX with the tuple encoding and competitive with CPLEX using the direct encoding for four out of six classes. Moreover, TOULBAR2 finds the best solutions for the largest number of instances in every class.

6 Conclusion and Future Work

It is now possible to model pseudo-Boolean linear constraints in deterministic and probabilistic graphical models. This provides greater modeling flexibility and allows a WCSP solver like TOULBAR2 to solve more problems, such as computational protein design problems with diversity guarantee or knapsack problems with conflict graphs. One of the weaknesses of our approach is that the algorithm fundamentally produces a suboptimal solution to the linear program, because it propagates the pseudo-Boolean linear constraints one by one and does not take into account other constraints (except at-most-one constraints). There are several ways to improve this, including adapting work previously done in this context on Lagrangian relaxation [30] or an approach closer to VAC [16]. It also opens up possibilities for other uses of linear constraints in the WCSP framework, such as the generation of cuts.

References

1. Allouche, D., et al.: Cost function networks to solve large computational protein design problems. In: Masmoudi, M., Jarboui, B., Siarry, P. (eds.) Operations Research and Simulation in Healthcare, pp. 81–102. Springer, Cham (2021). https://doi.org/10.1007/978-3-030-45223-0_4

2. Allouche, D., et al.: Tractability-preserving transformations of global cost functions. Artif. Intell. **238**, 166–189 (2016)

3. Allouche, D., et al.: Filtering decomposable global cost functions. In: Proceedings of AAAI-12. Toronto, Canada (2012)

4. Allouche, D., et al.: Computational protein design as an optimization problem. Artif. Intell. **212**, 59–79 (2014)

5. Ansótegui, C., et al.: Automatic detection of at-most-one and exactly-one relations for improved SAT encodings of pseudo-boolean constraints. In: Schiex, T., de Givry, S. (eds.) CP 2019. LNCS, vol. 11802, pp. 20–36. Springer, Cham (2019). https://doi.org/10.1007/978-3-030-30048-7_2

6. Bettinelli, A., Cacchiani, V., Malaguti, E.: A branch-and-bound algorithm for the knapsack problem with conflict graph. INFORMS J. Comput. **29**(3), 457–473 (2017)

7. Bofill, M., Coll, J., Suy, J., Villaret, M.: An MDD-based SAT encoding for pseudo-boolean constraints with at-most-one relations. Artif. Intell. Rev. **53**(7), 5157–5188 (2020)

8. Boussemart, F., Hemery, F., Lecoutre, C., Sais, L.: Boosting systematic search by weighting constraints. In: ECAI, vol. 16, p. 146 (2004)

9. Boussemart, F., Hemery, F., Lecoutre, C., Sais, L.: Boosting systematic search by weighting constraints. In: Proceedings of ECAI-04, vol. 16, p. 146 (2004)

10. Cabon, B., de Givry, S., Lobjois, L., Schiex, T., Warners, J.: Radio link frequency assignment. Constraints **4**(1), 79–89 (1999)

11. Coniglio, S., Furini, F., San Segundo, P.: A new combinatorial branch-and-bound algorithm for the knapsack problem with conflicts. Eur. J. Oper. Res. **289**(2), 435–455 (2021)

12. Cooper, M.C., de Givry, S., Sánchez, M., Schiex, T., Zytnicki, M., Werner, T.: Soft arc consistency revisited. Artif. Intell. **174**(7–8), 449–478 (2010)

13. Cooper, M.C., de Givry, S., Schiex, T.: Graphical models: queries, complexity, algorithms (tutorial). In: Proceedings of 37th International Symposium on Theoretical Aspects of Computer Science (STACS-20). LIPIcs, vol. 154, pp. 4:1–4:22. Montpellier, France (2020)

14. Cooper, M.C., de Givry, S., Schiex, T.: Valued constraint satisfaction problems. In: Marquis, P., Papini, O., Prade, H. (eds.) A Guided Tour of Artificial Intelligence Research, pp. 185–207. Springer, Cham (2020). https://doi.org/10.1007/978-3-030-06167-8_7

15. Dechter, R., Rish, I.: Mini-buckets: a general scheme for bounded inference. J. ACM (JACM) **50**(2), 107–153 (2003)

16. Dlask, T., Werner, T.: Bounding linear programs by constraint propagation: application to max-SAT. In: Simonis, H. (ed.) CP 2020. LNCS, vol. 12333, pp. 177–193. Springer, Cham (2020). https://doi.org/10.1007/978-3-030-58475-7_11

17. Dlask, T., Werner, T.: On relation between constraint propagation and block-coordinate descent in linear programs. In: Simonis, H. (ed.) CP 2020. LNCS, vol. 12333, pp. 194–210. Springer, Cham (2020). https://doi.org/10.1007/978-3-030-58475-7_12

18. Dyer, M.E.: An $o(n)$ algorithm for the multiple-choice knapsack linear program. Math. Program. **29**(1), 57–63 (1984)
19. Elffers, J., Nordström, J.: Divide and conquer: towards faster pseudo-boolean solving. In: Proceedings of IJCAI, Stockholm, Sweden, pp. 1291–1299 (2018)
20. de Givry, S., Schiex, T., Verfaillie, G.: Exploiting tree decomposition and soft local consistency in weighted CSP. In: Proceedings of AAAI-06, Boston, MA (2006)
21. de Givry, S., Heras, F., Zytnicki, M., Larrosa, J.: Existential arc consistency: getting closer to full arc consistency in weighted CSPs. In: Proceedings of IJCAI-05, Edinburgh, Scotland, pp. 84–89 (2005)
22. de Givry, S., Katsirelos, G.: Clique cuts in weighted constraint satisfaction. In: Beck, J.C. (ed.) CP 2017. LNCS, vol. 10416, pp. 97–113. Springer, Cham (2017). https://doi.org/10.1007/978-3-319-66158-2_7
23. Haller, S., Swoboda, P., Savchynskyy, B.: Exact map-inference by confining combinatorial search with LP relaxation. In: Proceedings of AAAI-18, New Orleans, Louisiana, USA, pp. 6581–6588 (2018)
24. Hebrard, E., Siala, M.: Explanation-based weighted degree. In: Salvagnin, D., Lombardi, M. (eds.) CPAIOR 2017. LNCS, vol. 10335, pp. 167–175. Springer, Cham (2017). https://doi.org/10.1007/978-3-319-59776-8_13
25. Hurley, B., et al.: Multi-language evaluation of exact solvers in graphical model discrete optimization. Constraints **21**(3), 413–434 (2016). https://doi.org/10.1007/s10601-016-9245-y
26. Jo Devriendt, A.G., Nordström, J.: Learn to relax: integrating 0–1 integer linear programming with pseudo-boolean conflict-driven search. In: Proceedings of CP-AI-OR 2020, Vienna, Austria (2020)
27. Johnson, E.L., Padberg, M.W.: A note of the knapsack problem with special ordered sets. Oper. Res. Lett. **1**(1), 18–22 (1981)
28. Koller, D., Friedman, N.: Probabilistic Graphical Models: Principles and Techniques. MIT Press, Cambridge (2009)
29. Kolmogorov, V.: Convergent tree-reweighted message passing for energy minimization. IEEE Trans. Pattern Anal. Mach. Intell. **28**(10), 1568–1583 (2006)
30. Komodakis, N., Paragios, N., Tziritas, G.: MRF energy minimization and beyond via dual decomposition. IEEE Trans. Pattern Anal. Mach. Intell. **33**(3), 531–552 (2010)
31. Larrosa, J.: On arc and node consistency in weighted CSP. In: Proceedings of AAAI 2002, Edmondton, CA, pp. 48–53 (2002)
32. Larrosa, J., Schiex, T.: In the quest of the best form of local consistency for weighted CSP. In: Proceedings of IJCAI-03, vol. 3, pp. 239–244 (2003)
33. Lee, J.H.M., Leung, K.L.: Consistency techniques for flow-based projection-safe global cost functions in weighted constraint satisfaction. J. Artif. Intell. Res. **43**, 257–292 (2012)
34. Lee, J.H., Leung, K.L., Shum, Y.W.: Consistency techniques for polytime linear global cost functions in weighted constraint satisfaction. Constraints **19**(3), 270–308 (2014)
35. Lee, J.H., Leung, K.L., Wu, Y.: Polynomially decomposable global cost functions in weighted constraint satisfaction. In: Proceedings of AAAI-12, Toronto, Canada (2012)
36. Marinescu, R., Dechter, R.: AND/OR branch-and-bound for graphical models. In: Proceedings of IJCAI-05, Edinburgh, Scotland, pp. 224–229 (2005)
37. Pisinger, D., Toth, P.: Knapsack problems. In: Du, D.Z., Pardalos, P.M. (eds.) Handbook of Combinatorial Optimization, pp. 299–428. Springer, Boston (1998). https://doi.org/10.1007/978-1-4613-0303-9_5

38. Prusa, D., Werner, T.: Universality of the local marginal polytope. In: Proceedings of the IEEE Conference on Computer Vision and Pattern Recognition, pp. 1738–1743 (2013)
39. Ruffini, M., Vucinic, J., de Givry, S., Katsirelos, G., Barbe, S., Schiex, T.: Guaranteed diversity & quality for the weighted CSP. In: 2019 IEEE 31st International Conference on Tools with Artificial Intelligence (ICTAI), pp. 18–25. IEEE (2019)
40. Ruffini, M., Vucinic, J., de Givry, S., Katsirelos, G., Barbe, S., Schiex, T.: Guaranteed diversity and optimality in cost function network based computational protein design methods. Algorithms **4**(6), 168 (2021)
41. Sakai, M., Nabeshima, H.: Construction of an ROBDD for a PB-constraint in band form and related techniques for PB-solvers. IEICE Trans. Inf. Syst. **98**(6), 1121–1127 (2015)
42. Sánchez, M., de Givry, S., Schiex, T.: Mendelian error detection in complex pedigrees using weighted constraint satisfaction techniques. Constraints **13**(1), 130–154 (2008)
43. Schiex, T.: Arc consistency for soft constraints. In: Dechter, R. (ed.) CP 2000. LNCS, vol. 1894, pp. 411–425. Springer, Heidelberg (2000). https://doi.org/10.1007/3-540-45349-0_30
44. Schlesinger, M.: Sintaksicheskiy analiz dvumernykh zritelnikh signalov v usloviyakh pomekh (Syntactic analysis of two-dimensional visual signals in noisy conditions). Kibernetika **4**, 113–130 (1976)
45. Sontag, D., Choe, D., Li, Y.: Efficiently searching for frustrated cycles in MAP inference. In: Proceedings of UAI, Catalina Island, CA, USA, pp. 795–804 (2012)
46. Sontag, D., Meltzer, T., Globerson, A., Weiss, Y., Jaakkola, T.: Tightening LP relaxations for MAP using message-passing. In: Proceedings of UAI, Helsinki, Finland, pp. 503–510 (2008)
47. Tourani, S., Shekhovtsov, A., Rother, C., Savchynskyy, B.: Taxonomy of dual block-coordinate ascent methods for discrete energy minimization. In: Proceedings of AISTATS 2020, Palermo, Sicily, Italy, pp. 2775–2785 (2020)
48. Traoré, S., et al.: A new framework for computational protein design through cost function network optimization. Bioinformatics **29**(17), 2129–2136 (2013)
49. Trösser, F., de Givry, S., Katsirelos, G.: Relaxation-aware heuristics for exact optimization in graphical models. In: Hebrard, E., Musliu, N. (eds.) CPAIOR 2020. LNCS, vol. 12296, pp. 475–491. Springer, Cham (2020). https://doi.org/10.1007/978-3-030-58942-4_31
50. Vaidya, P.: Speeding-up linear programming using fast matrix multiplication. In: 30th Annual Symposium on Foundations of Computer Science, pp. 332–337 (1989)
51. Werner, T.: A linear programming approach to max-sum problem: a review. IEEE Trans. Pattern Recogn. Mach. Intell. **29**(7), 1165–1179 (2007)
52. Zemel, E.: An $o(n)$ algorithm for the linear multiple choice knapsack problem and related problems. Inf. Process. Lett. **18**(3), 123–128 (1984)
53. Zytnicki, M., Gaspin, C., de Givry, S., Schiex, T.: Bounds arc consistency for weighted CSPs. J. Artif. Intell. Res. **35**, 593–621 (2009)

A Learning Large Neighborhood Search for the Staff Rerostering Problem

Fabio F. Oberweger[1]([⊠]), Günther R. Raidl[1], Elina Rönnberg[2], and Marc Huber[1]

[1] Institute of Logic and Computation, TU Wien, Vienna, Austria
fabio.oberweger@gmail.com, {raidl,mhuber}@ac.tuwien.ac.at
[2] Department of Mathematics, Linköping University, Linköping, Sweden
elina.ronnberg@liu.se

Abstract. To effectively solve challenging staff rerostering problems, we propose to enhance a large neighborhood search (LNS) with a machine learning guided destroy operator. This operator uses a conditional generative model to identify variables that are promising to select and combines this with the use of a special sampling strategy to make the actual selection. Our model is based on a graph neural network (GNN) and takes a problem-specific graph representation as input. Imitation learning is applied to mimic a time-expensive approach that solves a mixed-integer program (MIP) for finding an optimal destroy set in each iteration. An additional GNN is employed to predict a suitable temperature for the destroy set sampling process. The repair operator is realized by solving a MIP. Our learning LNS outperforms directly solving a MIP with Gurobi and yields improvements compared to a well-performing LNS with a manually designed destroy operator, also when generalizing to schedules with various numbers of employees.

Keywords: Staff rerostering · Large neighborhood search · Imitation Learning · Machine Learning

1 Introduction

Large neighborhood search (LNS) [33,38] is a powerful meta-heuristics for solving combinatorial optimization problems (COP). It has been successfully applied to many complex problems, including vehicle routing [38], facility location [18], and project scheduling [28]. A common LNS design is to define a neighborhood search by a destroy and repair operator pair that is applied in each iteration [33]. The destroy operator partially destructs the incumbent solution by freeing

This work is part of the pre-study *Decision support for railway crew planning* supported by KAJT (Capacity in the Railway Traffic System). It is also partially funded by the Doctoral Program *Vienna Graduate School on Computational Optimization*, Austrian Science Foundation (FWF), grant W1260-N35, and the Center for Industrial Information Technology (CENIIT), Project-ID 16.05.

© Springer Nature Switzerland AG 2022
P. Schaus (Ed.): CPAIOR 2022, LNCS 13292, pp. 300–317, 2022.
https://doi.org/10.1007/978-3-031-08011-1_20

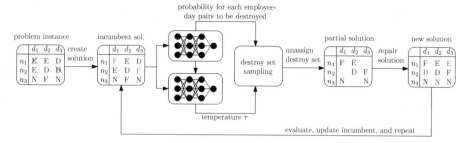

Fig. 1. Principle of the LNS applying trained ML models in the destroy operation.

a subset of the decision variables while fixing the others to their current values. A subproblem is hereby induced, and its space of feasible solutions forms a (large) neighborhood. By exactly or heuristically solving this subproblem, the repair operator tries to improve the previous solution by finding better assignments for these "destroyed" variables. If a new solution with an improved objective value is found, it becomes the new incumbent solution.

The destroy operator has a significant impact on the performance of an LNS. Frequently, a simple random selection of the variables is applied, which, however rarely gives the best results. Manually designing more effective destroy operators often is time-consuming and challenging. Recently, machine learning (ML) based techniques have been suggested to perform this task [2,14,39,40]. These techniques reduce or even eliminate the need for a manual design and have the potential to unveil connections that human experts might not see.

The staff rerostering problem (SRRP) is COP that deals with optimizing and reconstructing work schedules affected by disruptions, e.g., unplanned absences of employees or changes in demand for staff. Inspired by Sonnerat et al. [40], we propose an ML-based LNS consisting of a learning-based destroy operator and the use of a mixed-integer program (MIP) solver as repair method for heuristically solving challenging SRRP instances. Imitation learning is applied to train a conditional generative model predicting weights that indicate which elements of a solution are promising to destroy. Based on these weights, we propose a SRRP-specific sampling strategy for actually choosing the elements to destroy. Moreover, an additional ML model is used to obtain a suitable temperature parameter steering the sampling process. We employ graph neural networks (GNN) utilizing a SRRP-specific graph structure that enables efficient learning and inference for this highly constrained COP. Figure 1 shows our LNS scheme.

Experimental results show that our ML-based LNS outperforms both solving the respective MIP with Gurobi[1] and applying a well-performing LNS with a meaningful manually constructed destroy operator. While our approach is designed specifically for the SRRP, it may be generalized to other problems as well. The components requiring a problem-specific design are the graph structure for the GNN and the destroy set sampling process. Since these may be tailored

---

[1] https://www.gurobi.com.

to the needs of a particular problem, we believe that our approach might perform well on various types of COPs.

In Sect. 2, related work is reviewed and our approach is compared to existing ML-based LNS. The SRRP is introduced in Sect. 3 and the LNS framework is discussed in Sect. 4. Section 5 provides the details of our ML-based destroy operator. Experimental results are presented in Sect. 6 and concluding comments are given in Sect. 7. This work is based on the first author's master thesis [31].

2   Related Work

The SRRP is a generalization of the nurse rerostering problem (NRRP). Moz and Pato [25] were the first to formally define the NRRP, which deals with adapting an existing work schedule given employee absences on specific days. While the NRRP only considers employee-based disruptions, e.g., caused by illness, the SRRP also considers changes in demand. Already the NRRP with the single objective of minimizing the differences from the original schedule is NP-hard [27]. The SRRP and the NRRP are both extensions of the classical nurse rostering problem (NRP). For reviews of the NRP, we refer to Ernst et al. [8] and Van den Bergh et al. [5]. The NRRP is not as well studied as the NRP, but has been considered in several works [22,23,25–27,32,43].

Recently, many researchers have explored the application of ML in combinatorial optimization. ML techniques to learn heuristics for COPs in an end-to-end fashion [1,3,16,19,42] have been steadily improving. Nonetheless, they are usually still outperformed by state-of-the-art classical optimization methods, especially on problems with complex side constraints. For closing this performance gap, a new paradigm often referred to as "learning to search" [39] has evolved. This paradigm generally deals with the usage of NL-based heuristics within other methods. For example, learnable heuristics were used for guiding beam search [17,30], deciding on how to branch in branch-and-bound (B&B) algorithms [10,13,20,29], and learning destroy or repair operators in LNS [2,7,14,39,41].

Song et al. [39] proposed an ML-based destroy operator for a decomposition-based LNS solving general MIPs. In this LNS variant, one iteration consists of splitting all variables into disjoint sets and destroying and repairing each variable set one after the other. The authors use reinforcement learning (RL) and imitation learning to train a model performing the variable splits. To obtain data for the imitation learning, they randomly sample multiple decompositions for each solution and take the best. In contrast to the fixed variable subsets in the decomposition-based LNS, Addanki et al. [2] use RL to train a GNN for iteratively selecting one variable at a time until a destroy set of predetermined size is found. A current state in the optimization of a MIP is in [2] modeled by a graph structure called Constraint-Variable Incidence Graph (CVIG). A CVIG consists of one node for each variable and one node for each constraint, and edges represent the occurrence of variables in constraints. Sonnerat et al. [40] avoid the high overhead of one neural network (NN) inference step for each selection of a variable by training a conditional generative model with imitation learning. This model predicts a distribution over all nodes, which they use to sample a destroy set. The authors also employ a CVIG as input to their GNN. To generate training data, they use a local-branching [9] based mixed-integer programming approach that computes optimal destroy sets. All of the above approaches [2,39,40] rely on a MIP solver for repairing solutions.

We build on the approach of Sonnerat et al. [40] but modify it in the following ways. Since CVIGs are huge for the highly constrained SRRPs, they dramatically slow down training and inference already for instances of moderate sizes. Instead, we propose to use a more compact problem-specific graph that efficiently represents the choices to be made by the destroy operator, together with a generalized sampling strategy to choose the destroy set. Moreover, we employ an additional NN predicting suitable values of the temperature parameters for the destroy set sampling process.

3 Staff Rerostering Problem

The SRRP is defined on a set of employees N, a set of days D, and a set of shifts S, which we assume to be an early shift (7 am to 3 pm), the day shift (3 pm to 11 pm), the night shift (11 pm to 7 am), and the free shift modeling off-days. We treat the SRRP as an assignment problem, where each employee $n \in N$ shall be assigned to a shift $s \in S$ on each day $d \in D$. To represent a solution to an SRRP instance, we introduce decision variables $x_{nds} \in \{0,1\}$, where $x_{nds} = 1$ if and only if employee $n \in N$ is scheduled to work shift $s \in S$ on day $d \in D$. A solution is feasible if it satisfies to following hard constraints:

- Each employee must be assigned to exactly one working shift per day or the free shift.
- An employee has to rest at least 11 h after each working shift.
- An employee must not be assigned to less than a minimum or more than a maximum number of working shifts in the scheduling period.
- An employee must not be assigned to less than a minimum or more than a maximum number of consecutive working shifts.
- An employee must not have less than a minimum or more than a maximum number of assignments to a shift type in the scheduling period.
- An employee must not have less than a minimum or more than a maximum number of consecutive assignments to a shift type.
- Employees cannot be assigned to a working shift if they are absent for the time of this shift on this day.

Since the SRRP deals with disruptions to an existing schedule, such as absences of employees and changes in the demand for employees, the goal of the SRRP is to comply with the following soft constraints:

- The staffing requirements per day and shift should be met as well as possible.
- The original schedule should be modified as little as possible.

For a detailed formal definition of the SRRP, we refer to [31].

4 Large Neighborhood Search

We now define the repair operator used in each LNS applied in this work and propose a reasonable manually crafted destroy operator that serves as a baseline for comparison in our computational experiments. The construction heuristic (inspired by [35]) used to create an initial solution for the LNS simply takes the provided original schedule and, when an employee is absent on a working shift, changes the assignments to a free shift. The obtained initial solution will therefore typically be infeasible and this has to be considered in the LNS design.

4.1 Random Destroy Operator

Due to the constraints regulating the consecutive number of working shifts and the consecutive assignments per shift type, the repair operator has a greater chance to produce improvements if variables associated with consecutive days are unassigned in the destroy operator. Therefore, our baseline destroy operator randomly selects an employee $n \in N$ and a day $d \in D$ forming an employee-day pair (n, d). In addition, a period $P = \{\max(1, d - z_2), \ldots, d, \ldots, \min(|D|, d + z_2)\}$ is defined containing the days ranging from z_2 days before d to z_2 days after d, respecting that one is the index of the first and $|D|$ the index of the last day. Then, for the selected employee n and day d all variables $x_{nd's}$ for $d' \in P$ are destroyed. This process is repeated z_1 times for each application of the destroy operator; the selection is done without replacement. Both, $z_1 \in \mathbb{N}_0$ and $z_2 \in \mathbb{N}_0$ are fixed strategy parameters. The selection is done without replacement.

4.2 Repair Operator

The repair operator is to apply the Gurobi solver to a MIP representing an SRRP sub-instance induced by the destroy operator. Thus, a given partial solution is repaired by searching for best values for the unassigned variables while the other variables are considered fixed to their current values. As mentioned previously, the construction heuristic may return infeasible solutions. Moreover, the destroy operator might not select all the relevant variables required to turn the solution feasible with one repair operator application. As a consequence, we have to deal with infeasible solutions during the repair operation and the LNS in general. Therefore, an additional MIP is used, where a majority of the hard constraints are transformed into soft constraints. The constraints that are not relaxed are those that ensure that each employee is assigned to exactly one shift per day and that a working shift cannot be assigned to an absent employee. As a result, a solution to this relaxed model can violate the other hard constraints at the cost of additional penalization. The penalization is designed in such a way that infeasible solutions always have a worse objective values than feasible solutions. We refer to [31] for details on the MIP-formulations.

The repair operator uses the MIP with relaxed hard constraints as long as the incumbent solution before the destroy operation was infeasible. When the incumbent solution is feasible, the MIP with the regular hard constraints is employed. Thus, the two separate MIP in the repair operator put the focus effectively first on making an infeasible solution feasible and only then to further improve the objective value with respect to the remaining soft constraints.

5 Learning-Based Destroy Operator

The concept of our learning-based destroy operator is to utilize a NN that, given an SRRP instance and a current solution represented by features, returns weights to select promising employee-day pairs $(n, d) \in N \times D$ for which the respective decision variables x_{nds}, $s \in S$, are unassigned, i.e., "destroyed". Let π_θ represents the destroy set model, where θ is the learnable parameters. The model takes a featurized version of a state s^t at step t of an LNS run, consisting of an SRRP instance I and its current solution $x = (x_{nds})_{n \in N, d \in D, s \in S}$, as an input. The model π_θ outputs a value μ_{nd} for each $(n, d) \in N \times D$ indicating the probability that this employee-day pair is in an optimal destroy set, i.e., a destroy set that when realized yields, after an optimal repair, a solution with a minimum objective value. Note that we consider the size of the destroy set to be fixed to $z_1 \cdot (2z_2 + 1)$ employee-day pairs, just as in the random destroy operator from Sect. 4.1. More specifically, π_θ consists of two independently trained NNs: one handling states containing infeasible solutions and one dealing with states containing feasible solutions. This distinction is made as we observed that the behavior to learn can be quite different for feasible and infeasible solutions. Also, more training data is created for feasible solutions since the LNS spends more time in the feasible space. By considering two NNs, problems arising from the imbalance in training data are thus avoided. The only difference between the models in π_θ is the data used to train them. Hence, to improve readability, we will only refer to π_θ indicating the respective NNthroughout this whole section.

Another important aspect of our learning-based destroy operator is the temperature τ which is a parameter regulating the influence of π_θ's output in the destroy set sampling process. To choose a meaningful τ dynamically in dependence of the progress of the LNS, we use a second model π_ϕ^τ, which receives a state s^t and the output of π_θ as inputs and predicts a temperature τ for the current situation. Again, π_ϕ^τ consists of two independently trained NNs, one for infeasible and one for feasible solutions, and for the sake of readability, we will only refer to π_ϕ^τ.

The next step in our learning-based destroy operator is the destroy set sampling process. Here, we use π_θ's output and the predicted temperature τ to actually select the employee-day pairs to be unassigned in the current solution.

5.1 Markov Decision Process Formulation

A Markov decision process (MDP) [15] is represented by a 4-tuple consisting of the set of states ST, the set of actions A, a transition function T, and a reward function. In our setting, we define a state $s^t \in ST$ at step t of an episode to consist of an SRRP instance I and its current solution x^t. An action $a^t \in \{0, 1\}^{|N \times D|}$ at step t represents the selection of employee-day pairs to be destroyed. A positive assignment $a_{nd}^t = 1$ for an employee-day pair (n, d) indicates that it is chosen for the destroy set and, for all $s \in S$, the variables x_{nds}^t are unassigned in the current solution. Given a state s^t and an action a^t, the transition function $T : ST \times A \to ST$ determines the next state $s^{t+1} = T(s^t, a^t)$. Here, $T(s^t, a^t)$ is reached by destroying all variables associated with a_t in solution x^t and then repairing the partial solution with the repair operator. We do not define a reward function since it does not play a role in our method.

5.2 Destroy Set Prediction as Conditional Generative Modeling

Inspired by Nair et al. [29], we propose a conditional generative model representing the distribution of actions (i.e., destroy sets) in a current state. For step t in an LNS run, consider a state $s^t \in ST$, an action $a^t \in A$, and the transition function T. Moreover, let $c : ST \to \mathbb{R}$ represent a function returning the objective value of a current solution x^t of state $s^t \in ST$. Define the energy function

$$E(a^t; s^t) = \begin{cases} c(T(s^t, a^t)) & \text{if } c(T(s^t, a^t)) < c(s^t), \\ \infty & \text{otherwise,} \end{cases} \tag{1}$$

over the actions a^t of state s^t, which as in Nair et al. [29] defines the conditional distribution

$$\pi(a^t | s^t) = \frac{e^{-E(a^t; s^t)}}{\sum_{(a')^t} e^{-E((a')^t; s^t)}}. \tag{2}$$

Our learning efforts aim to approximate the conditional distribution in (2) utilizing a model $\pi_\theta(a^t | s^t)$ parameterized by θ. By using the unscaled energy function as presented in (1), destroy sets that leads to solutions with better (lower) objective values after being repaired get a higher probability. Furthermore, destroy sets not leading to improvements in objective value get zero probability. However, since our goal is to generate the best action in each state, we re-scale the energy function such that we assign probability $\pi((a^*)^t | s^t) = 1$ to an optimal action $(a^*)^t$ and a probability of zero to each other action in state s^t. Sonnerat et al. [40] adapted the conditional generative modeling design from Nair et al. [29] in the same way. As a consequence, we only have to consider training data containing optimal actions.

5.3 Sampling Destroy Sets

In Sect. 4.1, we introduced the baseline random destroy operator, which selects z_1 employee-day pairs $(n, d) \in N \times D$ and destroys all the variables associated with employee-day pairs within the range of z_2 days before to z_2 days after d. Remember that $z_1 \in \mathbb{N}_0$ and $z_2 \in \mathbb{N}_0$ are strategy parameters. Moreover, we described that selecting employee-day pairs without this range does not give good results since the SRRP contains constraints regarding consecutive working assignments. Although in our learning-based destroy operator, the NN π_θ outputs a value for each employee-day pair describing its probability to be in the destroy set, the same issues remain. Therefore, we propose a new destroy set sampling strategy, where blocks of consecutive employee-day pairs are selected. Each pair $(n, d) \in N \times D$ is assigned an aggregated weight

$$w_{nd} = \sum_{d'=\max(1,\, d-z_2)}^{\min(|D|,\, d+z_2)} (\pi_\theta(a_{nd'}^t = 1 \mid s^t) + \varepsilon)^{\frac{1}{\tau}} \cdot \mathbb{I}[(n, d') \in \mathcal{U}], \qquad (3)$$

where τ a temperature parameter to strengthen or weaken the influence of the NN, $\varepsilon > 0$ is an offset to give every pair a non-zero weight, and $\mathcal{U}$ is the destroy set selected so far. These weights can be interpreted as the sum of all NN outputs for employee n in a window of $2z_2 + 1$ consecutive days around day d, specifying the importance to add this range of employee-day pairs to the destroy set. We randomly select an employee-day pair (n, d) proportional to these weights w_{nd}, add pairs (n, d') for all $d' \in \{\max(1, d - z_2), \ldots, d, \ldots, \min(|D|, d + z_2)\}$ to $\mathcal{U}$, and repeat this process z_1 times. Eventually, for each $(n, d) \in \mathcal{U}$, we unassign variables $x_{nd's}^t$ for each shift $s \in S$ in the current solution.

5.4 Neural Networks

A GNN [12,36] is a NN architecture taking a graph as an input. It is typically independent of a specific graph size and able to represent underlying structural properties of a given graph by mapping it to a graph embedding expressed as vectors of values on the graph's nodes. We propose the following custom graph representation for the SRRP, which is not a reformulation of the SRRP as a graph problem but reflects a knowledge graph containing information for choosing suitable employee-day pairs. We define this graph as $G = (V, E, X)$, where V is the set of nodes, E the set of edges, and $X \in \mathbb{R}^{|V| \times p}$ a node feature matrix assigning each node $v \in V$ a p-dimensional feature vector $X_v = (X_{v,1} \ldots X_{v,p})$.

The set of nodes $V = V_{\text{emp}} \cup V_{\text{assign}} \cup V_{\text{day}}$ is composed of three different types which are employee $V_{\text{emp}} = \{n \mid n \in N\}$, assignment $V_{\text{assign}} = \{(n, d) \mid n \in N, d \in D\}$, and day $V_{\text{day}} = \{d \mid d \in D\}$ nodes. The assignment nodes represent the employee-day pairs. There are edges between an assignment node and its associated employee and day nodes. Moreover, since the days represent a sequence, we add edges between consecutive days.

There are thus $O(|N| \cdot |D|)$ nodes and edges, which is substantially less than in a CVIG for the underlying MIP (e.g., as used in [2]). A reasonably fast training and inference can therefore be expected. The interpretation of this representation is that an employee is involved in an assignment and this assignment takes place on a specific day in the planning horizon. A state $s^t \in ST$, consisting of an SRRP instance and its current solution x^t, holds all the required information to create such a graph structure. If we later use a state directly as an input to our NN, we implicitly assume that it is first transformed into such a graph.

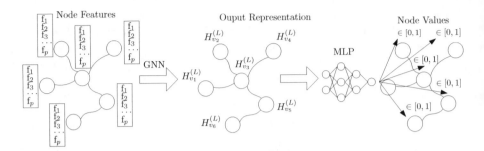

Fig. 2. Simplified representation of the NN architecture for the destroy set model π_θ; figure inspired by [6].

Naturally, we have different kinds of features with respect to employees, days, and assignments. For example, employee node features include the number of assignments to each shift $s \in S$ of an employee, day node features include the total number of assignments to each shift $s \in S$ on a given day, and assignment node features include the currently and originally assigned shift for an employee-day pair. For the complete list of the features used, we refer to [31]. To provide all these features to the GNN, one possibility would be to associate the individual features with the respective types of nodes in the graph representation and to apply the relational graph convolutional network (R-GCN) [37] for handling such an inhomogeneous graph with different feature types on different node types. However, similar to Chalumeau et al. [6], we use a simpler approach enabling us to employ more general GNNs for homogeneous graphs. Let $f_1^{emp}, \ldots, f_{q^{emp}}^{emp}$ be the employee features, $f_1^{assign}, \ldots, f_{q^{assign}}^{assign}$ the assignment features, and $f_1^{day}, \ldots, f_{q^{day}}^{day}$ the day node features, where $q^{emp}, q^{assign}, q^{day}$ are the number of employee, assignment, and day node features, respectively. Then each feature vector x_v of a node v, independent of the node type, is of the form

$$(f_1^{emp}, \ldots, f_{q^{emp}}^{emp}, f_1^{assign}, \ldots, f_{q^{assign}}^{assign}, f_1^{day}, \ldots, f_{q^{day}}^{day}, f_1^{enc}, f_2^{enc}, f_3^{enc}) \qquad (4)$$

where $f_1^{enc}, f_2^{enc}, f_3^{enc} \in \{0, 1\}$ is the mentioned one-hot encoding indicating whether the node is an employee, assignment, or day node, respectively. For example, if a node is an employee node, its feature vector contains the employee's values for $f_1^{emp}, \ldots, f_{q^{emp}}^{emp}$, zeros for the assignment and day features, and the associated one-hot encoding.

Architectures. So far, we have established the underlying graph structure. Figure 2 shows a simplified representation of the NN architecture of π_θ. First, we employ a GNN similar to the neural network for graphs from [24]. Our GNN updates the feature representation $H^{(l)}$ in a layer l by applying the update function

$$H^{(l)} = \sigma\left(H^{(l-1)}W_1^{(l)} + AH^{(l-1)}W_2^{(l)} + b^{(l)}\right), \qquad (5)$$

where $H^{(0)} = X$, A is the adjacency matrix, σ is a non-linear activation function, $b^{(l)} \in \mathbb{R}^q$ is the learnable bias, and $W_1^{(l)}, W_2^{(l)} \in \mathbb{R}^{m \times q}$ denote the weight matrices for layer l, where $m, q \in \mathbb{N}_0$ are the input and output feature dimensions, respectively. Applying the GNN to an input yields the node embedding $H^{(L)}$ of the graph, where $H_v^{(L)}$ is a vector for each node $v \in V$. These vectors from the GNN are further processed by a traditional MLP utilizing a sigmoid activation in the last layer to yield a value in $[0,1]$ for each node. We only require the final output for the assignment nodes.

To make the connection to our conditional generative modeling approach and the conditional distribution π from Eq. (2) in Sect. 5.2, let π_θ be our previously presented NN, where θ are all the learnable parameters, including the GNN and MLP weights. Remember that an action a_{nd}^t at step t indicates whether an employee-day pair $(n, d) \in N \times D$ is contained in the destroy set $(a_{nd}^t = 1)$ or not $(a_{nd}^t = 0)$. Also, remember that the sets V_{assign} and $N \times D$ are isomorphic, meaning that there is a one-to-one correspondence between each assignment node and employee day pair $(n, d) = v \in V_{\text{assign}}$. As Nair et al. [29] and Sonnerat et al. [40], we define π_θ to be a conditional-independent model of the form

$$\pi_\theta(a^t \mid s^t) = \prod_{(n,d) \in V_{\text{assign}}} \pi_\theta(a_{nd}^t \mid s^t), \tag{6}$$

which, given a state s^t, predicts the probability of an employee-day pair (n, d) being contained in an optimal destroy set independently of the other employee-day pairs. The probability $\pi_\theta(a_{nd}^t \mid s^t)$ is a Bernoulli distribution, and we compute its success probability μ_{nd} as

$$t_{nd} = \text{MLP}(H_{(n,d)}^{(L)}; \theta), \tag{7}$$

$$\mu_{nd} = \pi_\theta(a_{nd}^t = 1 \mid s^t) = \frac{1}{1 + e^{-t_{nd}}}. \tag{8}$$

As pointed out by Nair et al. [29], it is not possible to accurately model a multimodal distribution using the assumption of conditional independence. Despite that, they reported strong empirical results. Mathematically more accurate alternatives are autoregressive models [4] or inferring one employee-day pair at a time by repeatedly evaluating the NN. However, this increased accuracy comes at the cost of substantially slower inference times [29,40], which are not reasonable in our setting.

The architecture of the temperature NN π_ϕ^T is similar to the architecture of the destroy set model π_θ but there are some key differences. The temperature model π_ϕ^T shall predict how strongly π_θ shall influence the destroy set sampling process in a specific state s^t. Therefore, we add the output of π_θ as an additional feature to the node features of the graph representation. More specifically, we append the common node feature vector with $\mu_{nd} = \pi_\theta(a_{nd}^t = 1 \mid s^t)$ for assignment nodes $(n, d) = v \in V_{\text{assign}}$ and zero for every other node $v \in V_{\text{emp}} \cup V_{\text{day}}$. Another difference is that a read-out layer is applied on the final node representations $H_v^{\mathrm{T},(L)}$. This read-out layer aggregates the information over all nodes into a single vector by applying $\sum_{v \in V} H_v^{\mathrm{T},(L)}$. Finally, we employ an MLP with a softmax function in the last layer on this vector to return a probability for each temperature in a predefined set $\mathcal{T} = \{0.1, 0.2, 0.3, 0.4, 0.5, 0.6, 0.7, 0.8, 0.9, 1, 2, 3, 5\}$ to be the best selection. This classification-based approach turned out to be more effective in practice than a regression model.

5.5   Training

The training set $\mathcal{D}_{\text{train}} = \{(s_{(j)}^{1:T_j}, a_{(j)}^{1:T_j})\}_{j=1}^{M}$ for the destroy set model π_θ contains the data from M sampled trajectories of an expert strategy. For each such trajectory $j \in \{1, \ldots, M\}$ consisting of T_j steps, let $\{s_{(j)}^t\}_{t=1}^{T_j}$ be the states and $\{a_{(j)}^t\}_{t=1}^{T_j}$ be the corresponding expert actions in the form of optimal destroy sets. We learn the weights θ of our model π_θ by minimizing the loss function

$$\mathcal{L}(\theta) = -\sum_{j=1}^{M} \sum_{t=1}^{T_j} \log \pi_\theta(a_{(j)}^t \mid s_{(j)}^t), \tag{9}$$

which is the negative log likelihood of the expert actions.

The training set $\mathcal{D}_{\text{train}}^{\mathrm{T}} = \{(s_{(j)}^{1:T_j}, o_{(j)}^{1:T_j}, y_{(j)}^{1:T_j})\}_{j=1}^{M^{\mathrm{T}}}$ for the temperature model π_ϕ^{T} contains the outputs $\{o_{(j)}^t\}_{t=1}^{T_j}$ of π_θ in the respective states $\{s_{(j)}^t\}_{t=1}^{T_j}$ for each sampled trajectory $j = 1, \ldots, M^{\mathrm{T}}$. The associated labels $\{y_{(j)}^t\}_{t=1}^{T_j}$ consist of a one-hot encoding of the temperature found to be best by the expert for each time step t of a trajectory j. Eventually, we optimize the weights ϕ by minimizing the cross-entropy loss

$$\mathcal{L}^{\mathrm{T}}(\phi) = -\sum_{j=1}^{M^{\mathrm{T}}} \sum_{t=1}^{T_j} y_{(j)}^t \log \pi_\phi^{\mathrm{T}}(s_{(j)}^t, o_{(j)}^t). \tag{10}$$

We perform each training in mini-batches of size 32. In addition to $\mathcal{D}_{\text{train}}$ and $\mathcal{D}_{\text{train}}^{\mathrm{T}}$, we also create validation sets of the same form as the respective training sets containing about a fourth of the total generated trajectories. This data is hold out from $\mathcal{D}_{\text{train}}$ and $\mathcal{D}_{\text{train}}^{\mathrm{T}}$ to evaluate the progress on unseen data during training. Furthermore, we apply early stopping [11, p. 246] to avoid overfitting. As optimizer, we use ADAM [21] with a learning rate of 0.001 and an exponential decay rate of 0.9 for the first and 0.999 for the second momentum.

5.6   Training Data Generation

Our data generation process is inspired by the expert policy from Sonnerat et al. [40], which uses local branching [9] to create optimal destroy sets in a given state. In local branching, a constraint is added to the MIP that allows at most a certain number of decision variables to change compared to a given incumbent solution. In the following, we refer to the MIP extended with such a local branching constraints as extended or local branching-based MIP. If this extended MIP is solved to optimality in a current state s^t, an optimal destroy set can be derived by comparing the old solution x^t with the new solution x^{t+1} and collecting the variables with changed values. Since our destroy sets do not directly consist of decision variables but employee-day pairs, the local branching constraint is in our case

$$\sum_{(n,d,s)\in N\times D\times S:\, x^t_{nds}=0} x^{t+1}_{nds} + \sum_{(n,d,s)\in N\times D\times S:\, x^t_{nds}=1} (1 - x^{t+1}_{nds}) \le 2\eta, \quad (11)$$

which ensures that at most η employee-day pairs change. Note that one employee-day pair change always implies the change of two x_{nds} variables, since each employee must be assigned to exactly one shift, including the free shift, on each day. In the following, we refer to iteratively solving the local branching-based MIP and extracting the associated destroy set as our expert policy π^*.

To generate training samples for the destroy set model π_θ, we apply the Dataset Aggregation (DAGGER) algorithm which is an extension of classical behavior cloning [34]. In the first iteration, the expert policy is used to sample trajectories for training instances. Sampling a trajectory using a policy $\hat{\pi}$ means that in each state s^t at step t of an episode (LNS run), we store state s^t and the expert action $\pi^*(s^t)$ as a tuple in a dataset $\mathcal{D}$, use $\hat{\pi}$ to create a destroy set a^t, and move to the next state $s^{t+1} = T(s^t, a^t)$. Then, a model $\hat{\pi}_1$ is trained on the expert actions for all the encountered states in $\mathcal{D}$, to mimic the expert policy. In the second iteration, we use $\hat{\pi}_1$ to sample more trajectories and add more data to $\mathcal{D}$. For this learned policy $\hat{\pi}_1$, we apply the destroy set sampling strategy from Sect. 5.3 with temperature $\tau = 1$ to generate a destroy set a^t in a current state s^t. This process is in general iterated a certain number of times. Eventually, $\mathcal{D}$ is the final dataset which is used to train the destroy set model π_θ.

The idea behind this algorithm is that trained models may encounter very different states than the expert strategy and it is also important to train on those. Preliminary results showed in our case that DAGGER iterations slightly improve the result, although we ultimately decided to only perform two iterations as our expert policy is very time-consuming due to the iterative solving of the extended MIP. To further speed up the data generation, we took the following additional measures. First, we terminate the solving of the local branching-based MIP after a defined time limit, yielding a destroy set that is not guaranteed to be optimal. Second, we execute the trajectory sampling for each SRRP training instance in parallel. Lastly, we only create a training sample for every third visited state when sampling trajectories with a trained model if the solution is feasible. See [31] for more details on these refinements.

Concerning the temperature model π_ϕ^T, that steer the diversity of the randomized employee-day pair selection process by choosing a value of τ. A lower value increases the impact of π_θ, while a higher one decreases it. We may also interpret π_ϕ^T as an evaluator of π_θ. If the output of π_θ for a specific state s^t is highly promising/reliable, a lower temperature should be predicted for τ to increase the influence of π_θ. Vice versa, higher temperatures should be predicted if the current output of π_θ is not so likely to yield an improvement of the incumbent solution. To collect training data for π_ϕ^T, we produce multiple trajectories using π_θ, where each temperature $\tau \in \mathcal{T}$ is applied to sample three destroy sets in a state s^t at step t of an episode. Let τ_t^* be the temperature giving the best destroy sets on average in a state s^t. The best average performance is determined as follows: If the solution has been improved, we consider the objective value of the newly created solution, otherwise the objective value of the initially constructed solution is used. At each step t, we add s^t, $\pi_\theta(s^t)$, and τ_t^* to a dataset $\mathcal{D}^\mathrm{T}$. Then, we move to the next state $T(s^t, a_{\tau_t^*}^t)$ using an action sampled with τ_t^* and repeat this process until the termination of the LNS.

6   Experimental Evaluation

All algorithms were implemented in Julia[2] 1.6.1 and all MIPs were solved by Gurobi 9.1.0. The experiments were executed in single-threaded mode on a machine with an Intel Xeon E5-2640 processor with 2.40 GHz and a memory limit of 16 GB. Since it is essential to find high-quality solutions fast in practice, we worked with a time limit of 900 s to evaluate our optimization algorithms. We use %-gap, an optimality gap in percent, to evaluate computational performance. For an objective value o, the value of %-gap is $100 \cdot (o - lb)/o$, where lb is a lower bound for the optimal value obtained by solving the original MIP with Gurobi for three hours.

A time limit of five seconds is applied for solving the sub-MIP in a call of the repair operator of each LNS. For both the random and the learning-based destroy operator, the values $z_1 = 150$ and $z_2 = 2$ are used since they gave a good performance in preliminary experiments with the random destroy operator.

As baselines for our computational comparisons, we solved the MIP with Gurobi and used the LNS with the random destroy operator. In the following figures and tables, we will refer to these approaches as ILP and LNS_RND, respectively, while LNS_NN denotes the ML-based LNS. As the training data generation is time-consuming, we trained the NN of LNS_NN on data generated from a random single schedule with $|N| = 110$ employees and various sets of disruptions only. In total, we used 200 and 150 random sets of disruptions for this schedule to generate training trajectories for the destroy set and the temperature model, respectively. To accelerate the training data generation, a time limit of 30 minutes and $\eta = 375$ was used for solving the extended MIP.

In the following experiment, we aim to show the strong performance of LNS_NN and its ability to generalize to different schedules with different numbers of employees despite the rather restricted training data generation. For testing, we randomly created four schedules with 120, 130, 140, and 150 employees. For each of these schedules, 30 random sets of disruptions are sampled such that there is a total of 120 SRRP instances. The time horizon is four weeks, i.e., $|D| = 28$. The instance generation and training process are detailed in [31].

Table 1 presents the results including average %-gaps and respective standard deviations, average objective values, average numbers of performed iterations, grouped according to the instance size $|N|$. It is clear that LNS_NN performs

---
[2] https://julialang.org.

Table 1. Average results of the trained LNS_NN, the LNS_RND, and the ILP. For each number of employees $|N|$, there are 30 SRRP instances. Objective values are divided by 10^3 for improved readability.

	LNS_NN			LNS_RND			ILP	
$|N|$	%-gap	obj. val.	iter.	%-gap	obj. val.	iter.	%-gap	obj. val.
120	**3.42** ± 0.53	**926**	118.93	4.36 ± 0.57	935	127.27	34.18 ± 13.74	1,418
130	**4.45** ± 0.40	**1,079**	113.30	5.22 ± 0.51	1,088	128.10	28.51 ± 13.48	1,495
140	**8.04** ± 0.60	**1,314**	99.77	9.20 ± 0.58	1,331	105.43	31.31 ± 11.79	1,817
150	**5.74** ± 0.41	**1,427**	95.57	6.87 ± 0.46	1,446	100.97	26.71 ± 11.30	1,880

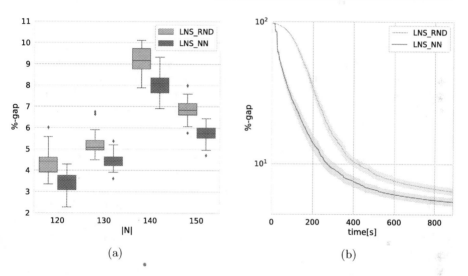

(a)   (b)

Fig. 3. (a) Boxplots for the %-gaps finally obtained by LNS_RND and LNS_NN in dependence of the numbers of employees $|N|$. (b) Solution quality development over time, aggregated over all runs/instances; note that %-gaps are here presented in log-scale.

better than LNS_RND for each of these instance classes even if it is trained only on the main schedule with $|N| = 110$. On average, LNS_NN improves the gap of LNS_RND by about one percent for each instance class. In total, LNS_RND only reached better solutions than LNS_NN for two of the 120 instances from this experiment. LNS_NN performs less iterations than LNS_RND. This decrease in iterations primarily comes from the fact that LNS_NN finds more meaningful destroy sets requiring more time to be repaired. These outcomes are even more substantial considering that already LNS_RND is a well-performing approach, which clearly surpassing the pure MIP solving. Considering optimality gap on average, LNS_NN outperforms ILP by factors between 3.89 and 9.99 across all instance groups.

Figure 3a visualizes the dominance of LNS_NN over LNS_RND and the significance of the differences with boxplots. Finally, Fig. 3b shows how the solution quality develops over time for LNS_NN and LNS_RND, aggregated over all runs for all benchmark instances. Most important to note is that in the beginning of the search, LNCS_NN finds improving solutions much faster than LNS_RND does. While LNS_NN reaches an optimality gap of less than 10% after 271 s on average, it takes LNS_RND 420 s. Also, LNS_NN finds feasible solutions after approximately 190 s and LNS_RND only after 310 s on average. Last but not least, remarkable are also the small standard deviations shown by the shaded areas, indicating the robustness of both LNS variants.

7   Conclusion and Future Work

We proposed a learning LNS for solving the SRRP, where the destroy operator is guided by a ML model trained upfront on representative problem instances with an imitation learning approach. More specifically, a conditional generative model predicts weights for all employee-day pairs, which are used in a randomized sampling strategy based on consecutive day selection to derive high-quality destroy sets. We use a custom graph structure modeling a current solution to the SRRP as an input to a GNN. Other so far proposed learning LNS approaches employ CVIG, resulting in prohibitively huge graphs for problems like the SRRP. In addition to the main NN predicting the employee-day pair weights, a second NN predicts a temperature value that controls the diversity of the destroy set sampling process. Experimental results clearly indicate the benefits of the learning LNS over a reasonably designed and already well working classical LNS in terms of solution quality and speed of solution improvement. Noteworthy also is the excellent generalization capability to unseen and even larger SRRP instances. The proposed learning LNS provides a scheme that may also be promising for other classes of highly constrained COPs.

In future work, training and testing of the approach on more diverse and even larger SRRP instances would be interesting. Investigating different variants of GNNs may lead to further improvements. Finally, finding ways to apply RL to optimize a performance metric directly instead of relying on the time-expensive imitation learning is a promising research direction.

References

1. Abe, K., Xu, Z., Sato, I., Sugiyama, M.: Solving NP-hard problems on graphs with extended AlphaGo zero. arXiv preprint arXiv:1905.11623 (2020)
2. Addanki, R., Nair, V., Alizadeh, M.: Neural large neighborhood search. In: Learning Meets Combinatorial Algorithms at Conference on Neural Information Processing Systems (2020)
3. Bello, I., Pham, H., Le, Q.V., Norouzi, M., Bengio, S.: Neural combinatorial optimization with reinforcement learning. In: Workshop Proceedings of the 5th International Conference on Learning Representations. OpenReview.net (2017)
4. Bengio, Y., Bengio, S.: Modeling high-dimensional discrete data with multi-layer neural networks. In: Advances in Neural Information Processing Systems, vol. 12, pp. 400–406. MIT Press (1999)

5. Van den Bergh, J., Beliën, J., De Bruecker, P., Demeulemeester, E., De Boeck, L.: Personnel scheduling: a literature review. Eur. J. Oper. Res. **226**, 367–385 (2013)
6. Chalumeau, F., Coulon, I., Cappart, Q., Rousseau, L.-M.: SeaPearl: a constraint programming solver guided by reinforcement learning. In: Stuckey, P.J. (ed.) CPAIOR 2021. LNCS, vol. 12735, pp. 392–409. Springer, Cham (2021). https://doi.org/10.1007/978-3-030-78230-6_25
7. Chen, M., Gao, L., Chen, Q., Liu, Z.: Dynamic partial removal: a neural network heuristic for large neighborhood search. arXiv preprint arXiv:2005.09330 (2020)
8. Ernst, A.T., Jiang, H., Krishnamoorthy, M., Sier, D.: Staff scheduling and rostering: a review of applications, methods and models. Eur. J. Oper. Res. **153**, 3–27 (2004)
9. Fischetti, M., Lodi, A.: Local branching. Math. Program. **98**, 23–47 (2003)
10. Gasse, M., Chetelat, D., Ferroni, N., Charlin, L., Lodi, A.: Exact combinatorial optimization with graph convolutional neural networks. In: Advances in Neural Information Processing Systems, vol. 32, pp. 15554–15566. Curran Associates, Inc. (2019)
11. Goodfellow, I., Bengio, Y., Courville, A.: Deep Learning. MIT Press, Cambridge (2016)
12. Gori, M., Monfardini, G., Scarselli, F.: A new model for learning in graph domains. In: 2005 Proceedings of the IEEE International Joint Conference on Neural Networks, vol. 2, pp. 729–734. IEEE (2005)
13. He, H., Daume, H., III., Eisner, J.M.: Learning to search in branch and bound algorithms. Adv. Neural. Inf. Process. Syst. **27**, 3293–3301 (2014)
14. Hottung, A., Tierney, K.: Neural large neighborhood search for the capacitated vehicle routing problem. In: Proceedings of the 24th European Conference on Artificial Intelligence. FAIA, vol. 325, pp. 443–450. IOS Press (2020)
15. Howard, R.A.: Dynamic Programming and Markov Processes. Wiley, Hoboken (1960)
16. Huang, J., Patwary, M., Diamos, G.: Coloring big graphs with AlphaGo zero. arXiv preprint arXiv:1902.10162 (2019)
17. Huber, M., Raidl, G.R.: Learning beam search: utilizing machine learning to guide beam search for solving combinatorial optimization problems. In: Machine Learning, Optimization, and Data Science - 7th International Conference, LOD 2021. LNCS, vol. 11943. Springer (2021, to appear)
18. Jatschka, T., Oberweger, F.F., Rodemann, T., Raidl, G.R.: Distributing battery swapping stations for electric scooters in an urban area. In: Olenev, N., Evtushenko, Y., Khachay, M., Malkova, V. (eds.) OPTIMA 2020. LNCS, vol. 12422, pp. 150–165. Springer, Cham (2020). https://doi.org/10.1007/978-3-030-62867-3_12
19. Khalil, E., Dai, H., Zhang, Y., Dilkina, B., Song, L.: Learning combinatorial optimization algorithms over graphs. In: Advances in Neural Information Processing Systems, vol. 30, pp. 6348–6358. Curran Associates, Inc. (2017)
20. Khalil, E.B., Bodic, P.L., Song, L., Nemhauser, G.L., Dilkina, B.N.: Learning to branch in mixed integer programming. In: Proceedings of the 30th AAAI Conference on Artificial Intelligence, pp. 724–731. AAAI Press (2016)
21. Kingma, D.P., Ba, J.: Adam: a method for stochastic optimization. In: Proceedings of the 3rd International Conference on Learning Representations (2015)
22. Maenhout, B., Vanhoucke, M.: An evolutionary approach for the nurse rerostering problem. Comput. Oper. Res. **38**, 1400–1411 (2011)
23. Maenhout, B., Vanhoucke, M.: Reconstructing nurse schedules: computational insights in the problem size parameters. Omega **41**, 903–918 (2013)

24. Micheli, A.: Neural network for graphs: a contextual constructive approach. IEEE Trans. Neural Netw. **20**, 498–511 (2009)
25. Moz, M., Pato, M.V.: An integer multicommodity flow model applied to the rerostering of nurse schedules. Ann. Oper. Res. **119**, 285–301 (2003)
26. Moz, M., Pato, M.V.: Solving the problem of rerostering nurse schedules with hard constraints: new multicommodity flow models. Ann. Oper. Res. **128**, 179–197 (2004)
27. Moz, M., Pato, M.V.: A genetic algorithm approach to a nurse rerostering problem. Comput. Oper. Res. **34**, 667–691 (2007)
28. Muller, L.F.: An adaptive large neighborhood search algorithm for the resource-constrained project scheduling problem. In: 2009 Proceedings of the VIII Metaheuristics International Conference (2009)
29. Nair, V., et al.: Solving mixed integer programs using neural networks. arXiv preprint arXiv:2012.13349 (2020)
30. Negrinho, R., Gormley, M.R., Gordon, G.J.: Learning beam search policies via imitation learning. In: Advances in Neural Information Processing Systems, vol. 31, pp. 10675–10684. Curran Associates Inc. (2018)
31. Oberweger, F.F.: A learning large neighborhood search for the staff rerostering problem. Diploma thesis, Institute of Logic and Computation, TU Wien, Austria (2021)
32. Pato, M.V., Moz, M.: Solving a bi-objective nurse rerostering problem by using a utopic pareto genetic heuristic. J. Heurist. **14**, 359–374 (2008)
33. Pisinger, D., Ropke, S.: Large neighborhood search. In: Gendreau, M., Potvin, J.Y. (eds.) Handbook of Metaheuristics. International Series in Operations Research & Management Science, vol. 146, pp. 399–419. Springer, Boston (2010). https://doi.org/10.1007/978-1-4419-1665-5_13
34. Pomerleau, D.A.: ALVINN: an autonomous land vehicle in a neural network. In: Advances in Neural Information Processing Systems, vol. 1, pp. 305–313. MIT Press (1988)
35. Rönnberg, E., Larsson, T., Bertilsson, A.: Automatic scheduling of nurses: what does it take in practice? In: Pardalos, P., Georgiev, P., Papajorgji, P., Neugaard, B. (eds.) Systems Analysis Tools for Better Healthcare Delivery. Springer Optimization and Its Applications, vol. 74, pp. 151–178. Springer, New York (2012). https://doi.org/10.1007/978-1-4614-5094-8_8
36. Scarselli, F., Gori, M., Tsoi, A.C., Hagenbuchner, M., Monfardini, G.: The graph neural network model. IEEE Trans. Neural Netw. **20**, 61–80 (2008)
37. Schlichtkrull, M., Kipf, T.N., Bloem, P., van den Berg, R., Titov, I., Welling, M.: Modeling relational data with graph convolutional networks. In: Gangemi, A., et al. (eds.) ESWC 2018. LNCS, vol. 10843, pp. 593–607. Springer, Cham (2018). https://doi.org/10.1007/978-3-319-93417-4_38
38. Shaw, P.: Using constraint programming and local search methods to solve vehicle routing problems. In: Maher, M., Puget, J.-F. (eds.) CP 1998. LNCS, vol. 1520, pp. 417–431. Springer, Heidelberg (1998). https://doi.org/10.1007/3-540-49481-2_30
39. Song, J., Lanka, R., Yue, Y., Dilkina, B.: A general large neighborhood search framework for solving integer linear programs. In: Advances in Neural Information Processing Systems, vol. 33, pp. 20012–20023. Curran Associates, Inc. (2020)
40. Sonnerat, N., Wang, P., Ktena, I., Bartunov, S., Nair, V.: Learning a large neighborhood search algorithm for mixed integer programs. arXiv preprint arXiv:2107.10201 (2021)

41. Syed, A.A., Akhnoukh, K., Kaltenhaeuser, B., Bogenberger, K.: Neural network based large neighborhood search algorithm for ride hailing services. In: Moura Oliveira, P., Novais, P., Reis, L.P. (eds.) EPIA 2019. LNCS (LNAI), vol. 11804, pp. 584–595. Springer, Cham (2019). https://doi.org/10.1007/978-3-030-30241-2_49

42. Vinyals, O., Fortunato, M., Jaitly, N.: Pointer networks. In: Advances in Neural Information Processing Systems, vol. 28, pp. 2692–2700. Curran Associates, Inc. (2015)

43. Wickert, T.I., Smet, P., Berghe, G.V.: The nurse rerostering problem: strategies for reconstructing disrupted schedules. Comput. Oper. Res. **104**, 319–337 (2019)

A MinCumulative Resource Constraint

Yanick Ouellet and Claude-Guy Quimper[✉]

Université Laval, Québec, Canada
yanick.ouellet.2@ulaval.ca, claude-guy.quimper@ift.ulaval.ca

Abstract. The CUMULATIVE constraint is the key to the success of Constraint Programming in solving scheduling problems with cumulative resources. It limits the maximum amount of a resource consumed by the tasks at any time point. However, there are few global constraints that ensure that a minimum amount of a resource is consumed at any time point. We introduce such a constraint, the MINCUMULATIVE. We show that filtering the constraint is NP-Hard and propose a checker and a filtering algorithm based on the fully elastic relaxation used for the CUMULATIVE constraint. We also show how to model MINCUMULATIVE using the SOFTCUMULATIVE constraint. We present experiments comparing the different methods to solve MINCUMULATIVE using Constraint Programming.

1 Introduction

Businesses and organizations often face scheduling problems where they must schedule tasks while satisfying various resources constraints. In the Constraint Programming community, with the CUMULATIVE constraint [1], significant work [3, 16, 17] has been devoted to scheduling problems where the resource usage of the tasks must not exceed the capacity of the resource. The reverse case, where a resource must have a minimum usage did not receive as much attention. However, many businesses and organizations need to solve scheduling problems where they need to ensure that a sufficient number of employees are working at any given time.

We introduce MINCUMULATIVE, a new global constraint that enforces that a minimum amount of the resource is used at any time. We show that applying domain or bounds consistency for this new constraint is NP-Hard. We propose a relaxed rule, the UnderloadCheck, to detect failures and a filtering algorithm related to this rule. We also show how to model the MINCUMULATIVE using the SOFTCUMULATIVE constraint, a soft version of the CUMULATIVE. This allows us to use the strong energetic reasoning rules of the CUMULATIVE.

We present relevant background in Sect. 2 and introduce the MINCUMULATIVE in Sect. 3. We present the UnderloadCheck rule and checker algorithm in Sect. 4 while Sect. 5 introduces the filtering algorithm. We show how to model the MINCUMULATIVE constraint using the SOFTCUMULATIVE constraint in Sect. 6. We compare the difference in strength between the different approaches in Sect. 7. Finally, experimental results are shown in Sect. 8 and Sect. 9 concludes the work.

© Springer Nature Switzerland AG 2022
P. Schaus (Ed.): CPAIOR 2022, LNCS 13292, pp. 318–334, 2022.
https://doi.org/10.1007/978-3-031-08011-1_21

2   Constraint Scheduling Background

A scheduling problem consists in scheduling a set $\mathcal{I}$ of n tasks over the time points $\mathcal{T} = 0..\text{hor} - 1$. Each task $i \in \mathcal{I}$ needs to execute *without preemption* for p_i units of *processing time* between its *earliest starting time* est_i and its *latest completion time* lct_i and consumes h_i units of a renewable resource, which are called height. A task i can be described using the tuple $\langle i, \text{est}_i, \text{lct}_i, p_i, h_i \rangle$. One can compute the *latest starting time* $\text{lst}_i = \text{lct}_i - p_i$ and the *earliest completion time* $\text{ect}_i = \text{est}_i + p_i$ of task i. We say that task i has a *compulsory part* in the interval $[\text{lst}_i, \text{ect}_i)$ if $\text{lst}_i < \text{ect}_i$. A task is necessarily executing during its compulsory part, regardless of its starting time.

One can model a scheduling problem using a starting time variable S_i for each task i with domain $\text{dom}(S_i) = [\text{est}_i, \text{lst}_i]$. Additional constraints can be added depending on the particularity of the problem (e.g., constraints limiting the amount of resource used, predecessor constraints, etc.). We say that task i is *fixed* if there is only one value in $\text{dom}(S_i)$. Thus we have $S_i = \text{est}_i = \text{lst}_i$.

Significant efforts have been made in the constraint programming community to efficiently solve scheduling problems for which the capacity of the resource is limited. The CUMULATIVE constraint [1] enforces that, at any time t, the sum of the heights of the tasks in execution does not exceed the capacity C of the resource. Deciding whether the CUMULATIVE constraint admits a solution is strongly NP-Complete [1]. Hence, checker and filtering algorithms for this constraint cannot apply domain or bounds consistency and must instead rely on rules to partially detect failures or partially filter the domains. Many such rules have been developed over the years, including the Overload Check [9,27], the Time Tabling [3], the Edge Finding [14,17,26] and the Energetic Reasoning [2,5,16,19,25]. By using lazy clause generations [10,18] with the CUMULATIVE constraint, Schutt et al. [23,24] closed many instances of hard scheduling problems.

Baptiste et al. [2] introduced a relaxation used in many rules for the CUMULATIVE constraint: the fully elastic relaxation. This relaxation allows the energy $e_i = p_i \cdot h_i$ of a task to be spent anywhere in the interval $[\text{est}_i, \text{lct}_i)$, regardless of the task's height or its non-preemption. For instance, on a resource of capacity 2, a task with $\text{est}_i = 0$, $\text{lct}_i = 4$, $p_i = 4$, and $h_i = 1$ could execute using 2 units of the resource at time 0, one unit at time 2 and one unit at time 3 for a total of $p_i \cdot h_i = 2 \cdot 2 = 4$ units of energy.

One of the strongest rules using the fully elastic relaxation is the energetic reasoning [2,16]. This rule, as the name suggests, is based on the notion of minimum energy in an interval $[l, u]$. The *left shift* of a task i in the interval, noted $\text{LS}(i, l, u) = h_i \cdot \max(0, \min(u, \text{ect}_i) - \max(l, \text{est}_i))$, is the amount of energy the task consumes in the interval when it is scheduled at its earliest. Conversely, the *right shift* of task i in interval $[l, u)$, noted $\text{RS}(i, l, u) = h_i \cdot \max(0, \min(u, \text{lct}_i) - \max(l, \text{lst}_i))$, is the amount of energy in the interval when the task is scheduled at its latest. The *minimum intersection* $\text{MI}(i, l, u) = \min(\text{LS}(i, l, u), \text{RS}(i, l, u))$ is the minimum between the left and right shift. It represents the minimum amount of energy that the task consumes in the interval regardless of when it starts. The sum of the minimum intersection of all tasks, noted $\text{MI}(\mathcal{I}, l, u) = \sum_{i \in \mathcal{I}} \text{MI}(i, l, u)$, is a fully elastic lower bound on the amount of energy consumed in an interval. The energetic reasoning detection rule (1) states that if there exists an interval such that the minimum intersection is greater than the energy available on the resource, i.e. $C \cdot (u - l)$, the CUMULATIVE constraint cannot be satisfied. Baptiste et al. [2] showed that it is sufficient to consider a subset of $O(n^2)$ intervals said *of interest* to apply the rule.

$$\exists [l, u) \mid \text{MI}(\mathcal{I}, l, u) > C \cdot (u - l) \implies \texttt{fail} \tag{1}$$

Many industrial problems require that a resource has a minimum usage instead of, or in addition to, a maximum. This is the case for the shifts scheduling [8] and the nurse rostering [4] problems. Both problems consist in scheduling the shifts of employees or nurses to satisfy a demand while minimizing the cost of exceeding it.

2.1 Global Cardinality Constraint

The Global Cardinality Constraint [13, 15, 22] $\text{GCC}([X_1, \ldots, X_n], [v_1, \ldots, v_m], [l_1, \ldots, l_m], [u_1, \ldots, u_m])$ ensures that each value v_i is assigned to at least l_i and at most u_i variables in X. If all tasks have a processing time $p_i = 1$ and share the same height, the GCC can be used to model the minimum usage of a resource. The variable X_i represents the starting time of the task i. The values are the time points. When the parameters u_j are set to infinity, the GCC forces each time point to be assigned a minimum number of times. However, there is currently no global constraint to handle the general case where tasks have distinct processing times or distinct heights.

2.2 Generalized Cumulative

Beldiceanu and Carlsson [3] introduced a generalization of the CUMULATIVE constraint, presented in (2). The GENERALIZEDCUMULATIVE constraint supports tasks with negative heights and an operator op $\in \{\leq, \geq\}$ that allows the capacity of the resource to be either a maximum that must not be exceeded or a minimum that must be reached. One can use this generalization to model a problem where tasks must meet a demand vector $d = [d_0, \ldots, d_{\text{hor}-1}]$. The demand d_t is the minimum amount of energy that must be spent by the tasks at time t. We show how this can be done in Sect. 3. We describe the portion of their algorithm that ensures that a minimum is reached.

$$\text{GENERALIZEDCUMULATIVE}(S, p, h, C, \text{op}) \overset{\text{def}}{\iff}$$
$$\sum_t \sum_i h_i \cdot \texttt{boolToInt}(S_i \leq t < S_i + p_i) \text{ op } C \tag{2}$$

To filter the GENERALIZEDCUMULATIVE constraint, Beldiceanu and Carlsson proposed a sweep algorithm that performs a Time-Tabling reasoning. The main idea behind the algorithm is to compute an upper bound of the resource usage at each time point and check whether that upper bound satisfies the minimum capacity.

Let $\text{GOOD}(t) = \{i \in \mathcal{I} \mid h_i > 0 \land \text{est}_i \leq t < \text{lct}_i\}$ be the set of tasks with positive height that *can* execute at time point t. These tasks increase the usage of the resource and thus can contribute to meet the minimum capacity. Let $\text{BAD}(t) = \{i \in \mathcal{I} \mid h_i < 0 \land \text{lst}_i \leq t < \text{ect}_i\}$ be the set of tasks with a negative height that have a compulsory part at time point t and therefore *must* execute at that time point. These tasks decrease the usage and thus make the minimum capacity harder to reach. The checker rule for Beldiceanu and Carlsson's algorithm is presented in (3). Recall that C is the constant representing the capacity of the resource.

$$\exists t \quad \sum_{i \in \text{GOOD}(t)} h_i + \sum_{i \in \text{BAD}(t)} h_i < C \implies \text{Failure} \tag{3}$$

The optimistic scenario occurs where, at a time point t, all $\text{GOOD}(t)$ tasks that can execute do so, and only the $\text{BAD}(t)$ tasks that *must* execute due to their compulsory parts do so. This means that if, at any time point, the sum is not enough to satisfy the capacity, the constraint cannot be satisfied. Note that this rule does not take the processing time of the tasks into account.

The filtering rule (4) filters GOOD tasks based on this idea. Consider a task j and a time point t. If the lower bound on the resource usage at time point t is insufficient to meet the minimum capacity without the contribution of task j, then task j must be executing at time t. The domain of S_j is changed for $\text{dom}(S_j) \cap [t - p_i + 1, t]$.

$$\forall t \, \forall j \in \text{GOOD}(t) \quad \sum_{i \in \text{GOOD}(t)} h_i + \sum_{i \in \text{BAD}(t)} h_i - h_j < C \implies \text{lst}_j \leq t < \text{ect}_j \tag{4}$$

2.3   SoftCumulative

De Clerc et al. [7] and Ouellet and Quimper [20] proposed checker and filtering algorithms for the SOFTCUMULATIVE constraint, a version of the CUMULATIVE constraint where it is possible to overload the resource but at a cost. The definition of SOFTCUMULATIVE (5) generalizes the CUMULATIVE constraint by adding the *overcost* variable Z, an upper bound on the cost incurred by overloading the resource. Note that it is not an equality. Generally, the cost is either minimized or subject to another constraint.

Ouellet and Quimper [20] introduced a generic cost function, but, for the sake of simplicity, we assume the cost function is linear in this paper. That is, overloading the resource by x units of resources always costs x units.

$$\text{SOFTCUMULATIVE}(S, p, h, C, Z) \overset{\text{def}}{\iff}$$
$$Z \geq \sum_t \max(0, \sum_i h_i \cdot \text{boolToInt}(S_i \leq t < S_i + p_i) - C) \tag{5}$$

Ouellet and Quimper adapted the energetic reasoning for the SOFTCUMULATIVE constraint. Instead of searching for one interval with an overload, their algorithm searches for a partition of the time line into contiguous intervals such that the sum of the overload in each interval is maximized. That sum is a lower bound on the overcost. The partition can be computed in $O(|T|^2)$ steps using dynamic programming, where T is the set of time points to partition. If the set T corresponds to the lower and upper bounds of the intervals of interest, there is $O(n^2)$ time points and thus, computing the partition is in $O(n^4)$. This is not reasonable for an algorithm that is called thousands of times during the search. However, Ouellet and Quimper proposed to consider a set with only $4n$ time points, which correspond to the est, ect, lst, and lct of the tasks. By doing so, the algorithm enforces a weaker filtering, but the complexity is better. Nevertheless, they showed that, when used on the hard version of the CUMULATIVE constraint (with $Z = 0$), the algorithm using only $4n$ time points applies the Time-Tabling and the Edge-Finding rules of the CUMULATIVE constraint.

3   Min-Cumulative

We introduce the MINCUMULATIVE constraint that ensures that the sum of the heights of the tasks in execution at each time point t is greater than or equal to the demand d_t.

$$\text{MINCUMULATIVE}([S_1,\ldots,S_n], [p_1,\ldots,p_n], \\ [h_1,\ldots,h_n], [d_0,\ldots,d_{\text{hor}-1}]) \iff \forall t \in \mathcal{T} \sum_{i \in \mathcal{I}: S_i \leq t < S_i + p_i} h_i \geq d_t$$

(6)

This constraint is a special case of the GENERALIZEDCUMULATIVE constraint presented by Beldiceanu and Carlsson, but a specialized version allows the design of more specialized and stronger checker and filtering algorithms. One can model the MIN-CUMULATIVE with GENERALIZEDCUMULATIVE by adding fixed tasks of negative heights to the problem to represent the demand. For each interval $[l, u)$ of maximal length such that $d_t = d_{t+1} \; \forall t \in \{l..u-2\}$, we created a fixed task i with $\text{est}_i = l$, $\text{lct}_i = u$, $p_i = u - l$, and $h_i = -d_l$. For instance, a demand of $[1, 1, 2, 2]$ in MINCU-MULATIVE can be represented by two tasks with $\text{est}_1 = 0$, $\text{lct}_1 = \text{est}_2 = 2$, $\text{lct}_2 = 4$, $p_1 = p_2 = 2$, $h_1 = -1$, and $h_2 = -2$. By fixing the capacity $C = 0$ and using operator $\text{op} = \geq$, the GENERALIZEDCUMULATIVE encodes the MINCUMULATIVE.

We can generalize the constraint where processing times and heights are variables rather than constants. The algorithms we present can substitute the height and the processing time by the maximum value in the domain of these variables.

Theorem 1. *Deciding whether the* MINCUMULATIVE *constraint admits a solution is NP-Complete even when domains are intervals.*

Proof. Deciding whether MINCUMULATIVE is feasible is a special case of the strongly NP-Complete unrestricted-output 3-Partition problem (3-Part-UO) [12]. Recall that 3-Part-UO is the problem of deciding whether it is possible to partition the multiset R containing $n = 3$ m integers with a total sum of $m \cdot T$ into m multisets, each of sum T. For each integer R_i, we declare a task with $\text{dom}(S_i) = [1, m]$, $p_i = 1$, $h_i = R_i$. The demand is T at each of the m time points. The starting times of the tasks correspond to the index of the set in which the matching integer is partitioned. The sum of the heights of all tasks is $m \cdot T$ and there is m time points of demand T. Hence, each of the m time points must have a usage of exactly T and corresponds to one of the m multisets.   □

Since deciding the feasibility of MINCUMULATIVE is NP-Complete even when domains are intervals, propagators cannot apply either bounds or domain consistency and must instead rely on relaxed rules to detect failures or filter variables.

To enforce the MINCUMULATIVE, it is sufficient to use a decomposition with summations and inequality constraints, as in (6). This method has the disadvantage of requiring a number of constraints and variables that is function of the horizon. In problems with a large horizon, this solution may not scale. We propose solutions that scale with the number of time points and that filters more than the decomposition.

4   Underload Check

We introduce the *underload check* rule that detects when MINCUMULATIVE is unfeasible. This new rule is inspired from the overload check [27] and the fully elastic relaxation. As with the overload check, the underload check is not sufficient to enforce the MINCUMULATIVE. It needs to be paired with another rule, such as the time-tabling.

The *underload detection rule* is based on the rule for the lower-bound constraint, a special case of the GCC constraint [22] where values must occur a minimum number of times within a vector of variables. The detection rule finds a non-empty subset $U \subseteq \mathcal{T}$ of time points for which the energy of the tasks that can be spent at these time points is not enough to satisfy the demand. When such a set exists, too much energy is spent in $\mathcal{T} \backslash U$ and not enough in U. We say that $\mathcal{T} \backslash U$ is *overloaded* while U is *underloaded*.

Underload Detection Rule: The underload detection rule fails iff there exists a set of time points U whose demand exceeds the energy of the tasks that can be executed during U:

$$\exists U \subseteq \mathcal{T} \ \sum_{t \in U} d_t > \sum_{i \in \mathcal{I}:[\text{est}_i, \text{lct}_i) \cap U \neq \emptyset} e_i \implies \text{Failure} \qquad (7)$$

To apply the rule, we design a greedy algorithm (Algorithm 2) that schedules the energy of the tasks, in non-decreasing order of lct, as early as possible. The algorithm wastes the demand at a given time point only if there is no other option. We call that waste *overflow*. Once all tasks are scheduled, time points where the demand was not met are included in the set U. If there is none, U is empty. If $U \neq \emptyset$, the constraint cannot be satisfied.

The algorithm uses the time line data structure [9] which efficiently schedules tasks according to the fully elastic relaxation. We begin by presenting a slightly modified version of the time line. Since the time line was designed for the CUMULATIVE, we replace the maximum capacity of the resource by the demand. We also prevent the algorithm from allocating energy that is not used to fulfill the demand.

The time line is first initialized with a vector of *critical time points* $\mathcal{T}^C$ that contains, in increasing order, the est and lct of the tasks without duplicates. We define a vector of demand Δ of dimension $|\mathcal{T}^C| - 1$. A critical time point $\mathcal{T}^C[j]$ has an associated demand $\Delta[j] = \sum_{k=\mathcal{T}^C[j]}^{\mathcal{T}^C[j+1]-1} d_k$ equal to the sum of the demand of the time points in the semi-open interval $[\mathcal{T}^C[j], \mathcal{T}^C[j+1])$. Two vectors, M_{est} and M_{lct} map a task i to the index of its est and lct in $\mathcal{T}^C$ such that $\text{est}_i = \mathcal{T}^C[M_{\text{est}}[i]]$ and $\text{lct}_i = \mathcal{T}^C[M_{\text{lct}}[i]]$. The time line uses a disjoint set (also called union-find) data structure [11] S upon the integers $1..|\mathcal{T}^C|$ containing the indexes of the critical time points. The operation $S.\texttt{union}(i, j)$ merges the set containing i with the set containing j. The operation $S.\texttt{findGreatest(i)}$ returns the greatest element of the set containing i. It has the same complexity as the classic $\texttt{find}$ operation and is implemented by keeping a map of the greatest element of each set and updating it when $\texttt{union}$ is called. The time line supports the operation $\texttt{ScheduleTask}(i)$ (Algorithm 1), which schedules task i as early as possible. Since the data structure is based on the fully elastic relaxation, the task can be preempted and can take more than its height at any given time point. It overflows at a time point only if it has to.

ScheduleTask(i) is implemented differently than in [9]. The algorithm computes the energy $e_i = h_i \cdot p_i$ that needs to be scheduled (line 1). Using the disjoint set S, line 2 finds the first time point j that is greater than or equal to est_i whose the demand is not yet satisfied. The while loop schedules as much energy as possible at that time point, but no more than the demand. If the demand of the current critical time point is fulfilled ($\Delta_t = 0$), the algorithm merges, on line 3, the sets of indices containing j with the one containing $j + 1$. This allows for future calls to S.findGreatest to return the next time point with unfulfilled demand. The process is repeated until all the task's energy is scheduled or the demand in the interval $[\text{est}_i, \text{lct}_i)$ is fulfilled.

Algorithm 1: ScheduleTask(i)	**Algorithm 2:** UnderloadCheck($\mathcal{I}, d$)
1 $\quad e \leftarrow h_i \cdot p_i$; $\quad j \leftarrow M_{\text{est}}[i]$; $\quad t_1 \leftarrow -\infty$; $\quad$**while** $e > 0 \wedge t_1 < \text{lct}_i$ **do** 2 $\quad\quad j \leftarrow S.\texttt{findGreatest}(j)$; $\quad\quad t_1 \leftarrow \mathcal{T}^C[j]$; $\quad\quad \tau \leftarrow \min(\Delta[j], e)$; $\quad\quad e \leftarrow e - \tau$; $\quad\quad \Delta[j] \leftarrow \Delta[j] - \tau$; $\quad\quad$**if** $\Delta[j] = 0$ **then** 3 $\quad\quad\quad \lfloor\ S.\texttt{union}(j, j + 1)$;	InitializeTimeline($\mathcal{I}, d$); **for** $i \in \mathcal{I}$ *sorted by* *non-decreasing* lct **do** $\quad \lfloor$ ScheduleTask(i); 1 $\quad$**return** $\quad S.\texttt{findGreatest}(1) = |\mathcal{T}^C|$;

We present the UnderloadCheck (Algorithm 2). Once the time line is initialized, the algorithm schedules each task in order of non-decreasing lct. The algorithm greedily spends the task's energy $e = p_i \cdot h_i$ as early as possible, but no more than the demand at each time point. If the loop stops because of the condition $t_1 < \text{lct}$, the unscheduled energy of the task cannot be used to fill the demand, because it would be scheduled after the latest completion time of the task. Thus, the energy is lost and we say that the task *overflowed*. Once all tasks are processed, the algorithm checks, on line 1, if all the sets have been merged together. If so, the demand of all time points is met and the check passes. Otherwise, the demand of at least one time point is not met and the check fails.

Example 1. Recall a task is defined by $\langle i, \text{est}_i, \text{lct}_i, p_i, h_i \rangle$. Given the three tasks $\langle 1, 2, 4, 2, 1 \rangle$, $\langle 2, 2, 5, 2, 1 \rangle$, $\langle 3, 0, 6, 2, 1 \rangle$ and a demand of 1 for each time point in $[0, 6)$, the UnderloadCheck computes the vector $\mathcal{T}^C = [0, 2, 4, 5, 6]$ of critical time points. In the disjoint sets data structure S, all these time points are in separate sets: $\{0\}, \{2\}, \{4\}, \{5\}, \{6\}$ (in this example, we use the time point values instead of indexes for the sake of clarity). The demand vector is initialized to $\Delta = [2, 2, 1, 1]$.

The algorithm begins by scheduling task 1, which has the smallest lct, in interval $[2, 4)$ (see Fig. 1). Since all the demand for critical time point 2 is met, the disjoint set $\{2\}$ is merged with disjoint set $\{4\}$. The UnderloadCheck then processes task 2, which has the second smallest lct. The algorithm finds the greatest time point greater than or

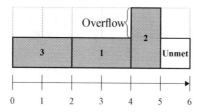

Fig. 1. Visual representation of the time line after the execution of the UnderloadCheck. One unit of demand has not been met in $[5, 6)$.

equal to $\text{est}_2 = 2$, which is 4. Hence, it schedules one unit at time point 4 and merges the set $\{2, 4\}$ with $\{5\}$. However, since $\text{lct}_2 = 5$, the remaining unit of energy of task 2 is lost, causing an overload. The UnderloadCheck then processes the last task, 3. As the greatest time point greater than or equal to $\text{est}_3 = 0$ is 0, the algorithm schedules 2 units of energy of task 3 in the interval $[0, 2)$, before merging the set $\{0\}$ with $\{2, 4, 5\}$. Once all tasks have been scheduled, we have $S = \{0, 2, 4, 5\}, \{6\}$. There is more than one disjoint set in S, which means that the demand has not been completely met, leading to a failure.

Theorem 2. UnderloadCheck *returns* true *if and only if the underload detection rule* (7) *does not fail.*

Proof. ($\Longrightarrow$) When the algorithm decrements $\Delta[j]$, it schedules the task i to spend its energy within $[\mathcal{T}^C[j], \mathcal{T}^C[j + 1]) \subseteq [\text{est}_i, \text{lct}_i)$. Once all tasks are scheduled, if $\Delta = \mathbf{0}$ (or equivalently $S.\texttt{findGreatest(1)} = |\mathcal{T}^C|$), all the demand is satisfied and the elastic schedule built by the algorithm disproves the existence of a non-empty set U that satisfies the underload check detection rule.

($\Longleftarrow$) If the algorithm returns *false*, we show that there exists a set U that satisfies the condition in (7). We associate a time interval τ_i to every task i. If task i does not overflow, we set $\tau_i = \emptyset$. If task i overflows, the critical index $M_{\text{lct}}[i]$ belongs to a set A in S forming an interval $\tau_i = [\mathcal{T}^C[\min(A)], \mathcal{T}^C[\max(A)])$. The demand is completely fulfilled in the time interval τ_i since $\Delta[j] = 0$ for all critical points $\mathcal{T}^C[j] \in \tau_i$. Whether task i overflows or not, we have $\sum_{j:[\text{est}_j, \text{lct}_j) \subseteq \tau_i} e_j \geq \sum_{t \in \tau_i} d_t$.

Let $U = [1, m] \backslash \bigcup_{i \in \mathcal{I}} \tau_i$. Consider a task j such that $[\text{est}_j, \text{lct}_j) \cap U \neq \emptyset$. Its energy was fully spent inside U. Indeed, all the non-empty intervals τ_i such that $\text{lct}_i < \text{lct}_j$ were fulfilled before j was processed. Its energy was therefore spent before or after τ_i. For non-empty intervals τ_i such that $\text{lct}_i \geq \text{lct}_j$, we necessarily have $\text{est}_j < \min(\tau_i)$ and since the task is scheduled at its earliest, it was first scheduled outside of τ_i. Moreover, it could not be scheduled in τ_i as the time point $M_{\text{est}}[i]$ would have been merged to the set A that defined τ_i, which contradicts $\text{est}_j < \min(\tau_i)$. Therefore, none of the energy of j was scheduled within τ_i. Since all the energy of the tasks j such that $[\text{est}_j, \text{lct}_j) \cap U \neq \emptyset$ was scheduled in U and that this energy does not fulfill the demand, the condition in (7) holds. $\qquad\square$

Theorem 3. UnderloadCheck *has a complexity of* $\Theta(n)$ *provided that the tasks are already sorted by their latest completion times.*

Proof. Fahimi et al. [9] showed that `ScheduleTask` executes in $O(1)$ amortized time when $|\mathcal{T}^C| \in O(n)$. As our modifications only make it run faster since the energy of a task is not always fully scheduled, the complexity remains unchanged. `ScheduleTask` is called n times hence $\Theta(n)$. $\square$

5   Underload Filtering

We can derive a filtering rule from the `UnderloadCheck` by fixing the starting time of a task, as shown in (8). If the `UnderloadCheck` fails when task i starts at time t, then task i obviously cannot start at time t and we can remove t from $\mathrm{dom}(S_i)$.

$$\neg\texttt{UnderloadCheck}(\mathcal{I}\backslash\{i\} \cup \{\langle t, t + p_i, p_i, h_i\rangle\}) \implies S_i \neq t \qquad (8)$$

By studying the set of time points U that makes the underload check fail, one can derive a filtering rule that prunes more than a single value from the domain of S_i.

$$\exists U \subseteq \mathcal{T} : \& D > E$$
$$\implies \mathrm{ect}_i \geq \min\{t \in U \mid t \geq \mathrm{ect}_i\} + \left\lceil \frac{D - E}{h_i} \right\rceil$$
$$\textbf{where } D = \sum_{t \in U} d_t \qquad (9)$$
$$E = \sum_{j \in \mathcal{I}\backslash\{i\}:[\mathrm{est}_j, \mathrm{lct}_j)\cap U \neq \emptyset} e_j$$

To filter a given task i, the rule (9) uses a set of time points U for which the sum of the demand $D = \sum_{t \in U} d_t$ exceeds the energy $E = \sum_{j \in \mathcal{I}\backslash\{i\}:[\mathrm{est}_j, \mathrm{lct}_j)\cap U \neq \emptyset} e_j$ of the tasks that can spend energy in U (except task i). If the energy of the tasks without i is insufficient to cover the demand for U by $D - E$ units, then task i must spend $D - E$ units of energy in U. Thus, we can filter the earliest completion time such that i ends at its earliest in U by the missing units.

5.1   Naive Algorithm

Algorithm 3 naively applies rule (8). For each task i, it fixes the starting time to est_i, then it executes the `UnderloadCheck`. If the check fails, it increases the starting time by one and executes the `UnderloadCheck` again. It repeats until it finds a starting time t for which the check passes or there is a starting time that exceeds the latest starting time of the task. In the first case, it filters the earliest starting time of task i to t. In the second case, it returns a failure. Filtering the latest completion time is symmetric.

This algorithm can fail even if the `UnderloadCheck` passes since the filtering algorithm, contrary to the `UnderloadCheck`, fixes the current task, preventing it from having elastic energy. Hence, the check performed by the filtering algorithm is stronger.

Algorithm 3: NaiveFiltering($\mathcal{I}, D$)

---

for $i \in \mathcal{I}$ **do**
  $t \leftarrow \text{lct}_i$;
  $\text{lct}_i \leftarrow \text{ect}_i$ //Temporarily fix i to its earliest;
  while $\text{est}_i \leq \text{lst}_i \wedge$
    $\neg \textit{UnderloadCheck}(\mathcal{I})$ **do**
      | $\text{est}_i \leftarrow \text{est}_i + 1$;
      | $\text{lct}_i \leftarrow \text{lct}_i + 1$;
  if $\text{est}_i \leq \text{lst}_i$ **then**
    | $\text{lct}_i \leftarrow t$ //Restore i to its original lct_i;
  else
    | **return** false // Failure;

return true // Consistent;

We call this algorithm naive because, although it is simple, it requires for each task, in the worst case, $\text{lst}_i - \text{est}_i$ calls to the UnderloadCheck, leading to a worst-case time complexity of $O(|\mathcal{T}| \cdot n^2)$, which depends on the number of time points. However, this does not make the algorithm irrelevant in practice. The best-case complexity is $\Theta(n^2)$ and it happens when there is no filtering done. Furthermore, for a given task i, the algorithm executes the UnderloadCheck $1 + k$ times, where k is the number of values removed from the domain of S. Hence, the complexity of the algorithm increases only if it filters, which is generally not frequent though essential.

5.2 Overflow Algorithm

We improve upon the naive algorithm by filtering the earliest starting time of a task i by more than one unit at a time, allowing the algorithm to directly apply rule (9). Let $\delta = \sum\limits_{j \in \mathcal{T}^C | \mathcal{T}_j^C \geq est_i} \Delta[\mathcal{T}_j^C]$ be the amount of demand not met at or after est_i after a call to UnderloadCheck. Let j be the task with the smallest est_j that overflowed such that $\text{est}_j \geq \text{est}_i$. If such a task exists, let $\tau = \text{est}_j$ otherwise let $\tau = \text{est}_i$. The starting time of i can be filtered such that $\text{est}_i \geq \tau + \left\lceil \frac{\delta}{h_i} \right\rceil$.

We now explain the intuition behind that filtering. In the first case, $\tau = \text{est}_i$. We know that there are δ units to fill after est_i. Moving the task by one unit can *free* at most h_i units of energy (it can come from the overflow of either i or another task that will its place). We need to move i by $\left\lceil \frac{\delta}{h_i} \right\rceil$ to gain enough energy to satisfy the demand.

The second case occurs when $\tau > \text{est}_i$. In that case, we move i after the est_j of the task that overflowed, by an amount corresponding to the missing energy δ. This allows j to take the place of i. The energy of i can then be used to fulfill the missing demand.

Example 2. Given two tasks $\langle 1, 0, 6, 3, 1 \rangle$ and $\langle 2, 0, 4, 3, 1 \rangle$ and a demand of 1 for each time point in $[0, 6)$, the overflow algorithm begins by filtering task 1. It temporarily fixes task 1 such that $\text{lct}_1 = 3$ before running the UnderloadCheck. Since the temporary

lct_1 is smaller than $lct_2 = 4$, the UnderloadCheck starts by scheduling task 1 in $[0, 3)$ (see Fig. 2). Then, it schedules task 2. There are 3 units of energy to schedule and the task can only be scheduled in $[0, 4)$. The algorithm schedules one unit at time 3 and the two remaining units are wasted. All tasks are scheduled, but two units of demand are not met ($\delta = 2$), as shown on Fig. 2. Since the UnderloadCheck fails, the overflow algorithm filters task 1. As task 2 overflowed, we have $\tau = est_2 = 0$. The algorithm filters est_1 to $\tau + \left\lceil \frac{\delta}{h_1} \right\rceil = 0 + \frac{2}{1} = 2$. Hence, the overflow algorithm increased the est_1 by two units with a single call to UnderloadCheck.

The algorithm iteratively applies this rule until the UnderloadCheck passes (or a failure is detected). Its worst-case complexity is as the naive algorithm's but it increments the est_i by more than one unit, reducing the number of calls to the UnderloadCheck.

6   Model with SoftCumulative

We show how to encode the MINCUMULATIVE using the SOFTCUMULATIVE. By doing so, we can use the strong energetic reasoning rules from the algorithms introduced by Ouellet and Quimper [20] for the SOFTCUMULATIVE.

Let $\mathcal{I}'$ be a set of fixed tasks that cover the horizon. For each interval $[l, u)$ of maximal length such that $d_j = d_{j+1} \; \forall j \in \{l..u - 2\}$, we create a task $i \in \mathcal{I}'$ with $est_i = l$, $lct_i = u$, $p_i = u - l$, and $h_i = \max_t(d_t) - d_l$. We have this equivalence.

$$\text{MINCUMULATIVE}([S_i \mid i \in \mathcal{I}], [p_i \mid i \in \mathcal{I}], [h_i \mid i \in \mathcal{I}], d) \iff$$
$$\text{SOFTCUMULATIVE}([S_i \mid i \in \mathcal{I} \cup \mathcal{I}'], [p_i \mid i \in \mathcal{I} \cup \mathcal{I}'], [h_i \mid i \in \mathcal{I} \cup \mathcal{I}'], \quad (10)$$
$$\max_t(d_t), \sum_i e_i - \sum_t d_t)$$

Lemma 1. *The* MINCUMULATIVE *can be encoded as a* SOFTCUMULATIVE *as in* (10).

Proof. The SOFTCUMULATIVE is associated to a resource of capacity $\max_t(d_t)$ over a horizon hor. The constraint allows an overflow of at most $\sum_{i \in \mathcal{I}} e_i - \sum_t d_t$ units of

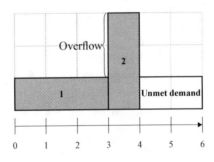

Fig. 2. Visual representation of the time line after the first call to the UnderloadCheck by the overflow algorithm, when filtering task 1. Task 1 is scheduled in $[0, 3)$. Task 2 has one unit scheduled in $[3, 4)$ and two units overflowed. Two units of demand are not met in $[4, 6)$.

energy. So at most $\max_t(d_t) \cdot \mathsf{hor} + \sum_{i \in \mathcal{I}} e_i - \sum_t d_t$ units of energy can be scheduled. The energy of the tasks matches this bound meaning that the resource is fully used.

$$\sum_{i \in \mathcal{I}} e_i + \sum_{i \in \mathcal{I}'} e_i = \sum_{i \in \mathcal{I}} e_i + \sum_l (\max_t(d_t) - d_l) = \sum_{i \in \mathcal{I}} e_i + \mathsf{hor} \cdot \max_t(d_t) - \sum_l d_l$$

Since the tasks in $\mathcal{I}'$ consume $\max_t(d_t) - d_l$ units of energy at time l, the tasks in $\mathcal{I}$ consume the remaining d_l (and can overflow) as required by the MINCUMULATIVE. The converse holds using the same argument.   □

7 Comparing the Rules

We compare the check rules of the Time-Tabling (TT), the Underload Check (UC), and the Energetic Reasoning (ER). The Underload Check rule, as mentioned earlier, is insufficient to enforce the MINCUMULATIVE constraint, as illustrated in Example 3.

Example 3. Consider an instance with two fixed tasks such that $\mathsf{est}_1 = 0$, $\mathsf{est}_2 = 1$, $\mathsf{lct}_1 = 2$, $\mathsf{lct}_2 = 3$, $p_{1,2} = 2$, $h_{1,2} = 1$ and a demand $d = [1, 1, 2]$. The Underload Check rule finds no underloaded set since it schedules the first unit of energy of task 1 at time point 1, the second unit at time point 2 and the two units of task 2 at time point 3. Even when the tasks are fixed, the Underload Check relaxes the heights of the tasks. On the other hand, the Time-Tabling finds that there are not enough tasks at time point 3 to satisfy the demand of 2. Similarly, the energetic reasoning finds that the minimum intersection in the interval $[1, 3)$ is 3. Since the demand in the interval is 2, there is one unit of overcost. The sum of the energy of the tasks is 4 and the sum of the demand is 4 so the SOFTCUMULATIVE allows no overcost, hence the detection of the failure.

While the weakness of the Underload Check is to relax the height of the tasks, the weakness of the Time-Tabling is that it does not take the processing time of the tasks into account, as demonstrated by Example 4

Example 4. Consider a task ($\mathsf{est}_1 = 0$, $\mathsf{lct}_1 = 5$, $p_1 = 1$, $h_1 = 1$) and a demand $d = [1, 1, 1, 1]$. The Time-Tabling notices that task 1 can cover the demand of each time point individually and passes. The Underload Check and the Energetic Reasoning quickly find that there is not enough energy in task 1 to satisfy the overall demand.

Examples 3 and 4 show that the Underload Check and the Time-Tabling are incomparable when comes the time to detect infeasibility, as it is the case for the CUMULATIVE. Lemma 2 shows that the energetic reasoning is stronger than the Time-Tabling.

Lemma 2. *If the Time-Tabling checker fails, then the* SOFTCUMULATIVE *model using the energetic reasoning checker also fails.*

Proof. Suppose the Time-Tabling fails because the total height of the tasks at time t is less than d_t. Consider the partition $\mathcal{T} = [0, t) \cup [t, t + 1) \cup [t + 1, \mathsf{hor})$. From Lemma 1, the resource is fully used. If it is underused in interval $[t, t+1)$, it is overused in $[0, t) \cup [t+1, \mathsf{hor})$. The energetic reasoning will therefore detect an overflow greater than $\sum_i e_i - \sum_t d_t$ in the intervals $[0, t) \cup [t + 1, \mathsf{hor})$ and thus, detect failure.   □

Lemma 3. *If the Underload Check fails, then the* SOFTCUMULATIVE *model using the energetic reasoning checker also fails.*

Proof. Suppose that the Underload Check returns a failure when checking the set U. The same reasoning as with Lemma 2 applies; the intervals in $T \setminus U$ have an overcost greater than $\sum_i e_i - \sum_t d_t$. Thus the SOFTCUMULATIVE fails. $\qquad \square$

The SOFTCUMULATIVE with its energetic checker is stronger than the Underload Check and Time-Tabling. In fact, it is strictly stronger, as shown in Example 5.

Example 5. Consider a demand $d = [1, 3, 1]$ and two tasks: $est_1 = est_2 = 0$, $lct_1 = 2$, $lct_2 = 3$, $p_1 = 2$, $p_2 = 1$, $h_1 = 2$, and $h_2 = 1$. The Time-Tabling passes since, when considered individually, there can be three units of height at time points 0 and 1 and one unit at time point 2. The Underload Check passes since the four units of energy of task 1 cover the demand of the first two time points and the single unit of task 2 covers the demand of the last time point. However, the SOFTCUMULATIVE model fails. Indeed, we have $MI(\mathcal{I}, 0, 1) = 2$, but the demand in $[0, 1)$ is only 1. Hence, we have an overcost of 1 in that interval, but no overcost is allowed since $\sum_{i \in \mathcal{I}} e_i - \sum_t d_t = (4 + 1) - 5 = 0$. Hence the SOFTCUMULATIVE detects the failure.

8 Experiments

We tested our algorithm on two benchmarks. The first contains randomly generated instances of a simple problem with one MINCUMULATIVE constraint. The tasks have a variable height in $\{0, 1\}$ indicating whether a task is activated or not. The goal is to satisfy the demand while minimizing the number of activated tasks. The instances are available on Github[1] and upon request to the second author. The Work Shift Scheduling Benchmark is an industrial benchmark introduced by [6]. The problem consists in scheduling the work shifts of employees while satisfying the demand. The benchmark has instances with up to 10 activities, but we only use the instances with one activity. Employee $e \in E$ can work between 6 to 8 hours. An employee e starts at time $S_{e,1}$, takes a 15-min break, resumes at S_{e_2}, takes a 1-hour lunch, resumes at $S_{e,3}$, takes a 15-min break, and resumes at $S_{e,4}$. The 4 work periods are at least one hour. The demand can vary at each 15-min time step. We minimize the number of employees. The model for the decomposition of our constraint in smaller binary constraints is given in (11)–(17). We replace (15) by the MINCUMULATIVE constraint for the other configurations. We break symmetries with (16) and (17). The upper limit on the number of employees is set to $|E| = 10$. This gives two unsatisfiable instances, which allows us to observe the behaviour of the algorithms on both satisfiable and unsatisfiable instances. The search heuristic branches on the minimum value of H_e, then P_e, and finally $S_{e,p}$.

$$\text{minimize} \sum_{e \in E} H_e \text{ subject to} \qquad (11)$$

$$S_{e,p} + P_{e,p} + \delta = S_{e,p+1} \qquad \forall e \in E, (p, \delta) \in \{(1, 1), (2, 4), (3, 1)\} \quad (12)$$

$$S_{e,4} + P_{e,4} + 4 \leq |T| \qquad \forall e \in E \quad (13)$$

---

[1] https://github.com/yanickouellet/min-cumulative-paper-public.

$$24 \leq \sum_{p \in \{1..4\}} P_{e,p} \leq 32 \qquad\qquad \forall e \in E \quad (14)$$

$$\sum_{e \in E, p \in \{1..4\}} H_e \cdot (S_{e,p} \leq t < S_{e,p} + P_{e,p}) \geq d_t \qquad\qquad \forall t \in T \quad (15)$$

$$H_e \leq H_{e-1} \wedge S_{e-1,2} \leq S_{e,2} \qquad\qquad \forall e \in 1..|E| - 1 \quad (16)$$

$$H_e = 0 \implies S_{e,p} = s \quad \forall e \in E, (p, s) \in \{(1,0), (2,5), (3,13), (4,18)\} \quad (17)$$

We implemented our algorithms in Java using Choco solver version 4.10.6 [21]. We ran the experiments on an Intel Xeon Silver 4110 (2.10 GHz). All models were implemented in MiniZinc. Experiments for the random benchmark were done with a timeout of 20 min. Since there are fewer instances, experiments for the Work Shift Scheduling Benchmark were done with a timeout of 1 h.

We report the time taken to solve instances of the Work Shift Scheduling Benchmark in Table 1. We experimented two versions of the problem, one with variable processing time and one with processing times fixed to 6 periods. For solved instances, the difference between the algorithms were similar, but we report only the latter since few instances were solved to optimality in the former. We tested the following configurations: the Decomposition (D), Time-Tabling filtering (TT), Time-Tabling with Underload Check (TT + UC), Time-Tabling with Underload Filtering (TT + UF), Time-Tabling with Energetic Reasoning Check (TT + EC) and Energetic Filtering (EF). We did not test the Underload algorithms alone since they are not sufficient to enforce the MINCUMULATIVE. Note that instances 7 and 10 are unsatisfiable instances.

The decomposition (D) is the configuration with the slowest solving times and it can solve to optimality only half of the instances within one hour. Time-Tabling alone is faster, but worse than when it is combined with Underload Check or Underload Filtering. The combination of Time-Tabling and Underload Check is clearly the best configuration. It solves all instances to optimality and is faster on every instance than all other configurations. The combination of Time-Tabling and Underload Filtering is the second-best configuration. However, the added cost of filtering does not seem to be worth the reduction in the search space it provides. Both Energetic Reasoning configurations are slower on this benchmark. They reduce the search space more than the Underload Check configurations, but their high complexity is not worth it.

We report the results of the random benchmark in Table 2. The first number in the name of each instance indicates the number of tasks. We report only instances for which at least one configuration found the optimality within 20 min. We can see that the Decomposition and Time-Tabling struggle on this benchmark, having difficulties solving instances of more than 20 tasks. On smaller instances, Time-Tabling combined with Underload Check performs better than other configurations, while, on larger instances, Time-Tabling combined with Energetic Check is better. The latter is the only configuration able to solve all instances. We conclude that, as the problem becomes harder, the combination of Time-Tabling and Energetic Checker becomes more interesting, even with a slower complexity. However, as for the Work Shift Scheduling Benchmark, the added cost of the Energetic Filtering is not worth the increased filtering.

Table 1. Time (s) and thousands of backtracks (bt) to optimally solve the work shift scheduling problem. A dash (-) means that the optimal was not proved within 1 h. Instances 7 and 10 are unsatisfiable.

Instance	D		TT		TT + UC		TT + UF		TT + EC		EF	
	time (s)	bt	time (s)	bt	time (s)	bt	time (s)	bt	time (s)	bt	time (s)	bt
1	3.7	5.5	0.8	6.3	0.7	3.3	0.8	1.6	1.6	3.3	3.3	2.3
2	729.8	139.9	45.4	1,528.2	4.4	128.9	15.6	86.7	39.2	93.3	131.7	54.6
3	-	-	3,106.7	103,275.4	560.7	21,395.7	2,798.7	16.3	-	-	-	-
4	-	-	2,503.2	114,540.2	242.3	9,592.2	1,035.7	6,564.2	1,637.4	3,668.9	-	-
5	542.3	1,083.1	32.0	1,154.5	6.7	221.0	23.7	145.1	54.6	146.8	188.3	77.2
6	-	-	945.2	41,763.3	144.9	5,635.7	671.6	4,067.2	2,731,462.0	1.1	-	-
7	-	-	-	-	488.3	20,286.2	2,140.0	14,867.0	1,255.1	3,928.8	-	
8	87.0	146.0	6.2	156.3	1.7	30.4	4.0	18.4	9.7	27.1	29.5	15.2
9	1.4	550.0	0.2	0.7	0.3	0.3	0.4	0.2	0.5	0.3	0.6	0.2
10	-	-	-	-	28.8	1,113.0	114.5	649.9	44.7	144.4	181.2	57.5

Table 2. Time (s) and thousands of backtracks (bt) to optimally solve random instances. (-) means the optimal solution was not proved within 20 min. The first number in the name of the instance is the number of tasks.

Instance	D		TT		TT + UC		TT + UF		TT + EC		EF	
	time (s)	bt	time (s)	bt	time (s)	bt	time (s)	bt	time (s)	bt	time (s)	bt
20_1	87.4	3,776.6	125.5	12,900.0	0.3	12.6	0.9	12.6	1.3	8.7	3.9	8.4
20_2	0.5	0.9	0.4	7.9	0.1	2.4	0.4	2.4	0.2	1.3	1.3	1.2
20_3	0.3	0.3	0.2	1.1	0.1	1.0	0.3	1.0	0.2	0.8	0.9	0.8
20_4	31.8	1,463.6	248.0	20,035.1	0.2	6.2	0.5	6.3	0.4	3.7	2.3	2.8
20_5	512.9	2,1434.8	1,205.6	138,626.8	0.2	2.2	0.4	2.2	0.3	2.0	455.0	1,080.3
20_6	0.0	0.1	0.2	7.1	0.4	8.5	1.1	8.5	0.7	7.4	3.6	7.3
20_7	0.2	0.5	0.9	34.8	0.6	26.6	2.1	26.2	0.9	9.0	3.9	7.7
20_8	0.1	0.2	0.2	2.0	0.2	2.1	0.4	2.1	0.3	1.8	1.0	1.8
20_9	0.1	1.0	0.5	25.1	0.5	30.2	0.8	8.3	1.3	6.7	3.6	7.0
20_10	4.0	170.3	7.1	302.1	0.3	1.5	0.4	1.5	0.4	1.1	1.3	1.1
30_1	-	-	-	-	3.6	309.4	29.0	285.1	8.4	5.4	119.3	52.7
30_2	-	-	-	-	2.1	210.5	16.3	204.1	9.1	68.6	110.8	53.4
30_3	15.8	523.1	-	-	-	-	-	-	3.5	25.6	44.1	19.7
30_4	-	-	-	-	0.4	7.8	2.2	7.7	0.9	4.4	6.8	4.2
30_5	-	-	-	-	0.8	27.7	4.2	27.7	0.5	2.3	4.3	2.3
30_6	-	-	-	-	1,390.1	142,143.5	-	-	213.8	2,340.3	43.7	23.6
30_7	-	-	-	-	20.4	1,393.0	178.4	1,371.7	23.7	186.0	263.6	181.9
30_8	-	-	-	-	2.1	116.4	16.1	115.9	13.3	97.1	178.4	94.1
30_9	-	-	-	-	2.5	151.9	18.3	151.5	17.3	147.2	231.3	143.2
30_10	2.2	59.0	406.0	17,834.6	5.9	336.9	34.1	336.7	7.6	42.5	67.5	39.4
40_1	-	-	-	-	-	-	-	-	668.9	3,028.2	-	-
40_2	-	-	-	-	724.1	35,924.8	-	-	36.5	106.5	435.6	77.3
40_3	-	-	-	-	1,973.2	191,817.2	-	-	1,423.0	6,603.5	-	-
40_4	-	-	-	-	-	-	-	-	4.1	11.6	59.0	10.1
40_5	-	-	-	-	1,312.8	127,012.2	-	-	1,112.2	4,452.5	-	-
40_6	-	-	-	-	-	-	-	-	9.7	33.6	108.3	21.3
40_7	-	-	-	-	-	-	-	-	2,573.3	8,849.2	-	-
40_8	-	-	-	-	-	-	-	-	2,899.0	12,032.3	-	-

9   Conclusion

The MINCUMULATIVE enforces tasks to cover a minimum demand. It is NP-Complete to test for feasibility. We proposed a checker algorithm, the UnderloadCheck, and two filtering algorithms based on the checker: the Naive filtering algorithm, and the Overflow filtering algorithm. MINCUMULATIVE can be encoded with a SOFTCUMULATIVE. We compared the strength of the different checking rules. The combination of Time-Tabling and Underload Check performs best in practice.

References

1. Aggoun, A., Beldiceanu, N.: Extending chip in order to solve complex scheduling and placement problems. Math. Comput. Model. **17**(7), 57–73 (1993)
2. Baptiste, P., Le Pape, C., Nuijten, W.: Constraint-Based Scheduling. Kluwer Academic Publishers (2001)
3. Beldiceanu, N., Carlsson, M.: A new multi-resource cumulatives constraint with negative heights. In: Van Hentenryck, P. (ed.) CP 2002. LNCS, vol. 2470, pp. 63–79. Springer, Heidelberg (2002). https://doi.org/10.1007/3-540-46135-3_5
4. Burke, E.K., De Causmaecker, P., Berghe, G.V., Van Landeghem, H.: The state of the art of nurse rostering. J. Sched. **7**(6), 441–499 (2004)
5. Carlier, J., Sahli, A., Jouglet, A., Pinson, E.: A faster checker of the energetic reasoning for the cumulative scheduling problem. Int. J. Prod. Res. 1–16 (2021)
6. Côté, M.C., Gendron, B., Quimper, C.G., Rousseau, L.M.: Formal languages for integer programming modeling of shift scheduling problems. Constraints **16**(1), 54–76 (2011)
7. De Clercq, A., Petit, T., Beldiceanu, N., Jussien, N.: A soft constraint for cumulative problems with over-loads of resource. In: Doctoral Programme of the 16th International Conference on Principles and Practice of Constraint Programming (CP 2010), pp. 49–54 (2010)
8. Ernst, A.T., Jiang, H., Krishnamoorthy, M., Owens, B., Sier, D.: An annotated bibliography of personnel scheduling and rostering. Ann. Oper. Res. **127**(1–4), 21–144 (2004)
9. Fahimi, H., Ouellet, Y., Quimper, C.-G.: Linear-time filtering algorithms for the disjunctive constraint and a quadratic filtering algorithm for the cumulative not-first not-last. Constraints **23**(3), 272–293 (2018). https://doi.org/10.1007/s10601-018-9282-9
10. Feydy, T., Stuckey, P.J.: Lazy clause generation reengineered. In: Gent, I.P. (ed.) CP 2009. LNCS, vol. 5732, pp. 352–366. Springer, Heidelberg (2009). https://doi.org/10.1007/978-3-642-04244-7_29
11. Gabow, H.N., Tarjan, R.E.: A linear-time algorithm for a special case of disjoint set union. J. Comput. Syst. Sci. **30**(2), 209–221 (1985)
12. Garey, M.R., Johnson, D.S.: Computers and Intractability, vol. 174. Freeman, San Francisco (1979)
13. Jean-Charles, R.E.: Generalized arc consistency for global cardinality constraint. In: American Association for Artificial Intelligence (AAAI 1996), pp. 209–215 (1996)
14. Kameugne, R., Fotso, L.P., Scott, J., Ngo-Kateu, Y.: A quadratic edge-finding filtering algorithm for cumulative resource constraints. Constraints **19**(3), 243–269 (2014)
15. Katriel, I., Thiel, S.: Complete bound consistency for the global cardinality constraint. Constraints **10**(3), 191–217 (2005)
16. Lopez, P., Esquirol, P.: Consistency enforcing in scheduling: A general formulation based on energetic reasoning. In: 5th International Workshop on Project Management and Scheduling (PMS 1996) (1996)

17. Mercier, L., Van Hentenryck, P.: Edge finding for cumulative scheduling. INFORMS J. Comput. **20**(1), 143–153 (2008)
18. Ohrimenko, O., Stuckey, P.J., Codish, M.: Propagation via lazy clause generation. Constraints **14**(3), 357–391 (2009)
19. Ouellet, Y., Quimper, C.-G.: A $O(n \log^2 n)$ checker and $O(n^2 \log n)$ filtering algorithm for the energetic reasoning. In: van Hoeve, W.-J. (ed.) CPAIOR 2018. LNCS, vol. 10848, pp. 477–494. Springer, Cham (2018). https://doi.org/10.1007/978-3-319-93031-2_34
20. Ouellet, Y., Quimper, C.G.: The softcumulative constraint with quadratic penalty. In: AAAI Conference on Artifical Intelligence proceeding (2022, to appear)
21. Prud'homme, C., Fages, J.G., Lorca, X.: Choco solver documentation. TASC, INRIA Rennes, LINA CNRS UMR 6241 (2016)
22. Quimper, C.G., Golynski, A., López-Ortiz, A., Van Beek, P.: An efficient bounds consistency algorithm for the global cardinality constraint. Constraints **10**(2), 115–135 (2005)
23. Schutt, A., Feydy, T., Stuckey, P.J.: Explaining time-table-edge-finding propagation for the cumulative resource constraint. In: Gomes, C., Sellmann, M. (eds.) CPAIOR 2013. LNCS, vol. 7874, pp. 234–250. Springer, Heidelberg (2013). https://doi.org/10.1007/978-3-642-38171-3_16
24. Schutt, A., Feydy, T., Stuckey, P.J., Wallace, M.G.: Explaining the cumulative propagator. Constraints **16**(3), 250–282 (2011)
25. Tesch, A.: A nearly exact propagation algorithm for energetic reasoning in $\mathcal{O}(n^2 \log n)$. In: Rueher, M. (ed.) CP 2016. LNCS, vol. 9892, pp. 493–519. Springer, Cham (2016). https://doi.org/10.1007/978-3-319-44953-1_32
26. Vilím, P.: Timetable edge finding filtering algorithm for discrete cumulative resources. In: Achterberg, T., Beck, J.C. (eds.) CPAIOR 2011. LNCS, vol. 6697, pp. 230–245. Springer, Heidelberg (2011). https://doi.org/10.1007/978-3-642-21311-3_22
27. Wolf, A., Schrader, G.: $O(n \log n)$ overload checking for the cumulative constraint and its application. In: Umeda, M., Wolf, A., Bartenstein, O., Geske, U., Seipel, D., Takata, O. (eds.) INAP 2005. LNCS (LNAI), vol. 4369, pp. 88–101. Springer, Heidelberg (2006). https://doi.org/10.1007/11963578_8

Practically Uniform Solution Sampling in Constraint Programming

Gilles Pesant[1], Claude-Guy Quimper[2], and Hélène Verhaeghe[1(✉)]

[1] Polytechnique Montréal, Montreal, Canada
{gilles.pesant,helene.verhaeghe}@polymtl.ca
[2] Université Laval, Quebec City, Canada
claude-guy.quimper@ift.ulaval.ca

Abstract. The ability to sample solutions of a constrained combinatorial space has important applications in areas such as probabilistic reasoning and hardware/software verification. A highly desirable property of such samples is that they should be drawn uniformly at random, or at least nearly so. For combinatorial spaces expressed as SAT models, approaches based on universal hashing provide probabilistic guarantees about sampling uniformity. In this short paper, we apply that same approach to CP models, for which hashing functions take the form of linear constraints in modular arithmetic. We design an algorithm to generate an appropriate combination of linear modular constraints given a desired sample size. We evaluate empirically the sampling uniformity and runtime efficiency of our approach, showing it to be near-uniform at a fraction of the time needed to draw from the complete set of solutions.

1 Introduction

Many important yet difficult problems can be expressed on a constrained combinatorial space: finding a *satisfying* element (i.e. a solution) in that space and looking for an *optimal* one according to some criterion are perhaps the most common tasks but *counting* how many solutions there are and *sampling* solutions uniformly at random also have important applications. Examples of the latter occurs in probabilistic planning [3], Bayesian inference [10], code verification [4], as well as other applications [5].

Model counters which follow an approach based on universal hashing benefit from probabilistic guarantees about the quality of the approximate count they provide. In the context of SAT models expressed through Boolean variables, such universal hashing takes the form of randomly generated parity (XOR) constraints [2]. In the wider context of finite-domain variables and CP models the latter generalize to randomly generated linear equality constraints in modular arithmetic, as recently investigated by Pesant et al. [9]. Because hashing-based model counting operates by partitioning the solution space into nearly-equal-size cells, it also offers an approach to solution *sampling*, which has been exploited for SAT [7]. Even though in principle sampling the solutions of a CP model could be achieved by first translating the model into SAT, it would be much preferable to do it directly in CP.

Vavrille et al. [11] recently proposed a solution sampler for CP inspired by that idea of splitting the search space into cells and focusing on one such cell—adding random

P. Schaus (Ed.): CPAIOR 2022, LNCS 13292, pp. 335–344, 2022.
https://doi.org/10.1007/978-3-031-08011-1_22

TABLE constraints to a CP model—but unfortunately without any theoretical guarantee about the quality of the sampling nor admittedly any generally-observed sampling uniformity in practice. In this short paper, we add certain linear modular constraints—sharing the same probabilistic guarantees as XOR constraints for SAT—in order to sample the solutions of a CP model. We show that in practice we achieve almost uniform sampling. As in [11] our approach is independent of the actual CP model being sampled and can be applied as long as the underlying CP solver supports such constraints. Additionally it is very simple to use, without any parameters to tune.

The literature is scarce on the problem of sampling CP models. Gogate and Dechter achieve uniform sampling on a CSP by first expressing it as a factored probability distribution over its solutions, which requires that each constraint be given as a set of allowed tuples [6]. Perez and Régin sample solutions according to a given probability distribution but require that the model be encoded as a single MDD constraint [8]. As previously mentioned Vavrille et al. add randomly generated TABLE constraints to a CP model thereby reducing the search space in a controlled manner prior to sampling [11]. Parameters v and p respectively control the arity of a table and the probability of a tuple being included. In practice v is chosen much smaller than the number of variables in order to keep in check the exponential growth of the tables, thus sacrificing uniformity in sampling.

In the rest of the paper Sect. 2 describes how a system of linear modular constraints can be used to partition the search space and offer guarantees on the quality of the sampling. We also give an algorithm to generate these constraints. Section 3 presents an empirical evaluation and a comparison to the state of the art in constraint programming.

2   Solution Sampling

The core idea of our sampling algorithm, borrowed from an approach to approximate model counting, is to employ pairwise-independent hash functions to partition the solution space into roughly equal-size cells of solutions. By aiming for a cell size corresponding to the desired number of samples, we then focus on one such cell and exhaustively enumerate its solutions. Note that while we could in principle enumerate solutions up to the desired number from a much larger cell, this would either introduce a bias in the sampling process from the branching heuristic used or require that we use a totally random branching heuristic to the detriment of better-performing ones and with poor uniformity as observed in [11]. Enumerating from an appropriately-sized cell also means a smaller search space, which could bring computational savings. Finally, because our samples come from the enumeration of a single smaller solution space, we are guaranteed there will be no duplicates.

Such near-uniform partitioning of the solutions can be achieved by adding a system of m linear modular equalities in n integer finite-domain variables

$$A\boldsymbol{x} = \boldsymbol{b} \quad (\text{mod } p)$$

where $\boldsymbol{x}$ is a vector of n integer finite-domain variables, A an $m \times n$ matrix whose elements belong to $[p] \triangleq \{0, 1, \ldots, p-1\}$, $\boldsymbol{b}$ a vector of m elements from $[p]$, and p the modulus. Linear modular equalities are closely related to universal hash functions—we recall two important properties [1]. Let modulus p be a *prime* number, $x_1, x_2 \in [p]^n$, A, b be filled uniformly at random, and $\Pr[e]$ denote the probability of event e:

uniform partitioning of solutions: $\Pr[Ax_1 = b \ (\text{mod } p)] = \frac{1}{p^m}$

pairwise independence: $\Pr[Ax_1 = b \ (\text{mod } p) \mid Ax_2 = b \ (\text{mod } p)] = \frac{1}{p^m}$

Another advantage of choosing p to be prime is that we can use Gauss-Jordan Elimination (GJE) to simplify and solve our system of linear equations: when p is prime every element of the finite field F_p has a multiplicative inverse, which is required in order to apply that algorithm. The system of linear equalities $Ax = b \ (\text{mod } p)$ is thus rewritten in parametric form. Then we encode its set of solutions—obtained by iterating through the domains of the parametric variables—as a TABLE constraint on all of x. This way we can efficiently achieve domain consistency for the system by using e.g. the compact table propagator. Details of the whole filtering algorithm can be found in [9].

Because the number of tuples for the TABLE constraint may be impractically large, we postpone adding the constraint until certain conditions are met. In this work we used the following:

– the size of the Cartesian product of the domains of the parametric variables falls below a given threshold (here, 1000);
– the likelihood that the non-parametric variables (denoted as x'') support a given combination of parametric values, estimated as $\prod_{x \in x''} |\text{domain}(x)|/p$, falls below a given threshold (here, 0.5).

Whenever either one of these conditions is met during search, a TABLE constraint is added dynamically, and retracted upon backtracking.

So ultimately for filtering we use TABLE constraints but, in contrast to Vavrille et al. [11]: (i) we express them over the whole set of variables instead of some randomly-chosen small subset; (ii) our tuples originate from hash functions with theoretical guarantees instead of being randomly selected according to some probabilistic threshold; (iii) we generate TABLE constraints dynamically during search.

While linear modular constraints provide us with a way to distribute solutions fairly evenly and independently among the partition of the search space into cells, we still need to control the size of a cell so that it likely contains the desired number of solutions. Simply adding an integral number m of linear modular equalities only provides limited control on cell size especially when p is large: each cell amounts to $1/p^m$ of the search space. To correct this, linear modular *inequalities* can be added as well since they provide a finer control on cell size.

2.1   Systems of Linear Modular Inequalities

As we just saw, we need to handle systems of linear modular inequalities $Ax + b \leq c$ $(\text{mod } p)^1$ in order to produce a cell containing about as many solutions as the desired

---

[1] i.e. each congruent to one of $\{0, 1, \ldots, c[i]\}$. We need to add b in the inequalities because otherwise the probability that e.g. the null solution $x = 0$ satisfies the inequality would be equal to 1.

number of samples. While the probabilistic guarantees of such systems were given in
Lemma 1 of Pesant et al. [9], namely

$$\Pr[Ax + b \le c] = \Pr[Ax + b \le c \mid Ay + b \le c] = \frac{\prod_{i=1}^{m}(c[i]+1)}{p^m},$$

no filtering algorithm was provided for them. We now outline one that essentially recasts
this in terms of equalities. For notational convenience, we first drop b by considering
that it has been appended to A and a variable fixed to 1 appended to x. Transform
system $Ax \le c$, seen as a conjunction of disjunctions of equalities (one conjunct per
constraint), into the equivalent disjunction of systems of equalities

$$\bigvee_{0 \le c'[i] \le c[i],\ 1 \le i \le m} Ax = c'$$

by pushing in the conjunction. There will be $\prod(c[i]+1)$ disjuncts that we can represent
compactly as Ax "=" C with C being the $m \times \prod(c[i]+1)$ matrix representing the
right-hand side of each disjunct. We then proceed as in the case of equalities, applying
Gauss-Jordan Elimination on this augmented system and eventually enumerating tuples
to be fed to a TABLE constraint.

2.2 Sampling Algorithm

Algorithm 1 describes our sampling algorithm. It takes as input the desired size of the
sample expressed as a fraction λ of the number of solutions and a set of variables x
spanning the search space of the CP model. The latter are typically the branching variables, which totally determine a solution, and not necessarily all the model variables. In
particular auxiliary (i.e. dependent) variables need not be included. The algorithm first
selects an appropriate value for p (e.g. from a precomputed table of prime numbers): p
should be large enough both to achieve (i) the previously-mentioned probabilistic guarantees and (ii) the desired cell size by offering at least a few possibilities in the choice
of right-hand side c for inequalities (with $p = 3$ the only option is $c = 1$). It then adds
to the CP model a suitable mix of m linear modular equality and inequality constraints
on x so that

$$\frac{\prod_{i=1}^{m'}(c[i]+1)}{p^m} \approx \lambda, \quad m' \le m,$$

as directed by Algorithm 2 described below. Finally it enumerates and returns all the
solutions in the resulting cell (i.e. the original CP model with the added linear modular
constraints). Note that it is sufficient to branch on the parametric variables of the system
of equality constraints since assigning these fixes the non-parametric variables as well.

   We chose to specify the sample size relative to the number of solutions because it
does not require some approximation of the latter. If such an approximation is available, generating some given absolute number of samples can easily be done as well by
deriving the corresponding fraction.

   The success of our approach relies on two factors: an even partition of solutions
into cells, provided by the linear modular constraints we use, and a relative cell size

---

Algorithm 1: Sampling algorithm

Input: sample fraction λ, model variables x
Output: set of sampled solutions
1   $\ell \leftarrow$ largest domain value among x
2   $p \leftarrow$ smallest prime $\geq \max(\ell, 5)$
3   $m, F \leftarrow$ partition(λ, p)
4   $m_\leq \leftarrow |F|$
5   $m_= \leftarrow m - m_\leq$
6   **if** $m_= > 0$ **then**
7   **for** $i \leftarrow 1$ **to** $m_=$ **do**
8          $b[i] \leftarrow \mathbb{U}_{[p]}$ // choose uniformly at random from [p]
9   **for** $j \leftarrow 1$ **to** $|x|$ **do** $A[i][j] \leftarrow \mathbb{U}_{[p]}$
10   post $Ax = b \pmod{p}$
11       $x' \leftarrow$ parametric variables of $Ax = b \pmod{p}$ // from GJE solved form
12   **else** $x' \leftarrow x$
13   **if** $m_\leq > 0$ **then**
14   **for** $i \leftarrow 1$ **to** $m_\leq$ **do**
15   remove a factor f from F
16          $c[i] \leftarrow f - 1$
17          $b[i] \leftarrow \mathbb{U}_{[p]}$
18   **for** $j \leftarrow 1$ **to** $|x'|$ **do** $A[i][j] \leftarrow \mathbb{U}_{[p]}$
19   post $Ax' + b \leq c \pmod{p}$
20   **return** all solutions of the resulting CP model, branching on x'

---

close to sample fraction λ. For the latter, we seek an expression whose denominator is determined by m (given p) and whose numerator can be decomposed into at most m factors all less than p (Algorithm 3), each giving rise to a corresponding inequality. Algorithm 2 computes a suitably accurate combination of linear modular equalities and inequalities to restrict the search space to a cell of the desired size. It repeatedly attempts to reach the target accuracy (given by parameter ϵ, set to 0.01 in our experiments) while keeping the numerator of our approximation to λ below threshold $\nu^{\max}$ (set to 100) because it corresponds to the number of disjuncts in the translation of the system of inequalities we will generate (Sect. 2.1), doubling parameter ϵ after each attempt until we succeed. In practice it only requires a few attempts. In each attempt we consider, for increasing values of m, a potential factorization of integral numerators $\lfloor \nu \rfloor$ and $\lceil \nu \rceil$, where ν is the exact rational numerator such that $\nu/p^m = \lambda$, provided that the resulting approximation of λ would be accurate enough. For example *partition*$(\lambda = 0.02, p = 11)$ requires two attempts, settling on $m = 3$ and $F = \langle 9, 3 \rangle$ with a relative error below 0.015.

     Algorithm 2 always terminates. Indeed, the inner loop increases ν on line 7 and eventually exits the loop on line 9 if ν becomes too large. The outer loop increases ϵ on line 12 and terminates on line 8 if it becomes too large.

     Algorithm 3 decomposes integer a into the fewest factors (and no more than d), none of them larger than f. Failing this, it returns an empty list. It is an adaptation of

---

Algorithm 2: partition(λ, p)

    Input: sample fraction λ, modulus p
    Output: number of constraints m and list of factors F

1 $F \leftarrow \langle \rangle$
2 **repeat**
3 | $m \leftarrow 0$
4 | $\nu \leftarrow \lambda$
5 | **while** $F = \langle \rangle$ **do**
6 | | $m \leftarrow m + 1$
7 | | $\nu \leftarrow \nu \times p$
8 | | **if** $\frac{|\nu - 1|}{\nu} \leq \epsilon$ **then return** $m, \langle \rangle$ `// equalities are sufficient`
9 | | **if** $\nu > \nu^{\max}$ **then break**   `// numerator too large; try bigger` ϵ
10 | | **if** $\frac{\nu - \lfloor \nu \rfloor}{\nu} \leq \epsilon$ **then** $F \leftarrow$ factorize$(\lfloor \nu \rfloor, p - 1, m)$
11 | | **if** $F = \langle \rangle \wedge \frac{\lceil \nu \rceil - \nu}{\nu} \leq \epsilon$ **then** $F \leftarrow$ factorize$(\lceil \nu \rceil, p - 1, m)$
12 | $\epsilon \leftarrow 2\epsilon$
13 **until** $\nu \leq \nu^{\max}$
14 **return** m, F

Algorithm 3: factorize(a, f, d)

    Input: integer a to factorize, largest possible factor f, maximum number of factors d
    Output: list F of factors

1 $F \leftarrow \langle \rangle$
2 **while** $a > 1 \wedge f > 1$ **do**
3 | **while** $a \bmod f = 0$ **do**
4 | | add f to F
5 | | $a \leftarrow a \div f$
6 | $f \leftarrow f - 1$
7 **if** $a = 1 \wedge |F| \leq d$ **then return** F
8 **else return** $\langle \rangle$

---

the simple *Trial division* algorithm[2] in which we instead try factors in *decreasing* order to minimize the number of factors and thus maximize the number of equality constraints we will use, which contributes to reduce the number of branching variables x'. Because $a \leq \nu^{\max}$ and $f < p$ using a more efficient algorithm is not worthwhile.

3 Experiments

In this section we evaluate the sampling uniformity and computational efficiency of our approach using benchmark instances from the literature. The `java.util.Random` random number generator is used in all our experiments. The random table app-

---

[2] https://en.wikipedia.org/wiki/Trial_division.

roach [11] is evaluated using the authors' own code. The code for our approach is available as well.[3]

3.1 Benchmark Problems

For all our experiments, we based ourselves on four problems. The N-Queens problem (used in [11]), the Feature Models problem (software configuration problem, used in [11]), a Synthetic_n_d problem (n variables of domain size d, composed of two sub-problems, the total number of solutions is computable analytically, used and defined in [9]), and the Myciel problem (graph coloring, used in [9]). For the instances with a low number of solutions (Feature Models, 9-Queens, Synthetic_10_5) we aim to generate as many samples as 30 times the number of solutions. For our approach we use fraction $\lambda \in \{0.01, 0.05, 0.25\}$ (it is hard to go below 1% with that few solutions). For the other instances (Myciel 4, 15-Queens, Synthetic_10_10) we perform 100 runs with $\lambda = 10^{-5}$. For the random table approach we use the best combination of parameters reported for instances appearing in [11] and their generally-recommended values ($\kappa = 16, p = 1/32$) otherwise.

3.2 Quality of Sampling

As in [11], we evaluate the statistical quality of the sampling by computing the p-value of Pearson's χ^2 statistic. Given the number of solutions to the problem $nSol$, the number of samples $nSample$, the number of occurrences $nOcc_k$ of a solution k in the samples and P the expected number of occurrences of a given solution assuming a uniform distribution ($P = \frac{nSample}{nSol}$), we compute

$$z_{exp} = \sum_{k=1}^{nSol} \frac{(nOcc_k - P)^2}{P}.$$

The p-value is computed based on this expression.[4] The closer to 1 the p-value is, the closer to a uniform distribution the sampling is.

Figure 1 shows the evolution of this p-value as samples are drawn. We compare our approach (*Lin Mod*) to that of [11] (*Rnd table*) as well as to an oracle (*Oracle*) that simulates random sampling from the pool of all solutions by generating a sequence of random integers between 1 and $nSol$. As pointed out in [11], this generator is not perfect. Our study of the evolution of the p-value confirms it.

For *9-Queens* the random table approach had achieved very good uniformity for some combinations of its parameters [11]. We achieve even nearer uniformity for all three values of λ. On *Synthetic*, the random tables over 2 variables perform very well (however, not as well over 4 variables) but our approach is practically perfect. *Feature Models* is an instance for which it was reported that the random table approach did not

---

[3] The sampling method is implemented in the MiniCPBP solver https://github.com/PesantGilles/MiniCPBP and our examples on how to use the method are available https://github.com/363734/UniformSampling.

[4] The statistical library used is "Apache Commons Mathematics Library" (https://commons.apache.org/proper/commons-math/).

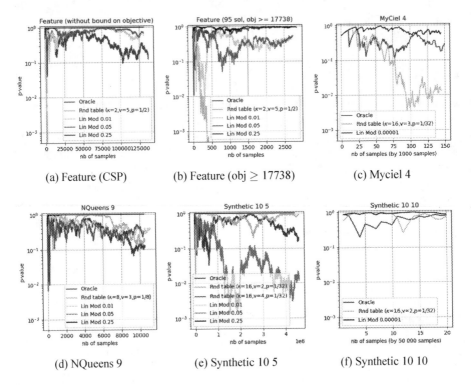

Fig. 1. Evolution of the p-value for competing sampling approaches on six benchmark instances.

perform well. On the contrary we see that our approach yields very good uniformity, even outperforming the oracle on the version with an unbounded objective. There is also a marked difference in quality on *Myciel*, whose model features 21 variables.

So on these benchmarks our sampling approach generally leads to better uniformity than the random table approach and even the oracle. Uniformity generally increases with λ, which is expected since the number of samples drawn at a time increases as well and these are necessarily distinct by design (we enumerate solutions in a more constrained space) whereas the oracle sampler may generate the same solution multiple times. A typical λ would be closer to 0.01 than to 0.25.

3.3 Runtime Efficiency of Our Sampling Approach

In the previous section we gave empirical evidence that our approach newly contributes near-uniform sampling. But it remains to show that it is less time-consuming than the brute-force approach of enumerating all solutions and then sampling uniformly at random from them (i.e. the oracle). When the number of solutions is small the latter may indeed be sufficient but in a realistic setting we sample from a large pool of solutions. Table 1 reports for the larger instances the runtime ratio between enumerating the solutions in the whole search space and in a cell approximately 10^{-5} of its size (our approach). In each case we observe computational savings, even by a few orders

Table 1. Runtime ratio between the other two approaches considered and ours to sample about 0.001% of solutions (for random table we use $\kappa = 16$, $p = 1/32$).

Instance	#solutions	$\frac{\#\text{solutions}}{\lvert\text{search space}\rvert}$	Oracle	Runtime ratio		
				Random table		
				$v = 2$	$v = 3$	$v = 4$
15-Queens	2.28e+6	5.2e-12	5.4	2e-3	3e-3	1e-2
Myciel 4	1.42e+8	3.0e-07	12.6	0.9	1.3	4.8
Synthetic_10_5	1.53e+5	1.6e-02	105.1	4.7	5.3	5.3
Synthetic_10_10	9.92e+8	1.0e-01	1183.9	2.7	13.3	137.9

of magnitude, without being close to an ideal 10^5 ratio from the corresponding reduction in search space because there is an overhead in handling linear modular constraints and recall that active domain filtering through TABLE constraints occurs lower in the search tree once they are posted. Better ratios appear to have more to do with the higher density of solutions than the actual number of solutions (see Table 1, second and third column). We also report the runtime ratio with the random table approach even though its sampling is not nearly as uniform. Except for 15-Queens, it is not faster either.

4   Conclusion

We described a novel approach to sample the set of solutions of any CP model. It is based on adding linear modular constraints to the model, thereby reducing its search space while preserving the proportion of solutions, and then exhaustively exploring that reduced search space. The amount of reduction being applied corresponds to the fraction of solutions one wishes to sample. Experiments on benchmark problems provided empirical evidence that our approach offers computationally-efficient near-uniform sampling.

Acknowledgements. We thank the anonymous reviewers for their constructive criticism which helped us improve the original version of the paper. Financial support for this research was provided in part by NSERC Discovery Grants 218028/2017 and 05953/2016.

References

1. Chakraborty, S., Meel, K.S., Mistry, R., Vardi, M.Y.: Approximate probabilistic inference via word-level counting. In: Schuurmans, D., Wellman, M.P. (eds.) Proceedings of the Thirtieth AAAI Conference on Artificial Intelligence, 12–17 February 2016, Phoenix, Arizona, USA, pp. 3218–3224. AAAI Press (2016)
2. Chakraborty, S., Meel, K.S., Vardi, M.Y.: A scalable approximate model counter. In: Schulte, C. (ed.) CP 2013. LNCS, vol. 8124, pp. 200–216. Springer, Heidelberg (2013). https://doi.org/10.1007/978-3-642-40627-0_18
3. Domshlak, C., Hoffmann, J.: Probabilistic planning via heuristic forward search and weighted model counting. J. Artif. Intell. Res. **30**, 565–620 (2007)

4. Dutra, R.T.: Efficient sampling of SAT and SMT solutions for testing and verification. Ph.D. thesis, University of California, Berkeley (2019)
5. Fichte, J.K., Hecher, M., Hamiti, F.: The model counting competition 2020. J. Exp. Algorithmics (JEA) **26**, 1–26 (2021)
6. Gogate, V., Dechter, R.: A new algorithm for sampling CSP solutions uniformly at random. In: Benhamou, F. (ed.) CP 2006. LNCS, vol. 4204, pp. 711–715. Springer, Heidelberg (2006). https://doi.org/10.1007/11889205_56
7. Meel, K.S., et al.: Constrained sampling and counting: universal hashing meets SAT solving. In: Darwiche, A. (ed.) Beyond NP, Papers from the 2016 AAAI Workshop, Phoenix, Arizona, USA, 12 February 2016, volume WS-16-05 of AAAI Workshops. AAAI Press (2016)
8. Perez, G., Régin, J.-C.: MDDs: sampling and probability constraints. In: Beck, J.C. (ed.) CP 2017. LNCS, vol. 10416, pp. 226–242. Springer, Cham (2017). https://doi.org/10.1007/978-3-319-66158-2_15
9. Pesant, G., Meel, K.S., Mohammadalitajrishi, M.: On the usefulness of linear modular arithmetic in constraint programming. In: Stuckey, P.J. (ed.) CPAIOR 2021. LNCS, vol. 12735, pp. 248–265. Springer, Cham (2021). https://doi.org/10.1007/978-3-030-78230-6_16
10. Sang, T., Beame, P., Kautz, H.A.: Performing Bayesian inference by weighted model counting. In: AAAI, vol. 5, pp. 475–481 (2005)
11. Vavrille, M., Truchet, C., Prud'homme, C.: Solution sampling with random table constraints. In: Michel, L.D. (ed.) 27th International Conference on Principles and Practice of Constraint Programming, CP 2021, Montpellier, France (Virtual Conference), 25–29 October 2021. LIPIcs, vol. 210, pp. 56:1–56:17. Schloss Dagstuhl - Leibniz-Zentrum für Informatik (2021)

Training Thinner and Deeper Neural Networks: Jumpstart Regularization

Carles Riera[1](✉), Camilo Rey[1](✉), Thiago Serra[2](✉), Eloi Puertas[1](✉), and Oriol Pujol[1](✉)

[1] Universitat de Barcelona, Barcelona, Spain
crieramo8@alumnes.ub.edu, camilorey@gmail.com,
{epuertas,oriol_pujol}@ub.edu
[2] Bucknell University, Lewisburg, USA
thiago.serra@bucknell.edu

Abstract. Neural networks are more expressive when they have multiple layers. In turn, conventional training methods are only successful if the depth does not lead to numerical issues such as exploding or vanishing gradients, which occur less frequently when the layers are sufficiently wide. However, increasing width to attain greater depth entails the use of heavier computational resources and leads to overparameterized models. These subsequent issues have been partially addressed by model compression methods such as quantization and pruning, some of which relying on normalization-based regularization of the loss function to make the effect of most parameters negligible. In this work, we propose instead to use regularization for preventing neurons from dying or becoming linear, a technique which we denote as *jumpstart regularization*. In comparison to conventional training, we obtain neural networks that are thinner, deeper, and—most importantly—more parameter-efficient.

Keywords: Deep learning · Model compression · ReLU networks

1 Introduction

> *Leap, and the net will appear.*
>
> Anonymous

Artificial neural networks are inspired by the simple, yet powerful idea that predictive models can be produced by combining units that mimic biological neurons. In fact, there is a rich discussion on what should constitute each unit and how the units should interact with one another. Units that work in parallel form a layer, whereas a sequence of layers transforming data unidirectionally define a feedforward network. Deciding the number of such layers—the *depth* of the network—is yet a topic of debate and technical challenges.

A neural network is trained for a particular task by minimizing the loss function associated with a sample of data in order for the network to learn a

© Springer Nature Switzerland AG 2022
P. Schaus (Ed.): CPAIOR 2022, LNCS 13292, pp. 345–357, 2022.
https://doi.org/10.1007/978-3-031-08011-1_23

function of interest. Although several universal approximation results show that mathematical functions can generally be approximated to arbitrary precision by single-layer feedforward networks, these results rely on using a very large number of units [12,26,43]. Moreover, simple functions such as XOR cannot be exactly represented with a single layer using the most typical units [45].

In fact, it is commonly agreed that depth is important in neural networks [7,38]. In the popular case of feedforward networks in which each unit is a Rectified Linear Unit (ReLU) [18,21,38,47], the neural network models a piecewise linear function [3]. Under the right conditions, the number of such "pieces"—the *linear regions*—may grow exponentially on the depth of the network [46,48,67]. Depending on the total number of units and size of the input, the number of linear regions is maximized with more or less layers [58]. Similarly, there is an active area of study on bounding the number of layers necessary to model any function that a given type of network can represent [3,14,20,27,45,69].

Although shallow networks present competitive accuracy results in some cases [4], deep neural networks have been established as the state-of-the-art over and again in areas such as computer vision and natural language processing [13, 24,29,30,37,39,62,70] thanks to the the development and popularization of backpropagation [41,54,71]. However, Stochastic Gradient Descent (SGD) [53]—the training algorithm associated with backpropagation—may have difficulties to converge to a good model due to exploding or vanishing gradients [6,28,35,49].

Exploding and vanishing gradients are often attributed to excessive depth, inadequate choice of parameters for the learning algorithm, or inappropriate scaling between network parameters, inputs, and outputs [17,32]. This issue has also inspired unit augmentations [25,44,60], additional connections across layers [23,30], and output normalization [32,52]. Indeed, it is somewhat intuitive that gradient updates, depth, and parameter scaling may affect one another.

In lieu of reducing depth, we may also increase the number of neurons per layer [22,64–66,72]. That leads to models that are considerably more complex, and which are often trained with additional terms in the loss function such as weight normalization to induce simpler models that hopefully generalize better. In turn, that helps model compression techniques such as network pruning methods to remove several parameters with only minor impact to model accuracy.

Nonetheless, vanishing gradients may also be caused by *dead* neurons when using ReLUs. If dead, a ReLU only outputs zero for every sample input. Hence, it does not contribute to updates during training and neither to the expressiveness of the model. To a lesser but relevant extent, similar issues can be observed with a RELU which never outputs zero, which we refer to as a *linear* neuron.

In this work, we aim to reverse neurons which die or become linear during training. Our approach is based on satisfying certain constraints throughout the process. For a margin defined for each unit, at least one input from the sample is above and another input is below. For each layer and input from the sample, at least one unit in the layer has that input above such a margin and another unit has it below. In order to use SGD for training, these constraints are dualized as part of the loss function and thus become a form of regularization that would prevent converging with the original loss function to spurious local minima.

2   Background

We consider a feedforward neural network modeling a function $\hat{y} = f_\theta(x)$ with an input layer $x = h^0 = [h_1^0 \; h_2^0 \ldots h_{n_0}^l]^T$, L hidden layers, and each layer $\ell \in \mathbb{L} = \{1, 2, \ldots, L\}$ having n_ℓ units indexed by $i \in \mathbb{N}_\ell = \{1, 2, \ldots, n_\ell\}$. For each layer $\ell \in \mathbb{L}$, let W^ℓ be the $n_\ell \times n_{\ell-1}$ matrix in which the j-th row corresponds to the weights of neuron j in layer ℓ and b^ℓ be vector of biases of layer ℓ. The preactivation output of unit j in layer ℓ is $g_i^\ell = W_j^\ell h^{\ell-1} + b_j^\ell$ and the output is $h_j^\ell = \sigma(g_j^\ell)$ for an activation function σ, which if not nonlinear would allow hidden layer ℓ to be removed by directly connecting layers $\ell - 1$ and $\ell + 1$ [55]. We refer to $g^\ell(\chi)$ and $h^\ell(\chi)$ as the values of g^ℓ and h^ℓ when $x = \chi$.

For the scope of this work, we consider the ReLU activation function $\sigma(u) = \max\{0, u\}$. Typically, the output of a feedforward neural network is produced by a softmax layer following the last hidden layer [10], $\hat{y} = \rho(h^L)$ with $\rho(h^L)_j = e^{h_j^L} / \sum_{k=1}^{n_L} e^{h_k^L} \; \forall j \in \{1, \ldots, n_L\}$, which is a peripheral aspect to our study.

The neural network is trained by minimizing a loss function $\mathcal{L}$ over a parameter set $\theta := \{(W^\ell, b^\ell)\}_{\ell=1}^L$ based on the N samples of a training set $\mathbb{X} := \{(x^i)\}_{i=1}^N$ to yield predictions $\{\hat{y}^i := f_\theta(x^i)\}_{i=1}^N$ that approximate the sample labels $\{y^i\}_{i=1}^N$ using metrics such as least squares or cross entropy [19, 59]:

$$\min_\theta \quad \mathcal{L}\left(\theta, \{(\hat{y}^i, y^i)\}_{i=1}^N\right) \tag{1}$$

$$\text{s.t.} \quad \hat{y}^i = f_\theta(x^i) \qquad \forall i \in \{1, 2, \ldots, N\} \tag{2}$$

whereas a neural network is not typically trained through constrained optimization, we believe that our approach is more easily understood under such a mindset, which aligns with further work emerging from this community [8, 15, 31].

3   Death, Stagnation, and Jumpstarting

Every ReLU is either *inactive* if $g_i^\ell \leq 0$ and thus $h_i^\ell = 0$ or *active* if $g_i^\ell > 0$ and thus $h_i^\ell = g_i^\ell > 0$. If a ReLU does not alternate between those states for different inputs, then the unit is considered *stable* [68] and thus the neural network models a less expressive function [56]. In certain cases, those units can be merged or removed without affecting the model [55, 57]. We consider in this work a superset of such units—those which do not change state at least for the training set:

Definition 1. *For a training set* $\mathbb{X}$*, unit j in layer ℓ is* dead *if $h_j^\ell(x^i) = 0 \; \forall i \in \{1, 2, \ldots, N\}$,* linear *if $h_j^\ell(x^i) > 0 \; \forall i \in \{1, 2, \ldots, N\}$, or* nonlinear *otherwise. Layer ℓ* dead *or* linear *if all of its units are dead or linear, respectively.*

Figures 1a to 1c illustrate geometrically the classification of the unit based on the training set. If dead, a unit impairs the training of the neural network because it always outputs zero for the inputs in the training set. Unless the units preceding a dead unit are updated in such a way that the unit is no longer dead, then the gradients of its output remain at zero and the parameters of the

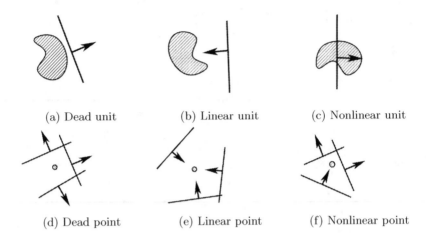

(a) Dead unit   (b) Linear unit   (c) Nonlinear unit

(d) Dead point   (e) Linear point   (f) Nonlinear point

Fig. 1. A unit j in layer ℓ separates the input space $\boldsymbol{h}^{\ell-1}$ into an open half-space $\boldsymbol{W}_j^{\ell}\boldsymbol{h}^{\ell-1} + b_j^{\ell} > 0$ in which the unit is active and a closed half-space $\boldsymbol{W}_j^{\ell}\boldsymbol{h}^{\ell-1} + b_j^{\ell} \leq 0$ in which the unit is inactive. The arrow in each case points to the active side. The unit is dead if the inputs from training set $\mathbb{X}$ lie exclusively on the inactive side (a); linear if exclusively on the active side (b); and nonlinear otherwise (c). In turn, an input is considered a dead point if it is in the closed half-space $\boldsymbol{W}_j^{\ell}\boldsymbol{h}^{\ell-1} + b_j^{\ell} \leq 0$ in which each and every unit $j \in \mathbb{N}_{\ell}$ is inactive (d); a linear point if it is in the open half-space $\boldsymbol{W}_j^{\ell}\boldsymbol{h}^{\ell-1} + b_j^{\ell} > 0$ in which each and every unit $j \in \mathbb{N}_{\ell}$ is active (e); and a nonlinear point otherwise (f).

dead unit are no longer updated [42,61], which effectively reduces the modeling capacity. If a layer dies, then the training stops because the gradients are zero.

For an intuitive and training-independent discussion, we consider incidence of dead layers at random. If the probability that a unit is dead upon initialization is p, as reasoned in [42], then layer ℓ is dead with probability p^{n_ℓ} and at least one layer is dead with probability $1 - \prod_{\ell=1}^{L}(1-p)^{n_\ell}$. If a layer is too thin or the network is too deep, then the network is more likely to be untrainable. We may discard dead unit initializations, but that ignores the impact on the training set:

Definition 2. *For a hidden layer $\ell \in \mathbb{L}$, an input x is considered a dead point if $\boldsymbol{h}^{\ell}(x) = 0$, a linear point if $\boldsymbol{h}^{\ell}(x) > 0$, and a nonlinear point otherwise.*

Figures 1d to 1f illustrate geometrically the classification of a point based on the activated units. If $x^i \in \mathbb{X}$ is a dead point at layer ℓ, then there is no backpropagation associated with x^i to the hidden layers 1 to $\ell - 1$. Hence, its contribution to training is diminished unless a subsequent gradient update at a preceding unit reverts the death. If $\ell = L$, then x^i is effectively not part of the training set. If all points die, regardless of the layer, then training halts.

If we also associate a probability q for x^i not activating a unit, then $\boldsymbol{x}^i$ is dead for layer ℓ with probability q^{n_ℓ} and for at least one layer of the neural network with probability $1 - \prod_{\ell=1}^{L}(1-q)^{n_\ell}$. Unlike p, q is bound to be significant.

We may likewise regard linear units and linear points as less desirable than nonlinear units and nonlinear points. A linear unit limits the expressiveness of the model, since it always contributes the same linear transformation to every input in the training set. A linear point can be more difficult to discriminate from other inputs, in particular if those inputs are also linear points.

Inspired by the prior discussion, we formulate the following constraints:

$$\max_{x^i \in \mathbb{X}} g_j^\ell(x^i) \geq 1 \qquad\qquad \forall \ell \in \mathbb{L}, j \in \mathbb{N}_\ell \qquad (3)$$

$$\min_{x^i \in \mathbb{X}} g_j^\ell(x^i) \leq -1 \qquad\qquad \forall \ell \in \mathbb{L}, j \in \mathbb{N}_\ell \qquad (4)$$

$$\max_{j \in \mathbb{N}_\ell} g_j^\ell(x^i) \geq 1 \qquad\qquad \forall \ell \in \mathbb{L}, x^i \in \mathbb{X} \qquad (5)$$

$$\min_{j \in \mathbb{N}_\ell} g_j^\ell(x^i) \leq -1 \qquad\qquad \forall \ell \in \mathbb{L}, x^i \in \mathbb{X} \qquad (6)$$

Dead and linear units are respectively prevented by the constraints in (3) and (4). Dead and linear points are prevented by the constraints in (5) and (6). Then we dualize those constraints and induce their satisfaction through the objective:

$$\min_\theta \quad \mathcal{L}\left(\theta, \{(\hat{y}^i, y^i)\}_{i=1}^N\right) + \lambda \mathcal{P}(\xi^+, \xi^-, \psi^+, \psi^-) \qquad (7)$$

$$\text{s.t.} \quad \hat{y}^i = f_\theta(x^i) \qquad\qquad \forall i \in \{1, 2, \ldots, N\} \qquad (8)$$

$$\xi_{j\ell}^+ = \max\left\{0, 1 - \max_{x^i \in \mathbb{X}} g_j^\ell(x^i)\right\} \qquad \forall \ell \in \mathbb{L}, j \in \mathbb{N}_\ell \qquad (9)$$

$$\xi_{j\ell}^- = \max\left\{0, -1 - \min_{x^i \in \mathbb{X}} g_j^\ell(x^i)\right\} \qquad \forall \ell \in \mathbb{L}, j \in \mathbb{N}_\ell \qquad (10)$$

$$\psi_{i\ell}^+ = \max\left\{0, 1 - \max_{j \in \mathbb{N}_\ell} g_j^\ell(x^i)\right\} \qquad \forall \ell \in \mathbb{L}, x^i \in \mathbb{X} \qquad (11)$$

$$\psi_{i\ell}^- = \max\left\{0, -1 - \min_{j \in \mathbb{N}_\ell} g_j^\ell(x^i)\right\} \qquad \forall \ell \in \mathbb{L}, x^i \in \mathbb{X} \qquad (12)$$

We denote by ξ^+, ξ^-, ψ^+, and ψ^- the nonnegative deficits associated with the corresponding constraints in (3)–(6) which are not satisfied. These deficits are combined and weighted against the original loss function $\mathcal{L}$ through a function $\mathcal{P}$, for which we have considered the arithmetic mean as well as the 1 and 2-norms.

We can apply this to convolutional neural networks [16,39] with only minor changes, since they are equivalent to a feedforward neural network with parameter sharing and which is not fully connected. The main difference to work with them directly is that the preactivation of the unit is a matrix instead of a scalar. We compute the margin through the maximum or minimum over those values.

4   Computational Experiments

Our first experiment (Fig. 2) is based on the MOONS dataset [51] with 85 points for training and 15 for validation. We test every width in $\{1, 2, 3, 4, 5, 10, 15, 20, 25\}$ with every depth in $\{1, 2, 3, 4, 5, 10, 15, 20, 25, 30, 35, 40, 45, 50, 60, \ldots, 150\}$. We chose a simpler dataset to limit the inference of factors such as overfitting, under-fitting, or batch size issues. The networks are implemented in Tensorflow [1] and Keras [11] with Glorot uniform initialization [17] and trained using Adam [34] for 5000 epochs, learning rate of $\epsilon = 0.01$, and batch size of 85. For each depth-width pair, we train a baseline network and a network with jumpstart using 1-norm as the aggregation function $\mathcal{P}$ and loss coefficient $\lambda = 10^{-4}$.

With jumpstart, we successfully train networks of width 3 with a depth up to 60 instead of 10 for the baseline and width 25 with a depth of up to 100 instead of 30. Hence, there is an approximately 5-fold increase in trainable depth.

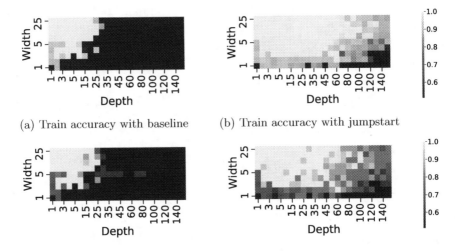

(a) Train accuracy with baseline   (b) Train accuracy with jumpstart

Fig. 2. Heatmap contrasting accuracy for neural networks trained on MOONS with depth between 1 and 150 and width between 1 and 25. The left plot is the baseline and the right plot shows the results when using jumpstart. The accuracy ranges from a low of 0.5 (black) to a high of 1.0 (beige), with the former corresponding to random guessing since the dataset has two balanced classes. (Color figure online)

Our second experiment (Table 1) evaluates convolutional neural networks trained on the MNIST dataset [40]. We test every depth from 2 to 68 in increments of 4 with every width in $\{2, 4, 8\}$, where the width refer to the number of filters per layer. The networks are implemented as before, but with a learning rate of 0.001 over 50 epochs, batch size of 1024, kernel dimensions $(3, 3)$, padding to produce an output of same dimensions as the input, Glorot uniform initialization [17], flattening before the output layer and using a baseline and a jumpstart network with 1-norm as the aggregation function $\mathcal{P}$ and loss coefficient $\lambda = 10^{-8}$.

Table 1. Summary of the results for the convolutional neural networks trained on the MNIST dataset without jumpstart (baseline) and with jumpstart.

	Baseline		Jumpstart	
	Training	Validation	Training	Validation
Best overall accuracy	0.999467	0.9885	0.999533	0.9911
Successful model	18	18	54	54
Best for depth-width pair	8	11	45	41

With jumpstart, we successfully train networks combining all widths and depths in comparison to only up to depth 12 for widths 2 and 4 and only up to depth 24 for width 8 in the baseline. In other words, only 18 baseline network trainings converge, which we denote as the successful models in Table 1.

Our third experiment (Figs. 3 and 4) evaluates convolutional networks trained on CIFAR-10 and CIFAR-100 [36]. For CIFAR-10, we test every depth in $\{10, 20, 30\}$ with every width in $\{2, 8, 16, 32, 64, 96, 192\}$. For CIFAR-100, we test depths in $\{10, 20\}$ with widths in $\{8, 16, 32, 64\}$. The networks are implemented in Pytorch [50], with learning rates $\varepsilon \in \{0.001, 0.0001\}$ over 400 epochs, batch size of 128, same kernel dimensions and padding, Kaiming uniform initialization [24], global max-avg concat pooling before the output layer, and jumpstart with 2-norm $(\mathcal{P} = L^2)$ and $\lambda \in \{0.001, 0.1\}$ or mean $(\mathcal{P} = \bar{x})$ and $\lambda \in \{0.1, 1\}$.

With jumpstart, we successfully train networks for CIFAR-10 with depth up to 30 in comparison to no more than 20 in the baseline. The best performance—0.766 for jumpstart and 0.734 for baseline—is observed for both with $\varepsilon = 0.001$, where the validation accuracy of each jumpstart experiment exceeds the baseline in 18 out of 21 depth-width pairs in one case and 20 out of 21 in another. The baseline is comparatively more competitive with $\varepsilon = 0.0001$, but the overall validation accuracy drops significantly. For CIFAR-100, the jumpstart experiments exceed the baseline in 12 out of 16 combinations of depth, width, and learning rate. The accuracy improves by 1 point in networks with 10 layers and 7.8 points in networks with 20 layers. The maximum accuracy attained is 0.37 for the baseline and 0.38 with jumpstart. The training time becomes 1.33 times greater in CIFAR-10 and 1.47 in CIFAR-100. The use of the precomputed pre-activations on the forward pass involves a similar memory cost: around 50% more.

The source code is at https://github.com/blauigris/jumpstart-cpaior.

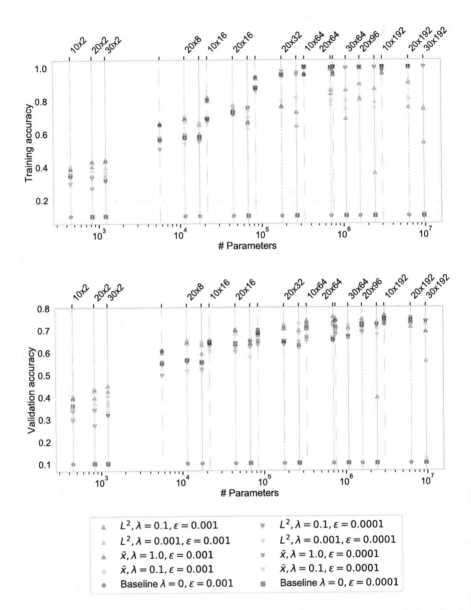

Fig. 3. Scatter chart of the number of parameters by accuracy for training (top) and validation (bottom) of convolutional neural networks trained on CIFAR-10. Some depth-width pairs are shown above the plots for reference and the gridlines are solid for depth 30, dashed for 20, and dotted for 10. The results of this experiment are plotted in this format due to their greater variability in comparison to the second experiment, which permits evaluating parameter efficiency. With same number of units but fewer parameters, the results for 20×8 are better than 10×16 and likewise for 20×32 when compared with 10×64.

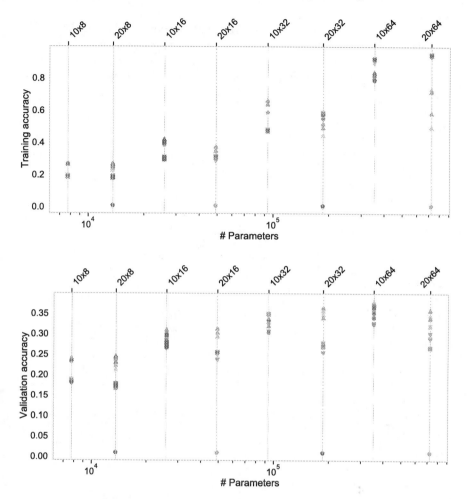

Fig. 4. Scatter chart of number of parameters by accuracy for training (top) and validation (bottom) of convolutional neural networks trained on CIFAR-100. Some depthwidth pairs are shown above the plots for reference and the gridlines are dashed for depth 20 and dotted for 10. Once certain capacity is reached at 640 units, we find that the performance for 20×32 is competitive with that of 10×64 while using less parameters.

5 Conclusion

We have presented a regularization technique for training thinner and deeper neural networks, which leads to a more efficient use of the dataset and to neural networks that are more parameter-efficient. Although massive models are currently widely popular in theory [33] and practice [2], their associated economical barriers and environmental footprint [63] as well as societal impact [5] are known concerns. Hence, we present a potential alternative to lines of work such as model compression [9] by avoiding to operate with larger models. Whereas deeper networks are often pursued, trainable thinner networks are surprisingly not.

Acknowledgements. Thiago Serra was supported by the National Science Foundation (NSF) grant IIS 2104583.

References

1. Abadi, M., et al.: TensorFlow: large-scale machine learning on heterogeneous systems (2015). https://www.tensorflow.org/
2. Amodei, D., Hernandez, D., Sastry, G., Clark, J., Brockman, G., Sutskever, I.: AI and compute (2018). https://openai.com/blog/ai-and-compute/. Accessed 23 Dec 2020
3. Arora, R., Basu, A., Mianjy, P., Mukherjee, A.: Understanding deep neural networks with rectified linear units. In: ICLR (2018)
4. Ba, J., Caruana, R.: Do deep nets really need to be deep? In: NeurIPS (2014)
5. Bender, E.M., Gebru, T., McMillan-Major, A., Shmitchell, S.: On the dangers of stochastic parrots: can language models be too big? In: FAccT (2021)
6. Bengio, Y., Simard, P., Frasconi, P.: Learning long-term dependencies with gradient descent is difficult. IEEE Trans. Neural Netw. **5**(2), 157–166 (1994)
7. Bengio, Y., Courville, A., Vincent, P.: Representation learning: a review and new perspectives (2014)
8. Bienstock, D., Muñoz, G., Pokutta, S.: Principled deep neural network training through linear programming. CoRR abs/1810.03218 (2018)
9. Blalock, D., Ortiz, J., Frankle, J., Guttag, J.: What is the state of neural network pruning? In: MLSys (2020)
10. Bridle, J.S.: Probabilistic interpretation of feedforward classification network outputs, with relationships to statistical pattern recognition. In: Soulié, F.F., Hérault, J. (eds.) Neurocomputing. NATO ASI Series, vol. 68, pp. 227–236. Springer, Berlin Heidelberg, Berlin, Heidelberg (1990). https://doi.org/10.1007/978-3-642-76153-9_28
11. Chollet, F., et al.: Keras (2015). https://keras.io
12. Cybenko, G.: Approximation by superpositions of a sigmoidal function. Math. Control Signals Syst. (MCSS) **2**(4), 303–314 (1989). https://doi.org/10.1007/BF02551274, http://dx.doi.org/10.1007/BF02551274
13. Devlin, J., Chang, M.W., Lee, K., Toutanova, K.: BERT: pre-training of deep bidirectional transformers for language understanding, 13 p. (2018). http://arxiv.org/abs/1810.04805
14. Eldan, R., Shamir, O.: The power of depth for feedforward neural networks (2016)
15. Fischetti, M., Stringher, M.: Embedded hyper-parameter tuning by simulated annealing. CoRR abs/1906.01504 (2019)
16. Fukushima, K., Miyake, S.: Neocognitron: a self-organizing neural network model for a mechanism of visual pattern recognition. In: Amari, S.I., Arbib, M.A. (eds.) Competition and Cooperation in Neural Nets. Lecture Notes in Biomathematics, vol. 45, pp. 267–285. Springer, Heidelberg (1982). https://doi.org/10.1007/978-3-642-46466-9_18
17. Glorot, X., Bengio, Y.: Understanding the difficulty of training deep feedforward neural networks. In: Proceedings of the International Conference on Artificial Intelligence and Statistics (AISTATS 2010). Society for Artificial Intelligence and Statistics (2010)

18. Glorot, X., Bordes, A., Bengio, Y.: Deep sparse rectifier neural networks. In: AIS-TATS (2011)

19. Goodfellow, I., Bengio, Y., Courville, A.: Deep Learning. MIT Press (2016). http://www.deeplearningbook.org

20. Gribonval, R., Kutyniok, G., Nielsen, M., Voigtlaender, F.: Approximation spaces of deep neural networks (2020)

21. Hahnloser, R., Sarpeshkar, R., Mahowald, M., Douglas, R., Seung, S.: Digital selection and analogue amplification coexist in a cortex-inspired silicon circuit. Nature **405**, 947–951 (2000)

22. Hasanpour, S.H., Rouhani, M., Fayyaz, M., Sabokrou, M., Adeli, E.: Towards principled design of deep convolutional networks: introducing SimpNet. CoRR abs/1802.06205 (2018). http://arxiv.org/abs/1802.06205

23. He, K., Zhang, X., Ren, S., Sun, J.: Deep residual learning for image recognition. CoRR abs/1512.03385 (2015). http://arxiv.org/abs/1512.03385

24. He, K., Zhang, X., Ren, S., Sun, J.: Delving deep into rectifiers: surpassing human-level performance on ImageNet classification. 2015 IEEE International Conference on Computer Vision (ICCV), pp. 1026–1034 (2015)

25. He, K., Zhang, X., Ren, S., Sun, J.: Delving deep into rectifiers: surpassing human-level performance on ImageNet classification. CoRR abs/1502.01852 (2015). http://arxiv.org/abs/1502.01852

26. Hecht-Nielsen, R.: Kolmogorov's mapping neural network existence theorem. In: Proceedings of the International Conference on Neural Networks, vol. 3, pp. 11–14. IEEE Press, New York (1987)

27. Hertrich, C., Basu, A., Summa, M.D., Skutella, M.: Towards lower bounds on the depth of ReLU neural networks (2021)

28. Hochreiter, S.: Untersuchungen zu dynamischen neuronalen netzen. Diploma Tech. Univ. München **91**(1) (1991)

29. Hochreiter, S., Schmidhuber, J.: Long short-term memory. Neural Comput. **9**(8), 1735–1780 (1997)

30. Huang, G., Liu, Z., van der Maaten, L., Weinberger, K.Q.: Densely connected convolutional networks. In: CVPR, pp. 2261–2269. IEEE Computer Society (2017). http://dblp.uni-trier.de/db/conf/cvpr/cvpr2017.html#HuangLMW17

31. Toro Icarte, R., Illanes, L., Castro, M.P., Cire, A.A., McIlraith, S.A., Beck, J.C.: Training binarized neural networks using MIP and CP. In: Schiex, T., de Givry, S. (eds.) CP 2019. LNCS, vol. 11802, pp. 401–417. Springer, Cham (2019). https://doi.org/10.1007/978-3-030-30048-7_24

32. Ioffe, S., Szegedy, C.: Batch normalization: accelerating deep network training by reducing internal covariate shift. CoRR abs/1502.03167 (2015). http://arxiv.org/abs/1502.03167

33. Jacot, A., Gabriel, F., Hongler, C.: Neural tangent kernel: convergence and generalization in neural networks. In: Proceedings of the 32nd International Conference on Neural Information Processing Systems, NIPS 2018, pp. 8580–8589. Curran Associates Inc., Red Hook (2018)

34. Kingma, D.P., Ba, J.: Adam: a method for stochastic optimization. CoRR abs/1412.6980 (2014). http://arxiv.org/abs/1412.6980

35. Kolen, J.F., Kremer, S.C.: Gradient flow in recurrent nets: the difficulty of learning long-term dependencies, pp. 237–243. Wiley-IEEE Press (2001). https://doi.org/10.1109/9780470544037.ch14

36. Krizhevsky, A.: Learning multiple layers of features from tiny images, pp. 32–33 (2009). https://www.cs.toronto.edu/~kriz/learning-features-2009-TR.pdf

37. Krizhevsky, A., Sutskever, I., Hinton, G.E.: ImageNet classification with deep convolutional neural networks. In: Pereira, F., Burges, C.J.C., Bottou, L., Weinberger, K.Q. (eds.) Advances in Neural Information Processing Systems, vol. 25, pp. 1097–1105. Curran Associates, Inc. (2012). http://papers.nips.cc/paper/4824-imagenet-classification-with-deep-convolutional-neural-networks.pdf

38. LeCun, Y., Bengio, Y., Hinton, G.: Deep learning. Nature **521**(7553), 436–444 (2015). https://doi.org/10.1038/nature14539, http://dx.doi.org/10.1038/nature14539

39. LeCun, Y., Bottou, L., Bengio, Y., Haffner, P.: Gradient-based learning applied to document recognition. In: Proceedings of the IEEE, vol. 86, pp. 2278–2324 (1998). http://citeseerx.ist.psu.edu/viewdoc/summary?doi=10.1.1.42.7665

40. LeCun, Y., Cortes, C.: MNIST handwritten digit database (2010). http://yann.lecun.com/exdb/mnist/

41. LeCun, Y., Touresky, D., Hinton, G., Sejnowski, T.: A theoretical framework for back-propagation. In: Proceedings of the 1988 Connectionist Models Summer School, vol. 1, pp. 21–28 (1988)

42. Lu, L., Shin, Y., Su, Y., Karniadakis, G.E.: Dying ReLU and initialization: theory and numerical examples. arXiv preprint arXiv:1903.06733 (2019)

43. Lu, Z., Pu, H., Wang, F., Hu, Z., Wang, L.: The expressive power of neural networks: a view from the width (2017)

44. Maas, A.L., Hannun, A.Y., Ng, A.Y.: Rectifier nonlinearities improve neural network acoustic models. In: in ICML Workshop on Deep Learning for Audio, Speech and Language Processing (2013)

45. Minsky, M., Papert, S.: Perceptrons: An Introduction to Computational Geometry. MIT Press, Cambridge (1969)

46. Montúfar, G., Pascanu, R., Cho, K., Bengio, Y.: On the number of linear regions of deep neural networks. In: NeurIPS (2014)

47. Nair, V., Hinton, G.: Rectified linear units improve restricted Boltzmann machines. In: ICML (2010)

48. Pascanu, R., Montúfar, G., Bengio, Y.: On the number of response regions of deep feedforward networks with piecewise linear activations. In: ICLR (2014)

49. Pascanu, R., Mikolov, T., Bengio, Y.: On the difficulty of training recurrent neural networks (2013)

50. Paszke, A., et al.: PyTorch: an imperative style, high-performance deep learning library. In: Wallach, H., Larochelle, H., Beygelzimer, A., d'Alché-Buc, F., Fox, E., Garnett, R. (eds.) Advances in Neural Information Processing Systems, vol. 32, pp. 8024–8035. Curran Associates, Inc. (2019). http://papers.neurips.cc/paper/9015-pytorch-an-imperative-style-high-performance-deep-learning-library.pdf

51. Pedregosa, F., et al.: Scikit-learn: machine learning in Python. J. Mach. Learn. Res. **12**, 2825–2830 (2011)

52. Pooladian, A., Finlay, C., Oberman, A.M.: Farkas layers: don't shift the data, fix the geometry. CoRR abs/1910.02840 (2019). http://arxiv.org/abs/1910.02840

53. Robbins, H., Monro, S.: A stochastic approximation method. Ann. Math. Stat. **22**(3), 400–407 (1951)

54. Rumelhart, D.E., Hinton, G.E., Williams, R.J.: Learning representations by back-propagating errors. Nature **323**, 533–536 (1986)

55. Serra, T., Kumar, A., Ramalingam, S.: Lossless compression of deep neural networks. In: Hebrard, E., Musliu, N. (eds.) CPAIOR 2020. LNCS, vol. 12296, pp. 417–430. Springer, Cham (2020). https://doi.org/10.1007/978-3-030-58942-4_27

56. Serra, T., Ramalingam, S.: Empirical bounds on linear regions of deep rectifier networks. In: AAAI (2020)

57. Serra, T., Kumar, A., Yu, X., Ramalingam, S.: Scaling up exact neural network compression by ReLU stability (2021)
58. Serra, T., Tjandraatmadja, C., Ramalingam, S.: Bounding and counting linear regions of deep neural networks (2018)
59. Shalev-Shwartz, S., Ben-David, S.: Understanding Machine Learning: From Theory to Algorithms. Cambridge University Press, USA (2014)
60. Shang, W., Sohn, K., Almeida, D., Lee, H.: Understanding and improving convolutional neural networks via concatenated rectified linear units. CoRR abs/1603.05201 (2016). http://arxiv.org/abs/1603.05201
61. Shin, Y., Karniadakis, G.E.: Trainability and data-dependent initialization of over-parameterized ReLU neural networks. CoRR abs/1907.09696 (2019). http://arxiv.org/abs/1907.09696
62. Simonyan, K., Zisserman, A.: Very deep convolutional networks for large-scale image recognition. CoRR abs/1409.1556 (2014). http://arxiv.org/abs/1409.1556
63. Strubell, E., Ganesh, A., McCallum, A.: Energy and policy considerations for deep learning in NLP. In: ACL (2019)
64. Szegedy, C., et al.: Going deeper with convolutions. CoRR abs/1409.4842 (2014). http://arxiv.org/abs/1409.4842
65. Tan, M., Le, Q.V.: EfficientNet: rethinking model scaling for convolutional neural networks. CoRR abs/1905.11946 (2019). http://arxiv.org/abs/1905.11946
66. Tan, M., Le, Q.V.: EfficientNetV2: smaller models and faster training. CoRR abs/2104.00298 (2021). https://arxiv.org/abs/2104.00298
67. Telgarsky, M.: Representation benefits of deep feedforward networks. CoRR abs/1509.08101 (2015)
68. Tjeng, V., Xiao, K., Tedrake, R.: Evaluating robustness of neural networks with mixed integer programming. In: ICLR (2019)
69. Vardi, G., Reichman, D., Pitassi, T., Shamir, O.: Size and depth separation in approximating benign functions with neural networks (2021)
70. Vaswani, A., et al.: Attention is all you need. In: Guyon, I., et al. (eds.) Advances in Neural Information Processing Systems, vol. 30. Curran Associates, Inc. (2017). https://proceedings.neurips.cc/paper/2017/file/3f5ee243547dee91fbd053c1c4a845aa-Paper.pdf
71. Werbos, P.J.: Applications of advances in nonlinear sensitivity analysis. In: Proceedings of the 10th IFIP Conference, 31.8 - 4.9, NYC, pp. 762–770 (1981)
72. Zagoruyko, S., Komodakis, N.: Wide residual networks. CoRR abs/1605.07146 (2016). http://arxiv.org/abs/1605.07146

Hybrid Offline/Online Optimization for Energy Management via Reinforcement Learning

Mattia Silvestri$^{(\boxtimes)}$, Allegra De Filippo, Federico Ruggeri, and Michele Lombardi

DISI, University of Bologna, Bologna, Italy
{mattia.silvestri4,allegra.defilippo,federico.ruggeri6,
michele.lombardi2}@unibo.it

Abstract. Constrained decision problems in the real world are subject to uncertainty. If predictive information about the stochastic elements is available offline, recent works have shown that it is possible to rely on an (expensive) parameter tuning phase to improve the behavior of a simple online solver so that it roughly matches the solution quality of an anticipative approach but maintains its original efficiency. Here, we start from a state-of-the-art offline/online optimization method that relies on optimality conditions to inject knowledge of a (convex) online approach into an offline solver used for parameter tuning. We then propose to replace the offline step with (Deep) Reinforcement Learning (RL) approaches, which results in a simpler integration scheme with a higher potential for generalization. We introduce two hybrid methods that combine both learning and optimization: the first optimizes all the parameters at once, whereas the second exploits the sequential nature of the online problem via the Markov Decision Process framework. In a case study in energy management, we show the effectiveness of our hybrid approaches, w.r.t. the state-of-the-art and pure RL methods. The combination proves capable of faster convergence and naturally handles constraint satisfaction.

Keywords: Deep reinforcement learning · Offline/online optimization · Uncertainty · Constrained optimization

1 Introduction

Real world constrained decision problems *often mix offline and online elements*. In many cases, a substantial amount of information about the uncertainty (e.g. in the form of historical solutions, event logs or probability distributions) is available before it is revealed, i.e. before the online execution starts. This information generally allows to make both strategic (offline) and operational (online) decisions: in production scheduling, for example, we may devise an initial plan to be revised at run time in case of disruptions; or in Energy Management Systems

© Springer Nature Switzerland AG 2022
P. Schaus (Ed.): CPAIOR 2022, LNCS 13292, pp. 358–373, 2022.
https://doi.org/10.1007/978-3-031-08011-1_24

(EMS) the electrical load should be planned the day ahead, while power flow balance should be maintained hour by hour.

The interplay of these offline and online phases has received attention in the last years [8]. Recent works [8,9] show that whenever distinct offline and online phases are present, a tighter integration can lead to substantial improvements in terms of both solution quality and computational costs. In particular, since in many application domains, efficient suboptimal algorithms for online optimization are already available or easy to design (e.g. greedy heuristics or myopic declarative models), such works exploit the available offline information to rely on a (typically expensive) parameter tuning phase to improve the behavior of the online solver, maintaining its original efficiency. [9] is based on the idea of injecting knowledge of a (convex) online approach into an offline solver. This is achieved by formulating the Karush-Kuhn-Tucker (KKT) optimality conditions for the online solver and adding them as constraints in a (offline) Mixed-Integer Programming (MIP) problem. The resulting model can be used to perform (offline expensive) parameter tuning. However, formulating optimality conditions is not trivial and requires operations research expertise. Moreover, KKT conditions introduce non-linearity to the initial model which dramatically reduces scalability. Finally, in this method, the uncertainty is managed by sampling, introducing approximations.

In this paper, we explore the idea of using learning-based approximations to lift this limitation. In particular, we employ Deep Reinforcement Learning (DRL) approaches as black-box solvers to perform (instance-specific) parameter tuning in a simpler integration scheme without requiring convexity for the online optimization problem.

We propose two hybrid approaches that combine both learning and optimization. The first one selects the parameters all at once, while the second approach exploits the sequential nature of the online problem by using the Markov Decision Process (MDP) framework.

Based on an Energy Management System case study, we show the effectiveness of our hybrid approaches, both compared to the (tuning) optimization problem from [7,9]. To demonstrate the advantages over full RL-based solutions, we have developed and compared RL end-to-end counterparts of our proposed methods. We show that the resulting hybrid approach benefits from powerful learning algorithms and is well suited to deal with operational constraints.

The rest of the paper is organized as follows. In Sect. 2 we provide a brief introduction on RL. Section 3 describes the proposed case study and the state-of-the-art approach for offline/online optimization grounded on it. Section 4 presents our proposed two hybrid approaches that combine both learning and optimization. Section 5 provides an analysis of results. Section 6 discusses the main approaches proposed in the literature focused on Deep RL, hybrid methods that combine both learning and optimization, and methods for hybrid offline/online optimization. Concluding remarks are in Sect. 7.

2   Background

Reinforcement Learning (RL) is a paradigm to solve sequential decision-making problems, defined on top of the MDP mathematical framework. Formally, a fully-observable MDP is defined by a tuple (S, A, p, r, γ), where S is the set of states, A is the set of actions, $p(\cdot|s, a)$ is the probability distribution of next states, $r(\cdot|s, a)$ is the probability distribution of the reward and $\gamma \in [0, 1]$, called discount factor, controls the impact of future rewards.

The sequential decision making problem is then cast down to a recurrent process where the RL agent interacts with the environment by performing actions according to its behavior policy $\pi_\theta(a_t|s_t)$. Consequently, these actions provoke the agent's state transitions. Based on the action outcome, a reward signal may be attributed to the RL agent. The learning process is then formulated as the maximization problem of cumulative rewards along state-action trajectories τ, dictated by $\pi_\theta(a_t|s_t)$.

$$J(\theta) = \mathbb{E}_{\tau \sim p_\theta(\tau)} \left[\sum_{t=1}^{T} \gamma^t r(s_t, a_t) \right] \tag{1}$$

$$p_\theta(\tau) = p_\theta(s_1, a_1, \ldots, s_T, a_T) = p(s_1) \prod_{t=1}^{T} \pi_\theta(a_t|s_t) p(s_{t+1}|s_t, a_t) \tag{2}$$

where T is the trajectory time horizon.

RL algorithms can be classified into two main categories: model-free and model-based RL. Model-free RL algorithms try to find the optimal policy π^* such that the expected cumulative discounted reward from the initial state $s_{t=1}$ is maximized. The idea of model-based RL is to learn the model of the environment, i.e. the transition probabilities $p(\cdot|s, a)$, rather than the optimal policy and then use the learned model to choose the optimal actions.

Within the model-free family, Policy Gradient algorithms are widely used when the actions space is continuous. One such an example is REINFORCE [20]: given a parametric policy π_θ, the parameters θ are optimized by gradient ascent to directly maximize $J(\theta)$.

$$\nabla_\theta J(\theta) = E_{\tau \sim \pi_\theta(\tau)} \left[\left(\sum_{t=1}^{T} \nabla_\theta \log \pi_\theta(a_t|s_t) \right) \left(\sum_{t=1}^{T} r(s_t, a_t) \right) \right] \tag{3}$$

Policy gradient algorithms are known to suffer from high variance. several non-mutually exclusive solutions can be employed to mitigate this issue, such as baseline subtraction to correctly isolate positive actions. Among the possible baselines, Actor-Critic (AC) methods are particularly effective in reducing variance. Instead of using a state-dependent baseline, one can reduce the variance by computing the advantage of taking an action a_t in state s_t. The advantage is defined as $A(s_t, a_t) = r + \gamma V(s_{t+1}) - V(s_t)$, where $V_\pi(s_t) = \mathbb{E}_\pi [J(\tau)|s = s_t]$ is the value function, r is the reward and s_{t+1} is the next state. Thus, the actor is represented by the policy, whereas the value function acts as the critic.

Modern RL approaches take advantage of deep learning models as power-ful tools for representation learning [15]. More precisely, neural networks are employed to approximate the policy π_θ and $V_\theta(\cdot)$. The scientific community usually refers to this research field as Deep Reinforcement Learning (DRL).

3 Problem Description

In this section, we present the details of the application scenario of an Energy Management System, and we illustrate the state-of-the-art offline/online approach grounded on it.

3.1 Energy Management Case Study

As a practical use case, we consider an Energy Management System (EMS) that requires allocating the minimum-cost power flows from different Distributed Energy Resources (DERs). The uncertainty stems from uncontrollable deviations from the planned loads of consumption and the presence of Renewable Energy Sources (RES). Based on actual energy prices and on the availability of DERs, the EMS decides: 1) how much energy should be produced; 2) which generators should be used for the required energy; 3) whether the surplus energy should be stored or sold to the energy market. Unlike in most of the existing litera-ture, we acknowledge that in many practical cases [8] *some parameters can be tuned offline*, while the energy balance should be maintained online by managing energy flows among the grid, the renewable and traditional generators, and the storage systems. Intuitively, handling these two phases in an integrated fashion should lead to some benefits, thus making the EMS a good benchmark for our integrated approach.

In our case study, it is desirable to encourage the online heuristic to store energy in the battery system when the prices of the Electricity Market are cheap and the loads are low, in anticipation of future higher users' demand. Storing energy has no profit so the online (myopic) solver always ends up in selling all the energy on the market. However, by defining a *virtual cost* parameter related to the storage system, it is possible to associate a profit (negative cost) to storing energy, which enables addressing this greedy limitation. Then, based on day-ahead RES generation and electric demand forecasts, we can find the optimal virtual costs related to the storage system to achieve better results in terms of solution quality (management costs of the energy system).

3.2 State-of-the-Art Offline/Online Approach

We refer to the integrated offline/online optimization method proposed in [7,9] that assumes *exogenous uncertainty*, and that is composed of two macro steps: an offline two-stage stochastic optimization model based on sampling and scenarios; and an online parametric algorithm, implemented within a simulator, that tries to make optimal online choices, by building over the offline decisions. The authors assume that *the online parametric algorithm is based on a convex optimization model*. Based on some configuration parameters of the online model, an offline parameter tuning step is applied. In this way, the authors can take advantage of the convexity of the online problem to obtain guaranteed optimal parameters. In particular, convexity implies that any local minimum must be a global minimum. Local minima can be characterized in terms of the KKT optimality conditions. Essentially, *those conditions introduce a set of constraints that must be satisfied by any solution that is compatible with the behavior of the online heuristic*. They can exploit this property by formulating the tuning phase as a Mathematical Program that is not a trivial task for every constrained real-world problem.

The online step is composed by a greedy (myopic) heuristic that minimizes the cost and covers the energy demand by manipulating the flows between the energy sources. We underline that this is a typical approach to handle the online optimization of an EMS [1]. The heuristic can be formulated as an LP model:

$$\min \sum_{k=1}^{n} \sum_{g \in G} c_g^k x_g^k \tag{4}$$

$$\text{s.t. } \tilde{L}^k = \sum_{g \in G} x_g^k \tag{5}$$

$$0 \le \gamma_k + \eta x_0^k \le \Gamma \tag{6}$$

$$\underline{x}_g \le x_g^k \le \overline{x}_g \tag{7}$$

For each stage k up to n, the decision variables x_g are the power flows between nodes in $g \in G$ and c_g are the associated costs. All flows must satisfy the lower and upper physical bounds $\underline{x}_g$ and $\overline{x}_g$. Index 0 refers to the storage system and the index 1 to the RES generators. Hence the virtual costs associated with the storage system are c_0^k. The battery charge, upper limit and efficiency are γ, Γ and η. The EMS must satisfy the user demand at each stage k referred to as $\tilde{L}^k$.

The baseline offline problem is modeled via MIP and relies on the KKT conditions to define a model for finding the optimal values of c_0^k for the set of sampled scenarios $\omega \in \Omega$. Such model is given by:

$$\min \frac{1}{|\Omega|} \sum_{\omega \in \Omega} \sum_{g \in G} \sum_{k=1}^{n} c_g^k x_{g,\omega}^k \tag{8}$$

$$\text{s.t. } \tilde{L}_\omega^k = \sum_{g \in G} x_{g,\omega}^k \qquad \forall \omega \in \Omega, \forall k = 1, \cdots, n \tag{9}$$

$$\underline{x}_g \le x_{g,\omega}^k \le \overline{x}_g \qquad \forall \omega \in \Omega, \forall k = 1, \cdots, n \tag{10}$$

$$0 \le \gamma_\omega^k \le \Gamma \qquad \forall k = 1, \cdots, n \tag{11}$$

$$\gamma_\omega^{k+1} = \gamma_\omega^k + \eta x_{0,\omega}^k \qquad \forall \omega \in \Omega, \forall k = 1, \cdots, n-1 \tag{12}$$

$$x_{1,\omega}^{k+1} = \hat{R}_k + \xi_{R,\omega}^k \qquad \forall \omega \in \Omega, \forall k = 1, \cdots, n \tag{13}$$

$$\tilde{L}_\omega^{k+1} = \hat{L}_k + y_k + \xi_{L,\omega}^k \qquad \forall \omega \in \Omega, \forall k = 1, \cdots, n \tag{14}$$

$\hat{R}_k$ and $\hat{L}_k$ are the estimated RES production and load, and ξ_R^k and ξ_L^k are the corresponding random variables representing the prediction errors. y_k are optimal load shifts and are considered as fixed parameters. The authors assume that the errors follow roughly a Normal distribution $N(0, \sigma^2)$ and that the variance σ^2 is such that 95% confidence interval corresponds to $\pm 10\%$ of the estimated value. $\tilde{L}_\omega^k$ is the observed user load demand for stage k of the scenario ω. Equations (12) to (14) model the transition functions.

The above formulation is free to assign variables (as long as the constraints are satisfied), whereas all decisions that are supposed to be made by the heuristic can not rely on future information. We account for this limitation by introducing, as constraints, the KKT optimality conditions for our convex online heuristic. The model achieves integration at the cost of offline computation time, because of the additional variables introduced and the presence of non-linearities.

In the following we show the KKT conditions formulation for the online heuristic in a single scenario:

$$-c_g^k = \lambda_\omega^k + \mu_{g,\omega}^k - \nu_{g,\omega}^k \qquad \forall g \in G \qquad (15)$$

$$\mu_{g,\omega}^k(x_{g,\omega}^k + \overline{x}_g) = 0 \qquad \forall g \in G \qquad (16)$$

$$\nu_{i,\omega}^k(\underline{x}_g - x_{g,\omega}^t) = 0 \qquad \forall g \in G \qquad (17)$$

$$\hat{\mu}_\omega^k(\eta x_{0,\omega}^k + \gamma^k - \Gamma) = 0 \qquad (18)$$

$$\hat{\nu}_\omega^k(\eta x_{0,\omega}^k + \gamma^k) = 0 \qquad (19)$$

$$\mu_{g,\omega}^k, \nu_{g,\omega}^k \geq 0 \qquad \forall g \in G \qquad (20)$$

$$\hat{\mu}_\omega^k, \hat{\nu}_\omega^k \geq 0 \qquad (21)$$

where $\mu_{g,\omega}^k$ and $\nu_{g,\omega}^k$ are the multipliers associated to the physical flow bounds, while $\hat{\mu}_\omega^k$ and $\hat{\nu}_\omega^k$ are associated to the battery capacity bounds. Injecting the conditions in the offline model yields:

$$\min \frac{1}{|\Omega|} \sum_{\omega \in \Omega} \sum_{g \in G} \sum_{k=1}^n c_g^k x_{g,\omega}^k$$

s.t. Eq. (9)–(14)   – offline problem constraints –

Eq. (15)–(21) $\forall \omega \in \Omega, \forall k = 1, \ldots n$ – KKT conditions –

where the decision variables are $x_{g,\omega}^k$, $\mu_{g,\omega}^k$, $\nu_{g,\omega}^k$, $\hat{\mu}_\omega^k$, $\hat{\nu}_\omega^k$. To those, the authors add the cost c_0^k associated with the flow from and to the storage system (the only parameter they allow the solver to adjust). This method allows the offline solver to associate a virtual profit for storing energy, which enables addressing the original limitation at no online computational cost.

4   Proposed Methods

Due to the limitations in terms of convexity assumption and scalability presented in Sect. 3, we devise an alternative to the TUNING approach of [7,9]. Decision-focused learning approaches are not directly applicable since the cost function

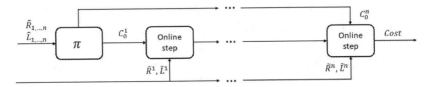

Fig. 1. In the SINGLE-STEP version, the policy π provides the set of $\{C_0^k\}_{k=1}^n$ all at once.

employed in the optimization problem takes also into account the (virtual) costs related to the storage system c_0^k, whereas the real cost to be minimized does not. In particular, we propose a hybrid method that employs DRL as a black-box tool to find the optimal c_0^k. The major benefit over the TUNING version of [7,9] is that *we do no longer require the greedy heuristic to be convex* but we are still able to compensate for its myopic behavior.

In the following sections, we will first describe our RL-based version of the TUNING algorithm. Then, to show the benefits of a hybrid approach that combines both learning and optimization, we will outline an alternative end-to-end RL method that directly provides the power flows.

4.1   RL-Based TUNING

We devise two viable ways to formulate the Reinforcement Learning problem. As shown in Fig. 1, in the first formulation (referred to as SINGLE-STEP), the policy $\pi : \mathbb{R}^{n\times 2} \to \mathbb{R}^n$ maps the day-ahead photovoltaic generation $\hat{R}^k$ and electric demand forecasting $\hat{L}^k$ to the set of all the virtual costs c_0^k for $k = \{1,\ldots,n\}$. Once c_0^k are provided, a solution $\{x_g^k\}_{k=1}^n$ is found solving the online optimization problem defined in Eqs. (4) to (7) and the reward is the negative real cost computed as:

$$-\sum_{k=1}^{n}\sum_{\substack{g\in G\\g\neq 0}} c_g^k x_g^k$$

The second formulation (referred to as MDP) exploits the sequential nature of the online step and fits the MDP framework and it is shown in Fig. 2. The policy π is a function $\pi : \mathbb{R}^{n\times 3+1} \to \mathbb{R}$. The state s_k keeps track of the battery charge γ^k and it is updated accordingly to the input and output storage flows. At each stage k, the agent's action a_k is the virtual cost c_0^k and the corresponding online optimization problem is solved. Then the environment provides as observations the battery charge γ^k, the set of forecasts $\tilde{R}_{1,\ldots,n}$ and $\tilde{L}_{1,\ldots,n}$, and a one-hot encoding of the stage k. The reward is again the negative real cost but for the only current stage k:

$$-\sum_{\substack{g\in G\\g\neq 0}} c_g^k x_g^k$$

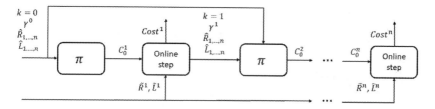

Fig. 2. In the MDP formulation, the agent sets the cost associated to the storage step-by-step, for each stage in the range from 1 to n.

The two formulations have complementary advantages and drawbacks. The single-step version is less prone to find suboptimal behaviors since it is rewarded with the actual real cost at the end of all the optimization steps. Instead, MDP receives a reward for each stage which makes it challenging to find a tradeoff between maximizing both immediate and far-in-time rewards. On the other hand, the task to be learned for MDP is simpler than for SINGLE-STEP because it has only to set one virtual cost at a time rather than deciding them all at once.

4.2   End-to-End RL

As for the hybrid approaches, we have developed both the single episode and sequential versions of the RL problem. Directly providing a feasible solution is extremely hard because the actions must satisfy all the constraints. To simplify the task, we make some architectural choices that allow for reducing the actions space.

For both the formulations, the observations are the same as for the corresponding counterparts described in Sect. 4.1. In the version equivalent to SINGLE-STEP, the output of the policy is a vector of dimension $n \times (|G| - 1)$ corresponding to the power flows x_g^k for each stage k from 1 to n. Since one of the power flows has no upper bound $\overline{x}_g$, we have set its value so that the power balance constraint of Eq. (9) is satisfied, reducing the actions space and making the task for RL easier. We refer to this decision variable as x_2^k. In the MDP version, the policy provides a $(|G| - 1)$-dimensional vector corresponding to the power flows for a single stage. The actions are clipped in the range $[-1, 1]$ and then rescaled in their feasible ranges $[\underline{x}_g, \overline{x}_g]$.

Despite adopting these architectural constraints, the actions provided by the agent may still be infeasible: the storage constraint of Eq. (12) and the lower bound $\underline{x}_2$ can be violated. Since the solutions' cost is in the range $[0, 3000]$, the policy network is rewarded with a value of -10000 when infeasible actions are selected to encourage the search for feasible solutions. The full RL version of SINGLE-STEP has the same reward of SINGLE-STEP itself. Unfortunately, this approach never founds a feasible solution during training. This is reasonable since the actions space is huge and the task extremely hard. Due to its poor performance, we do not consider this method for further investigation. In the MDP version, instead, the reward is non-zero only for the last stage and it is computed as the negative cumulative real cost. In following of the paper, we only consider this full RL method and refer to it as RL.

5 Experimental Results

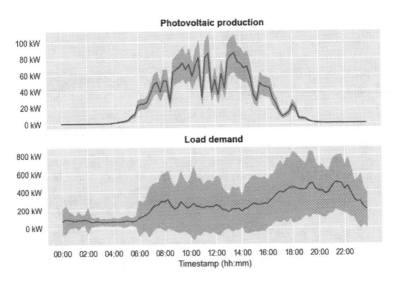

Fig. 3. Mean and standard deviation of the photovoltaic production and load demand forecasts obtained from the Public Dataset.

Training and test of the methods are performed on real data based on a Public Dataset[1]. From this dataset, we assume electric load demand and photovoltaic production forecasts, upper and lower limits for generating units and the initial status of storage units. During training of all the methods with an RL component, $\tilde{R}$ and $\tilde{L}$ are obtained from the forecasts by adding noise from a normal distribution as described in Sect. 3.2. The dataset presents individual profiles of load demand with a time step of 5 min resolution from 00:00 to 23:00. We consider aggregated profiles with a timestamp of 15 min and use them as forecasted load. The photovoltaic production is based on the same dataset with profiles for different sizes of photovoltaic units but the same solar irradiance (i.e. the same shape but different amplitude due to the different sizes of the panels used). Also in this case photovoltaic production is adopted as forecast.

To assess the variability of the dataset, in Fig. 3 we show the mean and standard deviation of photovoltaic production and user load demand regarding the hour of the day. Photovoltaic production has not a high variance and this is reasonable since it mainly depends on the solar irradiance. On the other hand, the load demand is extremely variable proving the robustness of the benchmark.

---

[1] www.enwl.co.uk/lvns.

The electricity demand hourly prices have been obtained based on data from the Italian national energy market management corporation[2] (GME) in €/MWh. The diesel price is taken from the Italian Ministry of Economic Development[3] and is assumed as a constant for all the time horizon (one day in our model) as assumed in literature [1] and from [11].

In the following, we will refer to the version of TUNING based on perfect information (i.e. without scenario sampling) as ORACLE. For SINGLE-STEP, MDP and RL, we have employed the Advantage Actor Critic (A2C) algorithm[4] since it is robust and it can deal with a continuous actions space. All the code and dataset to reproduce the results are publicly available at the following link[5]. Both training and evaluation were performed on a laptop with an Intel i7 CPU with 4 cores and 1.5 GHz clock frequency.

Since hyperparameter search was outside the scope of this paper, we employ a quite standard architecture. The policy is represented by a Gaussian distribution for each action dimension, parametrized by a feedforward fully-connected Neural Network with two hidden layers, each of 32 units and a hyperbolic tangent activation function. The critic is again a deep neural network with the same hidden architecture of the policy. Parameters are updated using Adam optimizer with a learning rate of 0.01, which is larger than the usual 0.001: we choose this value because it improves the speed of convergence without compromising the final results for our use case. Observations are rescaled in the same range $[0, 1]$ dividing by their maximum values. We have used a batch size of 100 for all the methods but MDP for which we have preferred a larger batch size of 9600 to have a comparable number of episodes for each training epoch.

For the evaluation, we randomly select 100 pairs of load demand and photovoltaic production forecasts, referred to as instances in the following of the section. Each method with a learning component (i.e. SINGLE-STEP, MDP and RL) is trained on each instance individually. Here we focus on probing the effectiveness of the proposed method and we intend to investigate the generalization capabilities in future work.

5.1   Cost Value over Computation Time

We start by comparing the mean cost on the generated realization during each training epoch as a function of the computation time, averaging the results considering the set of 100 instances. Since the optimal cost may be different among the instances, we normalize it by the best value found (i.e. the one provided by ORACLE). To make a fair comparison with TUNING, we train the methods with a learning phase choosing a number of epochs such that the computation time is similar. The mean epoch duration on the 100 instances, the number of epochs

---

[2] http://www.mercatoelettrico.org/En/Default.aspx.

[3] http://dgsaie.mise.gov.it/.

[4] A2C algorithm was implemented with the TensorFlow version of the **garage** [5] library.

[5] https://github.com/matsilv/rl-offline-online-opt.

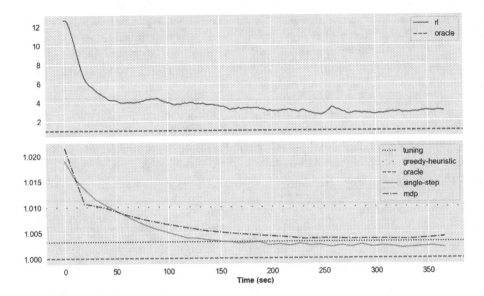

Fig. 4. Cost comparison of the methods w.r.t. the computational time.

and the total computation time required for all the methods are reported in Table 1.

Table 1. Mean epoch duration, number of epochs and total duration for the methods.

Method	Epoch duration (sec)	Num. of epochs	Total duration (sec)
SINGLE-STEP	9.85	37	364.45
MDP	19.28	19	366.32
RL	0.33	1085	358.05
ORACLE	–	–	74.13
TUNING	–	–	360.21
HEURISTIC	–	–	0.66

In the upper part of Fig. 4, RL is compared to ORACLE: despite the agent is actually minimizing the cost, it is far from being optimal. The results are so poor that we do not make a further comparison with the other methods. In the lower part of Fig. 4, our devised approaches (SINGLE-STEP and MDP) are compared with the greedy heuristic, TUNING and ORACLE. Since there is no learning for these last three methods, the solution found is used as a reference value and a simple horizontal line is plotted. Despite being extremely fast, the greedy heuristic provides considerably worse results than the oracle, due to its myopic behavior. Among our proposed methods, SINGLE-STEP provides better results and a faster convergence; in addition, it also outperforms the state-of-the-art TUNING in almost the same computation time and without requiring a convex online optimization problem. ORACLE finds the optimal solution and it is faster than our proposed methods. On the other hand, it requires perfect information so it is not applicable to real-world problems and here it is only used as a reference value to evaluate the performance of the other methods.

5.2 Decision Variables

Next, we proceed by comparing the power flows and storage capacity for each method (shown in Fig. 5). Since we introduce a virtual cost related to the battery system, our discussion focuses on storage usage. The end-to-end RL approach is only learning to satisfy the power balance and storage constraints and it does not take smarter actions to reduce the cost. One possible reason for this poor performance is the challenging exploration of the huge actions space. As one would expect, the greedy (and myopic) heuristic uses all the available energy in the storage and does not further charge the battery since it is not directly profitable. The hybrid approaches (SINGLE-STEP and MDP) have similar behaviors and extensively use the storage whereas TUNING focuses on the only hours close to the users demand peaks. The smartest decisions are taken by ORACLE which frequently resorts to the battery system but keeps the storage fully loaded for the first part day when the load demand is low.

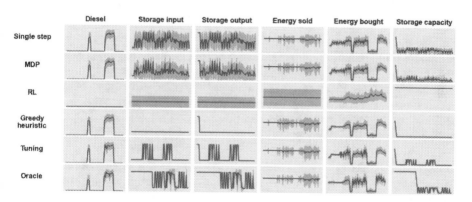

Fig. 5. Mean and standard deviation of the power flows and storage capacity w.r.t. the time for all the described methods.

6 Related Work

In this section we initially describe some recent Deep Reinforcement Learning approaches to solve combinatorial optimization problems. Then we illustrate the predict-then-optimize framework which has several properties in common with the method we have devised. The section ends with a brief overview of hybrid offline/online optimization approaches.

6.1 Deep Reinforcement Learning for Combinatorial Optimization

Recently, there has been an increasing interest in combining learning and optimization [13] with particular emphasis on DRL to solve combinatorial optimization problems [14]. Handcrafted heuristics are often used in place of exact solvers to find high-quality solutions in a reasonable time, but they require expert knowledge to be designed. Instead, DRL can learn its heuristic from a simple reward signal without any supervision. In the following, we describe the state-of-the-art methods adopting the same taxonomy proposed in [14].

Principal Learning. In principal learning, the agent directly provides a solution or takes actions that are part of the solution. One of the first attempts to solve Combinatorial Optimization problems with DRL has been made in [2] and mainly addresses the Traveling Salesman Problem. In particular, a Pointer Network [19] iteratively builds a tour by choosing the most probable remaining city at each step. The network is trained with an Actor-Critic algorithm using the negative tour length as a reward. Experimental results show that this method can achieve near-optimal solutions for tours with up to 100 nodes. In [12], the authors improve the results of [2] by replacing the Pointer Network with the Transformer architecture [18]. Similarly, [16] further extends [2] to the family of Vehicle Routing Problem (VRP). In [6], the authors develop a meta-algorithm to solve combinatorial optimization problems defined over graphs. The state is a partial solution and the set of actions is represented by the set of all possible nodes that can be added.

Generally, *DRL approaches have trouble dealing with combinatorial structures*: this issue could be addressed by injecting knowledge of the online solver into the policy itself, either by making the solver part of the environment, or by using Differentiable Programming to embed the online solver in the structure of the deep neural network. In this perspective, here we take advantage of the powerful learning framework provided by RL and rely on Declarative Optimization to deal with operational constraints.

Joint Training. Alternatively to principal learning, the policy can be jointly trained with an off-the-shelf solver to improve the solution quality or other performance metrics. For instance, in [3] DRL is employed as a value-selection heuristic to improve Constraint Programming searching strategy. Rather than *constructing* a solution, one can think of using RL to improve an already existing one, similarly to Local Search. For example, NeuRewriter [4] is an Actor Critic algorithm that learns a solution improvement heuristic by iteratively re-writing part of the solution until convergence is reached. Two policies are learned simultaneously: the region-picking and rule-picking policies. The region-picking policy chooses which part of the solution should be re-written, whereas the rule-picking policy selects the re-writing rule.

Despite the RL algorithm being trained end-to-end with the online solver, our proposed method is different from the approaches described above: *the agent is not directly integrated into a step of the solutions process of a pre-defined solver.*

Instead, it applies a parameters tuning phase separated from the optimization step and that guides the solver.

6.2   Predict-then-Optimize

Our approach is related to the family of decision-focused learning. Many real-world problems require a predictive model whose predictions are given as input to a combinatorial optimization problem. In decision-focused learning, the training of the predictive model is improved by taking into account the solutions of the optimization problem.

One such example is the Smart "Predict, then Optimize" (SPO) framework [10]: rather than simply minimizing the prediction error, the model is trained to provide estimates such that optimal solutions are found. Training is usually performed in a supervised fashion and the major challenge of this kind of approach is finding a differentiable and computational-efficient loss function, like the SPO+ that was proposed in [10].

Our method differs from decision-focused learning since *we allow for a discrepancy between the true cost that needs to be minimized and the cost function technically employed in the optimization problem.* As an additional benefit, we do not require differentiability on the cost function. This is the reason why we adopt RL rather than a supervised method in the learning stage.

6.3   Hybrid Offline/Online Optimization

Stochastic optimization problems are usually solved via offline or online methods. *Offline* approaches find a robust solution taking into account future uncertainty in advance but they are computationally expensive. On the other side, *online* algorithms take decisions once uncertainty is revealed but the solution quality is strictly affected by the available amount of computation time.

In many real-world cases, a large amount of information about the stochastic variables is available before the uncertainty is revealed. For example, in the energy management case study, historical data about past user demands can be used to model the uncertainty. This motivates the interest in developing hybrid offline/online approaches and taking advantage of both worlds to improve in terms of solution quality and computational cost.

If we model an n-stage stochastic optimization problem as a Markov Decision Process [17] then Dynamic Programming can be seen as a hybrid offline/online optimization approach. The policy and its corresponding value-function are iteratively improved offline, simulating executions, and then the resulting policy can be efficiently executed online.

When fast but sub-optimal, online algorithms are available (e.g. greedy heuristic), their behavior can be improved via a parameter tuning procedure without introducing additional computational cost during the online phase.

In [7,9], the authors propose a method to inject knowledge about a convex online solver in the offline problem. In practice, this is achieved by adding the KKT optimality conditions for the online solver as constraints in the offline problem. The method achieves positive results in cost/quality tradeoff by taking advantage of the offline/online integration. However, formulating optimality conditions is not trivial requiring experience and domain knowledge. Moreover, KKT conditions introduce non-linearity to the initial model and dramatically reduce scalability. Finally, in this method, the uncertainty is managed by sampling, introducing approximations.

The major benefit of our learning/optimization hybrid methods over the tuning version of [7,9] is that *we do no longer require the greedy heuristic to be convex* but we are still able to compensate for its myopic behavior.

7   Conclusions

This paper makes a significant step towards hybrid learning/optimization approaches for offline/online optimization under uncertainty.

We start from a state-of-the-art offline/online optimization method that makes offline parameter tuning by relying on optimality conditions to inject knowledge of a (convex) online approach into an offline solver. Then, we propose two approaches to replace this offline parameter tuning phase, by using DRL as a black-box solver. We present two hybrid methods that combine both learning and optimization: the first one optimizes all the parameters at once, whereas the second approach exploits the sequential nature of the online problem via the MDP framework.

In a case study in energy management, we show the effectiveness of our hybrid approaches w.r.t. the state-of-the-art methods. We also experimentally assess that a full RL-based approach struggles to find feasible solutions and its performance are poor compared to the state-of-the-art and our devised methods. The combination of RL and optimization proves capable of faster convergence and naturally handles constraint satisfaction. In contrast to current state-of-the-art approaches for offline/online optimization, our hybrid method has the potential to generalize: we leave probing generalization as an open question and future research direction.

Acknowledgements. This work has been partially supported by European ICT-48-2020 Project TAILOR (g.a. 952215). We thank professor Michela Milano (University of Bologna) for the valuable discussions.

References

1. Aloini, D., Crisostomi, E., Raugi, M., Rizzo, R.: Optimal power scheduling in a virtual power plant. In: 2011 2nd IEEE PES International Conference and Exhibition on Innovative Smart Grid Technologies, pp. 1–7 (2011)
2. Bello, I., Pham, H., Le, Q.V., Norouzi, M., Bengio, S.: Neural combinatorial optimization with reinforcement learning. arXiv preprint arXiv:1611.09940 (2016)

3. Cappart, Q., Moisan, T., Rousseau, L.M., Prémont-Schwarz, I., Cire, A.: Combining reinforcement learning and constraint programming for combinatorial optimization. arXiv preprint arXiv:2006.01610 (2020)
4. Chen, X., Tian, Y.: Learning to perform local rewriting for combinatorial optimization (2019)
5. Garage contributors, T.: Garage: a toolkit for reproducible reinforcement learning research (2019). https://github.com/rlworkgroup/garage
6. Dai, H., Khalil, E.B., Zhang, Y., Dilkina, B., Song, L.: Learning combinatorial optimization algorithms over graphs. arXiv preprint arXiv:1704.01665 (2017)
7. De Filippo, A., Lombardi, M., Milano, M.: Methods for off-line/on-line optimization under uncertainty. In: IJCAI, pp. 1270–1276 (2018)
8. De Filippo, A., Lombardi, M., Milano, M.: The blind men and the elephant: integrated offline/online optimization under uncertainty. In: IJCAI (2020)
9. De Filippo, A., Lombardi, M., Milano, M.: Integrated offline and online decision making under uncertainty. J. Artif. Intell. Res. **70**, 77–117 (2021)
10. Elmachtoub, A.N., Grigas, P.: Smart "predict, then optimize". Manag. Sci. **68**, 9–26 (2021)
11. Espinosa, A., Ochoa, L.: Dissemination document "low voltage networks models and low carbon technology profiles." University of Manchester, Technical report (2015)
12. Kool, W., Van Hoof, H., Welling, M.: Attention, learn to solve routing problems! arXiv preprint arXiv:1803.08475 (2018)
13. Lodi, A., Zarpellon, G.: On learning and branching: a survey. TOP **25**(2), 207–236 (2017). https://doi.org/10.1007/s11750-017-0451-6
14. Mazyavkina, N., Sviridov, S., Ivanov, S., Burnaev, E.: Reinforcement learning for combinatorial optimization: a survey. Comput. Oper. Res. **134**, 105400 (2021)
15. Mnih, V., et al.: Human-level control through deep reinforcement learning. Nature **518**(7540), 529–533 (2015)
16. Nazari, M., Oroojlooy, A., Snyder, L.V., Takáč, M.: Reinforcement learning for solving the vehicle routing problem. arXiv preprint arXiv:1802.04240 (2018)
17. Puterman, M.L.: Markov Decision Processes: Discrete Stochastic Dynamic Programming. John Wiley & Sons, Hoboken (2014)
18. Vaswani, A., et al.: Attention is all you need. In: Advances in Neural Information Processing Systems, pp. 5998–6008 (2017)
19. Vinyals, O., Fortunato, M., Jaitly, N.: Pointer networks. arXiv preprint arXiv:1506.03134 (2015)
20. Williams, R.J.: Simple statistical gradient-following algorithms for connectionist reinforcement learning. Mach. Learn. **8**(3), 229–256 (1992)

Enumerated Types and Type Extensions for MiniZinc

Peter J. Stuckey$^{(\boxtimes)}$ and Guido Tack

Department of Data Science and Artificial Intelligence, Monash University,
Melbourne, Australia
{peter.stuckey,guido.tack}@monash.edu

Abstract. Discrete optimisation problems often reason about finite sets of objects. While the underlying solvers will represent these objects as integer values, most modelling languages include enumerated types that allow the objects to be expressed as a set of names. Data attached to an object is made accessible through given arrays or functions from object to data. Enumerated types improve models by making them more self documenting, and by allowing type checking to point out modelling errors that may otherwise be hard to track down. But a *frequent modelling pattern* requires us to add new elements to a finite set of objects to represent extreme or default behaviour, or to combine sets of objects to reason about them jointly. Currently this requires us to map the extended object sets into integers, thus losing the benefits of using enumerated types. In this paper we introduce enumerated type extension, a restricted form of discriminated union types, to extend enumerated types without losing type safety, and default expressions to succinctly capture cases where we want to access data of extended types. The new language features allow for more concise and easily interpretable models that still support strong type checking and compilation to efficient solver-level models.

1 Introduction

Discrete optimisation models often reason about a set of given objects, and make use of data defined on those objects. In MiniZinc [9] (and other CP modelling languages) the core way of representing this information is as an *enumerated type* defining the objects, and arrays indexed by the enumerated type to store the data. Given that debugging constraint models can be quite difficult, particularly if the solver simply fails after a large amount of computation, an important role of enumerated types in modelling languages is to provide *type safety*. Many subtle errors can be avoided if we use strong type checking based on the enumerated types. Indeed MiniZinc and other languages such as Essence [4] provide strong type checking of enumerated types.

One of the greatest strengths of constraint programming modelling languages is the use of variable index lookups, i.e., looking up an array with a decision variable, supported in CP solvers by the `element` constraint. Variables in CP models are often declared specifically for this purpose. Accessing an array with an

P. Schaus (Ed.): CPAIOR 2022, LNCS 13292, pp. 374–389, 2022.
https://doi.org/10.1007/978-3-031-08011-1_25

incorrect index is one of the most common programming mistakes, and replacing integer index sets with enumerated types is a powerful technique that turns these mistakes into static compiler errors. This means that in order to index arrays with variables, we require variables that range over an enumerated type. Note that in the relational semantics [3] used by MiniZinc, the undefinedness from looking up an array at a non-existing index leads to falsity rather than a runtime abort, which may be difficult to detect if it does not occur in a root context (where the constraints have to hold). Hence, enumerated types and type checking are arguably even more important for constraint modelling languages.

However, real models are usually more complex, and quickly reach the limits of the current support for enumerated types. Although we define a set of objects to reason about, often modellers need to (a) add additional objects to the set to represent extreme or exceptional cases, and/or (b) reason about two sets of objects jointly. Currently we can resolve this problem by mapping the objects to integers and reasoning about index sets which are subsets of integers. But in doing so we lose the advantages of strong type checking.

In this paper we introduce mechanisms into MiniZinc that enable enumerated types to be defined by extending or joining other enumerated types, in a type-safe way. A review of the MiniZinc benchmark library reveals that many models can benefit from these new features.

2 Preliminaries

We give a brief introduction to MiniZinc in order to help parse the example code in the paper. A model consists of a set of variable declarations, constraints and predicate/function definitions, as well as an optional objective. Basic types (for our purposes) are integers, Booleans and enumerated types. MiniZinc also supports sets of these types. MiniZinc uses the notation $l..u$ to indicate the integer interval from l to u including the endpoints. We can define parameters and variables of these types, using a declaration [var] T: *varname* [= *value*] where T is a basic or set type or interval. Variable sets must be over integers or enumerated types. The optional *value* part can be used to initialise a parameter or variable. We can define multi-dimensional arrays in the form `array`[$indexset_1, indexset_2, .., indexset_n$] `of` [Var] T: *arrayname* where each *indexset_i* must be a range of either integers or an enumerated type.

One of the most important constructs in MiniZinc are array comprehensions written [e | *generator(s)*], where e is the expression to be generated. Generator expressions can be i `in` S where i is a new iterator variable and S is a set, or i `in` a where a is an array. These cause i to take values in order from the set or array. Optionally they can have a `where` *cond* expression which limits the generation to iterator values that satisfy the condition *cond*. We concatenate one dimensional arrays together using the `++` operator.

Generator call expressions of the form f (*generators*) (e) are syntactic sugar for f ([e | *generators*]). The most important functions used in generator call expressions are `forall` (conjoining the elements of the array), `exists` (disjoining the elements of the array) and `sum` (summing up the array elements).

Conditionals are of the form if *cond* then *thenexpr* else *elseexpr* endif. They evaluate as *thenexpr* if *cond* is true and *elseexpr* otherwise. Note that *cond* need not be a fixed Boolean expression, but may be decided by the solver [10].

Finally we occasionally use array slicing notation. In a two dimensional array a the expression $a[i, ..]$ returns the one dimensional array $[a[i, j] \mid j \in indexset_2]$ where $indexset_2$ is the second (declared) index set of array a.

3 Enumerated Types

An enumerated type is a simple type consisting of a named set of objects. Enumerated types are common to almost all programming languages as well as many modelling languages. They can be just syntactic sugar for integers, as in C; or they can be treated as distinct types by the type checker, as in Haskell, TypeScript, or MiniZinc, giving stronger checking of programs and models. Enumerated types are a special case of discriminated union types.

This section gives an overview of the existing enumerated type support in MiniZinc. Enumerated types are declared using the keyword enum. For example an enumerated type of colours might be

```
enum COLOUR = { Red, Orange, Yellow, Green, Blue, Violet };
```

which declares not only the type COLOUR, but six constant colour identifiers. These identifiers can then be used throughout the model. A model can also simply define the name of an enumerated type:

```
enum COLOUR;
```

which is then specified in a data file as

```
COLOUR = { Red, Orange, Yellow, Green, Blue, Violet };
```

Alternatively an *anonymous* enumerated type may be constructed using anon_enum. For example, imagine we are colouring a graph with n colours, we may use

```
COLOUR = anon_enum(n);
```

to specify the colours.

Definition 1. *In MiniZinc, an enumerated type is defined using the syntax*

$$\langle enum\text{-}declaration\rangle \rightarrow \textbf{enum } \textbf{\textit{id}} \; [= \langle enum\rangle \;]$$
$$\langle enum\rangle \rightarrow \{ \; \langle list\text{-}of\text{-}id\rangle \; \}$$
$$\langle enum\rangle \rightarrow \textbf{anon\_enum} \; (\; \langle expr\rangle \;)$$
$$\langle list\text{-}of\text{-}id\rangle \rightarrow \langle list\text{-}of\text{-}id\rangle \; , \; \textbf{\textit{id}}$$
$$\langle list\text{-}of\text{-}id\rangle \rightarrow \textbf{\textit{id}}$$

where **id** *is a MiniZinc identifier and expr is an (integer) expression. The identifiers defined in different enumerated types are required to be distinct.* □

Values of an enumerated type naturally represent a set of different, ordered (as given in the list) objects. Operators such as =, !=, <, >= and functions such as min and max have the natural definition on enumerated types. MiniZinc also supports the partial function enum_succ (the next element in the type) and enum_prev (the previous element in the type).

The most common use of enumerated types is as a set to iterate over in constraints. For example a simple knapsack problem can be defined by

```
enum PRODUCT;  int: budget;
array[PRODUCT] of int: price;
array[PRODUCT] of int: profit;
array[PRODUCT] of var bool: chosen;
constraint sum(i in PRODUCT)(price[i]*chosen[i]) <= budget;
solve maximize sum(i in PRODUCT)(profit[i]*chosen[i]);
```

One of the great strengths of CP modelling is the use of global constraints. While global constraints are defined on integers, we often want to apply them to enumerated types. In MiniZinc this is accomplished by treating enumerated types as subtypes of integers, and automatically coercing them to integers when required. For example in the model where we are ordering people in a line

```
enum PERSON;
enum ORDER = anon_enum(card(PERSON));
array[ORDER] of var PERSON: x;
constraint alldifferent(x);
```

the alldifferent constraint acts on PERSON which are automatically coerced into the integers $\{1, \ldots, n\}$ where n is the number of elements in PERSON. This also applies when we apply arithmetic operations, e.g. the successor function is similar in effect to $x + 1$, which has the effect of taking an enumerated type value x, coercing it to an integer and adding one, returning an integer. In order to map back from integers, MiniZinc supports the to_enum partial function which maps an integer back to an enumerated value (when possible), e.g. x = to_enum(PERSON, y+1) returns the successor of PERSON y. According to the relational semantics of MiniZinc, to_enum will become false in the enclosing Boolean context if the given integer is outside of the valid values of the enumerated type.

One of the reasons that enumerated types are critically important to CP modelling languages is the use of variable array lookups. Frequently CP models make use of the fact that we can build constraints where array lookups depend on variables (implemented in solvers by the element constraint). Consider an alternate knapsack model where we are restricted to take exactly k items:

```
int: k;  int: budget;
enum PRODUCT;
array[PRODUCT] of int: price;
array[PRODUCT] of int: profit;
array[1..k] of var PRODUCT: chosen;
```

```
constraint alldifferent(chosen);
constraint sum(i in 1..k)(price[chosen[i]]) <= budget;
solve maximize sum(i in 1..k)(profit[i]);
```

Note the second last line where we use a variable of enumerated type to look up the price of a product. This is a powerful feature of CP modelling languages.

In a language without strict type checking for enumerated types, this model will run and give seemingly meaningful answers (as long as there are more than k products). With strong type checking, a type error is reported, illustrating that the last line should read

```
solve maximize sum(i in 1..k)(profit[chosen[i]]);
```

4   Type Extensions

Enumerated types are a powerful modelling tool, and strict type checking has significant benefits, since the kind of errors that can arise without it may not necessarily be obvious to track down during the solving of the model. But together they may make it hard to express some reasonably common modelling patterns. One such modelling pattern is that we often want to reason about two or more sets of objects in the same way.

Example 1. Consider the usual objects for a vehicle routing problem:

```
enum CUSTOMER;   % set of customers to be served
enum TRUCK;      % set of trucks to deliver
```

The common *grand tour* modelling of such problems constructs a set of nodes: one node for each customer and two for each truck, a start node representing its leaving the depot, and an end node representing its return. Currently to represent such nodes we are forced to use integers, e.g.

```
set of int: NODE = 1..card(CUSTOMER)+2*card(TRUCK);
array[NODE] of var NODE:  next;  % next node after this one
array[NODE] of var TRUCK: truck; % truck visiting node
```

This means we give up on type checking, risking the possibility of subtle modelling errors, particularly when doing arithmetic to access the truck nodes.   □

In order to avoid moving to integers, we propose *enumerated type extensions*. They allow us to create new enumerated types by mapping existing enumerated types using type constructors and possibly adding new elements.

4.1   Syntax and Examples

Definition 2. *An enumerated type extension is defined by extending the syntax*

$$\langle enum \rangle \rightarrow \textbf{\textit{id}} \; (\; \langle enum \rangle \;)$$
$$\langle enum \rangle \rightarrow \langle enum \rangle \; \text{++} \; \langle enum \rangle$$

The first rule builds a new enumerated type from an existing one via a constructor function, while the second rule allows concatenation of enumerated types.   □

Example 2. To express the node type using type extension we would write

```
enum NODE = C(CUSTOMER) ++ S(TRUCK) ++ E(TRUCK);
```

The new enumerated type has one element per customer and two per truck. We can access the names of the elements using the constructor functions, so e.g. the node for customer c is C(c) and the end node for truck t is E(t).   □

The order of the elements in the extension types is given by the order in the definition. In the example, customer nodes are before start nodes, which are before end nodes. The definition automatically creates the constructor function and its inverse, e.g. C(.) and $C^{-1}(.)$.[1] The inverse functions are partial. For example, $C^{-1}(S(t))$, which attempts to map the start node of truck t back to a customer, will become false in its enclosing Boolean context. We extend the constructor function to also work on sets of the base type, e.g. C(CUSTOMER) returns all the customer nodes.

Example 3. Given the NODE type defined in Example 2, we can set up the constraints on the trucks visiting each node as follows:

```
constraint forall(n in NODE diff E(TRUCK))
                  (truck[next[n]] = truck[n]);
constraint forall(t in TRUCK)(truck[S(t)] = t /\ truck[E(t)] = t);
```

That is, the truck visiting a node also visits its successor for all but the end nodes. And each truck visits its own start and end nodes.   □

Note how the rules for type extension support concatenation of arbitrary enumerated types, not just the new constructor functions. This allows us to add "extra" elements to an enumerated type, as shown in the following example.

Example 4. A common modelling trick for vehicle routing problems where not every customer needs to be visited is to add a dummy truck, and all non-visited customers are "visited" by this truck. We can extended the enumerated type as

```
enum TRUCKX = T(TRUCK) ++ { DUMMYT };
```

The NODE type would then use TRUCKX instead of TRUCK. Now imagine we need to check that the individual trucks each visit between *mincust* and *maxcust* customers, and no more than *misscust* are not visited.

```
int: mincust;  % minimum customers visited by each truck
int: maxcust;  % maximum customers visited by each truck
int: misscust; % maximum missed customers
array[NODE] of var TRUCKX: truck; % truck or dummy visiting node
constraint global_cardinality_low_up([ truck[C(c)] | c in CUSTOMER],
                       TRUCKX,
                       [ mincust | t in TRUCK ] ++ [0],
                       [ maxcust | t in TRUCK ] ++ [misscust]);
```

---

[1] This can be written both in ASCII as C^-1 or using the Unicode character for $^{-1}$.

The global cardinality constraint restricts the lower and upper bounds of the number of customers visited by each truck (including the dummy).   □

Type extension is also useful for anonymous enumerated types, in particular if we have two or more anonymous enumerated types that we need to treat both separately and together.

Example 5. Consider a model for a graceful bipartite graph [5] defined as:

```
int: left;   enum LEFT = anon_enum(left);
int: right;  enum RIGHT = anon_enum(right);
array[LEFT,RIGHT] of var bool: e;      % edges
var int: m = count(e);                 % number of edges
enum NODE = L(LEFT) ++ R(RIGHT);
array[NODE] of var 0..left*right: label; % node label
constraint forall(n in NODE)(label[n] <= m);
constraint alldifferent(label);
constraint alldifferent_except_0([ e[l,r]*abs(label[L(l)] - label[R(r)])
                                | l in LEFT, r in RIGHT ]);
```

The left and right nodes in the bipartite graph are separate anonymous enumerated types. The graph itself is represented by a 2D array of Booleans indicating which edges exist. We need to label nodes with different values from 0 to m where m is the number of edges. But the nodes are from two different classes, LEFT and RIGHT, so the NODE type is instrumental to defining the model. Finally each (existing) edge should be labelled with a different number from 1 to m.   □

4.2   Pattern Matching and Range Notation

We extend the generator syntax of MiniZinc to include *pattern matching*, to make it easier to reason about different cases. In MiniZinc one can write a generator x in a where a is an array, so x takes the value of all elements of the array in turn. Once we have extended enumerated types it is worth extending this syntax to allow pattern matching: $P(x)$ in a iterates through all elements of the form $P(b)$ in a, setting pattern variable x to b for each such element.

Example 6. Consider a model for scheduling search and rescue teams of up to size members made up of humans, robots, and dogs. Each dog must be paired with their handler, and a robot requires a team member qualified to run them.

```
int: size;
set of int: TEAM = 1..8;
enum PERSON;
enum DOG;
array[DOG] of PERSON: handler;
enum ROBOT;
array[PERSON] of set of ROBOT: skills;
enum MEMBER = P(PERSON) ++ D(DOG) ++ R(ROBOT) ++ { NOONE };
enum ZONE; % Zone to be searched
```

```
array[ZONE,TEAM] of var MEMBER: x;
% Each dog is paired with their handler
constraint forall(z in ZONE, D(d) in x[z,..])
                  (exists(t in TEAM)(x[z,t] = P(handler[d])));
% Each robot is in a team with the skills to run it
constraint forall(z in ZONE, R(r) in x[z,..])
                  (exists(P(p) in x[z,..])(r in skills[p]));
```

The constraints for dogs iterate over the zones and apply a constraint to team members matching the pattern $D(d)$. The robot constraints use pattern matching twice: to match the robots in a team, and to find the matching person.   □

4.3   Implementing Enumerated Type Extension

Enumerated type extension allows for type safe construction of new types. Interestingly we can implement this feature entirely as syntactic sugar, i.e., by automatically rewriting extended enumerated types into standard MiniZinc.

In order to implement this feature, the MiniZinc lexer and parser need to be extended so that they recognise the new syntax. The type checking phase of the compiler is extended to introduce the new enumerated type, the constructor functions and inverse constructors. It makes use of the fact that we can always map enumerated types to integers.

Example 7. Consider the NODE type defined in Example 2. This is translated to a series of definitions:

```
int: nc = card(CUSTOMER);
int: nt = card(TRUCK);
enum NODE = anon_enum(nc + nt + nt);
function var NODE: C(var CUSTOMER: c) = to_enum(NODE,c);
function var NODE: S(var TRUCK: t) = to_enum(NODE,nc + t);
function var NODE: E(var TRUCK: t) = to_enum(NODE,nc + nt + t);
function var CUSTOMER: C⁻¹(var NODE: n) = to_enum(CUSTOMER,n);
function var TRUCK: S⁻¹(var NODE: n) = to_enum(TRUCK,n - nc);
function var TRUCK: E⁻¹(var NODE: n) = to_enum(TRUCK,n - nc - nt);
```

We introduce a new anonymous enumerated type of the right size. Each of the constructor functions coerces the original enumerated types to nodes using the to_enum function. Each of the inverse constructor functions performs the reverse coercion. Note that $to\_enum(E,i)$ is a partial function which is undefined if the integer second argument i is outside $1..card(E)$. This gives exactly the right behaviour for the partial inverse constructors. In the implementation we extend the constructors to also work on sets, and return sets, and generate specialized versions for when the input argument is fixed at compile time.   □

Pattern matching expressions are again treated as syntactic sugar. The expression [$g(x)$ | P(x) in a] where x has type T is mapped to

[if e in P(T) then g(P⁻¹(e)) else <> endif | e in a]

The absent value <> acts as an identity element for the operator applied to the array, for more details see [8]. For special cases, in particular where a has par type (i.e., the test whether an element in a has the constructor P can be performed at compile time), we can avoid the creation of an array containing <> elements, but we leave out the details for brevity.

Example 8. The pattern matching in Example 6 is translated to

```
constraint forall(z in ZONE, e in x[z,..])
    ( if e in D(DOG) then
        exists(t in TEAM)(x[z,t] = P(handler[D⁻¹(e)]))
      else true endif);
constraint forall(z in ZONE, e in x[z,..])
    ( if e in R(ROBOT) then
        exists(f in x[z,..])
            ( if f in P(PERSON) then R⁻¹(e) in skills[P⁻¹(f)]
              else false endif )
      else true endif);
```

where the compiler has replaced the absent value <> by the correct identity elements, false for the exists, and true for the two forall functions.   □

5   Defaults

In MiniZinc, objects represented as enumerated types are usually implemented via arrays indexed by an object identifier. Type safety will check that we only access these arrays with the correct type. But this will often require us to guard the access to avoid undefinedness.

Example 9. In the vehicle routing problem, a critical part of the model is deciding arrival times at each node, based on some travel time matrix. Given data on customers and locations

```
enum LOCATION;  % set of locations of interest
array[CUSTOMER] of LOCATION: loc;    % location of customer
LOCATION: depot;                     % depot location
array[LOCATION,LOCATION] of int: tt; % travel time loc -> loc
array[CUSTOMER] of int: service;     % service time at customer
```

A model could decide the arrival time at each node as follows.

```
int: maxtime; set of int: TIME = 0..maxtime;
array[NODE] of var TIME: arrival;    % arrival time at node
constraint forall(t in TRUCK)(arrival[S(t)] = 0);  % start nodes
constraint forall(n in NODE diff E(TRUCK))(
    arrival[next[n]] >= arrival[n] +
    if n in C(CUSTOMER) then service[C⁻¹(n)] else 0 endif +
    tt[if n in C(CUSTOMER) then loc[C⁻¹(n)] else depot endif,
        if next[n] in C(CUSTOMER) then loc[C⁻¹(next[n])] else depot endif]
);
```

Note that we need to guard the lookup of the service time and location arrays to check that the node represents a customer, and then extract the customer from the node name.   □

5.1   The `default` Operator

The guarding of data lookups, as well as the use of the inverse constructor to extract the subtype information is verbose. In order to shorten models and make them more readable we introduce *default* expressions into MiniZinc. Default expressions are not *directly related* to type extensions, rather they are a way of capturing undefinedness. However, they become particularly useful due to the addition of (partial) inverse enum constructors.

In MiniZinc expressions can be undefined, and take the value ⊥, as a result of division by zero, or by accessing an array out of bounds. The undefined value percolates up the expression, making all enclosing expressions also undefined ⊥ until a Boolean expression is reached where the undefinedness is interpreted as *false*; thus following the relational semantics treatment of undefinedness in modelling languages [3].

Many languages feature similar functionality. For example, C/C++ programmers may use a ternary operator to guard against `nullptr`. Haskell programmers would use the `maybe` function, and in Rust you might use `unwrap_or`.

Definition 3. *The default expression* x **default** y *takes the value* x *if* x *is defined (not equal to* ⊥*) and* y *otherwise. If* x *and* y *are both* ⊥ *the expression evaluates to* ⊥*.*   □

Example 10. With default expressions we can drastically shorten the arrival time reasoning shown in Example 9:

```
constraint forall(n in NODE diff E(TRUCK))(
    arrival[next[n]] >= arrival[n] +
    service[C⁻¹(n)] default 0 +
    tt[loc[C⁻¹(n)] default depot, loc[C⁻¹(next[n])] default depot]
)
```

The partial function $C^{-1}(n)$ results in undefinedness when node `n` is not a customer node. This also makes the resulting array lookup undefined, which is then replaced by the default value.   □

Default expressions are also useful for guarding other undefinedness behaviour. For example to calculate the minimum positive value occurring in a list, or return 0 if there are none, we can write

```
var int: minval = min([x | x in xs where x > 0]) default 0;
```

Defaults can also be useful for simplifying integer reasoning.

Example 11. A frequent idiom in constraint models over 2D representations of space is to use a matrix indexed by ROW and COL(umn). But then care has to be taken when indexing into the matrix. Imagine choosing k different positions in a matrix where the sum of (orthogonally) adjacent positions is non-negative. A model encoding this is

```
int: nrow; set of int: ROW = 1..nrow;
int: ncol; set of int: COL = 1..ncol;
array[ROW,COL] of int: m;  % given matrix
array[1..k] of var ROW: y; % row position chosen
array[1..k] of var COL: x; % col position chosen
constraint alldifferent([y[i]*ncol + x[i] | i in 1..k]);
constraint forall(i in 1..k)
                (sum(dr in -1..1, dc in -1..1 where abs(dr)+abs(dc) = 1)
                    (if y[i]+dr in ROW /\ x[i]+dc in COL
                     then m[y[i]+dr,x[i]+dc] else 0 endif) >= 0);
```

Notice that the model has to guard against the possibility that the position chosen is on one of the extreme rows or columns, e.g. `y[i] = 1`, since when `dr = -1` the lookup of `m` will fail and the relational semantics [3] will make the sum false. We can replace the sum if-then-else-endif expression simply by `m[y[i]+dr,x[i]+dc] default 0`.   □

5.2   Implementing Defaults

A naive implementation would simply replace the expression x `default` y by

`if defined(x) then x else y endif`

given a suitable built-in function `defined`. Internally, the MiniZinc compiler already evaluates each expression into a pair of values: the result value of the expression, and a Boolean that signals whether the result is defined. We therefore chose to implement the `default` operator as a special built-in operation that can directly access the partiality component.

For the use case where the undefinedness arises from array index value out of bounds, the motivating case we consider, the MiniZinc compiler can choose to implement the default in a more efficient way than using if-then-else-endif.

For an expression $a[i]$ `default` y where i may possibly be outside the index set I of a we can build an extended array ax over the index set $lb(i)..ub(i)$, where $ax[i] = y$ for $i \notin I$, where $lb(i)$ ($ub(i)$) is the least (greatest) value in the declared domain of i.

We can extend this rewriting also to expressions of the form $a[f(i)]$ `default` y where f is a (possibly partial) function, by building an array ax over the index set $lb(i)..ub(i)$ where $ax[i] = y$ for $f(i) \notin I$ (including the case that $f(i)$ is not defined) and $ax[i] = a[f(i)]$ otherwise.

Example 12. This is particularly useful for undefinedness that results from the use of inverse constructors. Here we extend the array type to the full supertype NODE. Consider the arrival time constraint shown in Example 10. The automatic translation of defaults as extended arrays would then be

```
array[NODE] of int: servicex = array1d(NODE,
    [ if n in C(CUSTOMER) then service[C⁻¹(n)] else 0 endif | n in NODE]);
array[NODE] of LOCATION: locx = array1d(NODE,
    [ if n in C(CUSTOMER) then loc[C⁻¹(n)] else depot endif | n in NODE]);
constraint forall (n in NODE where not (n in E(TRUCK))) (
    arrival[next[n]] >= arrival[n] + servicex[n] +
                        tt[locx[n],locx[next[n]]]);
```

This is essentially equivalent to how an expert might write the model using integer indices.   □

We can use the same approach for higher-dimensional arrays (as in Example 11). Note that if the bounds of the index variable i are substantially larger than the original index set of the array, the compilation approach may produce very large arrays (particularly for multi-dimensional arrays). Currently we limit the compilation of default expressions on arrays to no more than double the size of the original array, otherwise the if-then-else-endif interpretation is used.

6 Experiments

The first experiment is qualitative, examining how valuable the language extensions we propose here are likely to be. Considering all the models used in the MiniZinc challenge[2] as a representation of a broad range of constraint programming models, we examined each of the models to determine (a) if the model could be improved with (more) enumerated types; and (b) if the model could benefit from extensions and defaults. Note that some models used in the challenge were written before enumerated types were available in MiniZinc. In addition expert modellers (particularly those used to modelling directly for solvers) who submit models to the challenge often use integer domains even when an enumerated type might be suggested from the problem.

Of the 129 models used in the challenge over its history we find 15 that could make use of enumerated type extensions to improve type safety. Another 64 models could improve type safety simply by using enumerated types. Clearly the extensions we develop here are not restricted to a very special class of models.

As an example of a model that could be improved using enumerated type extensions we illustrate parts of the **freepizza** model. In the problem you must purchase a set of pizzas each with a given price, but you have vouchers that can be used, e.g. buy 2 get 1 free. A voucher is enabled by buying enough pizzas for it, then it can be used to get some free pizzas, but the free pizzas must always be no more expensive than the enabling bought pizzas. The key decisions in the original model are how you bought each pizza, expressed as follows.

```
int: m; % no of vouchers
set of int: VOUCHER = 1..m;
set of int: ASSIGN = -m .. m; % -i pizza is used to buy voucher
                              %  i pizza is for free using
                              %  0 no voucher used on pizza
array[PIZZA] of var ASSIGN: how;
```

[2] https://github.com/minizinc/minizinc-benchmarks.

A key constraint in the model ensures that pizzas that enable a voucher are no less expensive than pizzas obtained for free:

```
constraint forall(p1, p2 in PIZZA)
                ((how[p1] < how[p2] /\ how[p1] = -how[p2])
                -> price[p2] <= price[p1]);
```

The `ASSIGN` set used in this model is an ideal case for an extended enumerated type. We can rewrite the model in a type-safe way as

```
int: m; % no of vouchers
enum VOUCHER = anon_enum(m);        % strong type check for VOUCHER
set of int: ASSIGN = Buy(VOUCHER) ++ % pizza is used to buy voucher v
                    { NOVOUCHER } ++ % no voucher used on pizza
                    Free(VOUCHER);   % pizza is for free using voucher v
array[PIZZA] of var ASSIGN: how;
```

The critical constraint is now simply

```
constraint forall(p1, p2 in PIZZA)
                (Free(Buy⁻¹(how[p1]))) = how[p2]
                -> price[p2] <= price[p1]);
```

The partiality of the inverse constructors is used to trivially satisfy the implication. We would argue that the resulting model is far easier to understand than the original, and compared to the set $-m..m$, the extended type is self-documenting. Indeed a version of the original model has been used as a debugging exercise, since it is quite hard to reason about it. Note that because the original model uses negation to indicate that a voucher is bought, it represents those vouchers in the reverse order compared to the extended enum. The solver may therefore perform a different search, which results in different runtime behaviour (faster for some instances, slower for others). If negation $-v$ in the original model is replaced by $v-m-1$, the two models behave identically.

Our second experiment demonstrates that translating array access expressions with defaults by extending the array with the default elements can lead to improvements in solving time. We ran a version of the capacitated vehicle routing problem from the MiniZinc benchmarks repository,[3] which we modified to use enumerated types and defaults. Table 1 shows the solving time and number of variables of defaults implemented as if-then-else-endif expressions[4] versus the extended arrays as explained in Example 12. For the experiments, we used the Chuffed solver with a timeout of 10 min, `A-n64-k9` and `B-n45-k5` data files, reduced to 8 and 9 customers to enable complete solving within the timeout. The results show an average improvement in solving time of 20%–30%, and a small reduction in the number of generated variables.

---

[3] https://github.com/minizinc/minizinc-benchmarks.

[4] Compiled as described in [10].

Table 1. Solving times and number of generated variables for Chuffed on several CVRP instances with 8 and 9 customers, extended arrays (x[y]) versus if-then-else expressions (i-t-e).

Instance/Customer set	Solving time (sec)		No. of variables	
	x[y]	i-t-e	x[y]	i-t-e
B-n45-k5/1–8	1.724	2.060	39 794	40 122
B-n45-k5/9–16	1.776	2.217	39 722	40 050
A-n37-k5/1–8	6.997	8.251	38 372	38 700
A-n37-k5/17–24	9.432	10.726	37 894	38 222
B-n45-k5/25–32	9.545	12.340	41 262	41 590
A-n37-k5/9–16	10.115	14.727	33 104	33 432
B-n45-k5/17–24	13.290	24.691	43 556	43 884
A-n37-k5/25–32	20.608	35.316	33 834	34 162
B-n45-k5/1–9	31.266	43.143	47 683	48 065
B-n45-k5/19–27	125.936	177.199	54 928	55 183
B-n45-k5/28–36	159.009	209.935	50 427	50 809
A-n37-k5/1–9	174.749	223.807	46 499	46 881
B-n45-k5/10–18	169.007	229.181	45 787	46 169
A-n37-k5/19–27	189.714	265.302	42 611	42 993
A-n37-k5/10–18	254.691	346.922	47 647	48 029
A-n37-k5/28–36	262.204	366.466	42 347	42 729

7   Related Work

Most programming languages support enumerated types in some form, it being a critical feature to avoid "magic constants". Enumerated type extension corresponds to using discriminated unions, for languages where those are available. No modelling language we are aware of except Zinc [7] supports such types, but Zinc does not support variables of such types, defeating one of the key purposes for introducing enumerated type extension.

AMPL [2] supports using sets of strings to define a form of enumerated types. Since the strings are only ever used as fixed parameters (there are no variables of type string) the language checks correct array lookups for arrays indexed by sets of strings during model compilation.

Similarly, OPL [11] does not support enumerated types, rather it supports the string data type, and the effect of enumerated types is mimicked by using sets of strings. Again since there are no variables of string type, the array index lookup for string indices is restricted to fixed parameters and checked during model compilation. Note that using strings to encode enumerated types has the advantage that one can simply build an array indexed by the union of two sets of strings, but this is not that helpful in the NODE example where we want to associate two nodes to each TRUCK. OPL does support arrays indexed by more complex types such as tuples which can significantly improve some models.

Essence [4] supports enumerated types that are very similar to MiniZinc's. They can be explicitly defined by sets of identifiers, in the model or the data, or defined as anonymous new types by size. Enumerated types can be used almost anywhere in the complex type language of Essence which includes parametric types for sets, multisets, functions, tuples, relations, partitions and matrices. Enumerated types support equality, ordering, and successor and predecessor functions. Essence is strongly typed, ensuring that all uses of enumerated types are correct. Currently there is no way to coerce an enumerated value to an integer within Essence. In order to make use of global constraints on enumerated types the mapping of enumerated types to integers is performed during the translation of Essence to Essence' by Conjure. Because of this restriction there is no way to write an Essence model for the VRP using enumerated types, since one cannot associate enumerated types with (even integer) node values. This means an Essence model for VRP will be forced to use integers for all types CUSTOMER, TRUCK and NODE, thus losing strong type checking. We believe that the Essence type system can be extended to support the concepts presented here.

There are a number of constraint modelling languages with a focus on object orientation, where complex data is given as sets of objects as opposed to arrays indexed by enumerated types, and subclassing provides another approach to effectively reason about multiple different types of objects simultaneously.

In s-COMMA [1] one can define classes which include constraints across their fields, and (single inheritance) subclassing. Enumerated types are supported as base types (which cannot be subclasses). There are no variables that range across objects, meaning that the issues we address here don't arise.

ConfSolve [6] is an object-oriented modelling language aimed at specifying configuration problems. Again it supports enumerated types as base types that cannot be extended. The class system supports reference types which allow for powerful modelling of complicated relationships. This allows for similar kinds of subclass reasoning as extended enumerated types. It is not clear exactly how much type checking is applied to ConfSolve models. Interestingly the models are compiled to MiniZinc to actually run, essentially mapping object identifiers to integers and using arrays to represent fields and pointers to other objects.

8   Conclusion

Enumerated types are critical for type safety of models that manipulate objects. Type extension allows us to have the same safety properties for models that manipulate two sets of objects together, or need to extend a set of objects to define extreme cases. This is a frequent modelling pattern in complex constraint programming models. Hence we believe all CP modelling languages should support them. In this paper we show how they are implemented in MiniZinc, with enough detail so that other modelling language authors can translate the ideas to their own language.

We believe the use of enumerated types by modellers should be strongly encouraged, since we know that debugging models can be very challenging, and strong type checking of array access and function arguments can prevent very subtle errors when the model is solved.

Future work. The concept of enumerated type extension should generalise to tuple and record types, although the interactions of these types with arrays and decision variables are more difficult to handle in the compiler. Such an extension would make it much easier to interface MiniZinc models with object-oriented programming languages and data sources.

Acknowledgments. This research was partially supported by the OPTIMA ARC training centre IC200100009.

References

1. Chenouard, R., Granvilliers, L., Soto, R.: Model-driven constraint programming. In: Proceedings of the 10th International ACM SIGPLAN Symposium on Principles and Practice of Declarative Programming, PPDP 2008 (2008). https://doi.org/10.1145/1389449.1389479
2. Fourer, R., Kernighan, B.: AMPL: A Modeling Language for Mathematical Programming. Duxbury (2002)
3. Frisch, A.M., Stuckey, P.J.: The proper treatment of undefinedness in constraint languages. In: Gent, I.P. (ed.) CP 2009. LNCS, vol. 5732, pp. 367–382. Springer, Heidelberg (2009). https://doi.org/10.1007/978-3-642-04244-7_30
4. Frisch, A.M., Harvey, W., Jefferson, C., Hernández, B.M., Miguel, I.: Essence: a constraint language for specifying combinatorial problems. Constraints **13**(3), 268–306 (2008)
5. Golomb, S.W.: How to number a graph. In: Graph Theory and Computing. Academic Press (1972)
6. Hewson, J.A.: Constraint-based specification for system configuration. Ph.D. thesis, University of Edinburgh (2013)
7. Marriott, K., Nethercote, N., Rafeh, R., Stuckey, P., Garcia de la Banda, M., Wallace, M.: The design of the Zinc modelling language. Constraints **13**(3), 229–267 (2008). https://doi.org/10.1007/s10601-008-9041-4
8. Mears, C., Schutt, A., Stuckey, P.J., Tack, G., Marriott, K., Wallace, M.: Modelling with option types in MiniZinc. In: Simonis, H. (ed.) CPAIOR 2014. LNCS, vol. 8451, pp. 88–103. Springer, Cham (2014). https://doi.org/10.1007/978-3-319-07046-9_7
9. Nethercote, N., Stuckey, P.J., Becket, R., Brand, S., Duck, G.J., Tack, G.: MiniZinc: towards a standard CP modelling language. In: Bessière, C. (ed.) CP 2007. LNCS, vol. 4741, pp. 529–543. Springer, Heidelberg (2007). https://doi.org/10.1007/978-3-540-74970-7_38
10. Stuckey, P.J., Tack, G.: Compiling conditional constraints. In: Schiex, T., de Givry, S. (eds.) CP 2019. LNCS, vol. 11802, pp. 384–400. Springer, Cham (2019). https://doi.org/10.1007/978-3-030-30048-7_23
11. Van Hentenryck, P.: The OPL Optimization Programming Language. MIT Press (1999)

A Parallel Algorithm for GAC Filtering of the Alldifferent Constraint

Wijnand Suijlen$^{(\boxtimes)}$ (iD), Félix de Framond, Arnaud Lallouet (iD),
and Antoine Petitet

Huawei Technologies France, Paris Research Center, CSI, Boulogne-Billancourt,
France
{wijnand.suijlen,arnaud.lallouet,antoine.petitet}@huawei.com

Abstract. In constraint programming the Alldifferent constraint is one of the oldest and most used global constraints. The algorithm by Régin enforces generalized arc-consistency, which is the strongest level of consistency for a single constraint. It is also the most time consuming despite several optimizations that were developed by others.

This paper parallelizes the Alldifferent generalized arc-consistent filtering algorithm, which is one of the first attempts for any global constraint. It does so by using a parallel graph search algorithm for two major parts of Régin's algorithm: finding a maximum matching and finding the strongly connected components. Most effective known optimizations are also ported. Experiments solving a large N-queens problem or a resource constrained scheduling problem show that generalized arc-consistent filtering can be significantly sped-up on a 64-core shared-memory system and on a 200-core distributed-memory system. We discuss also several scenarios where this algorithm should be applied or not.

1 Introduction

A major strength of Constraint Programming is the availability of *global constraints*. First, they provide a meaningful modeling element natural to the user, because it either occurs frequently in models or because it is strongly related to a specific application. The *Alldifferent* constraint falls in the first category, while, for example, the *Cumulative* constraint falls in the second one. Also, they speed up the programming process, because they encapsulate complex algorithms.

Alldifferent is one of the most useful and widely used global constraints. It ensures that a vector of variables will take different values. Stating Alldifferent(x_1, x_2, x_3) is equivalent to stating the conjunction $x_1 \neq x_2$, $x_2 \neq x_3$, and $x_1 \neq x_3$. Such conditions often arise as pigeonhole argument in optimization problems where resources are assigned exclusively. The particular interest of this constraint as modeling brick has attracted a lot of attention to its filtering algorithm [16], from the simple decomposition into a clique of differences to the full, generalized arc-consistent filtering algorithm by Régin [28]. In combination with precedence constraints, more effective filtering algorithms also exist [3]. For an extensive survey of the various consistencies and algorithms, we refer the reader to [16] and the original works [22–24, 27, 28]. In practice, choosing a proper level

P. Schaus (Ed.): CPAIOR 2022, LNCS 13292, pp. 390–407, 2022.
https://doi.org/10.1007/978-3-031-08011-1_26

of consistency can reduce the resolution time considerably, but this is highly problem dependent and out of the scope of this paper.

Régin [28] describes a GAC filtering algorithm which operates on the bipartite graph connecting variables and values. It can be rephrased in three major steps: find a maximum matching to get a support, interpret this matching as a flow and compute its residual, find its strongly connected components and remove inter-component arcs. This algorithm is a major contribution to Constraint Programming and has paved the way to a vast literature on global constraints [17]. Yet, while subsequent authors have developed several optimizations [11,26,37], all previous work only considers a sequential or a very limited (SIMD) parallel [21] environment, which begs the question what a more general parallel environment can add, focusing only on the filtering algorithm itself.

This paper proposes to parallelize the algorithm's two main components: maximal matching (MM) and strongly connected components (SCC). We use the Ford-Fulkerson algorithm with Breadth-First Search (BFS) as basis of the matching algorithm because it was found to be faster than the original Hopcroft-Karp algorithm by [11]. Both MM and SCC can be expressed in terms of *reachability queries* which can be parallelized. Our first version of the parallel algorithm uses a regular, single source BFS and we present two optimizations. The first one replaces BFS by a different traversal we call *Local-First Search* that gives priority to local data. The second one performs multiple search queries in parallel; i.e., a multi-source query. In the experimental evaluation, that studies the algorithm in isolation from other parallelization techniques, such as parallel search or parallel propagation, we observe the algorithm performance on two applications that are varied in size, vertex degree, and the amount of backtracking required.

The structure of the paper is as follows. Section 2 summarizes Régin's GAC filtering algorithm whose sequential algorithm forms the basis of ours. As introduction to parallel programming, Sect. 3 briefly treats the Bulk Synchronous Parallel model. Then, Sect. 4 develops the parallel algorithm. Experiments are conducted in Sect. 5 whose results are discussed and related to other work in Sect. 6.

2   Régin's Alldifferent Filtering Algorithm

The filtering algorithm by [28] represents the problem as the bipartite variable-value graph $G = (X \cup D, E)$. It has a vertex for each variable and for each domain value. Between every variable $x \in X$ and domain value $d \in D_x$, there is an edge $(x, d) \in E$. Now, the act of filtering coincides with removing edges from this graph. In order to decide which edges to remove, the algorithm proceeds in four steps:

1. Find a maximum matching (Fig. 1a). In filtering rounds after initial propagation, the maximum matching from the previous round can be reused. If none of the variables lost a value that was also in the previous matching, the same matching can be reused straight away. Otherwise, the matching should be repaired by finding augmenting paths. If one or more variables cannot be matched, the domain is inconsistent (Fig. 1b).

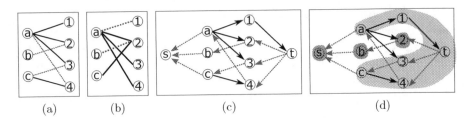

Fig. 1. Example graphs for variables $X = \{a, b, c\}$ and domain $D = \{1, 2, 3, 4\}$: (a) Maximum matching, (b) domain failure, (c) residual flow graph, (d) strongly connected components

2. Interpret the maximum matching as a maximum flow through the digraph $G_f = (X \cup D \cup \{s, t\}, A)$ with arcs from the source vertex s to all variables, from all domain values to the sink vertex t, and from each variable to each domain value that is in its domain. Now, the maximum matching defines a maximum flow through G_f and so the residual flow graph G_r (Fig. 1c) can be constructed by reversing those arcs that appear in the maximum flow.
3. Find all strongly connected components (SCCs) in G_r (Fig. 1d).
4. The arcs which are not in the matching while their endpoints lie in different SCCs, cannot be part in any (other) maximum matching. Thus, these can be removed.

For a correctness proof the reader is referred to [11] and [16].

Several algorithms exist that can find a maximum matching. Empirical study [11] showed that an application of the Ford-Fulkerson algorithm in breadth-first search order, denoted by the abbreviation FF-BFS, is often the quickest. A BFS has time complexity $\mathcal{O}(X + D + E)$ in the worst case and FF-BFS requires X of those traversals. Hence, the time complexity for finding a maximum matching initially is $\mathcal{O}(X(X + D + E))$, while each incremental reparation for m removed edges from the matching requires $\mathcal{O}(m(X + D + E))$ time. For finding SCCs, Tarjan's algorithm [33] is a very efficient sequential algorithm with a $\mathcal{O}(X + D + E)$ time complexity. It requires only one depth-first search (DFS) graph traversal.

The main workload of both algorithms is the graph traversal itself and is mainly bottlenecked by memory access. By reducing these memory footprint to a minimum, e.g. by exploiting the particular structure of the graph and by encoding the graph efficiently, raw processing speed can be optimized. In order to evaluate our parallel Alldifferent filtering algorithm, we implemented the sequential algorithm too. Not only did we focus on good raw processing efficiency, but we also applied more sophisticated optimizations as described in [11]. In particular, we implemented *incremental matching* and *exploiting strongly connected components*. Pseudo code of the sequential filtering algorithm appears below as Procedure 1 (SeqAllDiffFilter).

Exploiting strongly connected components is a form of decomposition. It recognizes that during the next incremental matching an augmenting path cannot cross SCCs, and that therefore SCCs can only split and never join. Effectively, each SCC can be regarded as an independent Alldifferent constraint on only the variables and domain values that constitute that SCC.

Procedure 1: SeqAllDiffFilter(X, U, D, M, $\mathcal{S}$)

In: Variables X, updated variables $U \subseteq X$.
In/Out: Domains $(D_x)_{x \in X}$ (GAC output), matching M, SCC G_r partition $\mathcal{S}$.
Data: G_r: the residual flow graph that matches the matching M

1 **foreach** *SCC* $S \in \mathcal{S}$ *with* $S \cap U \neq \emptyset$ **do**
2   **foreach** $x \in S$ *not in* M **do**   /* Repair matching with FF-BFS */
3   BFS of augmenting path in G_r starting from x;
4   If unsuccessful, **return** *"Inconsistent"*;
5   Otherwise, update M and G_r with the augmenting path;
6   **end**
7   **foreach** $x \in S$ *still unvisited by Tarjan* **do**   /* Tarjan's algorithm */
8   DFS traversal of S in G_r starting from x;
9   Split SCC S into the SCCs that were found by above DFS ;
10   Remove edges from E that cross SCCs and are not in matching M;
11   Update variable domains D_x for each $x \in S$ and update G_r;
12   **end**
13 **end**
14 **return** *"Consistent"*;

---

3   Bulk Synchronous Parallel Model

Since the main motivation of parallel computing is to increase processing speed, it is essential to have a model that can analyze the time complexity of a parallel algorithm. We choose the Bulk Synchronous Parallel (BSP) bridging model [35], because it is simple and accurate up to small constant factors. It describes how a parallel workload can be mapped to any parallel computer.

The BSP model assumes a BSP computer with three types of components: P processors with local memory, a communications network that can route messages between individual processors, and a synchronization facility that can synchronize all processors at once. Operations on these components cost time. The model counts g time units for the transmission of each word, while it counts ℓ time units for the synchronization of all processors at once.

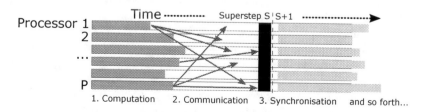

Fig. 2. Time-activity diagram of any BSP program

Such a BSP computer can be programmed in *direct mode* with a BSP program, whose execution proceeds through *supersteps*. Each superstep consists of two phases: a computation phase, in which each processor s works W_s time alone, and a communication phase, in which processors send and receive at most H_s data. Between each superstep, the BSP computer synchronizes all processors (see Fig. 2 for an illustration of activity in a BSP program over time). The time to complete a superstep is hence $\max_s W_s + \max_s H_s g + \ell$.

Notations. The algorithms that follow in subsequent sections are listed as *Single Program Multiple Data* (SPMD) procedures, meaning that the same code is executed on every process using s as process identifier ($0 \leq s < P$). All data held by variables indexed with s and iteration variables are local to the process. The s index is omitted for data which are known to be the same on all processes, even if their implementation is done by a different variable on each process. Data can only be shared by explicit communication.

4   A Parallel Alldifferent Filtering Algorithm

We propose a parallel filtering algorithm for the Alldifferent constraint that follows the same ideas as Régin's sequential algorithm [28]. Main workload of his algorithm is finding a maximum matching and finding all SCCs.

Observe that maximum matching and SCCs can both be found through *reachability* queries. The Ford-Fulkerson method relies on identification of augmenting paths, which is a reachability problem. For finding SCCs, Tarjan's algorithm is avoided, because depth-first search is hard to parallelize efficiently. Instead, the membership of a vertex to the SCC of another vertex can be decided by reachability queries in both forward and backward direction of the arcs. Then, in a divide-and-conquer fashion, all SCCs in a graph are identified by repeatedly removing SCCs.

ParAllDiffFilter. The basis of our parallel Alldifferent filtering algorithm is outlined as Procedure 2 (ParAllDiffFilter). This SPMD procedure engages all processors in the same computation, that removes all inconsistent domain values, while each processor stores an equal part of the graph. It traverses each SCC that contains a modified variable, picks an unmatched variable vertex (line 2), restores the matching in parallel using FindAugmentingPath (Procedure 3) and finds the SCCs with the parallel procedures ForwardReachable (Procedure 7) and BackwardReachable. Note that the definition of BackwardReachable is identical to ForwardReachable but follows arcs in the reverse direction. The initial propagation calls this procedure with ParAllDiffFilter($X, X, D, \emptyset, \{X\}$).

Some improvements to ParAllDiffFilter are not explored. SCCs are not processed in parallel (the foreach-loop lines 1–12), because the Search procedure is already parallel and experiments suggest that there is usually one non-trivial SCC, which contains all the undecided variables. For the same reason, the possibility to find multiple SCCs (lines 7–11) simultaneously, see e.g. [5], is not explored.

Procedure 2: ParAllDiffFilter(X, U, D, M, $\mathcal{S}$)

In: Variables X evenly assigned to all processes, updated variables $U \subseteq X$.
In/Out: Domains $(D_x)_{x \in X}$ (GAC output), matching M, SCC G_r partition $\mathcal{S}$.
Data: G_r residual flow graph from matching M; t sink vertex; $\mathcal{S}'$ new SCCs

1 **foreach** $SCC\ S \in \mathcal{S}$ with $S \cap U \neq \emptyset$ **do**
2   **foreach** $x \in S \cap X \backslash M$ **do**   /* Repair matching */
3   path $A \leftarrow$ FindAugmentingPath(G_r, x, t);
4   **if** $A = \emptyset$ **then return** *"Inconsistent"*;
5   update M and G_r with the (augmenting) path A
6   **end**
7   **while** $S \neq \emptyset$ **do**   /* Find SCCs and filter domains */
8   Pick $v \in S$;
9          $T \leftarrow$ ForwardReachable(G_r, v) $\cap$ BackwardReachable(G_r, v);
10          $\mathcal{S}' \leftarrow \mathcal{S}' \cup \{T\}$;
11          $S \leftarrow S \backslash T$; **foreach** $x \in T \cap X$ **do** $D_x \leftarrow D_x \cap T$;
12   **end**
13 **end**
14 $\mathcal{S} \leftarrow \mathcal{S}'$;
15 **return** *"Consistent"*;

FindAugmentingPath. FindAugmentingPath is used to restore a matching for a variable x. It proceeds in two steps. First it calls Procedure 4 (Search) that returns a tree rooted by x in which appears the target t. Then, starting from the target, it repeatedly gets the parent node from the process which owns it until the augmenting path a is gathered (lines 3 to 7).

Search. The Search procedure builds a tree in T_s whose arcs are again distributed over the processes. For this, it performs a parallel breadth-first search traversal through the residual graph from the root r, until the target t is found. Parallel BFS is a work-efficient algorithm because its total number of operations is proportional to the best sequential algorithm. Parallel algorithms with lower time

---

Procedure 3: FindAugmentingPath(G, r, t)

In: Graph G, source vertex r, target vertex t
Out: The augmenting path a to the target t if it was found or $\emptyset$ otherwise
1 $T_s \leftarrow \emptyset$; $a \leftarrow \emptyset$; $v \leftarrow t$;
2 **if** Search(G, r, t, T_s) **then**
3   **while** $v \neq r$ **do**
4   **if** *process s owns v* **then** pick w_s s.t. $(v, w_s) \in T_s$ and broadcast it;
5   Synchronize;
6   Receive broadcast as w;
7          $a \leftarrow a \cup \{(w, v)\}$; $v \leftarrow w$;
8   **end**
9 **end**
10 **return** a;

Procedure 4: Search(G, r, t, T_s)

In: Graph G, root vertex r, target vertex t
Out: Boolean *found* whether vertex t was found; local arcs T_s of spanning tree
Data: A visited status $visited_s[v] \in \{0, 1\}$ for each local vertex v, the
        parent-child vertex pair sets J_s and K_s for the current and next levels

1 *found* $\leftarrow$ 0; $visited_s[v] \leftarrow 0$ for each local vertex v;
2 **if** r *is local vertex* **then** $J_s \leftarrow \{(\bot, r)\}$ **else** $J_s \leftarrow \emptyset$;
3 **repeat**
4 | $K_s \leftarrow \emptyset$; $f_s \leftarrow 0$;
5 | **foreach** $(i, j) \in J_s$ **do** VisitLocal(G, i, j, t, T_s, $visited_s[]$, f_s, K_s);
6 | **foreach** $(j, k) \in K_s$ **do** Send (j, k) to process assigned to k ;
7 | Start global logical-or *found* $\leftarrow \bigvee_{r=0}^{P-1} f_r$;
8 | Synchronize;
9 | $J_s \leftarrow$ union of all received child-parent vertex pairs;
10 **until** *found or* $\bigcup_{0 \leq s < P} J_s = \emptyset$;
11 **return** *found*;

---

complexity exist, but require many more (super-linear) processors. The result of a BFS on a graph is a tree. The paths from the leaves to the root of that tree coincide with the minimum length paths between the respective vertices in the graph. The presence of the sink vertex as a leaf testifies, therefore, its reachability from the root.

The level-synchronous BFS starts by visiting the root vertex and then continues to visit all its children in parallel. From then on, this pattern repeats itself: all of its grandchildren are visited in parallel, all of its great grandchildren, and so on. Of course, when the number of available processors is limited, some of its grandchildren will have to be visited after some of the others, but that does not matter in a level-synchronous BFS. The Boolean $visited_s[v]$ is used to keep track of the visited vertices. While the whole tree is returned in T_s, the Search procedure uses two intermediate sets: J_s contains the current level to be explored and K_s contains the child vertices. Procedure VisitLocal is used to fill K_s. It is implemented as classical BFS in Procedure 5 and with an internal local BFS we call Local-First Search (LFS) in Procedure 6.

Procedure 5: VisitLocal(G, i, j, t, T_s, $visited_s$, f_s, K_s) for BFS

In: Graph G, parent vertex i, child vertex j, target vertex t
In/Out: $visited_s[]$ status, target found f_s, child edges for next round K_s
Out: Local arcs T_s of spanning tree

1 **if** $\neg f_s \wedge \neg visited_s[j]$ **then**
2 | $T_s \leftarrow T_s \cup \{(i, j)\}$; $visited_s[j_s] \leftarrow 1$;
3 | **if** $j = t$ **then** $f_s \leftarrow 1$;
4 | **else foreach** *neighbor* k *of* j *s.t.* $\neg visited_s[k]$ **do** $K_s \leftarrow K_s \cup \{k\}$;
5 **end**

Any measure to reduce communication may speed up the search. Recall that the variables and domain values, and therefore also the graph's vertices, are evenly assigned to all processes. The idea is that if a vertex is visited, all reachable vertices on the *same* process are visited in BFS order first, before continuing the traversal from the adjacent vertices on other processes. This may effectively reduce the number of synchronizations in practice, although it does not change the worst case complexity.

When the target vertex has been found by a local process, the local Boolean f_s is set to true and the search has to be stopped. This is done by broadcasting these Booleans and gathering their global OR.

Procedure 6: VisitLocal(G, i, j, t, T_s, $visited_s$, f_s, K_s) for LFS

   In: Graph G, parent vertex i, child vertex j, target vertex t
   In/Out: $visited_s$[] status, target found f_s, child edges for next round K_s
   Out: Local arcs T_s of spanning tree
1 Initialize queue $J_s \leftarrow \{(i,j)\}$;
2 **while** $\neg f_s \wedge J_s \neq \emptyset$ **do**
3   Dequeue (a, b) from J_s;
4   **if** $\neg visited_s[b]$ **then**
5        $T_s \leftarrow T_s \cup \{(a,b)\}$; $visited_s[b] \leftarrow 1$;
6   **if** $b = t$ **then**
7           $f_s \leftarrow 1$;
8   **break**
9   **end**
10      $N_s \leftarrow \{(b,c) \mid c \in \text{neighbors of } b \wedge \neg visited_s[c]\}$;
11      $R_s \leftarrow \{(b,c) \in N_s \mid c \text{ not assigned to process } s\}$;
12      $K_s \leftarrow K_s \cup R_s$;
13   Enqueue $N_s \backslash R_s$ to J_s;
14   **end**
15 **end**

ForwardReachable and BackwardReachable. These procedures are used in the SCC computation and reuse BFS Procedure 4.

Procedure 7: ForwardReachable(G, r)

   In: Graph G, set of source vertices r
   Out: All vertices U that are reachable from *root*
1 $T_s \leftarrow \{r\}$;
2 Search(G, r, T_s);
3 Broadcast T_s;
4 Synchronize ;
5 $T \leftarrow$ gather broadcast of all visited vertices;
6 **return** $\{child \mid (parent, child) \in T\}$;

Complexity. On a graph with diameter δ and V vertices with maximum degree d, Procedure 4 (Search) requires at most δ rounds of the outer loop (lines 3–10) and hence δ synchronizations. Each process has no more than V/P vertices, because variables and domain values are evenly distributed. Each vertex has at most d neighbors and, therefore, K_s will have no more than Vd/P entries. Hence, the number of outgoing messages is bounded by Vd/P. Likewise, J_s will not receive more than Vd/P entries. Therefore, the total time is $T_{\mathrm{BFS}}(V,\delta) = \mathcal{O}(\delta(\frac{Vd}{P}g + \ell))$. This bound is tight for the unlucky but possible case where neighbors are on another process each time. Complexity of Procedure 2 ParAllDiffFilter is $T_{\mathrm{ParAllDiffFilter}} = \sum_{S \in \mathcal{S}} T_{\mathrm{BFS}}(V_S, \delta_S)$.

Parallel Multi-source LFS. At initial propagation a new augmenting path has to be discovered for every variable, which costs XT_{BFS} time. Again, this time will be dominated by synchronization. To reduce this cost, we have developed a multi-source search which finds multiple vertex disjoint augmenting paths simultaneously. Of course, it will generally not be possible to find a maximum matching in one pass, but it is guaranteed to find at least one augmenting path, so that it can be repeated a few times until the matching is indeed maximum. The algorithm is not included here for space reasons.

5   Experimental Evaluation

The performance characteristics of the presented parallel algorithm is quite different from the original sequential algorithm. Firstly, the parallel algorithm is more susceptible to data-dependent performance variations. The diameter δ and the maximum degree d of the bipartite graph directly affect the inter-process communication necessary to perform a graph traversal. Since communication is almost all that it does, it strongly affects the total running time. Secondly, there is a special penalty associated with each call of the propagator, namely a synchronization of all processes, which is, even on the tightest parallel machines, an order of magnitude slower than a local memory access. We use for experiments a multicore system and a cluster as described in Table 1.

Table 1. Parallel systems used for testing

Feature	Epyc64	Ivy200
OS/Compiler	openEuler 20.03/GCC 7.3.0	CentOS 7/GCC 7.3.0
BSPlib	MulticoreBSP for C 2.0.4 [36]	Intel MPI 2018 &
		BSPonMPI 1.1 [32]
CPU	AMD Epyc 7452 @ 2.35GHz	Intel E5-2690v2 @ 3GHz
Architecture	Zen 2	Ivybridge
Memory	256 GB DDR4*	10× 256 GB DDR3*
Interconnect	Infinity Fabric	EDR Infiniband
P	$32 \times 2 \times 1 = 64$ [#]	$10 \times 2 \times 10 = 200$ [#]
Inverse bandwidth g	120 ns/word [†]	7600 ns/word [†]
Sync latency ℓ	37 μs [†]	4831 μs [†]

* Configured with maximal bandwidth [1, 19].
[#] P = Number of Cores × Sockets × Server Nodes.
[†] Estimated to 2 significant digits with 95% confidence by randomly probing communication patterns of up to 32KB with 16-byte messages [32].

Hence the experimental question presents itself how the total running time compares with respect to three different parameters of a given problem: the size, the amount of backtracking required, and the degree. Note that degree is correlated to the diameter.

The benchmark suite used by other Alldifferent studies [11,37], contains only CSPs where the Alldifferent constraints have small scopes and always require a considerable amount of backtracking search. Certainly, on those problems the parallel filtering algorithm is not competitive with the sequential algorithm. For that reason we choose different CSPs for our experiments: An N-Queens satisfaction problem to evaluate running time as function of problem size, an N-Queens counting problem to evaluate backtracking speed, and a scheduling problem to evaluate running time as function of degree.

5.1   Data Structures, Data Distribution, and CP Solver Integration

Our algorithms assume a distributed address space. Data on a remote process can only be modified through explicit communication. Since all algorithms can broadcast any modifications cheaply, it is possible to maintain exact copies of the entire solver state on all processes. However, that would mean that total memory use would increase by a factor P.

By far, most storage is needed for the variable domains, whose memory requirement tends to grow quadratically with the number of variables. As a simple solution, this paper proposes to store a variable domain only on the process that the variable is assigned to, analogous to a 1D distribution of the adjacency matrix in parallel graph computations. Variables are assigned to processes cyclically. All other data, which is either only proportional to the number of variables, domains, or processes, can remain replicated. The residual flow graph does not need separate storage, as it can be directly read and altered via the variable domains and the current matching. In our experiments, we store variable domains in Sparse Sets [6], the SCCs in a backtrackable variable set partition [11], and the matching in an array with backtrackable entries.

The CP-solver in which this filtering algorithm is integrated, called *HPCS*, is original but essentially not much different from its contemporaries. There are variables, domains, views, propagators, a state trail, an event queue, and search. Only the domains and their trail are distributed. The solver has only a few constraints so far, among which is the Alldifferent constraint. Like the propagators, the HPCS solver is also an SPMD-style program. All processes are executing the exact same source code all the time but on different parts of the data. To ensure that they remain on the same control-flow path, information has to be broadcast regularly, which implies regular synchronization. Indeed, in the variable and value selection heuristics each process can only view the domains of the variables that are assigned to it. All processes have to communicate in order to collectively agree on the chosen variable and value for every decision.

5.2   N-Queens

The N-Queens problem is to place N queens on an $N \times N$ chessboard so that no queen can attack another. The simplest variant starts with an empty chessboard and just asks to construct one solution. This is a constant time problem, because a solution exists for $N \geq 4$, while a solution can be constructed analytically. A hard variant is to count the number of solutions. The CSP can be defined with three Alldifferent constraints, as is shown in Fig. 3.

$$
\begin{aligned}
\text{Domain} \quad & D = \{0, 1, \ldots, N-1\} \\
\text{Variables} \quad & Q_0, Q_1, \ldots, Q_{N-1} \in D \\
\text{Constraints} \quad & \text{Alldifferent}(Q_0, Q_1, \ldots, Q_{N-1}) \\
& \text{Alldifferent}(Q_0 + 0, Q_1 + 1, \ldots, Q_{N-1} + N - 1) \\
& \text{Alldifferent}(Q_0 - 0, Q_1 - 1, \ldots, Q_{N-1} - N + 1).
\end{aligned}
$$

Fig. 3. CSP of N-queens problem. Each Q_i describes the column in which the queen on the i-th row is placed.

Problem Size. When N is a multiple of 6, the N-queen problem has an analytic solution [9] of the form:

$$
Q_i = \begin{cases}
2i + 1 & 0 \leq i < N/2 \\
2i - N & N/2 \leq i < N
\end{cases}
$$

In the first experiment, we use this solution as heuristic in order to reach the first solution without backtracking and only measure the propagation time. The parallel version is tested with the four parallel graph search traversals, namely single source (SS) or multi-source (MS) BFS or LFS. These are compared to our sequential implementation and to the sequential N-queens implementation in Gecode [10] using three `distinct` constraints. On this problem, Gecode's number of executed propagators, number of failures, and search tree depth match exactly with our solver. Results are presented in Fig. 4.

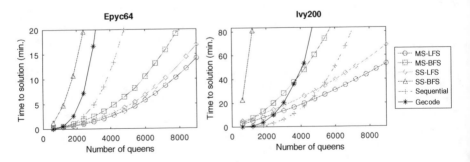

Fig. 4. Time required by Gecode and HPCS using either sequential or parallel algorithm to find the analytic solution to N-queens

In these experiments, we observe that BFS with single-source search performs the worst and is the only parallel setting slower than sequential. LFS is a critical improvement because it reduces the number of synchronizations. The high synchronization time in the cluster gives advantage to the sequential algorithm up to a large number of queens. The speed of our sequential implementation compared to Gecode has to be relativized by the fact that Gecode is far more generic.

Backtracking. To compare backtracking speed, the solvers are instructed to enumerate all solutions of the N-queens problems until a set number of failures has been reached. The total running time is measured for a range of failure number limits. Results are presented in Fig. 5.

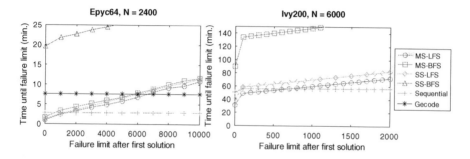

Fig. 5. Time required by Gecode and HPCS using either sequential or parallel algorithm to reach failure limit while counting N-queens solutions

We can see that Gecode time appears constant because it is really fast on backtracking and fails at a rate this granularity is not able to measure. In contrast, backtracking comes at a high synchronization cost for any version of our parallel algorithm.

5.3   Scheduling

As scheduling problem, a collection of tasks of unit time linked by precedence constraints is to be scheduled on a number of parallel machines. The goal is to find a schedule with a minimum makespan. The CSP is given in Fig. 6.

Let $\mathcal{T} = \{0, \ldots, n-1\}$ be a set of n tasks to be scheduled on m machines before a time horizon t, a set of dependencies $\mathcal{D} \subset \mathcal{T} \times \mathcal{T}$ and a compatibility set $\mathcal{M}_i \subseteq [0..m-1]$ that allows the task i to be executed only on a machine in $\mathcal{M}_i$.

Variables	Time	$T_1, T_2, \ldots, T_n \in [0, t-1]$
	Machine	$M_1, M_2, \ldots, M_n \in [0, m-1]$
	Slot	$S_1, S_2, \ldots, S_n \in [0, tm-1]$
Constraints	Compatibility	$\forall i \in \mathcal{T}, M_i \in \mathcal{M}_i$
	Precedence	$\forall (i,j) \in \mathcal{D}, T_i < T_j$
	Resource	$\forall i \in \mathcal{T}, S_i = T_i \times m + M_i$
		Alldifferent$(S_1, S_2, \ldots, S_t)$

Fig. 6. The CSP of the task scheduling problem

In this CSP, there are many precedence constraints to handle. If these were to be handled as separate binary constraints, the parallel solver would lose a lot of time in synchronizations. Instead, we choose to handle this network of less-than constraints as one global constraint. At initial propagation, the longest path between each pair of tasks is computed by repeatedly squaring the adjacency matrix, which costs $\lceil \log T \rceil (T^3/P + T^2 g + \ell)$, ignoring small constant factors. After that, global bounds of task start times can be found by scattering locally found bounds and then gathering them to processes that own the tasks that are concerned. If t variables per process changed, then, ignoring small constant factors, one round of propagation costs no more than $tT + Tg + \ell$. This is listed as Procedure 8.

Procedure 8: ParLessThanFilter(X, U_s, D, $\bar{A}$, $\bar{B}$,)

In: Variables X, updated local variables U_s, closures $\bar{A}$ and $\bar{B}$ of the less-than constraint network for lower and upper bound relations, respectively.

In/Out: Domains $(D_x)_{x \in X}$ (GAC output)

Data: Arrays $lb_s[]$ and $ub_s[]$ store locally found lower and upper bounds

1 **foreach** $x \in X$ **do**
2   $\quad lb_s[x] = \max_{y \in U_s} \{ \min D_y + \bar{A}_{y,x} \}$; $ub_s[x] = \min_{y \in U_s} \{ \max D_y + \bar{B}_{y,x} \}$
3 **end**
4 Scatter $lb_s[x]$ and $ub_s[x]$ to process to whom x is assigned;
5 Synchronize;
6 **foreach** *local* $x \in X$ **do**
7   $\quad$ Gather lbv $\leftarrow \max_{0 \le s < P} lb_s[x]$; Gather ubv $\leftarrow \min_{0 \le s < P} ub_s[x]$;
8   $\quad D_x \leftarrow \{ v \in D_x \mid \text{lbv} \le v \le \text{ubv} \}$;
9 **end**

Degree. This last experiment randomly samples generated task graphs with a fixed number of tasks n, a constant out degree $d = 2$ (except for the start and end vertices), and an expected square shape, which means that the longest dependency chain is as about as long as there are tasks that can run in parallel [34]. Figure 7 shows a sample of some such task graphs of small size.

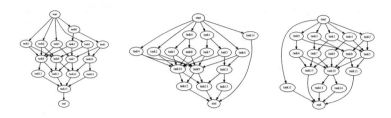

Fig. 7. A sample of three random task graphs with 16 tasks

Each task is compatible with μ randomly drawn machines. A small μ will result in a low-degree variable-value graph in the Alldifferent filtering algorithm, while a large μ results in a high-degree graph. A high-degree graph will have smaller diameter δ but also more arcs to process. While μ varies, the CP solver is instructed to find a solution with minimal makespan using the list-schedule heuristic [13]. The results are presented in Fig. 8 using $m = 10$ and $t = 660$. As can be expected, the sequential algorithm performs best on the low-degree problems, while the parallel algorithm also benefits from a low diameter.

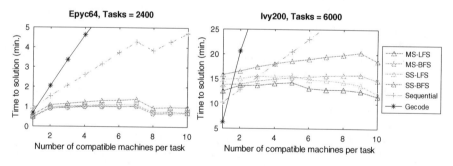

Fig. 8. Time required by Gecode and HPCS using either sequential or parallel algorithm to solve scheduling problems of various degree

6   Related Work and Discussion

Survey papers [12,29] describe the state of parallel inference and filtering. Most works [14,15,25,30,31] concern the execution of multiple propagators in parallel. This includes one work [7] which also implements a parallel propagator for the table constraint. Their propagator for Alldifferent is based on its binary decomposition, hence unable to guarantee GAC, and uses parallel propagation.

A SIMD algorithm is proposed by [21], which also parallelizes graph search in Régin's algorithm. Their proposed search algorithms [8] assume that the graphs are dense and, therefore, only parallelize the lookup of unvisited vertex neighbors, which may cause scalability issues on large problems. Like these two attempts, our solution does not exclude the possibility of running multiple propagators in parallel. Additionally, parallel search methodologies, like EPS, may also be combined, if sufficient memory is available. Parallel search requires much more memory, because it requires duplication of the search state for each search-tree branch that is explored in parallel.

Most of the aforementioned works mention that a major drawback of parallel propagation is its synchronization cost, which is also confirmed by our experiments. This is inevitable whenever multiple processes have to come to a collective conclusion. Synchronizations generally happen in two places: during fixed-point iteration of the propagators and when making a new decision. While P-completeness of arc-consistency on binary constraint networks [20] testifies the existence of a worst-case scenario, it is possible to avoid too many fixed-point iterations by grouping several constraints together in order to propagate domain updates more quickly. We used this trick with the global less-than constraint in the scheduling application. Alternatively, it may be possible to use hardware that has very low synchronization costs, such as the integrated vector instructions in modern CPUs. On the contrary, avoiding a synchronization for every decision seems less obvious. Backtracking search is therefore much slower than in a sequential environment. That is not a reason to avoid parallel filtering, however. Their main application area are large problems for which backtracking search is infeasible anyway. Large problems require good heuristics so that backtracking search is limited to a minimum.

Important algorithmic improvements have been studied by Gent et al. [11]. Our algorithm implements most of them, including the use of domain iterators as adjacency lists, although that last optimization does not improve performance much [26]. We did not include the most recently described improvement [37], which filters some domain values already while finding a maximum matching, so that SCC finding is less work.

Our parallel algorithms for maximum matching and SCCs are simplistic and are vulnerable to some worst-case inputs. Those worst-case inputs did not occur in our experiments. Much more robust algorithms are available for SCC finding [5] and maximum matching [2]. The best time bound for a work-efficient maximum matching algorithm remains Hopcroft-Karp [18]. The ideas for our maximum matching algorithm are based on [4, exercise 5.9].

7   Conclusion

The Alldifferent constraint is one of the most often used global constraints in constraint programs. CP solvers may choose to use one of several filtering algorithms, each of which reaches a particular level of consistency. The strongest algorithms today are based on Régin's GAC filtering algorithm. Its weak point, however, is that it is relatively time consuming.

This paper presents a parallel algorithm that can significantly speed up filtering for Alldifferent constraints with large scopes and domains. It uses a parallel multi-source graph search to parallelize the two main phases of Régin's algorithm: finding the maximum matching and finding the SCCs. Although worst-case time complexity is not better than the sequential algorithm, experiments demonstrate that CSPs with large Alldifferent constraints can be solved much faster, provided that a good heuristic is in place. The start-up cost for each invocation of the algorithm makes it less suitable for problems that require extensive backtracking search, if GAC is to be maintained at each decision.

There are many ways possible to reduce start-up cost, especially if one is willing to make system specific optimizations. Alternatively, it is possible to change to the sequential algorithm at runtime after the variable domains have been reduced below a certain threshold.

In any case, complete search is usually not a feasible strategy for large problems because of the exponential size of the search tree. Therefore, large problems are necessarily only feasible if a good search heuristic is available. Indeed, if a CP solver's time is mostly spent on Alldifferent constraint filtering, then the presented algorithm can reduce that time significantly.

References

1. AMD: Memory population guidelines for AMD EPYC processors. Tech. Rep. 56301, revision 1.1, Advanced Micro Devices (2018)
2. Azad, A., Buluç, A., Pothen, A.: Computing maximum cardinality matchings in parallel on bipartite graphs via tree-grafting. IEEE Trans. Parallel Distrib. Syst. **28**(1), 44–59 (2016). https://doi.org/10.1109/TPDS.2016.2546258
3. Bessiere, C., Narodytska, N., Quimper, C.-G., Walsh, T.: The alldifferent constraint with precedences. In: Achterberg, T., Beck, J.C. (eds.) CPAIOR 2011. LNCS, vol. 6697, pp. 36–52. Springer, Heidelberg (2011). https://doi.org/10.1007/978-3-642-21311-3_6
4. Bisseling, R.H.: Parallel Scientific Computing: A Structured Approach Using BSP. 2nd edn. Oxford University Press (2020)
5. Blelloch, G.E., Gu, Y., Shun, J., Sun, Y.: Parallelism in randomized incremental algorithms. In: Proceedings of the 28th ACM Symposium on Parallelism in Algorithms and Architectures, p. 467–478. SPAA 2016, Association for Computing Machinery, NY (2016). https://doi.org/10.1145/2935764.2935766
6. Briggs, P., Torczon, L.: An efficient representation for sparse sets. ACM Lett. Program. Lang. Syst. **2**(1–4), 59–69 (1993). https://doi.org/10.1145/176454.176484
7. Campeotto, F., Dal Palù, A., Dovier, A., Fioretto, F., Pontelli, E.: Exploring the use of GPUs in constraint solving. In: Flatt, M., Guo, H.-F. (eds.) PADL 2014. LNCS, vol. 8324, pp. 152–167. Springer, Cham (2014). https://doi.org/10.1007/978-3-319-04132-2_11
8. Cheriyan, J., Mehlhorn, K.: Algorithms for dense graphs and networks on the random access computer. Algorithmica **15**(6), 521–549 (1996). https://doi.org/10.1007/BF01940880
9. Erbas, C., Tanik, M.M.: Generating solutions to the N-queens problem using 2-circulants. Math. Mag. **68**(5), 343–356 (1995). http://www.jstor.org/stable/2690923

406   W. Suijlen et al.

10. Gecode Team: Gecode: generic constraint development environment (2019). http://www.gecode.org
11. Gent, I.P., Miguel, I., Nightingale, P.: Generalised arc consistency for the alldifferent constraint: an empirical survey. Artif. Intell. **172**(18), 1973–2000 (2008). https://doi.org/10.1016/j.artint.2008.10.006
12. Gent, I.P., et al.: A review of literature on parallel constraint solving. Theory Pract. Logic Program. **18**(5–6), 725–758 (2018). https://doi.org/10.1017/S1471068418000340
13. Graham, R.L.: Bounds on multiprocessing timing anomalies. SIAM J. Appl. Math. **17**(2), 416–429 (1969). https://doi.org/10.1137/0117039
14. Granvilliers, L., Hains, G.: A conservative scheme for parallel interval narrowing. Inf. Process. Lett. **74**(3–4), 141–146 (2000). https://doi.org/10.1016/S0020-0190(00)00048-X
15. Hamadi, Y.: Optimal distributed arc-consistency. Constraints **7**(3–4), 367–385 (2002). https://doi.org/10.1023/A:1020594125144
16. van Hoeve, W.-J.: The alldifferent constraint: a survey. CoRR cs.PL/0105015 (2001). https://arxiv.org/abs/cs/0105015
17. van Hoeve, W.-J., Katriel, I.: Global constraints. In: Rossi, F., van Beek, P., Walsh, T. (eds.) Handbook of Constraint Programming, Foundations of Artificial Intelligence, vol. 2, pp. 169–208. Elsevier (2006). https://doi.org/10.1016/S1574-6526(06)80010-6
18. Hopcroft, J.E., Karp, R.M.: A n5/2 algorithm for maximum matchings in bipartite. In: 12th Annual Symposium on Switching and Automata Theory, SWAT 1971, pp. 122–125 (1971). https://doi.org/10.1109/SWAT.1971.1
19. Huawei: Tecal RH2288H v2 rack server v100r002. Tech. rep., Huawei Technologies Co., Ltd. (2013), issue 01
20. Kasif, S.: On the parallel complexity of discrete relaxation in constraint satisfaction networks. Artif. Intell. **45**(3), 275–286 (1990). https://doi.org/10.1016/0004-3702(90)90009-O
21. Van Kessel, P., Quimper, C.-G.: Filtering algorithms based on the word-ram model. In: Hoffmann, J., Selman, B. (eds.) Proceedings of the Twenty-Sixth AAAI Conference on Artificial Intelligence, 22–26 July 2012, Toronto, Ontario, AAAI Press (2012). http://www.aaai.org/ocs/index.php/AAAI/AAAI12/paper/view/5135
22. Leconte, M.: A bounds-based reduction scheme for constraints of difference. In: Second International Workshop on Constraint-Based Reasoning, Key West, FL, p. 19–28 (1996)
23. López-Ortiz, A., Quimper, C.-G., Tromp, J., van Beek, P.: A fast and simple algorithm for bounds consistency of the alldifferent constraint. In: Gottlob, G., Walsh, T. (eds.) IJCAI-03, Proceedings of the Eighteenth International Joint Conference on Artificial Intelligence, Acapulco, 9–15 August 2003, pp. 245–250. Morgan Kaufmann (2003). http://ijcai.org/Proceedings/03/Papers/036.pdf
24. Mehlhorn, K., Thiel, S.: Faster algorithms for bound-consistency of the sortedness and the alldifferent constraint. In: Dechter, R. (ed.) CP 2000. LNCS, vol. 1894, pp. 306–319. Springer, Heidelberg (2000). https://doi.org/10.1007/3-540-45349-0_23
25. Nguyen, T., Deville, Y.: A distributed arc-consistency algorithm. Sci. Comput. Program. **30**(1–2), 227–250 (1998). https://doi.org/10.1016/S0167-6423(97)00012-9
26. Nightingale, P.: Are adjacency lists worthwhile in alldifferent. Tech. rep., Citeseer (2009)

27. Puget, J.: A fast algorithm for the bound consistency of alldiff constraints. In: Mostow, J., Rich, C. (eds.) Proceedings of the Fifteenth National Conference on Artificial Intelligence and Tenth Innovative Applications of Artificial Intelligence Conference, AAAI 98, IAAI 98, 26–30 July 1998, Madison, Wisconsin, pp. 359–366. AAAI Press/The MIT Press (1998). http://www.aaai.org/Library/AAAI/1998/aaai98-051.php

28. Régin, J.-C.: A filtering algorithm for constraints of difference in csps. In: Hayes-Roth, B., Korf, R.E. (eds.) Proceedings of the 12th National Conference on Artificial Intelligence, Seattle, WA, 31 July–4 August 1994, vol. 1, pp. 362–367. AAAI Press/The MIT Press (1994). http://www.aaai.org/Library/AAAI/1994/aaai94-055.php

29. Régin, J.-C., Malapert, A.: Parallel constraint programming. In: Handbook of Parallel Constraint Reasoning, pp. 337–379. Springer, Cham (2018). https://doi.org/10.1007/978-3-319-63516-3_9

30. Rolf, C.C., Kuchcinski, K.: Parallel consistency in constraint programming. In: Arabnia, H.R. (ed.) Proceedings of the International Conference on Parallel and Distributed Processing Techniques and Applications, PDPTA 2009, Las Vegas, Nevada, 13–17 July 2009, 2 vols, pp. 638–644. CSREA Press (2009)

31. Ruiz-Andino, A., Araujo, L., Sáenz-Pérez, F., Ruz, J.J.: Parallel arc-consistency for functional constraints. In: Sagonas, K. (ed.) Proceedings of the International Workshop on Implementation Technology for Programming Languages Based on Logic, held in conjunction with the Joint International Conference and Symposium on Logic Programming, Manchester, UK, 20 June 1998, pp. 86–100 (1998)

32. Suijlen, W.J.: BSPonMPI - BSPlib implementation on top MPI (2019). https://github.com/wijnand-suijlen/bsponmpi, version 1.1

33. Tarjan, R.E.: Depth-first search and linear graph algorithms. SIAM J. Comput. 1(2), 146–160 (1972). https://doi.org/10.1137/0201010

34. Topcuoglu, H., Hariri, S., Wu, M.: Performance-effective and low-complexity task scheduling for heterogeneous computing. IEEE Trans. Parallel Distrib. Syst. 13(3), 260–274 (2002). https://doi.org/10.1109/71.993206

35. Valiant, L.G.: A bridging model for parallel computation. Commun. ACM 33(8), 103–111 (1990)

36. Yzelman, A.N., Bisseling, R.H., Roose, D., Meerbergen, K.: MulticoreBSP for C: a high-performance library for shared-memory parallel programming. Int. J. Parallel Prog. 42(4), 619–642 (2013). https://doi.org/10.1007/s10766-013-0262-9

37. Zhang, X., Li, Q., Zhang, W.: A fast algorithm for generalized arc consistency of the alldifferent constraint. In: Lang, J. (ed.) Proceedings of the Twenty-Seventh International Joint Conference on Artificial Intelligence, IJCAI 2018, 13–19 July 2018, Stockholm, Sweden, pp. 1398–1403. ijcai.org (2018). https://doi.org/10.24963/ijcai.2018/194

Analyzing the Reachability Problem in Choice Networks

Piotr Wojciechowski[1], K. Subramani[1(✉)], and Alvaro Velasquez[2]

[1] LDCSEE, West Virginia University, Morgantown, WV, USA
{pwojciec,k.subramani}@mail.wvu.edu
[2] Information Directorate, AFRL, Rome, NY, USA
alvaro.velasquez.1@us.af.mil

Abstract. In this paper, we investigate the problem of determining $s-t$ reachability in **choice networks**. In the traditional $s-t$ reachability problem, we are given a weighted network tuple $\mathbf{G} = \langle V, E, \mathbf{c}, s, t \rangle$, with the goal of checking if there exists a path from s to t in G. In an optional choice network, we are given a choice set $S \subseteq E \times E$, in addition to the network tuple G. In the $s-t$ reachability problem in choice networks (OCR_D), the goal is to find whether there exists a path from vertex s to vertex t, with the caveat that at most one arc from each arc-pair $(e_i, e_j) \in S$ is used in the path. OCR_D finds applications in a number of domains including **routing in wireless networks** and **sensor placement**. We analyze the computational complexities of the OCR_D problem and its variants from a number of algorithmic perspectives. We show that the problem is **NP-complete** and its optimization version is **NPO PB-complete**. Additionally, we show that the problem is fixed-parameter tractable in the cardinality of the choice set S. We also consider weighted versions of the OCR_D problem and detail their computational complexities; in particular, the optimization version of the $WOCR_D$ problem is **NPO-complete**.

1 Introduction

This paper is concerned with the $s-t$ reachability in choice networks. Choice networks differ from traditional networks in that there is a choice set. The choice set is a binary relation on the arc set of the network. It consists of arc pairs. If one arc from an arc pair is used in the construction of a path, then the second arc cannot be used. The $s-t$ reachability problem in such networks is called the Optional Choice Reachability problem (OCR_D). It is important to note that paths are not required to intersect the choice set S at all.

The OCR_D problem is a variant of the resource-constrained shortest path problem (RCSP), which has been widely studied in the operations research and

---

K. Subramani—This research was supported in part by the Air-Force Office of Scientific Research through Grant FA9550-19-1-0177 and in part by the Air-Force Research Laboratory, Rome through Contract FA8750-17-S-7007.

P. Schaus (Ed.): CPAIOR 2022, LNCS 13292, pp. 408–423, 2022.
https://doi.org/10.1007/978-3-031-08011-1_27

theoretical computer science communities. Our interest in this problem stems from applications in wireless routing [18], sensor placement [25], air-force logistics [26] and program verification [15].

2   Statement of Problems

In this section, we define the problems studied in this paper. We will be studying the complexity of a variant of reachability known as optional choice reachability. This variant examines reachability in choice networks.

Definition 1 (Choice Network). *A Choice Network* $\mathbf{G} = \langle V, E, S, s, t, \mathbf{c} \rangle$ *consists of the following: 1. Vertex set* V*. 2. Arc set* $E \subseteq V \times V$*. 3. Choice set* $S \subseteq E \times E$*. 4. Start vertex* $s \in V$*. 5. Target vertex* $t \in V$*. 6. Arc cost vector* $\mathbf{c} \in \mathbb{Z}^{|E|}$*.

Throughout this paper, we use n to denote the cardinality of the vertex set V and we use m to denote the cardinality of the arc set E.

In a choice network, the choice set S is used to determine whether a path p is valid.

Definition 2 (Valid Path). *A path* p *in a choice network* $\mathbf{G}$ *is valid, if for each pair* $S_i \in S$*,* p *contains at most one arc in* S_i*.

We can now define the reachability problem in choice networks.

Definition 3 ($\mathbf{OCR}_D$). *The* Optional Choice Reachability *(OCR_D) problem: given a choice network* $\mathbf{G}$*, does* $\mathbf{G}$ *have a valid path from* s *to* t*?

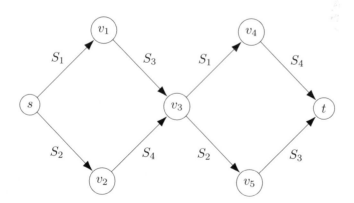

Fig. 1. Choice network $\mathbf{G}$. Each arc is labeled according to its pair

Example 1. Consider the choice network $\mathbf{G}$ in Fig. 1.

Consider a path p in $\mathbf{G}$. Any path that leaves s must go through v_1 or v_2. If p goes through v_1, then it includes the arcs (s, v_1) and (v_1, v_3). Thus, p cannot include arc (v_3, v_4) or (v_5, t). Consequently, there is no valid path from s to t going through v_1.

If p goes through v_2, then it includes the arcs (s, v_2) and (v_2, v_3). Thus, p cannot include arc (v_3, v_5) or (v_4, t). Consequently, there is no valid path from s to t going through v_2.

This means that there is no valid path in **G** from s to t.

In addition to the OCR$_D$ problem, we also study the complexity of finding shortest paths in choice networks. We do this for both weighted and unweighted networks.

Definition 4 (OCR$_{Opt}$). *The* Optional Choice Shortest Path *(OCR$_{Opt}$) problem: given a choice network* **G***, what is the valid path in* **G** *from s to t with the fewest arcs?*

Definition 5 (WOCR$_{Opt}$). *The* Weighted Optional Choice Shortest Path *(WOCR$_{Opt}$) problem: given a choice network* **G***, what is the valid path in* **G** *from s to t with the lowest total cost?*

In this paper we relate problems to the complexity classes **NPO** and **NPO PB**.

The principal contributions of this paper are as follows:

1. Establishing that the OCR$_D$ problem is **NP-complete** (see Sect. 4.1). 2. Designing an $O^*(1.42^{|S|})$ parameterized algorithm for the OCR$_D$ and OCR$_{Opt}$ problems (see Sect. 4.2). 3. A $O(2^n)$ time exact exponential algorithm for the OCR$_D$ problem (see Sect. 4.3). 4. Showing that the OCR$_{Opt}$ problem is **NPO PB-complete** (see Sect. 4.4). 5. A proof that there cannot be a $o(1.18^{|S|})$ algorithm for the OCR$_D$ problem unless the Strong Exponential Time Hypothesis (SETH) fails (see Sect. 6).

3   Motivation and Related Work

The $s - t$ reachability problem and hence the single-source shortest paths problem is one of the most well-studied problems in operations research and theoretical computer science [1,2,7]. Several algorithms have been proposed for various variants of the problem, including the case where the arcs are non-negatively weighted, the arcs are real-weighted, and so on. The $s - t$ reachability problem also plays a fundamental role in the modeling of complexity theoretic issues as evidenced by [21]. Likewise, constrained shortest path problems or more generally resource-constrained shortest path problems have been investigated for quite some time owing to their wide applicability [3]. One of the earliest works in constrained shortest paths is described in [22]. Since then, there have been multiple variants of constrained shortest paths that have been studied in the literature [5,23]. Needless to say, most resource constrained shortest path problems are **NP-hard** [11]. Variants of this problem are amenable to the approximation guarantees [12] and exact approaches [8]. Probabilistic versions of the constrained shortest path problem are discussed in [19]. Additionally, this problem has been studied for paths with forbidden pairs of vertices [10,16,17,27]. However, in this paper, we examine the problem for forbidden pairs of arcs. While these two problems are closely related, transforming one problem into another will involve increasing the size of the graph.

A restricted version of this problem, known as the reachability problem in graphs with forbidden transitions was studied in [24]. In this restriction, each pair of arcs must consist of two arcs such that the tail of one arc is the head of the other. This restricted version is **NP-complete** [24]. In [14], this result was extended and the reachability problem in graphs with forbidden transitions was shown to be **NP-complete** even for grid graphs. A variant of this problem, in which the arcs in a path must alternate direction, was shown to be **NP-complete** [4]. In this paper, we study a more general problem. However, we are able to establish results for even more restricted forms of graphs.

4   Reachability in Choice Networks

In this section, we study the OCR_D and OCR_{Opt} problems.

4.1   Computational Complexity

First, we show that the OCR_D problem is **NP-complete**. Given a path from s to t in a choice network $\mathbf{G}$, it is easy to check if the path is valid. Thus, the OCR_D problem is in **NP**. All that remains is to show **NP-hardness**. This will be done by a reduction from 3-SAT.

Let Φ be a 3-CNF formula. We construct the corresponding choice network $\mathbf{G}$ as follows:

1. For each variable x_i we create the structure shown in Fig. 2. Note that X_i is the set of arcs paired with the arcs in the upper path through the structure. Similarly, $\bar{X}_i$ is the set of arcs paired with the arcs in the lower path through the structure.

   In Fig. 2, the dashed lines connect the arcs which are in the same pair. If a valid path p traverses the lower path, then, since each of the arcs in the set $\bar{X}_i$ are paired with an arc in p, none of the arcs in set $\bar{X}_i$ can be in p. Similarly, if p traverses the upper path, then none of the arcs in the set X_i can be in p. Note that $|X_i|$ is the number of clauses in Φ with the literal x_i and $|\bar{X}_i|$ is the number of clauses in Φ with the literal $\neg x_i$. Note that the arcs in the sets X_i and $\bar{X}_i$ are not part of the variable gadget but are used to construct the clause gadgets.
2. Using the arcs in the sets X_i and $\bar{X}_i$ we now construct a gadget for each clause ϕ in Φ. The clause gadget corresponding to a clause of the form $(x_i, \neg x_j, x_k)$ is made using the arcs from the sets X_i, $\bar{X}_j$, and X_k. The resultant gadget is shown in Fig. 2.

   We construct the choice network $\mathbf{G}$ by combining the gadgets as shown in Fig. 3.

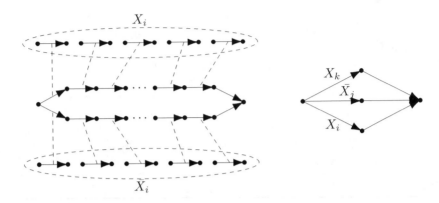

Fig. 2. Variable gadget corresponding to the variable x_i and clause gadget corresponding to the clause $(x_i, \neg x_j, x_k)$.

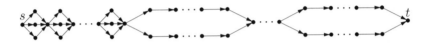

Fig. 3. Choice network corresponding to 3-CNF formula Φ.

Theorem 1. OCR_D is **NP-hard**.

Proof. Let Φ be a 3-CNF formula and let **G** be the corresponding choice network.

First, we assume that Φ is satisfiable. Let **x** be a truth assignment that satisfies Φ. We use **x** to construct the valid path p. For each variable x_i assigned a value of **true**, p will traverse the lower path of the corresponding variable gadget. This prevents any of the arcs in $\bar{X}_i$ from being on p. Similarly, for each variable x_i assigned a value of **false** the path will traverse the upper path of the corresponding variable gadget. This prevents any of the arcs in X_i from being on the path.

Let $(x_i, \neg x_j, x_k)$ be an arbitrary clause in the system. At least one literal in this clause is assigned a value of **true**. Since the three paths through the corresponding gadget correspond to the sets X_i, $\bar{X}_j$, and X_k, at least one path through the gadget can still be traversed by p. Thus, we can construct a valid path from s to t in **G**.

Now assume that there is a valid path p from s to t in **G**. We use p to construct the assignment **x** as follows: for each variable x_i, if p traverses the upper path of the gadget corresponding to x_i, then set x_i to **true**, otherwise set x_i to **false**.

Consider a clause ϕ of Φ and look at how p traverses the corresponding gadget in **G**. p either uses an arc in the set X_i for a literal x_i in ϕ, or p uses an arc in the set $\bar{X}_i$ for a literal $\neg x_i$ in ϕ. If p uses an arc in the set X_i, then p cannot traverse the lower path of the gadget corresponding to x_i. Thus, p must traverse the upper path of the gadget. Consequently, x_i is assigned a value of **true** and **x** satisfies the clause ϕ.

If p uses an arc in the set $\bar{X}_i$, then p cannot traverse the upper path of the gadget corresponding to x_i. Thus, p must traverse the lower path of the gadget. Consequently, x_i is assigned a value of **false** and **x** satisfies the clause ϕ. Since ϕ was an arbitrarily chosen clause, **x** satisfies Φ as desired. □

We now use structural properties of the graph **G** to strengthen Theorem 1.

Corollary 1. *The OCR$_D$ problem is* **NP-complete** *for bipartite, acyclic graphs with path-width 2.*

Proof. By construction **G** is both bipartite and acyclic. Additionally, **G** has path-width 2. Thus, the OCR$_D$ problem is still **NP-complete** even for such restricted networks. □

4.2 A Fixed-Parameter Algorithm

Now, we show that there exists a $O^*(1.42^{|S|})$ time **FPT** algorithm for the OCR$_D$ and OCR$_{Opt}$ problems.

First, we provide a $O^*(2^{|S|})$ time algorithm for solving these problems.

For each set $S_i \in S$, we know that no valid path can traverse both arcs in S_i. Thus, if a valid path exists and we consider every possible way to remove one arc from each pair S_i, t will be reachable from s in at least one resultant network.

This gives us the brute force approach in Algorithm 4.1.

Algorithm 4.1. Brute force algorithm for OCR$_D$.

   Input: Choice network **G**
   Output: **true** if **G** has a valid path from s to t, **false** otherwise.
1: **procedure** OCR-DEC(**G**)
2:     **for** $(C = 0$ to $(2^{|S|} - 1))$ **do**
3:         **for** $(i = 1$ to $|S|)$ **do**
4:             **if** $(C \mod 2^i \geq 2^{i-1})$ **then**   ▷ The ith bit of C is 1.
5:                 Remove the first arc of S_i from **G**.
6:             **else**
7:                 Remove the second arc of S_i from **G**.
8:         **if** $(t$ is reachable from s in **G**) **then**   ▷ **G** has a valid path from s to t.
9:             **return true**.
10: **return false**.

Algorithm 4.1 can be easily modified to find shortest paths by performing a shortest path check for each resultant network instead of a reachability check.

We now define several reduction rules which can be used to both reduce the size of **G** and the number of pairs in S. These rules are as follows: 1. Remove all vertices not reachable from s: If a vertex v_j is not reachable from s, then v_j cannot be part of a valid path from s to t. Thus, v_j can be removed from $\mathbf{G}$ without affecting any valid paths from s to t. 2. Remove all vertices from which t is not reachable: If t is not reachable from a vertex v_j, then v_j cannot be part of a valid path from s to t. Thus, v_j can be removed from $\mathbf{G}$ without affecting any valid paths from s to t.

In particular, these rules can be performed each time an arc is removed from $\mathbf{G}$ in Algorithm 4.1. If a reduction rule removes an arc belonging to a pair S_i, then that pair no longer matters since there is no way for any path from s to t to use both arcs in S_i. Thus, we can also remove the pair S_i from $\mathbf{G}$ without affecting the validity of any paths from s to t.

For choice network $\mathbf{G}$ and arc e in $\mathbf{G}$, let $R(e, \mathbf{G})$ be the number of pairs removed from $\mathbf{G}$ by the above reduction rules if e is removed from $\mathbf{G}$. Also let $G(e, \mathbf{G})$ be the choice network formed by applying the above reduction rules after arc e is removed from $\mathbf{G}$.

For each pair $S_i = (e_{S_{i,1}}, e_{S_{i,2}})$, we can calculate both $R(e_{S_{i,1}}, \mathbf{G})$ and $R(e_{S_{i,2}}, \mathbf{G})$. We now show that if $\mathbf{G}$ has no valid path from s to t, then there always exists either:

1. A pair S_i such that $R(e_{S_{i,1}}, \mathbf{G}) \geq 1$ and $R(e_{S_{i,2}}, \mathbf{G}) \geq 1$.
2. A pair S_i such that removing $e_{S_{i,1}}$ from $\mathbf{G}$ also removes $e_{S_{i,2}}$ from $\mathbf{G}$ (or vice versa).

Lemma 1. *If choice network $\mathbf{G}$ has no valid path from s to t, then there always exists a pair S_i such that either:*

1. *$R(e_{S_{i,1}}, \mathbf{G}) \geq 1$ and $R(e_{S_{i,2}}, \mathbf{G}) \geq 1$.*
2. *Removing $e_{S_{i,1}}$ from $\mathbf{G}$ also removes $e_{S_{i,2}}$ from $\mathbf{G}$ (or vice versa).*

Proof. Let $\mathbf{G}$ be a choice network with no valid paths from s to t. We can assume without loss of generality that s is reachable from t. Additionally, assume that there is no pair S_i that satisfies the desired conditions. We will show that this always results in a contradiction by induction on $|S|$.

If $|S| = 1$, then there is only one pair $S_1 = (e_{S_{1,1}}, e_{S_{1,2}})$ in S. Since $\mathbf{G}$ has no valid path from s to t, any path from s to t must use both $e_{S_{1,1}}$ and $e_{S_{1,2}}$. Thus, removing $e_{S_{1,1}}$ from $\mathbf{G}$, will also remove $e_{S_{1,2}}$ from $\mathbf{G}$ when applying the reduction rules to generate $G(e_{S_{1,1}}, \mathbf{G})$.

Now let us assume that the lemma holds when $|S| = k$. Let $\mathbf{G}$ with $|S| = k + 1$ be a choice network with no valid paths from s to t. If for every pair $S_i = (e_{S_{i,1}}, e_{S_{i,2}})$, $R(e_{S_{i,1}}, \mathbf{G}) = 0$ or $R(e_{S_{i,2}}, \mathbf{G}) = 0$, then either removing $e_{S_{i,1}}$ will not make any vertex unreachable from s or removing $e_{S_{i,2}}$ will not make any vertex unreachable from s. Thus, t is still reachable from s in $G(e_{S_{i,1}}, \mathbf{G})$, or in $G(e_{S_{i,2}}, \mathbf{G})$ for every i. Consequently, $\mathbf{G}$ has a valid path from s to t.   □

Lemma 1 lets us construct the improved parameterized algorithm Algorithm 4.2 for the OCR_D problem.

We can now analyze the running time of Algorithm 4.2. Note that Algorithm 4.2 only recurses on multiple graphs if for some $S_i \in S$, $R(e_{S_{i,1}}, \mathbf{G}) \geq 1$ and

Algorithm 4.2. Improved **FPT** algorithm for OCR$_D$.

Input: Choice network **G**, start vertex s, and target vertex t.
Output: true if **G** has a valid path from s to t, **false** otherwise.
1: **procedure** OCR-FPT($\mathbf{G}, s, t$)
2:     **if** ($|S| = 1$) **then**
3:         **if** ($t \notin G(e_{S_{1,1}}, \mathbf{G})$ and $t \notin G(e_{S_{1,2}}, \mathbf{G})$) **then** ▷ Any path from s to t must use both $e_{S_{1,1}}$ and $e_{S_{1,2}}$.
4:             **return false.**
5:         **return true.**
6:     **for** ($i = 1$ **to** $|S|$) **do**
7:         **if** ($R(e_{S_{i,1}}, \mathbf{G}) \geq 1$ and $R(e_{S_{i,2}}, \mathbf{G}) \geq 1$) **then**
8:             $res_1 := $ OCR-FPT($G(e_{S_{i,1}}, \mathbf{G}), s, t$).
9:             $res_2 := $ OCR-FPT($G(e_{S_{i,2}}, \mathbf{G}), s, t$).
10:             **return** $res_1 \vee res_2$.
11:         **if** ($e_{S_{i,2}} \notin G(e_{S_{i,1}}, \mathbf{G})$) **then**
12:             $res_1 := $ OCR-FPT($G(e_{S_{i,2}}, \mathbf{G}), s, t$).
13:             **return** res_1.
14:         **if** ($e_{S_{i,1}} \notin G(e_{S_{i,2}}, \mathbf{G})$) **then**
15:             $res_1 := $ OCR-FPT($G(e_{S_{i,1}}, \mathbf{G}), s, t$).
16:             **return** res_1.
17:     **return true.**   ▷ By Lemma 1, there is a valid path from s to t in **G**

$R(e_{S_{i,2}}, \mathbf{G}) \geq 1$. Observe that in $G(e_{S_{i,1}}, \mathbf{G})$ the size of the choice set has been reduced by $(R(e_{S_{i,1}}, \mathbf{G}) + 1)$ since S_i and at least one other element of the choice set have been removed. Similarly, in $G(e_{S_{i,2}}, \mathbf{G})$ the size of the choice set has been reduced by $(R(e_{S_{i,2}}, \mathbf{G}) + 1)$. Thus, in both $G(e_{S_{i,1}}, \mathbf{G})$ and $G(e_{S_{i,2}}, \mathbf{G})$ the size of the choice set has been reduced by at least 2.

Let $T(|S|)$ be the running time of Algorithm 4.2 on a choice network with choice set S. Note that for each i, $R(e_{S_{i,1}}, \mathbf{G})$, $R(e_{S_{i,2}}, \mathbf{G})$, $G(e_{S_{i,1}}, \mathbf{G})$, and $G(e_{S_{i,2}}, \mathbf{G})$ can be calculated in time $O(m+n)$. Since Algorithm 4.2 only recurses when the size of the choice set decreases by at least 2, the running time of Algorithm 4.2 is governed by the recurrence relation $T(|S|) \leq T(|S| - 2) + T(|S| - 2) + O(|S| \cdot (m + n))$.

From [6], Algorithm 4.2 runs in time $O(|S| \cdot (m + n) \cdot 1.42^{|S|})$. This is an improvement over the running time of the brute force approach of Algorithm 4.1.

We now prove the correctness of Algorithm 4.2.

Theorem 2. *Algorithm 4.2 returns* **true***, if and only if* **G** *has a valid path from* s *to* t.

Proof. First, assume that Algorithm 4.2 returns **false**. Algorithm 4.2 can return **false** on Line 4, Line 10, Line 13, or Line 16. We will prove that **G** has no valid path from s to t through induction on the size of the choice set S.

Assume that $|S| = 1$. In this case, Algorithm 4.2 can only return **false** on Line 4. Thus, the **if** statement on Line 3 must be satisfied. This means that

$t \notin G(e_{S_{1,1}}, \mathbf{G})$ and $t \notin G(e_{S_{1,2}}, \mathbf{G})$. Since $t \notin G(e_{S_{1,1}}, \mathbf{G})$, t is not reachable from s if $e_{S_{1,1}}$ is removed from $\mathbf{G}$. Thus, any path from s to t must use $e_{S_{1,1}}$. Similarly, since $t \notin G(e_{S_{1,1}}, \mathbf{G})$, any path from s to t must use $e_{S_{1,2}}$. Thus, any path from s to t must use both arcs in S_1. Consequently, $\mathbf{G}$ does not have a valid path from s to t.

Now assume that if Algorithm 4.2 returns **false** on a choice network $\mathbf{G}$ with $|S| < k$, then $\mathbf{G}$ has no valid path from s to t. Suppose we run Algorithm 4.2 on a choice network $\mathbf{G}$ with $|S| = k$.

If Algorithm 4.2 returns **false** on line 10, then, for some pair S_i, both OCR-FPT$(G(e_{S_{i,1}}, \mathbf{G}), s, t)$ and OCR-FPT$(G(e_{S_{i,2}}, \mathbf{G}), s, t)$ returned **false**. Since OCR-FPT$(G(e_{S_{i,1}}, \mathbf{G}), s, t)$ returned **false**, then by the inductive hypothesis, there is no valid path from s to t in $G(e_{S_{i,1}}, \mathbf{G})$. Note that $G(e_{S_{i,1}}, \mathbf{G})$ is made by removing $e_{S_{i,1}}$ from $\mathbf{G}$ and then removing all vertices made unreachable from s. Thus, any valid path from s to t in $\mathbf{G}$ must use $e_{S_{i,1}}$. Similarly, since OCR-FPT$(G(e_{S_{i,2}}, \mathbf{G}), s, t)$ returned **false**, any valid path from s to t in $\mathbf{G}$ must use $e_{S_{i,2}}$. However, by definition no valid path can use both $e_{S_{i,1}}$ and $e_{S_{i,2}}$. Thus, $\mathbf{G}$ has no valid path from s to t.

If Algorithm 4.2 returns **false** on line 13, then, for some pair S_i, the arc $e_{S_{i,2}}$ is not in the graph $G(e_{S_{i,1}}, \mathbf{G})$ and OCR-FPT$(G(e_{S_{i,2}}, \mathbf{G}), s, t)$ returned **false**. Since $e_{S_{i,2}} \notin G(e_{S_{i,1}}, \mathbf{G})$, then, after removing $e_{S_{i,1}}$ from $\mathbf{G}$, the arc $e_{S_{i,2}}$ is not on any path from s to t. Assume that $\mathbf{G}$ has a valid path p from s to t. If p uses $e_{S_{i,2}}$ then, by definition, p cannot use $e_{S_{i,1}}$. However, this path would remain in the graph after removing $e_{S_{i,1}}$. Since $e_{S_{i,2}} \notin G(e_{S_{i,1}}, \mathbf{G})$, this cannot happen. Thus, p cannot use $e_{S_{i,2}}$. Consequently, p is a valid path in $G(e_{S_{i,2}}, \mathbf{G})$. By the inductive hypothesis, OCR-FPT$(G(e_{S_{i,2}}, \mathbf{G}), s, t)$ could not have returned **false**. Since OCR-FPT$(G(e_{S_{i,2}}, \mathbf{G}), s, t)$ does return **false**, $\mathbf{G}$ has no valid path from s to t. Similarly, if Algorithm 4.2 returns **false** on line 16, then $\mathbf{G}$ has no valid path from s to t.

Now, assume that Algorithm 4.2 returns **true**. Algorithm 4.2 can return **true** on Line 5, Line 10, Line 13, Line 16, or Line 17. We will prove that $\mathbf{G}$ has a valid path from s to t through induction on the size of the choice set S.

Assume that $|S| = 1$. In this case, Algorithm 4.2 can only return **true** on Line 5. Thus, $t \in G(e_{S_{1,1}}, \mathbf{G})$ or $t \in G(e_{S_{1,2}}, \mathbf{G})$. If $t \in G(e_{S_{1,1}}, \mathbf{G})$ then there is a path from s to t in $\mathbf{G}$ that does not use $e_{S_{1,1}}$. Since $S_1 = (e_{S_{1,1}}, e_{S_{1,2}})$ is the only pair in $\mathbf{G}$ this path is, by definition, a valid path from s to t. Similarly, if $t \in G(e_{S_{1,2}}, \mathbf{G})$, then $\mathbf{G}$ has a valid path from s to t.

Now assume that if Algorithm 4.2 returns **true** on a choice network $\mathbf{G}$ with $|S| < k$, then $\mathbf{G}$ has a valid path from s to t. Suppose we run Algorithm 4.2 on a choice network $\mathbf{G}$ with $|S| = k$.

If Algorithm 4.2 returns **true** on Line 10, Line 13, or Line 16, then either OCR-FPT$(G(e_{S_{i,1}}, \mathbf{G}), s, t)$ or OCR-FPT$(G(e_{S_{i,2}}, \mathbf{G}), s, t)$ returns **true** for some pair S_i. If OCR-FPT$(G(e_{S_{i,1}}, \mathbf{G}), s, t)$ returns **true**, then by the inductive hypothesis, $G(e_{S_{i,1}}, \mathbf{G})$ has a valid path from s to t. Note that this path is also a valid path in $\mathbf{G}$. Thus, $\mathbf{G}$ has a valid path from s to t. Similarly, OCR-FPT$(G(e_{S_{i,2}}, \mathbf{G}), s, t)$ returns **true**, then $\mathbf{G}$ has a valid path from s to t.

If Algorithm 4.2 returns **true** on Line 17, then there is no pair S_i that satisfies either condition 1 or condition 2 in Lemma 1. Thus, by Lemma 1, **G** must have a valid path from s to t. □

4.3   Exact Exponential Algorithm

We now look at a different way of analyzing Algorithm 4.2.

Let $T'(n)$ be the running time of Algorithm 4.2 on a choice network with n vertices. Observe that the reduction rules in Sect. 4.2 only reduce the size of the choice set if vertices are removed from **G**. Additionally, Algorithm 4.2 only recurses if the size of the choice set is reduced by the reduction rules. Thus, if Algorithm 4.2 recurses on choice S_i, then both $G(e_{S_{i,1}}, \mathbf{G})$ and $G(e_{S_{i,2}}, \mathbf{G})$ have fewer vertices than **G**. Consequently, the running time of Algorithm 4.2 is governed by the recurrence relation $T'(n) \leq T'(n-1) + T'(n-1) + O(|S| \cdot (m+n))$.

From [6], Algorithm 4.2 runs in time $O(|S| \cdot (m + n) \cdot 2^n)$. Thus, Algorithm 4.2 is also an exact exponential algorithm for OCR_D.

4.4   Approximation Complexity

We now show that the OCR_{Opt} problem is **NPO PB-complete**.

Note that a valid path p in a choice network has at most n arcs. Additionally, recall that the validity of a path can be verified in polynomial time. Thus the OCR_{Opt} is in **NPO PB**. All that remains is to show **NPO PB-hardness**. This is done by a reduction from the Min-Ones problem.

The Min-Ones problem is defined as follows: Given a 3-CNF formula Φ, what is the satisfying assignment to Φ with the fewest variables set to **true**. This problem is known to be **NPO PB-complete** [13]. We will modify the network construction used in Sect. 4.1 to prove the inapproximability of OCR_{Opt}.

Let Φ be a 3-CNF formula with m' clauses over n' variables and let **G** be the choice network corresponding to Φ as per the construction in Sect. 4.1. From Sect. 4.1, **G** has the following properties: 1. In each variable gadget, the upper path has $(|X_i| + 1)$ arcs where $|X_i|$ is the number of clauses in Φ with the literal x_i. 2. In each variable gadget, the lower path has $(|\bar{X}_i| + 1)$ arcs where $|\bar{X}_i|$ is the number of clauses in Φ with the literal $\neg x_i$. 3. In each clause gadget, each path through the gadget has two arcs. From **G**, we construct the choice network **G'** by adding unpaired arcs to the end of each path through each variable gadget, until:

1. The upper path in each variable gadget has m' arcs.
2. The lower path in each variable gadget has $m' \cdot (n' + 3)$ arcs.

This gives the network **G'** the following properties:

1. The shortest path from s to t has $m' \cdot (n' + 2)$ arcs. This path uses the upper path in each variable gadget (a total of $m' \cdot n'$ arcs) and any path through each clause gadget (an additional $2 \cdot m'$ arcs). Note that this path is valid if and only if Φ is satisfied by the all **false** assignment.

2. Each time a path from s to t uses the lower path through a variable gadget, the length of the path increases by $m' \cdot (n' + 2)$ arcs.

We can now relate the length of paths in $\mathbf{G}'$ to the number of **true** variables in a satisfying assignment to Φ.

Theorem 3. $\mathbf{G}'$ *has a valid path p from s to t with* $(k+1) \cdot m' \cdot (n'+2)$ *arcs if and only if Φ has a satisfying assignment in which k variables are set to* **true**.

Proof. From Theorem 1, $\mathbf{G}'$ has a valid path p from s to t if and only if Φ has a satisfying assignment $\mathbf{x}$. In each variable gadget, p takes the lower path if and only if $\mathbf{x}$ assigns that variable to **true**. Note that every satisfying assignment to Φ corresponds to such a valid path.

Let $\mathbf{x}$ be a satisfying assignment to Φ with k variables set to **true**. Thus, $(n' - k)$ variables are set to **false**.

Let p the the corresponding valid path from s to t in $\mathbf{G}'$. Note that p takes the upper path through $(n' - k)$ variable gadgets and the lower path through k variable gadgets. This means that p has length

$$(n' - k) \cdot m' + k \cdot m' \cdot (n' + 3) + 2 \cdot m' = (k+1) \cdot m' \cdot (n' + 2).$$

Thus $\mathbf{G}'$ has a valid path p from s to t with $(k+1) \cdot m' \cdot (n'+2)$ arcs if and only if Φ has a satisfying assignment in which k variables are set to **true**. $\square$

Thus, the reduction from Min-Ones to OCR_{Opt} is approximation preserving. Since the Min-Ones problem is **NPO PB-complete** so is the OCR_{Opt} problem. Thus, the OCR_{Opt} problem cannot be approximated to within a polylogarithmic factor unless $\mathbf{P} = \mathbf{NP}$.

5   Weighted Reachability in Choice Networks

In this section, we study the WOCR_{Opt} problem.

5.1   Approximation Complexity

We now show that the WOCR_{Opt} problem is **NPO-complete**.

Recall that the validity of a path can be verified in polynomial time. Thus the WOCR_{Opt} problem is in **NPO**. All that remains is to show **NPO-hardness**. This is done by a reduction from the Weighted Min-Ones problem.

The Weighted Min-Ones problem is defined as follows: Given a 3-CNF formula Φ and variable weight function w, what is the satisfying assignment to Φ with the least weight of variables set to **true**. This problem is known to be **NPO-complete** [20]. We will modify the network construction used in Sect. 4.1 to prove the inapproximability of WOCR_{Opt}.

Let Φ be a 3-CNF formula with m' clauses over n' variables and let $\mathbf{G}$ be the choice network corresponding to Φ as per the construction in Sect. 4.1. We weight the arcs in $\mathbf{G}$ as follows: 1. In each clause gadget in $\mathbf{G}$, give every arc a cost of 0. 2. In the variable gadget corresponding to variable x_i, give every arc a cost of 0 except for the unpaired arc on the lower path, which gets a cost of $w(x_i)$. This is shown in Fig. 4.

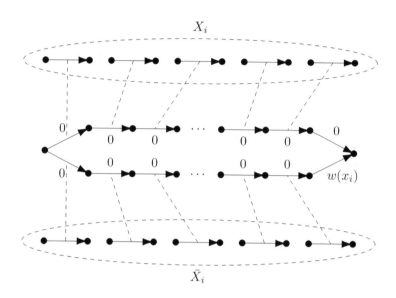

Fig. 4. Weighted variable gadget corresponding to the variable x_i.

We can now relate the costs of paths in $\mathbf{G}$ to the weight of **true** variables in satisfying assignments to Φ.

Theorem 4. $\mathbf{G}$ *has a valid path p from s to t with cost k if and only if Φ has a satisfying assignment in which the weight of variables set to* **true** *is k.*

Proof. From Theorem 1, $\mathbf{G}$ has a valid path p from s to t if and only if Φ has a satisfying assignment $\mathbf{x}$. In each variable gadget, p takes the lower path if and only if $\mathbf{x}$ assigns that variable to **true**. Note that every satisfying assignment to Φ corresponds to such a valid path.

Let $\mathbf{x}$ be a satisfying assignment to Φ in which the total weight of variables set to **true** is k. Let p the the corresponding valid path from s to t in $\mathbf{G}'$.

For every variable x_i set to **true** in $\mathbf{x}$, p has one arc of cost $w(x_i)$. All other arcs in p have cost 0. This means that p has total cost k.

Thus $\mathbf{G}$ has a valid path p from s to t with cost k if and only if Φ has a satisfying assignment in which the weight of variables set to **true** is k. $\square$

Thus, the reduction from Weighted Min-Ones to WOCR$_{Opt}$ is approximation preserving. Since the Weighted Min-Ones problem is **NPO-complete** so is the

WOCR$_{Opt}$ problem. Thus, the WOCR$_{Opt}$ problem cannot be approximated to within a polynomial factor unless **P = NP**.

6 Lower Bounds for OCR$_D$

We now use the Strong Exponential Time Hypothesis (SETH) to establish lower bounds on the running time of any algorithm for the OCR$_D$ problem.

Let Φ be a CNF formula with n' variables. We construct the corresponding choice network **G** as follows:

For each variable x_i we create the structure shown in Fig. 5.

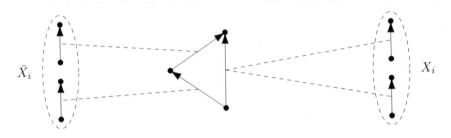

Fig. 5. Variable gadget corresponding to the variable x_i.

In Fig. 5 the dashed lines connect the arcs which are in the same pair. If a valid path p traverses the left path, then both of the arcs in the set $\bar{X}_i$ are paired with an arc in p. Thus, neither of the arcs in set $\bar{X}_i$ can be in p. Similarly, if p traverses the right path, then neither of the arcs in the set X_i can be in p.

Using the arcs in the sets X_i and $\bar{X}_i$ we now construct a gadget for each clause ϕ in Φ. The clause gadget corresponding to a k-literal clause ϕ_j has k possible paths through the gadget corresponding to the variables in ϕ_j. The path corresponding to the variable $x_i \in \phi_j$ is made using the arc $l_i \in X_i$ if the literal x_i is in ϕ_j, or the arc $l_i \in \bar{X}_i$ if the literal $\neg x_i$ is in ϕ_j. An example of such a gadget is shown in Fig. 6.

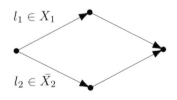

Fig. 6. Clause gadget corresponding to the clause $(x_1, \neg x_2)$.

We construct the choice network **G** by combining the gadgets as they were combined in Sect. 4.1. Thus, any valid path from s to t must traverse all the variable gadgets, followed by all the clause gadgets.

Theorem 5. **G** *has a valid path from s to t if and only if Φ is satisfiable.*

Proof. First, we assume that Φ is satisfiable. Let **x** be a truth assignment that satisfies Φ. We use **x** to construct the valid path p. For each variable x_i assigned a value of **true**, p will traverse the left path of the corresponding variable gadget. This prevents any of the arcs in $\bar{X}_i$ from being on p. Similarly, for each variable x_i assigned a value of **false** the path will traverse the right path of the corresponding variable gadget. This prevents any of the arcs in X_i from being on the path.

Let $(x_i, \neg x_j, x_k)$ be an arbitrary clause in the system. At least one literal in this clause is assigned a value of **true**. Since the three paths through the corresponding gadget correspond to the sets X_i, $\bar{X}_j$, and X_k, at least one path through the gadget can still be traversed by p. Thus, we can construct a valid path from s to t in **G**.

Now assume that there is a valid path p from s to t in **G**. We use p to construct the assignment **x** as follows: for each variable x_i, if p traverses the left path of the gadget corresponding to x_i, then set x_i to **true**, otherwise set x_i to **false**.

Consider a clause ϕ of Φ and look at how p traverses the corresponding gadget in **G**. p either uses an arc in the set X_i for a literal x_i in ϕ, or p uses an arc in the set $\bar{X}_i$ for a literal $\neg x_i$ in ϕ. If p uses an arc in the set X_i, then p cannot traverse the right path of the gadget corresponding to x_i. Thus, p must traverse the left path of the gadget. Consequently, x_i is assigned a value of **true** and **x** satisfies the clause ϕ.

If p uses an arc in the set $\bar{X}_i$, then p cannot traverse the left path of the gadget corresponding to x_i. Thus, p must traverse the right path of the gadget. Consequently, x_i is assigned a value of **false** and **x** satisfies the clause ϕ. Since ϕ was an arbitrarily chosen clause, **x** satisfies Φ as desired.   □

Note that **G** has $|S| = 4 \cdot n'$ pairs of arcs. Thus, if an algorithm solved the OCR_D problem in time $o(1.18^{|S|})$ time, then that algorithm would solve SAT in time $o(1.18^{4 \cdot n'}) \subseteq o(1.94^{n'})$. This would violate the Strong Exponential Time Hypothesis.

7   Conclusion

In this paper, we investigated algorithmic issues associated with the reachability problem in choice networks (OCR_D) and its variants. The OCR_D problem is a special case of the Resource-Constrained Shortest path (RCSP) problem. The literature on RCSP specializations is immense. As mentioned before, this problem finds applications in a number of different domains such as transportation logistics and program verification. The three principal contributions of this paper are showing that OCR_D is **NP-complete**, OCR_{Opt} is **NPO PB-complete** and that OCR_D/OCR_{Opt} are **fixed-parameter tractable** in the cardinality of the choice set. We also studied the $WOCR_{Opt}$ problem and derived algorithmic and complexity results for the same. From our perspective, the following problem is worth investigating: A lower bound on the algorithmic complexity of the $WOCR_{Opt}$ problem - We are in the process of showing that no algorithm for this problem can run in $o(n^2 \cdot 2^n)$ time unless the Traveling Salesman problem can be solved in $o(n^2 \cdot 2^n)$ time. The Heldman-Karp bound of $\Omega(n^2 \cdot 2^n)$ time has not seen improvement in the last five decades [9].

References

1. Ahuja, R.K., Magnanti, T.L., Orlin, J.B.: Network Flows: Theory, Algorithms and Applications. Prentice-Hall (1993)
2. Ahuja, R.K., Mehlhorn, K., Orlin, J.B., Tarjan, R.E.: Faster algorithms for the shortest path problem. J. ACM **37**(2), 213–223 (1990)
3. Bajaj, C.P.: Some constrained shortest-route problems. Unternehmensforschung **15**(1), 287–301 (1971)
4. Bang-Jensen, J., Bessy, S., Jackson, B., Kriesell, M.: Antistrong digraphs. J. Comb. Theory Ser. B **122**, 68–90 (2017)
5. Beasley, J.E., Christofides, N.: An algorithm for the resource constrained shortest path problem. Networks **19**(4), 379–394 (1989)
6. Cygan, M., et al.: Parameterized Algorithms. Springer, Cham (2015). https://doi.org/10.1007/978-3-319-21275-3
7. Deo, N., Pang, C.Y.: Shortest-path algorithms: taxonomy and annotation. Networks **14**(2), 275–323 (1984)
8. Ferone, D., Festa, P., Guerriero, F.: An efficient exact approach for the constrained shortest path tour problem. Optim. Methods Softw. **35**(1), 1–20 (2020)
9. Fomin, F.V., Kratsch, D., Woeginger, G.J.: Exact (exponential) algorithms for the dominating set problem. In: Hromkovič, J., Nagl, M., Westfechtel, B. (eds.) WG 2004. LNCS, vol. 3353, pp. 245–256. Springer, Heidelberg (2004). https://doi.org/10.1007/978-3-540-30559-0_21
10. Gabow, H.N., Maheswari, S.N., Osterweil, L.J.: On two problems in the generation of program test paths. IEEE Trans. Software Eng. **2**(3), 227–231 (1976)
11. Garey, M.R., Johnson, D.S.: Computers and Intractability: A Guide to the Theory of NP-Completeness. W.H. Freeman Company, San Francisco (1979)
12. Horváth, M., Kis, T.: Multi-criteria approximation schemes for the resource constrained shortest path problem. Optim. Lett. **12**(3), 475–483 (2017). https://doi.org/10.1007/s11590-017-1212-z
13. Kann, V.: Polynomially bounded minimization problems that are hard to approximate. Nordic J. Comput. **1**(3), 317–331 (1994)
14. Kanté, M.M., Moataz, F.Z., Momège, B., Nisse, N.: Finding paths in grids with forbidden transitions. In: Mayr, E.W. (ed.) WG 2015. LNCS, vol. 9224, pp. 154–168. Springer, Heidelberg (2016). https://doi.org/10.1007/978-3-662-53174-7_12
15. Büning, H.K., Wojciechowski, P., Subramani, K.: Finding read-once resolution refutations in systems of 2CNF clauses. Theor. Comput. Sci. **729**, 42–56 (2018)
16. Kolman, P., Pangrác, O.: On the complexity of paths avoiding forbidden pairs. Discret. Appl. Math. **157**(13), 2871–2876 (2009)
17. Kovác, J.: Complexity of the path avoiding forbidden pairs problem revisited. Discret. Appl. Math. **161**(10–11), 1506–1512 (2013)
18. Kukunuru, N.: Secure and energy aware shortest path routing framework for WSN. In: Balas, V.E., Sharma, N., Chakrabarti, A. (eds.) Data Management, Analytics and Innovation. AISC, vol. 808, pp. 379–390. Springer, Singapore (2019). https://doi.org/10.1007/978-981-13-1402-5_29
19. Hosseini Nodeh, Z., Babapour Azar, A., Khanjani Shiraz, R., Khodayifar, S., Pardalos, P.M.: Joint chance constrained shortest path problem with Copula theory. J. Comb. Optim. **40**(1), 110–140 (2020)
20. Orponen, P., Mannila, H.: On approximation preserving reductions: complete problems and robust measures. Technical report, Department of Computer Science, University of Helsinki (1987)

21. Christos, H.: Papadimitriou. Computational Complexity. Addison-Wesley, New York (1994)
22. Saigal, R.: Letter to the editor - a constrained shortest route problem. Oper. Res. **16**(1), 205–209 (1968)
23. Sivakumar, R.A., Batta, R.: The variance-constrained shortest path problem. Transp. Sci. **28**(4), 309–316 (1994)
24. Szeider, S.: Finding paths in graphs avoiding forbidden transitions. Discret. Appl. Math. **126**(2–3), 261–273 (2003)
25. Wang, N., Li, J.: Shortest path routing with risk control for compromised wireless sensor networks. IEEE Access **7**, 19303–19311 (2019)
26. Wojciechowski, P., Williamson, M., Subramani, K.: On finding shortest paths in arc-dependent networks. In: Baïou, M., Gendron, B., Günlük, O., Mahjoub, A.R. (eds.) ISCO 2020. LNCS, vol. 12176, pp. 249–260. Springer, Cham (2020). https://doi.org/10.1007/978-3-030-53262-8_21
27. Yinnone, H.: On paths avoiding forbidden pairs of vertices in a graph. Discret. Appl. Math. **74**(1), 85–92 (1997)

Model-Based Approaches
to Multi-attribute Diverse Matching

Jiachen Zhang$^{(\boxtimes)}$, Giovanni Lo Bianco$^{(\boxtimes)}$, and J. Christopher Beck$^{(\boxtimes)}$

Department of Mechanical and Industrial Engineering, University of Toronto,
Toronto, ON M5S 3G8, Canada
{jasonzjc,giolb,jcb}@mie.utoronto.ca

Abstract. Bipartite b-matching is a classical model that is used for utility maximization in various applications such as marketing, healthcare, education, and general resource allocation. Multi-attribute diverse weighted bipartite b-matching (MDWBM) balances the quality of the matching with its diversity. The recent paper by Ahmadi et al. (2020) introduced the MDWBM but presented an incorrect mixed integer quadra-tic program (MIQP) and a flawed local exchange algorithm. In this work, we develop two constraint programming (CP) models, a binary quadratic programming (BQP) model, and a quadratic unconstrained binary optimization (QUBO) model for both the unconstrained and constrained MDWBM. A thorough empirical evaluation using commercial solvers and specialized QUBO hardware shows that the hardware-based QUBO approach dominates, finding best-known solutions on all tested instances up to an order of magnitude faster than the other approaches. CP is able to achieve better solutions than BQP on unconstrained problems but under-performs on constrained problems.

Keywords: Specialized hardware · QUBO · Constraint programming · Digital Annealer · Diverse matching

1 Introduction

Bipartite matching problems assign an agent on one side of a market to an agent on the other side. Weighted bipartite b-matching generalizes such problems to the setting where matches have real-valued weights and agents on one side of the market can be matched to at most b agents on the other side. The weighted bipartite b-matching problem serves as the core of applications such as general resource allocation [6] and recommender systems [18].

The Multi-attribute Diverse Weighted Bipartite b-Matching (MDWBM) problem has been recently introduced to simultaneously maximize the quality and diversity of a bipartite b-matching [1]. The quality is measured by weighted costs of assignments and the diversity is calculated in terms of differences across multiple feature classes. Ahmadi et al. [1] proved that MDWBM is NP-hard and tackled it with a mixed integer quadratic programming (MIQP) model and an exact local exchange algorithm. However, there are flaws in both of

© Springer Nature Switzerland AG 2022
P. Schaus (Ed.): CPAIOR 2022, LNCS 13292, pp. 424–440, 2022.
https://doi.org/10.1007/978-3-031-08011-1_28

these approaches. In this work, we address MDWBM and its constrained variant with three model-based paradigms: Constraint Programming (CP), Binary Quadratic Programming (BQP), and Quadratic Unconstrained Binary Optimization (QUBO). We make three primary contributions:

1. We propose two CP models, a BQP model, and a QUBO model of MDWBM.
2. By adding several practical constraints, we introduce constrained MDWBM.
3. We obtain state-of-the-art results for the standard and constrained MDWBM on software-based solvers with the CP and BQP models and on specialized hardware with the QUBO model, demonstrating that recent hardware architectures can be harnessed for combinatorial optimization problems.

2   Diverse Matching

A matching market often aims to maximize quality subject to some fairness constraints, such as assuring equal opportunity amongst agents. Benabbou et al. [5] study the trade-off between social welfare and diversity for the Singapore housing allocation, modeling diversity with constraints added to a model of an extension of the classic assignment problem.

Diverse bipartite b-matching [2] represents the trade-off between efficiency and diversity, where a matching provides good coverage over different varieties of agents. Diversity has been generally measured by some expression of coverage of the space of possible variation. Mathematically, researchers have used submodular functions, which encode the diminishing returns of similarity. For example, submodular diversity metrics are used in information retrieval communities, including determinantal point processes [21] and Maximum Marginal Relevance (MMR) [7]. Multi-attribute diverse weighted bipartite b-matching is a more general problem as it deals with multiple classes of features. The goal is to form diverse matchings with respect to all feature classes.

To our knowledge, there is no work on *constrained* diverse matching in the literature other than that on conflict and degree constraints in non-diverse bipartite b-matching in e-commerce [8] and vehicular networks [14].

2.1   Multi-attribute Diverse Weighted Bipartite b-Matching

In the Multi-attribute Diverse Weighted Bipartite b-Matching problem (MDWBM) [1], there are two sets of nodes U and V in a bipartite graph. Every node in V has multiple features, each of which belongs to a different feature class. Let F be the set of feature classes, for example in a worker-team assignment context, $F = \{\text{Gender}, \text{Nationality}\}$. For each $f \in F$, we use G_f to denote the set of values for class f, such as $G_{\text{Nationality}} = \{\text{France}, \text{Canada}\}$. The set of nodes in V with feature value g_f for feature f is denoted by $F_f^{g_f}$. A b-matching on a bipartite graph allows a node from U to be connected to multiple (b) nodes in V via edges. An example of the bipartite b-matching is illustrated in Fig. 1. Each edge is weighted by a cost, hence connecting nodes comes with the cost. The purpose of MDWBM is to minimize the weighted sum of cost and diversity.

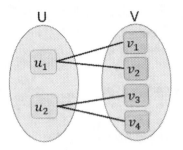

Fig. 1. Bipartite 2-matching.

For MDWBM, the objective function to be minimized is:

$$obj = S + W = \sum_{f \in F} \sum_{u \in U} \sum_{g_f \in G_f} \left(\lambda_f \cdot \left(c_{u,f,g_f}\right)^2 + \lambda_0 \cdot w_{u,f,g_f} \cdot c_{u,f,g_f} \right), \quad (1)$$

where S represents the similarity of the matching and W represents the weighted assignment cost. c_{u,f,g_f} denotes the number of nodes in V connected to node $u \in U$ having value g_f for feature class f. Accordingly, if S_f is the similarity of a matching w.r.t. feature class f, then S can also be represented by:

$$S = \sum_{f \in F} \lambda_f \cdot S_f = \sum_{f \in F} \lambda_f \cdot \sum_{u \in U} \sum_{g_f \in G_f} \left(c_{u,f,g_f} \right)^2. \quad (2)$$

where $\lambda_f \in \mathbb{Z}^+$ is a weight expressing the importance of feature f. Minimizing S, namely the supermodular similarity function[1] w.r.t. multiple features, has been proved to be NP-hard [1].

A weight is associated with a connection between a feature value g_f to a node $u \in U$. Specifically, the weight $w_{u,f,g_f} \in \mathbb{Z}^+$ represents the cost of assigning a node in V whose feature class f value is g_f to node $u \in U$. The costs are assumed to be integers [1]. The total cost of a matching is

$$W = \lambda_0 \sum_{f \in F} \sum_{u \in U} \sum_{g_f \in G_f} w_{u,f,g_f} \cdot c_{u,f,g_f} \quad (3)$$

where $\lambda_0 \in \mathbb{Z}^+$.

Each node $u \in U$ has a degree of d_u, specifying that the number of nodes in V connected to u is exactly d_u in a matching. Each node $v \in V$ can only be connected to at most one node in U. Ahmadi et al. address problems with these degree constraints, a special case of the general MDWBM.

Constraints. Assignment/allocation problems often contain constraints that, to our knowledge, have not been included in any formulations of MDWBM. In this paper, we address six types of constraints as follows.

---

[1] The negative submodular diversity is equivalent to the supermodular similarity.

- *Conflict (C)*: Two nodes $v_1, v_2 \in V$ cannot both be assigned to a node $u \in U$. Consider an example in the worker-team assignment context: two workers cannot be assigned to the same team due to a personal conflict.
- *Binding (B)*: Nodes $v_1, v_2 \in V$ must be assigned to the same node $u \in U$.
- *Conflict Assignment (CA)*: A node $v \in V$ cannot be assigned to a node $u \in U$. For example in the paper review context, a reviewer cannot be assigned to a paper due to the conflict of interest.
- *Binding Assignment (BA)*: A node $v \in V$ must be assigned to a node $u \in U$. For example, a particular reviewer must be assigned to a specific paper.
- *Must-Have (MH)*: A node $u \in U$ must be assigned at least one node $v \in V$ with a specific value $g_f \in G_f$. For instance, in an engineering course project, a team must have at least one student who is good at coding.
- *Not-Alone (NA)*: A node $u \in U$ is assigned either 0 or at least E $(E \geq 2)$ nodes in V, which have value $g_f \in G_f$. For example in engineering course projects, each group must have either 0 or at least two female students.

We call the multi-attribute diverse matching with degree constraints the *standard MDWBM* and the extension including any of the six practical constraints the *constrained MDWBM*.

2.2 Related Work

There is only one work in the literature that studied the MDWBM. Ahmadi et al. [1] proposed an mixed integer quadratic programming (MIQP) model for the standard MDWBM and also introduced an local exchange algorithm based on negative cycle detection. However, both the model and the algorithm are flawed.

The key problem of the MIQP model is that it uses c_{u,f,g_f} defined above as the decision variable. However, c_{u,f,g_f} does not represent an assignment but rather the number of nodes with a particular feature value assigned to a node. Thus, there is no bijection between the set of assignments and the set of solutions to the MIQP model. In fact, the decision variable choice decouples the combination of feature values from an assignment and hence can provide superoptimal solutions (i.e. the model is a relaxation of the true problem).

The local exchange algorithm uses the identification of negative cycles to improve a matching. A series of moves that leads to a decrease in the objective is called a negative cycle. In each iteration, the algorithm evaluates a neighborhood of solutions via node movement and detects the existence of negative cycles. The algorithm stops when it cannot find any negative cycle. The authors claimed and proved that the algorithm terminates at a global optimum [1]. However, the claim is false as there exists potential objective decrease that cannot be captured by the negative cycles.

The detailed information and counterexamples for both the model and the algorithm flaws are provided in the online appendix.[2]

---

[2] https://tidel.mie.utoronto.ca/pubs/Appendix_Matching_CPAIOR22.pdf.

3   Constraint Programming Models for MDWBM

We propose two constraint programming (CP) models for MDWBM. The first is based on integer assignment variables, like most CP models for bin packing problems. The second model manipulates a list of integer selection variables for each node $u \in U$, requiring more effort to link variables and parameters. For convenience, we use *assignment CP* (ACP) and *selection CP* (SCP), respectively, to represent the two models.

3.1   Assignment CP Model

The ACP is as follows:

$$\min_{x} W + S \tag{4a}$$

$$\text{s.t. } \text{cardinality}(\{x_1, ..., x_{|V|}\}, \{0, 1, ..., |U|\}, \{|V| - \sum_{u \in U} d_u, d_1, ..., d_{|U|}\}), \tag{4b}$$

$$\text{knapsack}(\mathbf{x}_{f,g_f}, \{c_{0,f,g_f}, ..., c_{|U|,f,g_f}\}, \{1, ..., 1\}), \ \forall f \in F, \forall g_f \in G_f, \tag{4c}$$

$$\text{spread}(\{c_{u,f,1}, ..., c_{u,f,|G_f|}\}, \frac{d_u}{|G_f|}, \sigma_{u,f}), \ \forall u \in U, \forall f \in F, \tag{4d}$$

$$W = \lambda_0 \sum_{v \in V} \mathbf{A}_{x_v, v}, \tag{4e}$$

$$S = \sum_{f \in F} \sum_{u \in U} \lambda_f \cdot \left(\sigma_{u,f}^2 \cdot |G_f| + \frac{d_u^2}{|G_f|} \right), \tag{4f}$$

$$x_v \in \{0, 1, ..., |U|\}, \ \forall v \in V, \tag{4g}$$

$$c_{u,f,g_f} \in \{0, 1, ..., d_u\}, \ \forall u \in U \cup \{0\}, \forall f \in F, \forall g_f \in G_f, \tag{4h}$$

$$x_{v_1} \neq x_{v_2}, \ \forall (v_1, v_2) \in C_C, \tag{4i}$$

$$x_{v_1} = x_{v_2}, \ \forall (v_1, v_2) \in C_B, \tag{4j}$$

$$x_v \neq u, \ \forall (u, v) \in C_{CA}, \tag{4k}$$

$$x_v = u, \ \forall (u, v) \in C_{CB}, \tag{4l}$$

$$c_{u,f,g_f} \geq 1, \ \forall (u, f, g_f) \in C_{MH}, \tag{4m}$$

$$c_{u,f,g_f} \in \{0, E, E+1, ..., d_u\}, \ \forall (u, f, g_f, E) \in C_{NA}. \tag{4n}$$

The integer decision variable $x_v = u$ if the node v is assigned to node u, and 0 if the node u is assigned to a dummy node $0 \notin U$. The dummy node does not have a fixed upper bound on degree and is used when some node in V is not matched to any node in U. The integer variable c_{u,f,g_f} represents the number of nodes in V connected to node $u \in U \cup \{0\}$ having value g_f for feature class f. In our implementation, the dummy node is assigned the last index instead of the first one to better fit the typical default search algorithm of CP solvers.

In the model, constraint (4b) is the *global cardinality constraint (gcc)* ensuring the number of nodes in V matched to node $u \in U$ is exactly d_u. The rest of the nodes are matched to the dummy node, which has degree $|V| - \sum_{u \in U} d_u$.

Constraint (4c) is the *multi-knapsack* constraint that links variables $\mathbf{x}$ and $\mathbf{c}$ together. The variable set $\mathbf{x}_{f,g_f}$ contains all the x_v, where node v has g_f as the feature value of feature class f. Constraint (4d) is the *spread* constraint to obtain the standard deviation $\sigma_{u,f}$ of c_{u,f,g_f} (excluding the dummy node) over $g_f \in G_f$. Note that the mean value of c_{u,f,g_f} over $g_f \in G_f$ is the constant $d_u/|G_f|$ as there are exactly d_u nodes in V that are matched to u.

Constraint (4e) represents the assignment cost of the matching, which is also the first component of the objective function to minimize.[3] The (u, v) entry of the matrix $\mathbf{A}$ is the cost of assigning v to u, which is pre-calculated by

$$\mathbf{A}_{u,v} = \sum_{(f,g_f) \text{ if } v \in F_f^{g_f}} w_{u,f,g_f}. \tag{5}$$

Constraint (4f) expresses the similarity of the matching, which is equivalent to term (2). Constraints (4g) and (4h) address the variable ranges.

The standard MDWBM is modeled by objective (4a) and constraints (4b)–(4h), while constraints (4i)–(4n) are for constrained MDWBM. Constraint (4i) is the conflict constraint where C_C contains pairs of conflict nodes in V. Constraint (4j) is the binding constraint where C_B contains pairs of binding nodes in V. Constraint (4k) is the conflict assignment constraint where C_{CA} contains node pairs that cannot be connected. Constraint (4l) is the binding assignment constraint where C_{BA} contains node pairs that must be connected by the assignment. Constraint (4m) is the must-have constraint where C_{MH} contains the 3-tuples ⟨node in U, feature class, feature value⟩. Constraint (4n) is the not-alone constraint where C_{NA} contains the 4-tuples ⟨node in U, feature class, feature value, the number of eligible nodes in V⟩. Note that constraints (4k) - (4n) are implemented via direct domain pruning in the model.

3.2   Selection CP Model

The SCP is as follows:

$$\min_x \ W + S \tag{6a}$$

$$\text{s.t.} \quad \text{alldifferent}(\{x_{u,k}, \forall u \in U, \forall k = 1, ..., d_u\}), \tag{6b}$$

$$\text{table}(\mathbf{T}, x_{u,k}, \{z_{u,k}^1, ..., z_{u,k}^{|F|}\}), \ \forall u \in U, \forall k = 1, ..., d_u, \tag{6c}$$

$$\text{cardinality}(\{\{z_{u,1}^f, ..., z_{u,d_u}^f\}, \{1, ..., |G_f|\}, \{c_{u,f,1}, ..., c_{u,f,|G_f|}\}\}),$$
$$\forall u \in U, \forall f \in F, \tag{6d}$$

$$\text{spread}(\{c_{u,f,1}, ..., c_{u,f,|G_f|}\}, \frac{d_u}{|G_f|}, \sigma_{u,f}), \ \forall u \in U, \forall f \in F, \tag{6e}$$

$$W = \lambda_0 \sum_{u \in U} \sum_{1 \le k \le d_u} \mathbf{A}_{u, x_{u,k}}, \tag{6f}$$

---

[3] We use (4e) instead of (3) according to the superior results in our experiments.

$$S = \sum_{f \in F} \sum_{u \in U} \lambda_f \cdot \left(\sigma_{u,f}^2 \cdot |G_f| + \frac{d_u^2}{|G_f|} \right), \tag{6g}$$

$$x_{u,k} \in \{1, ..., |V|\}, \ \forall u \in U, \forall k = 1, ..., d_u, \tag{6h}$$

$$c_{u,f,g_f} \in \{0, 1, ..., d_u\}, \ \forall f \in F, \forall g_f \in G_f, \tag{6i}$$

$$\text{count}(\{x_{u,1}, ..., x_{u,d_u}\}, v_1) + \text{count}(\{x_{u,1}, ..., x_{u,d_u}\}, v_2) \leq 1,$$
$$\forall u \in U, \forall (v_1, v_2) \in C_C, \tag{6j}$$

$$\text{count}(\{x_{u,1}, ..., x_{u,d_u}\}, v_1) = \text{count}(\{x_{u,1}, ..., x_{u,d_u}\}, v_2),$$
$$\forall u \in U, \forall (v_1, v_2) \in C_B, \tag{6k}$$

$$x_{u,k} \neq v, \ \forall k = 1, ..., d_u, \forall (u, v) \in C_{CA}, \tag{6l}$$

$$\text{count}(\{x_{u,1}, ..., x_{u,d_u}\}, v) = 1, \ \forall (u, v) \in C_{CB}, \tag{6m}$$

$$c_{u,f,g_f} \geq 1, \ \forall (u, f, g_f) \in C_{MH}, \tag{6n}$$

$$c_{u,f,g_f} \in \{0, E, E+1, ..., d_u\}, \ \forall (u, f, g_f, E) \in C_{NA}. \tag{6o}$$

The integer decision variable $x_{u,k} = v$ if v is the k-th node selected by node u. The integer variable c_{u,f,g_f} is the same as in ACP. In the model, constraint (6b) is the *all-different* constraint guaranteeing that nodes in U select distinct nodes in V. Constraint (6c) is the *table* constraint that links $\mathbf{x}$ and $\mathbf{z}$. $\mathbf{T}$ is the feature matrix, where the (v, f) entry is the value of feature f of node v. Then, $z_{u,k}^f$ represents the feature value of the k-th node matched to u. Constraint (6d) is the *gcc* constraint that links $\mathbf{c}$ and $\mathbf{z}$. Constraint (6e) is the same as (4d) and constraint (6f) represents the assignment cost of the matching, with the same cost matrix $\mathbf{A}$ as in ACP. Constraint (6g) is the same as (4f). Constraints (6h) and (6i) express the variable ranges.

The standard MDWBM is modeled by (6a)–(6i), while constraints (6j)–(6o) are for the constrained MDWBM. Similar to ACP, constraints (6j), (6k), (6l), (6m), (6n), and (6o) represent the conflict, binding, conflict assignment, binding assignment, must-have, and not-alone constraints, respectively.

4   Quadratic Models for MDWBM

In this section, we propose a Binary Quadratic Programming (BQP) model and a Quadratic Unconstrained Binary Optimization (QUBO) model for MDWBM.

4.1   BQP Model

We introduce a BQP model with binary assignment variables. The decision variable $x_{u,v} = 1$ if node $v \in V$ is assigned to node $u \in U$, and 0 otherwise. Based on the feature values of each node $v \in V$, we can generate a feature matrix $\mathbf{B} = \{b_{v,f,g_f}\}$ where $b_{v,f,g_f} = 1$ if node v has value g_f for feature f, and 0 otherwise. Similarly, we can generate a weighted cost matrix $\mathbf{C} = \{c_{u,v,f}\}$ based on weighted cost parameters w_{u,f,g_f}. We use $c_{u,v,f}$ to represent the cost for feature class f if node $v \in V$ is assigned to node $u \in U$. We set $c_{u,v,f} = w_{u,f,g_f}$ if node v has value g_f for feature class f and 0 otherwise. In addition, we also add a dummy node indexed by 0 to deal with situations where a node in V might not be matched to any node in U. Our BQP model is shown below.

$$\min_{x} \quad \lambda_0 \cdot \sum_{f \in F} \sum_{u \in U} \sum_{v \in V} c_{u,v,f} \cdot x_{u,v} + \tag{7a}$$

$$\sum_{f \in F} \lambda_f \cdot \sum_{u \in U} \sum_{g_f \in G_f} \left(\sum_{v \in V} b_{v,f,g_f} \cdot x_{u,v} \right)^2 \tag{7b}$$

$$\text{s.t.} \quad \sum_{v \in V} x_{u,v} = d_u, \ \forall u \in U, \tag{7c}$$

$$\sum_{u \in U \cup \{0\}} x_{u,v} = 1, \ \forall v \in V, \tag{7d}$$

$$x_{u,v_1} + x_{u,v_2} \leq 1, \ \forall u \in U, \forall (v_1, v_2) \in C_C, \tag{7e}$$

$$x_{u,v_1} = x_{u,v_2}, \ \forall u \in U, \forall (v_1, v_2) \in C_B, \tag{7f}$$

$$x_{u,v} = 0, \ \forall (u, v) \in C_{CA}, \tag{7g}$$

$$x_{u,v} = 1, \ \forall (u, v) \in C_{BA}, \tag{7h}$$

$$\sum_{v \in V} b_{v,f,g_f} \cdot x_{u,v} \geq 1, \ \forall (u, f, g_f) \in C_{MH}, \tag{7i}$$

$$\sum_{v \in V} b_{v,f,g_f} \cdot x_{u,v} \geq E \cdot y_{u,f,g_f}, \ \forall (u, f, g_f, E) \in C_{NA}, \tag{7j}$$

$$\sum_{v \in V} b_{v,f,g_f} \cdot x_{u,v} \leq d_u \cdot y_{u,f,g_f}, \ \forall (u, f, g_f, E) \in C_{NA}, \tag{7k}$$

$$x_{u,v} \in \{0, 1\}, \ \forall u \in U \cup \{0\}, \forall v \in V, \tag{7l}$$

$$y_{u,f,g_f} \in \{0, 1\}, \ \forall u \in U \cup \{0\}, \forall f \in F, \forall g_f \in G_f. \tag{7m}$$

Term (7a) represents the weighted cost of the assignment. Term (7b) represents the supermodular similarity w.r.t. all feature classes. Constraint (7c) guarantees that node $u \in U$ (excluding the dummy node) has a degree of d_u. Constraint (7d) ensures that each node in V is only assigned to one node in U. These components form the BQP model for the standard MDWBM. Constraints (7e) to (7i) model the conflict constraints to must-have constraints, respectively.

Constraints (7j) and (7k) represent the not-alone constraints. The variable $y_{u,f,g_f} = 0$ if node u is not matched to any node in V with feature value g_f of feature f and $y_{u,f,g_f} = 1$ if the number of such nodes matched to u is greater than or equal to E.

4.2 Quadratic Unconstrained Binary Optimization

The recent emergence of specialized hardware has opened up new ways to solve specific computational tasks, such as combinatorial optimization problems [26]. Including adiabatic and gate-based quantum computers [23] and CMOS annealers [3], these novel technologies represent a variety of designs and underlying models of computation. Many of the designs for combinatorial optimization target problems formulated as an Ising model or equivalently as a Quadratic Unconstrained Binary Optimization (QUBO) model [9], which is the following problem:

$$\min y = \frac{1}{2} \sum_i \sum_{j \neq i} W_{i,j} x_i x_j + \sum_i b_i x_i + c, \tag{8}$$

where $x \in \{0,1\}^n$ are binary decision variables, $W \in \mathbf{M}_{n,n}(\mathbb{R})$ is a symmetric weight matrix, $b \in \mathbb{R}^n$ is a bias vector, and $c \in \mathbb{R}$ is a constant [20]. QUBO has been used to represent problems in combinatorial scientific computing [26], machine learning [10], and finance [25]. With multiple feature classes, the supermodular similarity function of MDWBM is naturally quadratic, suggesting that it might be a good candidate for such novel hardware.

In our QUBO model, we use the same binary assignment variables $x_{u,v}$ as in the BQP model. The QUBO model of MDWBM is shown below.

$$\min_x \ \lambda_0 \cdot \sum_{f \in F} \sum_{u \in U} \sum_{v \in V} c_{u,v,f} \cdot x_{u,v} + \tag{9a}$$

$$\sum_{f \in F} \lambda_f \cdot \sum_{u \in U} \sum_{g_f \in G_f} \left(\sum_{v \in V} b_{v,f,g_f} \cdot x_{u,v} \right)^2 + \tag{9b}$$

$$p_1 \cdot \sum_{u \in U} \left(\sum_{v \in V} x_{u,v} - d_u \right)^2 + \tag{9c}$$

$$p_2 \cdot \sum_{v \in V} \left(\sum_{u \in U \cup \{0\}} x_{u,v} - 1 \right)^2 + \tag{9d}$$

$$p_3 \cdot \sum_{(v_1,v_2) \in C_C} \sum_{u \in U} x_{u,v_1} \cdot x_{u,v_2} + \tag{9e}$$

$$p_4 \cdot \sum_{(v_1,v_2) \in C_B} \sum_{u=1}^{|U|} (x_{u,v_1} - x_{u,v_2})^2 + \tag{9f}$$

$$p_5 \cdot \sum_{(u,v) \in C_{CA}} x_{u,v} + \tag{9g}$$

$$p_6 \cdot \sum_{(u,v) \in C_{BA}} (x_{u,v} - 1)^2 + \tag{9h}$$

$$p_7 \cdot \sum_{(u,f,g_f) \in C_{MH}} \left(\sum_{v \in V} b_{v,f,g_f} \cdot x_{u,v} - 1 - s_7 \right)^2 + \tag{9i}$$

$$p_8 \cdot \sum_{(u,f,g_f,E) \in C_{NA}} \left(\sum_{v \in V} b_{v,f,g_f} \cdot x_{u,v} + s_8 - d_u \cdot y_{u,f,g_f} \right)^2 . \tag{9j}$$

Term (9a) and (9b) are the same as (7a) and (7b). p_1 to p_8 are penalty coefficients that are set to $10 * |F|$. Terms (9c) to (9i) are the penalized terms of constraints (7c) to (7i). Take (7c) and (9c) as an example; as we are minimizing the overall objective function, we want (9c) to evaluate to 0 when constraint (7c) is satisfied and to a non-zero value proportional to its violation when it is not. In term (9i), s_7 is the non-negative slack variable. The lower bound of s_7 is 0 and the upper bound is $|V| - 1$. Thus, we use binary variables $z_1, ..., z_{|V|-1}$ to represent s_7 in QUBO, i.e., $s_7 = z_1 + \cdots + z_{|V|-1}$. Similarly, $s_8 = z_1 + ... + z_{d_u - E}$ is the slack variable in term (9j) for not-alone constraints. Note that y_{u,f,g_f} is the same indicator variable as in ACP.

5   Empirical Evaluation

In this section, we present our experimental results on standard and constrained MDWBM with the commercial constraint programming solver CP Optimizer (CPO) v20.1.0, the commercial mathematical programming solver Gurobi v9.5.0, a multistart tabu search algorithm for QUBO [24], and a computer architecture designed for QUBO: the Fujitsu Digital Annealer. CPO/Gurobi are the state-of-the-art for general purpose constraint/mathematical programming. Though the multistart tabu search was developed more than 10 years ago, according to recent work, it is still one of the best metaheuristic approaches to QUBO [13]. We use the software-based implementation of the multi-start tabu search version 2 by D-Wave [11].

5.1   Fujitsu Digital Annealer

The Fujitsu Digital Annealer (DA) is a recent computer architecture designed for solving QUBO problems [22]. The third generation DA (DA3), a hybrid system of hardware and software, can represent QUBOs with up to 100000 variables. For our DA environment,[4] the integer coefficients for the quadratic terms range from -2^{62} to 2^{62} and those for the linear terms range from -2^{73} to 2^{73} [16]. The DA algorithm is based on Simulated Annealing (SA), however it takes advantage of the massive parallelization provided by the custom CMOS hardware. The difference between the SA and DA algorithm are as follows:

- DA utilizes parallel tempering that runs a number of problem solving processes (*replicas*) in parallel with different temperatures [12]. Replicas swap temperatures to diversify the search. In each replica, each Monte Carlo step considers all possible one-bit flips in parallel [4].
- DA employs a dynamic offset to raise the energy of a state to escape local minima.
- DA supports a dedicated bit flip mechanism, over a subset of variables belonging to one-hot equivalent constraints when using DA3.
- DA can deal with inequality constraints that are not modeled in QUBO. As a consequence, the terms (9i) and (9j) are not included in the QUBO model when using DA3. Instead, they are represented as the following constraints:

---

[4] All experiments were conducted on the Digital Annealer environment prepared exclusively for the research at the University of Toronto.

$$1 - \sum_{v \in V} b_{v,f,g_f} \cdot x_{u,v} \leq 0, \; \forall (u, f, g_f) \in C_{MH}. \tag{9i'}$$

$$E \cdot y_{u,f,g_f} \leq \sum_{v \in V} b_{v,f,g_f} \cdot x_{u,v} \leq d_u \cdot y_{u,f,g_f}, \; \forall (u, f, g_f, E) \in C_{NA}. \tag{9j'}$$

In our experiments, we run the DA3 on a remote computer and do not include the communication time in our runtime limits and results. The programs (for running DA, CPO, Gurobi, and tabu search) are written in Python 3.7 and conducted on a Window PC with Intel(R) Core(TM) i7-8700K CPU @3.20 GHz with 16 GB RAM.

5.2 Experimental Setting

The proposed QUBO model is tested with three solvers (Gurobi, DA, and tabu search), while the proposed CP and BQP models are tested with CPO and Gurobi, respectively. The six model-solver combinations are each run for 600 s for each instance.

Since the runtime limits are the same for the six approaches, we use the best objective value, the time of finding the best objective, and the mean relative error as performance measures. Denote by $B_{i,t,a}$ the best solution attained by runtime t of approach a for instance i. The relative error at time t for approach a on instance i is given by

$$RE(i, t, a) = \frac{B_{i,t,a} - B_i}{B_i} \tag{11}$$

where B_i represents the best solution over all approaches at the end of runtime. For a minimization problem, this expression is always non-negative. The mean relative error of approach a at time t, $MRE(t, a)$, can be computed as

$$MRE(t, a) = \frac{1}{|\mathcal{I}|} \sum_{i \in \mathcal{I}} RE(i, t, a). \tag{12}$$

5.3 Experiments on Standard MDWBM

For standard MDWBM, we first run the paper review [17] benchmark dataset from UIUC [19] that were used by Ahmadi et al. It contains 73 papers accepted by SIGIR 2007, 189 prospective reviewers, and 25 major topics. For each paper, a 25-dimensional label is provided based on its relevance to those topics. Similarly for the 189 reviewers, a 25-dimensional expertise representation is provided.

Following Ahmadi et al., we first use spectral clustering to divide reviewers into five clusters based on their topic vectors. We treat the cluster label as the first feature. The assignment cost of a reviewer to a paper is calculated as the relevance of each cluster for each paper. We take the average cosine similarity of label vectors of reviewers in that cluster and the paper. The reviewer demand of each paper is set to 4 ($b = 4$) and no reviewer is assigned to more than 1 paper. Again, following the methodology of Ahmadi et al., to increase the dataset size and the feature number, we create a copy of each reviewer and invert the gender in the copy. The gender is considered as the second feature. We set $\lambda_0 = \lambda_1 = \lambda_2 = 1000$ and round the assignment costs after multiplying by λ_0 as DA only supports integral coefficients. The results are summarized in Table 1. The number of bits is the number of $x_{u,v}$ variables in the QUBO model.

Table 1. Objective results of UIUC paper review instances.

Parameters				Exact methods				Non-exact methods	
$|U|$	$|V|$	$|F|$	#bits	ACP	SCP	GBQP	GBQB	DA3	TABU
3	378	2	1512	**45911**	**45911**	**45911**	**45911**	**45911**	**45911**
13	378	2	5292	**201139**	**201139**	**201139**	**201139**	**201139**	**201139**
23	378	2	9072	**356652**	**356652**	**356652**	357927	**356652**	**356652**
33	378	2	12852	**512177**	**512177**	**512177**	513621	**512177**	**512177**
43	378	2	16632	**669264**	**669264**	**669264**	676947	**669264**	MemOut
53	378	2	20412	**824742**	**824742**	**824742**	MemOut	**824742**	MemOut
63	378	2	24192	**979525**	**979525**	**979525**	MemOut	**979525**	MemOut
73	378	2	27972	**1136424**	**1136424**	**1136424**	MemOut	**1136424**	MemOut

GBQP and GBQB represent Gurobi with the BQP and QUBO models, respectively. The results show that ACP, SCP, GBQP, and DA3 achieve the same solutions, but none proves optimality for any instance. TABU finds the same solutions for the first four instances, but runs out of memory on larger problems. GBQB is outperformed by all other approaches.

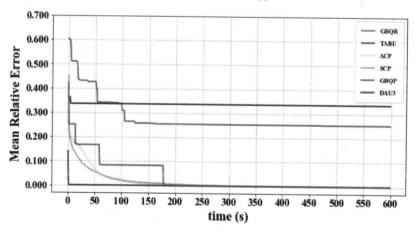

Fig. 2. MRE plot of UIUC paper review instances.

MRE comparisons are shown in Fig. 2. To avoid infinite MREs, for instances that induce a memory error, we use a simple heuristic to find an initial solution and use it for each time point during the generation of MRE plots. The heuristic assigns reviewer $\{1, 2, 3, 4\}$ to the first paper, reviewer $\{5, 6, 7, 8\}$ to the second paper, and so forth. From the MRE plot we see that the non-exact methods, TABU and DA3, reach solutions immediately and rarely improve the quality after 10 s, though DA3 is much better than TABU. The exact methods gradually increase their solution quality, with ACP, SCP, and GBQP eventually finding the same solutions as DA3. The performance of GBQB, however, never surpasses DA3 at 10 s. While we have shown that the two solution approaches of Ahmadi et al. are incorrect in the online appendix, we note that that their reported runtimes are up to 4 orders of magnitude longer than the runtime of DA3. These numeric results, therefore lead to very different conclusions w.r.t. solving MDWBM problems in practice.

Table 2. MRE of randomly generated instances.

	Parameters				Exact				Non-exact	
ID	$|U|$	$|V|$	$|F|$	#bits	ACP	SCP	GBQP	GBQB	DA3	TABU
1–5	25	100	10	2600	0.027	0.032	0.011	0.016	**0.000**	0.028
6–10	25	100	100	2600	0.030	0.035	0.006	0.008	**0.000**	0.014
11–15	50	200	10	10200	0.049	0.052	0.115	0.019	**0.000**	0.030
16–20	50	200	100	10200	0.073	0.065	MemOut	0.009	**0.000**	0.018

In the UIUC dataset, there are only two feature classes. In other application contexts such as machine learning [15], the number of feature classes can be very large. We hence uniformly randomly generated instances with more feature classes [27]. The newly generated instances are of the sizes in terms of $|V| \times |U|$: $\{100 \times 25, 200 \times 50\}$, and the number of feature classes $|F|$: $\{10, 100\}$. The assignment costs are uniformly distributed from 1 to 5. Each feature class randomly has 2 to 10 different values. Each node in U needs to have a degree of 4 ($b = 4$). We generate 5 instances for each size and set $\lambda_0 = \lambda_1 = \ldots = \lambda_{|F|} = 1$. The MRE results are shown in Table 2 and Fig. 3.

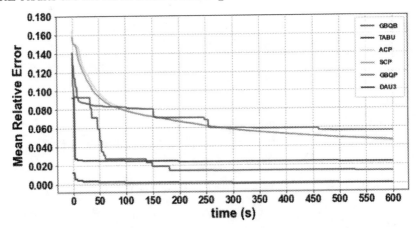

Fig. 3. MRE plot of randomly generated instances.

For the 20 randomly generated instances, DA3 remains the best solution approach. Unlike for UIUC instances, DA3 gradually improves its solution quality, though the improvement is small. For these instances, QUBO-based methods are much better than CP/BQP-based methods, as Gurobi and CPO at 600s do not produce a better solution than DA3/TABU/GBQB at 60s. CPO with both CP models is worse than GBQP initially, but achieves better performance after 280s. Also note that the performance of CP approaches degrades when the problem size or the number of feature classes increases.

5.4   Experiments on Constrained MDWBM

In this section, we test constrained MDWBM. Focusing more on the constraints, we consider the multi-class instances with IDs 1, 2, 6, and 7. Due to limited paper space, we select four different constraint patterns according to the number of each type of constraints, as shown in Table 3.

Table 3. Number of constraints in different patterns.

	Constraint					
Pattern	C	B	CA	BA	MH	NA
PT1	5	5	5	5	5	5
PT2	10	10	10	10	10	10
PT3	0	20	0	0	20	20
PT4	50	0	50	0	0	0

MDWBM with PT2 is more constrained than that with PT1. PT3 and PT4 are practically interesting constraint patterns. PT3 illustrates the situation that nodes in V have binding preferences while nodes in U need to meet specific requirements, while PT4 reflects the circumstances when there are only conflict constraints. The constraints are randomly generated. The results of the constrained MDWBM are shown in Table 4 and Fig. 4.

Table 4. MRE of instances with constraints.

	Constraint	Exact				Non-exact	
ID	Pattern	ACP	SCP	GBQP	GBQB	DA3	TABU
1,2,6,7	PT1	0.032	0.026	0.004	0.040	**0.000**	0.035
1,2,6,7	PT2	0.027	0.026	0.006	0.071	**0.000**	0.065
1,2,6,7	PT3	0.028	0.019	0.004	0.114	**0.000**	0.127
1,2,6,7	PT4	0.043	0.034	0.008	0.014	**0.000**	0.016

For constrained MDWBM, though DA3 with QUBO models is still the state-of-the-art, CP/BQP-based methods perform better than other QUBO-based methods. One of the reasons is that CP/BQP models can naturally deal with constraints while QUBO has to convert constraints to penalty terms. Surprisingly, though designed for constrained optimization, CP approaches are worse than GBQP and DA3. We speculate that CPO might not deal with the six types of constraints efficiently as they are not expressed in terms of global constraints with effective filtering algorithms. TABU and GBQB are worse than DA3 by around 6%. Though TABU and GBQB improve the solution quality during their runs, the solutions at 600s are still far from competitive.

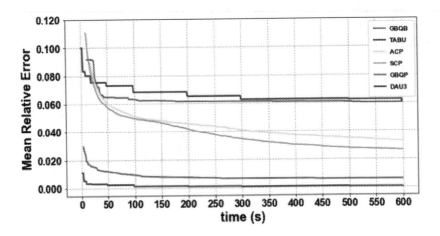

Fig. 4. MRE plot of instances with constraints.

6   Conclusions

We have developed two constraint programming, a binary quadratic programming, and a quadratic unconstrained binary optimization models for the multi-attribute diverse weighted bipartite b-matching problem and introduced practical constraints into the models. Experiments on the standard and constrained MDWBM show that novel hardware DA3 with the QUBO model has an advantage over CP Optimizer with the CP models, Gurobi with the BQP and QUBO models, and tabu search on the QUBO model. We have also identified flaws in existing approaches for standard MDWBM [1].

As traditional computers suffer from the end of Moore's law, it is increasingly important to understand how AI and OR problems can benefit from novel hardware and computation architectures. Our work has demonstrated that for multi-attribute diverse weighted bipartite b-matching, state-of-the-art performance can be delivered by such hardware using a natural model.

Acknowledgement. The authors would like to thank Fujitsu Ltd. and Fujitsu Consulting (Canada) Inc. for providing financial support and access to the Digital Annealer at the University of Toronto. Partial funding for this work was provided by Fujitsu Ltd. and the Natural Sciences and Engineering Research Council of Canada.

References

1. Ahmadi, S., Ahmed, F., Dickerson, J.P., Fuge, M., Khuller, S.: An algorithm for multi-attribute diverse matching. In: Proceedings of the 29th International Joint Conference on Artificial Intelligence (IJCAI), pp. 3–9. AAAI Press (2020)
2. Ahmed, F., Dickerson, J.P., Fuge, M.: Diverse weighted bipartite b-matching. In: Proceedings of the 26th International Joint Conference on Artificial Intelligence (IJCAI), pp. 35–41. AAAI Press (2017)
3. Aramon, M., Rosenberg, G., Valiante, E., Miyazawa, T., Tamura, H., Katzgraber, H.G.: Physics-inspired optimization for quadratic unconstrained problems using a digital annealer. Front. Phys. **7**, 48 (2019)
4. Bagherbeik, M., Ashtari, P., Mousavi, S.F., Kanda, K., Tamura, H., Sheikholeslami, A.: A permutational Boltzmann machine with parallel tempering for solving combinatorial optimization problems. In: Bäck, T., Preuss, M., Deutz, A., Wang, H., Doerr, C., Emmerich, M., Trautmann, H. (eds.) PPSN 2020. LNCS, vol. 12269, pp. 317–331. Springer, Cham (2020). https://doi.org/10.1007/978-3-030-58112-1_22
5. Benabbou, N., Chakraborty, M., Ho, X.V., Sliwinski, J., Zick, Y.: Diversity constraints in public housing allocation. In: 17th International Conference on Autonomous Agents and MultiAgent Systems, AAMAS 2018 (2018)
6. Bertsimas, D., Papalexopoulos, T., Trichakis, N., Wang, Y., Hirose, R., Vagefi, P.A.: Balancing efficiency and fairness in liver transplant access: tradeoff curves for the assessment of organ distribution policies. Transplantation **104**(5), 981–987 (2020)
7. Carbonell, J., Goldstein, J.: The use of MMR, diversity-based reranking for reordering documents and producing summaries. In: Proceedings of the 21st Annual International ACM SIGIR Conference on Research and Development in Information Retrieval, pp. 335–336 (1998)
8. Chen, C., Zheng, L., Srinivasan, V., Thomo, A., Wu, K., Sukow, A.: Conflict-aware weighted bipartite b-matching and its application to e-commerce. IEEE Trans. Knowl. Data Eng. **28**(6), 1475–1488 (2016)
9. Coffrin, C., Nagarajan, H., Bent, R.: Evaluating ising processing units with integer programming. In: Rousseau, L.-M., Stergiou, K. (eds.) CPAIOR 2019. LNCS, vol. 11494, pp. 163–181. Springer, Cham (2019). https://doi.org/10.1007/978-3-030-19212-9_11
10. Cohen, E., Senderovich, A., Beck, J.C.: An ising framework for constrained clustering on special purpose hardware. In: Hebrard, E., Musliu, N. (eds.) CPAIOR 2020. LNCS, vol. 12296, pp. 130–147. Springer, Cham (2020). https://doi.org/10.1007/978-3-030-58942-4_9
11. D-Wave System Inc.: D-wave tabu (2021). https://docs.ocean.dwavesys.com/projects/tabu/en/latest/, Accessed 21 July 2021
12. Dabiri, K., Malekmohammadi, M., Sheikholeslami, A., Tamura, H.: Replica exchange MCMC hardware with automatic temperature selection and parallel trial. IEEE Trans. Parallel Distrib. Syst. **31**(7), 1681–1692 (2020)

13. Dunning, I., Gupta, S., Silberholz, J.: What works best when? A systematic evaluation of heuristics for Max-Cut and QUBO. INFORMS J. Comput. **30**(3), 608–624 (2018)
14. Fazliu, Z.L., Chiasserini, C.F., Malandrino, F., Nordio, A.: Graph-based model for beam management in mmwave vehicular networks. In: Proceedings of the Twenty-First International Symposium on Theory, Algorithmic Foundations, and Protocol Design for Mobile Networks and Mobile Computing, pp. 363–367 (2020)
15. Fern, X.Z., Brodley, C.E., et al.: Cluster ensembles for high dimensional clustering: an empirical study. Technical Report CS06-30-02, Oregon State University (2006)
16. Fujitsu Limited: The third generation of the digital annealer (2021). https://www.fujitsu.com/jp/group/labs/en/documents/about/resources/tech/techintro/3rd-g-da_en.pdf, Accessed 20 Aug 2021
17. de Givry, S., Schiex, T., Schutt, A., Simonis, H.: Modelling the conference paper assignment problem. In: 19th Workshop on Constraint Modeling and Reformulation, ModRef-20 (2020)
18. Kadıoğlu, S., Kleynhans, B., Wang, X.: Optimized item selection to boost exploration for recommender systems. In: Stuckey, P.J. (ed.) CPAIOR 2021. LNCS, vol. 12735, pp. 427–445. Springer, Cham (2021). https://doi.org/10.1007/978-3-030-78230-6_27
19. Karimzadehgan, M., Zhai, C.: Constrained multi-aspect expertise matching for committee review assignment. In: Proceedings of the 18th ACM conference on Information and knowledge management, pp. 1697–1700 (2009)
20. Kochenberger, G., et al.: The unconstrained binary quadratic programming problem: a survey. J. Comb. Optim. **28**(1), 58–81 (2014). https://doi.org/10.1007/s10878-014-9734-0
21. Kulesza, A., Taskar, B.: Determinantal point processes for machine learning. arXiv preprint arXiv:1207.6083 (2012)
22. Matsubara, S., et al.: Digital annealer for high-speed solving of combinatorial optimization problems and its applications. In: 2020 25th Asia and South Pacific Design Automation Conference, ASP-DAC, pp. 667–672. IEEE (2020)
23. Mohseni, M., et al.: Commercialize quantum technologies in five years. Nature News **543**(7644), 171 (2017)
24. Palubeckis, G.: Multistart tabu search strategies for the unconstrained binary quadratic optimization problem. Ann. Oper. Res. **131**(1), 259–282 (2004)
25. Rosenberg, G., Haghnegahdar, P., Goddard, P., Carr, P., Wu, K., De Prado, M.L.: Solving the optimal trading trajectory problem using a quantum annealer. IEEE J. Sel. Top. Signal Process. **10**(6), 1053–1060 (2016)
26. Tran, T.T., et al.: Explorations of quantum-classical approaches to scheduling a mars lander activity problem. In: Workshops at the Thirtieth AAAI Conference on Artificial Intelligence (2016)
27. Zhang, J., Lo Bianco, G., Beck, J.C.: MDWBM Instances (2021). https://github.com/JasonZhangjc/mdwbm-instances, Accessed 11 Feb 2022

Author Index

Printed in the United States
by Baker & Taylor Publisher Services